Photoelectric Measurement and Sensing: New Technology and Applications

Photoelectric Measurement and Sensing: New Technology and Applications

Editors

Qibo Feng
Jiakun Li
Qixin He

MDPI

Basel • Beijing • Wuhan • Barcelona • Belgrade • Novi Sad • Cluj • Manchester

Editors
Qibo Feng
Beijing Jiaotong University
Beijing
China

Jiakun Li
Beijing Jiaotong University
Beijing
China

Qixin He
Beijing Jiaotong University
Beijing
China

Editorial Office
MDPI
St. Alban-Anlage 66
4052 Basel, Switzerland

This is a reprint of articles from the Special Issue published online in the open access journal *Sensors* (ISSN 1424-8220) (available at: https://www.mdpi.com/journal/sensors/special_issues/171A8C2SU0).

For citation purposes, cite each article independently as indicated on the article page online and as indicated below:

Lastname, A.A.; Lastname, B.B. Article Title. *Journal Name* **Year**, *Volume Number*, Page Range.

ISBN 978-3-0365-9252-7 (Hbk)
ISBN 978-3-0365-9253-4 (PDF)
doi.org/10.3390/books978-3-0365-9253-4

Contents

About the Editors

Qibo Feng

Qibo Feng obtained bachelor's and master's degrees from Hefei University of Technology in 1983 and 1986, respectively, and completed a Doctor of Philosophy (PhD) degree at Tsinghua University in 1993, majoring in Optical Engineering. Currently, he is chair professor in the Discipline of Optical Engineering of Beijing Jiaotong University. He was the dean of the Faculty of Science from 2005 to 2017. His research interests mainly focus on optical measurement technologies and instrumentation, and advanced monitoring technology and equipment for railways. He has published more than 200 journal papers and achieved more than 60 Chinese patents including 7 international patents for inventions. He has acted as the principal investigator (PI) for presiding over more than 100 cases of research funding, including one Major Scientific Instruments Project, one Key Project from the National Natural Science Foundation of China, two 863 Projects from the Ministry of Science and Technology, China, etc. He has won three first prizes and three second prizes of provincial and ministerial science and technology achievement awards. He was also awarded the national teaching achievement 2nd prize, as well as the 1st and 2nd prizes in teaching achievements in Beijing.

Jiakun Li

Jiakun Li received his Bachelor and Ph.D. degrees in 2009 and 2015, respectively, from the School of Optics and Photonics, Beijing Institute of Technology, PR China. Since 2020, he has been an associate professor at the School of Physical Science and Engineering, Bejing Jiaotong University. He is mainly engaged in research on optoelectronic detection technology, including laser multi-degree-of-freedom measurement, machine vision measurement, multi-band image processing, FPGA and embedded system design, optical instrument design, etc. He is responsible for and has participated in more than 30 projects, including Major Scientific Instruments Projects and other projects funded by the National Natural Science Foundation of China, as well as provincial-level projects with enterprises and institutions. He has published over 30 SCI/EI indexed papers and obtained 1 US invention patent and over 10 national invention patents. He has won first prize and second prize at provincial and ministerial science and technology achievement awards. He is recognized as an excellent graduation design supervisor in Beijing.

Qixin He

Qixin He received his received his M.S. and Ph.D. degrees in 2013 and 2018, respectively, from the College of Electronic Science and Engineering, Jilin University, PR China. Currently, he is an associate professor at the School of Physical Science and Engineering, Bejing Jiaotong University. His research interests include infrared gas detection (TDLAS, cavity-enhanced spectroscopy, and magnetic rotation spectroscopy) and machine vision measurement (structured light measurement and digital image processing). He has published more than 20 scientific journal articles in the above technical fields as a first author or corresponding author.

sensors

MDPI

Editorial

Photoelectric Measurement and Sensing: New Technology and Applications

Qibo Feng *, Jiakun Li and Qixin He

MoE Key Lab of Luminescence and Optical Information, Beijing Jiaotong University, No. 3 Shangyuancun, Beijing 100044, China; jkli@bjtu.edu.cn (J.L.); heqixin@bjtu.edu.cn (Q.H.)
* Correspondence: qbfeng@bjtu.edu.cn

Laser-based measurement and sensing technology has been paid more and more attention by academia and industry because of its incomparable advantages, such as high sensitivity, fast response, and no contact. Its application has penetrated various fields of scientific research and industrial production, including industrial measurements, material analysis, environmental monitoring, and more.

In industrial production, laser measurement technology can be used for precision measurement, quality control, and inspection. For example, laser interferometry can be used to measure the shape and surface quality of components, while machine vision technology can be used to measure the distance and position of objects [1–3]. These technologies can help improve product quality and production efficiency. In the field of material analysis, laser spectroscopy is an important tool. For instance, laser-induced breakdown spectroscopy (LIBS) can be used to analyze the elemental composition of samples, and laser-induced fluorescence spectroscopy (LIFS) can be used to analyze the chemical substances in samples [4–6]. Furthermore, laser measurement and sensing technologies can also be used in environmental monitoring and safety fields. For example, laser radar can be used to measure the concentration and distribution of atmospheric pollutants, while laser absorption spectroscopy technology can be used for trace gas concentration analysis [7–9]. In recent years, with the development of laser sources and measurement approaches, many new technologies and applications of laser measurement and sensing have appeared [10,11].

This Special Issue aims to collect original research papers and reviews on recent developments of laser measurement technologies and innovative applications. Potential topics include, but are not limited to, laser measurement and sensing, micro- and nano-photoelectric measurement, simultaneous measurement of multiple parameters, structured light measurement, online digital measurement, computational measurement, embedded photoelectric measurement, and laser spectroscopy analysis.

Geometric parameters are important basic quantities that reflect the physical properties of an object, and they are also some of the physical quantities requiring the highest measurement accuracy in modern industrial production. For example, the measurement accuracy of a modern 3-nanometer wire's width needs to be sub-nanometer, and laser measurement has become the only choice for such measurement [12]. Length measurement is the basis of geometric measurement. Based on the combination of length and angle measurement, the shape, size, position, and attitude of the target object can be measured with high precision. In today's highly developed information society, fast and accurate simultaneous acquisition of a variety of information is the inevitable trend of the development of future measuring instruments; laser multi-parameter simultaneous measurement is also one of the future important development trends [13]. There are 10 articles in this Special Issue, in which different laser measurement methods have been used to measure geometric parameters.

In contribution 1, an absolute-type four-degrees-of-freedom (four-DOF) grating encoder that can simultaneously measure the three-axis pose (θx, θy, θz) and one-axis out-of-plane position of an object with high accuracy was demonstrated. The presented grating

Citation: Feng, Q.; Li, J.; He, Q. Photoelectric Measurement and Sensing: New Technology and Applications. *Sensors* **2023**, *23*, 8584. https://doi.org/10.3390/s23208584

Received: 16 October 2023
Accepted: 18 October 2023
Published: 19 October 2023

encoder was composed of a stationary reading head and a movable grating reflector. The authors investigated the modeling and decoupling algorithms to guarantee an independent calculation of these four-DOF absolute positions. In addition, the authors constructed a prototype and verified that the proposed grating encoder could achieve the absolute measurement of four-DOF θx, θy, θz, and Z with an accuracy of sub-arcseconds and sub-micrometers. The encoder proposed in this research is the first one to achieve absolute simultaneous measurements of four-DOF position and pose with a large measurement range. The success of this new grating encoder can benefit various multi-DOF positioning applications, especially for large-scale synthetic aperture optics (SAO), including stitching off-axis parabolic mirrors and pulse compression grating.

In contribution 2, a precision 3D measurement instrument integrating multiple laser range sensors was designed, which fuses the information of multiple redundant laser range sensors to obtain the coordinates of a 3D position. The authors developed an identification model of laser beam position and orientation parameters based on redundant distance information and standard spherical constraint to reduce the requirement for the assembly accuracy of laser range sensors. In addition, they designed a hybrid identification algorithm of PSO-LM (particle swarm optimization Levenberg–Marquardt) to solve the high-order nonlinear problem of the identification model. Experiments of identification of position and orientation, verifications of the measuring accuracy, and calibration of industrial robots were conducted, which show the effectiveness of the proposed 3D measurement instrument and identification methods. Moreover, the proposed instrument is small and can be used in narrow industrial sites.

The authors of contribution 3 proposed an adaptive hybrid sampling method for free-form surfaces based on geodesic distance. The free-form surfaces are divided into segments, and the sum of the geodesic distance of each surface segment is taken as the global fluctuation index of free-form surfaces. The number and location of the sampling points for each free-form surface segment are reasonably distributed. Compared with the common methods, this method can significantly reduce the reconstruction error for the same sampling points. This method overcomes the shortcomings of the current commonly used method of taking curvature as the local fluctuation index of free-form surfaces, and provides a new perspective for the adaptive sampling of free-form surfaces.

Contribution 4 introduced a targetless and simultaneous measurement method of three-degrees-of-freedom (3-DOF) angular motion errors using digital speckle pattern interferometry (DSPI). Based on the analysis of the sensitivity mechanism of DSPI to DOF errors and the formation mechanism of the phase fringes, the relationship between the angular motion errors and the distribution of the interferometric phases was established, and a new simultaneous measurement model of 3-DOF angular motion errors was further proposed by the authors. Furthermore, repetitive tests, noise tests, and precision analysis were carried out to verify the performance of the system. The test results showed that the measurement resolution of the system was <1 μrad, which is capable of measuring the pitch angle, yaw angle, and roll angle at the submicron arc level simultaneously without target mirrors.

In contribution 5, an improved calibration method based on coplanar constraint was proposed for a camera with a large FOV. Firstly, with an auxiliary plane mirror provided, the positions of the calibration grid and the tilt angles of the plane mirror are changed several times to capture several mirrored calibration images. Secondly, the initial parameters of the camera are calculated based on each group of mirrored calibration images. Finally, adding the coplanar constraint between each group of calibration grids, the external parameters between the camera and the reference plane are optimized via the Levenberg–Marquardt algorithm (LM). The experimental results show that the proposed camera calibration method has good robustness and accuracy.

The authors of contribution 6 described a novel laser scattering instrument that measures mass concentration and particle size in real time over a wide concentration range. The instrument combines laser scattering and time-of-flight aerodynamics in one optical device.

In this study, two APD detectors were used to receive the forward-scattered light and the side-scattered light, respectively, which can increase the sensitivity greatly. In addition, a high-speed ADC and FPGA were combined to achieve an anti-overlap algorithm objective to measure the high concentrations of aerosol. It was verified in experiments that the anti-overlapping algorithm can effectively improve the applicability of the aerodynamic particle size spectrometer under high concentration conditions.

In contribution 7, a compact and high-precision three-degrees-of-freedom (DOF; X, Y, and Z directions) grating encoder based on quadrangular frustum pyramid (QFP) prisms was proposed in this paper to solve the insufficient installation space problem of the reading head of the multi-DOF in high-precision displacement measurement applications. The authors built a three-DOF measurement platform through the self-collimation function of the miniaturized QFP prism. The overall size of the reading head is $12.3 \times 7.7 \times 3$ cm^3, and it has the potential for further miniaturization. The measurement accuracy of the main displacement is below 500 nm on average; the minimum and maximum errors are 0.07% and 2.84%, respectively. This design will help further popularize the research and applications of multi-DOF grating encoders in high-precision measurements.

In contribution 8, an evaluation method based on fitting planes was proposed to evaluate laser plane attitude and determine the degree of laser coplanarity effectively. Real-time fitting of laser planes with three planar targets of different heights provides information about the laser plane attitude on both sides of the rails. On this basis, laser coplanarity evaluation criteria were developed to determine whether the laser planes on both sides of the rails are coplanar. Using the method in this study, the laser plane attitude can be quantified and accurately assessed on both sides, effectively resolving the problem with traditional methods that can only assess the laser plane attitude qualitatively and roughly, thereby providing a solid foundation for calibration and error correction of the measurement system.

In contribution 9, a high-precision real-time pose measurement method for the primary lens of a space telescope in orbit based on laser ranging was proposed. The measurement of the pose of the primary lens in real time and with high precision is one of the important techniques for a space telescope. Through this method, the pose change of the telescope's primary lens can be easily calculated through six high-precision laser distance changes. Analysis and experiments show that this method can accurately obtain the pose of the primary lens in real time. The rotation error of the measurement system is 2×10^{-5} degrees (0.072 arcsecs), and the translation error is 0.2 μm. This study will provide a scientific basis for high-quality imaging of a space telescope.

Contribution 10 proposed a LiDAR-based method for sensing the thickness of tunnel wet spray, which aims to improve efficiency and quality. The proposed method utilizes an adaptive point cloud standardization processing algorithm to address differing point cloud postures and missing data, and the segmented Lamé curve is employed to fit the tunnel design axis using the Gauss–Newton iteration method. This establishes a mathematical model of the tunnel section and enables the analysis and perception of the thickness of the tunnel to be wet-sprayed through comparison with the actual inner contour line and the design line of the tunnel. Experimental results show that the proposed method is effective in sensing the thickness of tunnel wet spray, with important implications for promoting intelligent wet spraying operations, improving wet spraying quality, and reducing labor costs in tunnel lining construction.

Besides geometric parameter measurement, laser technology also has a wide range of applications in vibration measurement, molecular concentration detection, and more.

The authors of contribution 11 studied the most relevant figures of merit (FoM) of a DNTT-based organic phototransistor as a function of the timing parameters of light pulses to assess the device's suitability for real-time applications. The dynamic response to light pulse bursts at ~470 nm (close to the DNTT absorption peak) was characterized at different irradiances under various working conditions, such as pulse width and duty cycle. Several bias voltages were explored to allow for a trade-off to be made between operating points. Amplitude distortion in response to light pulse bursts was also addressed.

In contribution 12, the measurement characteristics of speckles based on the photoinduced electromotive force (photo-emf) effect for high-frequency, small-amplitude, and in-plane vibration were theoretically and experimentally studied. The relevant theoretical models were utilized. A GaAs crystal was used as the photo-emf detector for experimental research, as well as to study the influence of the amplitude and frequency of the vibration, the imaging magnification of the measuring system, and the average speckle size of the measuring light on the first harmonic of the induced photocurrent in the experiments. The correctness of the supplemented theoretical model was verified, and a theoretical and experimental basis was provided for the feasibility of using GaAs to measure in-plane vibrations with nanoscale amplitudes.

Contribution 13 introduced a small-scale water quality detection instrument that can detect two representative water quality parameters: the permanganate index and total dissolved solids (TDS). The permanganate index measured through the laser spectroscopy method can show the approximate value of organic matter in the water, and the TDS measured through the conductivity method can show the approximate value of inorganic matter in the water. In addition, the authors proposed an evaluation method of water quality based on the percent scores to facilitate the popularization of civilian applications. The instrument designed in this paper has the advantages of high sensitivity, high integration, and small volume, which lays the foundation for the popularity of the detection instrument.

The authors of contribution 14 studied a thermal management system for battery modules (BTMS) of a hybrid train. The authors analyzed the flow rates in each branch and the pressure losses. Since many branches of this system are built inside the battery box of the hybrid train, flow rate measurements were conducted by means of an ultrasonic clamp-on flow sensor because of its minimal invasiveness and its ability to be quickly installed without modifying the system layout. Experimental data of flow rate and pressure drop were then used to validate a lumped parameter model of the system, realized in the Simcenter AMESim® environment.

Contribution 15 mainly focuses on the dust sensing system based on the light scattering method. The authors minimized the cost by replicating the particle count (PC) of an existing dust sensing device. The existing device uses multiple sensors to measure the number of particles according to the size of dust. In this study, the authors attempted to replicate the performance of a multi-sensor device through a single-sensor device to minimize the power consumption and reduce the cost of the dust sensing system.

Machine learning and deep learning have been readily adopted in laser measurement and sensing [14–16]. Using machine learning can improve the automation level and efficiency of optoelectronic detection systems. By training and optimizing machine learning algorithms with optoelectronic detection data, automatic processing and analysis of data can be achieved, reducing manual intervention time and error rates, and improving measurement accuracy and efficiency. This Special Issue includes two articles that combine optoelectronic detection with machine learning.

Contribution 16 introduced an improved blind/referenceless image spatial quality evaluator (BRISQUE) algorithm. The algorithm was formulated by using image characteristic extraction technology to obtain a characteristic vector (CV) that consisted of 36 characteristic values that could effectively reflect the defocusing condition of the corresponding image. The authors constructed an image database that contained a sufficient number of training samples. The trained model is trained to obtain the support vector machine (SVM) model by using the regression function of the SVM. The method of establishing the image definition evaluation model via SVM is feasible and yields higher subjective and objective consistency.

Contribution 17 introduced a polarization imaging device of cotton foreign fiber based on the difference in optical properties and polarization characteristics between cotton fibers. The authors proposed an object detection and classification algorithm based on an improved YOLOv5 to achieve small foreign fiber recognition and classification. In the algorithm, the lightweight network Shufflenetv2 with the Hard-Swish activation function was used as the backbone feature extraction network. The PANet network connection of YOLOv5 was modified to obtain a fine-grained feature map to improve the detection accuracy for small targets. A CA attention module was added to the YOLOv5 network to increase the weight of the useful features while suppressing the weight of invalid features to improve the detection accuracy of foreign fiber targets. The model volume, mAP@0.5, mAP@0.5:0.95, and FPS of the improved YOLOv5 were up to 0.75 MB, 96.9%, 59.9%, and 385 f/s, respectively, compared to YOLOv5, and the improved YOLOv5 increased by 1.03%, 7.13%, and 126.47%, respectively, which proves that the method can be applied to the vision system of an actual production line for cotton foreign fiber detection.

Contributions 18 and 19 are review papers. Contribution 18 focused on the wheel flat detection techniques and flat signal processing methods based on wayside deployment. The timely and accurate detection of wheel flats is of great significance to ensure the safety of train operation and reduce maintenance costs. Commonly used wheel flat detection methods, including sound-based methods, image-based methods, and stress-based methods, are introduced and summarized. The advantages and disadvantages of these methods are discussed, and conclusions are drawn. In addition, the flat signal processing methods corresponding to different wheel flat detection techniques are also summarized and discussed by the authors.

In contribution 19, the advances in research on oscillation principles and key technologies of the different kinds of dual-frequency solid-state lasers are reviewed, including birefringent dual-frequency solid-state lasers and biaxial and two-cavity dual-frequency solid-state lasers. The system composition, operating principle, and some main experimental results are briefly introduced. Several typical frequency difference stabilizing systems for dual-frequency solid-state lasers are introduced and analyzed. The main development trends of research on dual-frequency solid-state lasers are predicted.

In summary, This Special Issue presents a variety of advanced laser measurement techniques and their interesting applications in many areas. We hope that this SI will help researchers to better understand the state of the art of laser-based measurement and sensing technologies. We hope that the 19 published papers will also help researchers working in the field to disclose future perspectives.

List of Contributions

1. Wang, S.; Luo, L.; Zhu, J.; Shi, N.; Li, X. An Ultra-Precision Absolute-Type Multi-Degree-of-Freedom Grating Encoder. *Sensors* **2022**, *22*, 9047; https://doi.org/10.3390/s22239047.
2. Gao, G.; Kuang, L.; Liu, F.; Xing, Y.; Shi, Q. Modeling and Parameter Identification of a 3D Measurement System Based on Redundant Laser Range Sensors for Industrial Robots. *Sensors* **2023**, *23*, 1913; https://doi.org/10.3390/s23041913.
3. Chen, C.; Jia, H.; Lu, Y.; Zhang, X.; Chen, H.; Yu, L. An Adaptive Hybrid Sampling Method for Free-Form Surfaces Based on Geodesic Distance. *Sensors* **2023**, *23*, 3224; https://doi.org/10.3390/s23063224.
4. Shi, L.; Wu, S.; Yan, M.; Niu, H. A Targetless Method for Simultaneously Measuring Three-Degree-of-Freedom Angular Mo-tion Errors with Digital Speckle Pattern Interferometry. *Sensors* **2023**, *23*, 3393; https://doi.org/10.3390/s23073393.
5. Lu, R.; Wang, Z.; Zou, Z. Accurate Calibration of a Large Field of View Camera with Coplanar Constraint for Large-Scale Specular Three-Dimensional Profile Measurement. *Sensors* **2023**, *23*, 3464; https://doi.org/10.3390/s23073464.
6. Zhang, J.; Zhang, Z.; Hou, L.; Zhou, W. A Novel Optical Instrument for Measuring Mass Concentration and Particle Size in Real Time. *Sensors* **2023**, *23*, 3616; https://doi.org/10.3390/s23073616.

7. Wang, S.; Liao, B.; Shi, N.; Li, X. A Compact and High-Precision Three-Degree-of-Freedom Grating Encoder Based on a Quadrangular Frustum Pyramid Prism. *Sensors* **2023**, *23*, 4022; https://doi.org/10.3390/s23084022.
8. Wang, L.; Wang, H.; Han, Q.; Fang, Y.; Wang, S.; Wang, N.; Li, G.; Ren, S. A Laser Plane Attitude Evaluation Method for Rail Profile Measurement Sensors. *Sensors* **2023**, *23*, 4586; https://doi.org/10.3390/s23104586.
9. Shi, H.; Du, J.; Wang, L.; Bian, J.; Gao, G.; Liu, D.; Fan, B.; Yang, H. A High-Precision Real-Time Pose Measurement Method for the Primary Lens of Large Aperture Space Telescope Based on Laser Ranging. *Sensors* **2023**, *23*, 4833; https://doi.org/10.3390/s23104833.
10. Xu, D.; Song, Q.; Fang, S.; Guo, Y. Sensing Method for Wet Spraying Process of Tunnel Wall Based on the Laser LiDAR in Complex Environment. *Sensors* **2023**, *23*, 5167; https://doi.org/10.3390/s23115167.
11. Campajola, M.; Di Meo, P.; Di Capua, F.; Branchini, P.; Aloisio, A. Dynamic Photoresponse of a DNTT Organic Phototran-sistor. *Sensors* **2023**, *23*, 2386; https://doi.org/10.3390/s23052386.
12. Gao, J.; Zhang, B.; Feng, Q.; Shen, X.; Xue, Y.; Liu, J. Speckle Measurement for Small In-Plane Vibration Using GaAs. *Sensors* **2023**, *23*, 2724; https://doi.org/10.3390/s23052724.
13. Tian, Z.; Chen, H.; Ding, Q.; Che, X.; Bi, Z.; Wang, L. Research on Small-Scale Detection Instrument for Drinking Water Combined Laser Spectroscopy and Conductivity Technology. *Sensors* **2023**, *23*, 2985; https://doi.org/10.3390/s23062985.
14. De Rosa, R.; Romagnuolo, L.; Frosina, E.; Belli, L.; Senatore, A. Validation of a Lumped Parameter Model of the Battery Thermal Management System of a Hybrid Train by Means of Ultrasonic Clamp-On Flow Sensor Measurements and Hydronic Optimization. *Sensors* **2023**, *23*, 390; https://doi.org/10.3390/s23010390.
15. Lee, S.; Kwon, J.; Park, D. Optimized Replication of ADC-Based Particle Counting Algorithm with Reconfigurable Multi-Variables in Pseudo-Supervised Digital Twining of Reference Dust Sensor Systems. *Sensors* **2023**, *23*, 5557; https://doi.org/10.3390/s23125557.
16. Zhang, N.; Lin, C. The Image Definition Assessment of Optoelectronic Tracking Equipment Based on the BRISQUE Algo-rithm with Gaussian Weights. *Sensors* **2023**, *23*, 1621; https://doi.org/10.3390/s23031621.
17. Wang, R.; Zhang, Z.; Yang, B.; Xi, H.; Zhai, Y.; Zhang, R.; Geng, L.; Chen, Z.; Yang, K. Detection and Classification of Cotton Foreign Fibers Based on Polarization Imaging and Improved YOLOv5. *Sensors* **2023**, *23*, 4415; https://doi.org/10.3390/s23094415.
18. Fu, W.; He, Q.; Feng, Q.; Li, J.; Zheng, F.; Zhang, B. Recent Advances in Wayside Railway Wheel Flat Detection Techniques: A Review. *Sensors* **2023**, *23*, 3916; https://doi.org/10.3390/s23083916.
19. Jiao, M.; Jiang, F.; Xing, J.; Liu, Y.; Lian, T.; Liu, J.; Li, G. Advances of Research on Dual-Frequency Solid-State Lasers for Synthetic-Wave Absolute-Distance Interferometry. *Sensors* **2023**, *23*, 3206; https://doi.org/10.3390/s23063206.

Conflicts of Interest: The authors declare no conflict of interest.

References

1. Jiang, B. Low Noise Phase-locked Laser System for Atom Interferometry. *Appl. Phys. B* **2022**, *128*, 71. [CrossRef]
2. Bazaz, H.; Fatimah, M.; Asim, L.; Zabit, U.; Bernal, O. Integration of Zero Crossing Method in a Nonuniform Sampling System Using Optical Feedback Interferometry. *IEEE Sens. J.* **2023**, *23*, 14397. [CrossRef]
3. Chen, Y.; Li, Y.; Niu, G.; Zuo, M. Offline and Online Measurement of the Geometries of Train Wheelsets: A Review. *IEEE Trans. Instrum. Meas.* **2022**, *71*, 3523915. [CrossRef]
4. Harmon, R.; Senesi, G. Laser-Induced Breakdown Spectroscopy—A geochemical tool for the 21st century. *Appl. Geochem.* **2021**, *128*, 104929. [CrossRef]
5. Galbács, G. A Critical Review of Recent Progress in Analytical Laser-induced Breakdown Spectroscopy. *Anal. Bioanal. Chem.* **2015**, *407*, 7537. [CrossRef] [PubMed]

6. Li, J.; Chu, Y.; Zhao, N.; Zhou, R.; Yi, R.; Guo, L.; Li, J.; Li, X.; Zeng, X.; Lu, Y. Detection of Trace Elements in Active Luminescent Glass Using Laser-induced Breakdown Spectroscopy Combined with Laser-induced Fluorescence. *Chin. J. Anal. Chem.* **2016**, *44*, 1042. [CrossRef]
7. Jiang, Y.; Karpf, S.; Jalali, B. Time-stretch LiDAR as a Spectrally Scanned Time-of-flight Ranging Camera. *Nat. Photon.* **2020**, *14*, 14. [CrossRef]
8. Lin, S.; Chang, J.; Sun, J.; Xu, P. Improvement of the Detection Sensitivity for Tunable Diode Laser Absorption Spectroscopy: A Review. *Front. Phys.* **2022**, *10*, 853966. [CrossRef]
9. Bolshov, M.; Kuritsyn, Y.; Romanovskii, Y. Tunable Diode Laser Spectroscopy as a Technique for Combustion Diagnostics. *Spectrochim. Acta Part B At. Spectrosc.* **2015**, *106*, 45. [CrossRef]
10. Victor, T.; Weiner, A. Optical Frequency Comb Technology for Ultra-broadband Radio-frequency Photonics. *Laser Photon. Rev.* **2014**, *8*, 368.
11. Zhang, C.; Qu, F.; Ou, P.; Sun, H.; He, S.; Fu, B. Recent Advances and Outlook in Single-Cavity Dual Comb Lasers. *Photonics* **2017**, *10*, 221. [CrossRef]
12. Chen, Z.; Pu, H.; Liu, X.; Peng, D.; Yu, Z. A Time-Grating Sensor for Displacement Measurement with Long Range and Nanometer Accuracy. *IEEE Trans. Instrum. Meas.* **2015**, *64*, 3105. [CrossRef]
13. Feng, Q.; Zhang, B.; Cui, C.; Kuang, C.; Zhai, Y.; You, F. Development of a simple system for simultaneously measuring 6DOF geometric motion errors of a linear guide. *Opt. Express* **2015**, *21*, 25805.
14. Zuo, C.; Qian, J.; Feng, S.; Yin, W.; Li, Y.; Fan, P.; Han, J.; Qian, K.; Chen, K. Deep learning in optical metrology: A review. *Light Sci. Appl. Rev. Artic.* **2022**, *11*, 39. [CrossRef] [PubMed]
15. Cui, Y.; Chen, R.; Chu, W.; Chen, L.; Tian, D.; Li, Y.; Cao, D. Deep Learning for Image and Point Cloud Fusion in Autonomous Driving: A Review. *IEEE Trans. Intell. Transp. Syst.* **2022**, *23*, 722. [CrossRef]
16. Genty, G.; Salmela, L.; Dudley, J.; Brunner, D.; Kokhanovskiy, A.; Kobtsev, S.; Turitsyn, S. Machine learning and applications in ultrafast photonics. *Nat. Photon.* **2021**, *15*, 91. [CrossRef]

sensors

MDPI

Article

An Ultra-Precision Absolute-Type Multi-Degree-of-Freedom Grating Encoder

Shengtong Wang [1], Linbin Luo [1], Junhao Zhu [1], Ningning Shi [1] and Xinghui Li [1,2,*]

1 Tsinghua Shenzhen International Graduate School, Tsinghua University, Shenzhen 518055, China
2 Tsinghua-Berkeley Shenzhen Institute, Tsinghua University, Shenzhen 518055, China
* Correspondence: li.xinghui@sz.tsinghua.edu.cn

Abstract: An absolute-type four-degree-of-freedom (four-DOF) grating encoder that can simultaneously measure the three-axis pose (θx, θy, θz) and one-axis out-of-plane position (Z) of an object with high accuracy is demonstrated for the first time in this research. This grating encoder is composed of a stationary reading head and a movable grating reflector. A light beam from the reading head is projected onto the grating, and three diffracted beams (0th-, +1st-, and −1st-order) are generated, collimated, and received by three separate quadrant photodetectors (QPDs). The information of θx, θy, θz, and Z is coded into spot positions of these three diffracted beams on the QPDs. Thus, the modeling and decoupling algorithms were investigated, and an independent calculation of these four-DOF absolute positions was theoretically guaranteed. A prototype was then designed, constructed, and evaluated. Experimental results verified that the proposed grating encoder could achieve the absolute measurement of four-DOF θx, θy, θz, and Z with an accuracy of sub-arcseconds and sub-micrometers. To the best of our knowledge, the proposed encoder in this research is the first one to achieve absolute simultaneous measurements of four-DOF position and pose with a large measurement range. The success of this new grating encoder can benefit various multi-DOF positioning applications, especially for large-scale synthetic aperture optics (SAO), including stitching off-axis parabolic mirrors and pulse compression grating.

Keywords: precision positioning; multi-degree-of-freedom; absolute measurement; grating encoder; synthetic aperture optics; laser autocollimation

Citation: Wang, S.; Luo, L.; Zhu, J.; Shi, N.; Li, X. An Ultra-Precision Absolute-Type Multi-Degree-of-Freedom Grating Encoder. *Sensors* **2022**, 22, 9047. https://doi.org/10.3390/s22239047

Academic Editors: Qibo Feng, Jiakun Li and Qixin He

Received: 16 October 2022
Accepted: 21 November 2022
Published: 22 November 2022

1. Introduction

Large-scale synthetic aperture optics (SAO) systems are highly important in several applications, such as deep space exploration [1], high-energy laser physics, national defense security [2], and other basic research fields [3,4]. The position and pose monitoring of the sub-mirror affects its performance [5]. Therefore, highly accurate, absolute-type, multi-degree-of-freedom position and pose monitoring equipment is necessary for aligning or adjusting each sub-mirror to its ideal posture.

Currently, the multi-degree-of-freedom position and pose measurement schemes mainly include optical and electrical schemes. The electrical method can demonstrate a high accuracy but is limited in its applications [6,7] because of its short measurement range, the requirement of conductive measurement target, and complexity of expansion to multi-DOF. The optical scheme can provide both a nanometric accuracy and a large range, mainly represented by the laser interferometer [8,9] and the grating encoder [10–13]. However, the laser interferometer is vulnerable to environmental variation because of its long exposed optical path, which greatly influences the measurement standard, that is, light wavelength. In contrast, the optical path in the grating encoder is much shorter, and its primary measuring standard is the physical grating pitch [14–16]. Thus, the grating encoder is less affected by the environment, normally demonstrates a more stable performance, and has great potential for precision position and pose measurement [17–19]. However, the state-of-the-art grating encoders mainly focus either on the absolute measurement in one-axis, or on the multi-DOF incremental measurement [20–23], and there are few types of research on absolute-type multi-DOF measurement, and it remains a large challenge to combine them while keeping a compact and lower weight structure. Thus, there is a pressing need to develop a grating-based 6-DOF absolute position and pose measurement method with high stability, accuracy, and compactness.

To address the same problems, some effective measurement schemes were proposed for in-plane position measurement by phase detection [24]. Li et al. [25,26] proposed an absolute 2-DOF encoder with two probes, which uses the correlation of reference codes to obtain the absolute position information of the measurement target and achieves a 0.5 μm absolute position accuracy in X- and Y-directions. Furthermore, Shi et al. [27] proposed a hybrid-positioning methodology that combines a pulse signal generated from the correlation of reference codes as that in Ref. [25] and an incremental interference signal to improve the positioning accuracy of the reference position. The positioning repeatability was greatly improved and reached 10 nm for the motion range of several tens of millimeters. Due to the success of the in-plane X- and Y-direction absolute position measurement, the main research work of the six-DOF absolute measurement can focus on the out-of-plane, i.e., the 4-DOF absolute position and pose measurement of θx, θy, θz, and Z-direction.

Several related types of research have been proposed for out-of-plane measurement [28,29]. Gao et al. first demonstrated a three-DOF (θx, θy, θz,) autocollimator by using a grating reflector to replace the flat mirror in the conventional laser autocollimator [30,31]. Although this innovative autocollimator can provide a high resolution and a simultaneous measurement of three-DOF poses, this is mainly for incremental measurement and not Z-direction position measurement. Then, Liu et al. modified the proposal in Ref. [30] and proposed a four-DOF system, allowing simultaneous measurement of four error motions involving Z-direction [32], which was then expanded to be five-degrees-of-freedom approach [33]. These studies still focused on incremental measurement tests and did not discuss the possibility of absolute measurement or inevitable alignment error compensation, which is indispensable for these kinds of multi-DOF measurement systems. They also do not demonstrate each posture for the multi-DOF position and pose motion inputs, so these studies lack comprehensive evaluation of decoupling accuracy. The actual experiment ranges in these studies were not sufficient at less than 4 arcseconds and with a large relative measurement error, e.g., the 3×STDEV of yaw error is 0.4″ in the range of 0.6″ also prevented its advancement.

In order to meet the high accuracy and absolute measurement demand of the out-of-plane 4-DOFs of the sub-mirror, three improvement works based on the abovementioned multi-DOF out-of-plane research were carried out in this study. Firstly, an absolute zero-point of the QPD coordinate system was proposed to establish the absolute coordinate of the spot position, and the absolute position and pose of the grating reflector can be decoupled by the absolute coordinates of the three diffracted spots (+1st-, 0th-, and −1st-order beams). Secondly, a homogeneous error compensation matrix involving installation posture error and installation distance error was proposed, which can significantly reduce the crosstalk error and improve the measurement accuracy. Finally, a compact prototype system was designed and built in this study. For the first time, a verification experiment of simultaneous input of multiple main motions was used to obtain the optimal homogeneous error compensation matrix, and high-precision independent decoupling of the absolute position and pose of the out-of-plane 4-DOF was realized, fully verifying the excellent performance of the proposed grating encoder.

2. Principle and Method

As shown in Figure 1a, the change in the SAO focus and the error between the ideal and real-time posture of the sub-mirror decreases the SAO performance. The proposed encoder is used as a monitor in the system described in Figure 1b to help the actuator to adjust the sub-mirror to its ideal posture.

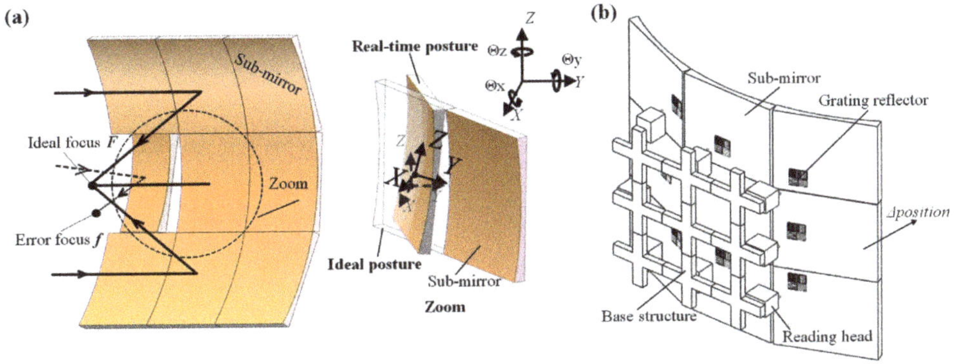

Figure 1. (**a**) The focus change of the synthetic aperture optics (SAO) and the ideal and real-time postures of the sub-mirror; (**b**) the application of the encoder.

Figure 2a shows the schematic of the grating encoder. This grating encoder is composed of a stationary reading head and a movable grating reflector. The 660 nm wavelength laser beam is emitted from the laser diode (LD) and illuminates the grating reflector through the beam splitter (BS). The diffraction beams are the +1st-, 0th-, and −1st-order beams, which are refracted and focus on three QPDs through three convex lenses (CLs). The position information regarding the light spot on the QPD is shown in Figure 2b. The specific positions (x_A, y_A, x_B, y_B, x_C, y_C) of the light spot are calculated according to the back-end photocurrent information, and the specific calculation formula can be expressed as formula (1). In this paper, the obtained coordinates are absolute coordinates; thus, the position coordinates on the QPD are all signed, and from the front of the QPD, right and up are positive directions.

$$\begin{cases} x_\alpha = \frac{(I_{\alpha 1}+I_{\alpha 2})-(I_{\alpha 3}+I_{\alpha 4})}{I_{\alpha 1}+I_{\alpha 2}+I_{\alpha 3}+I_{\alpha 4}}, \\ y_\alpha = \frac{(I_{\alpha 1}+I_{\alpha 3})-(I_{\alpha 2}+I_{\alpha 4})}{I_{\alpha 1}+I_{\alpha 2}+I_{\alpha 3}+I_{\alpha 4}}, \end{cases} \quad (1)$$

where I_α are the four-way current signal outputs of the QPD, α = A, B, or C.

Taking QPDA as an example, due to different light intensity distributions, the coordinate value can be calculated from Formula (1). The coordinate origin of the two-dimensional coordinate system is the symmetric center of the four photodetectors. Due to the mechanical structure, it is difficult to adjust the light spot to be perfectly located at the symmetrical center of the QPD. Therefore, at the beginning of the test, the light spot is located at the position (x_A, y_A) of the coordinate system of the QPD, and x_A and y_A are very close to 0. The point (x_A, y_A) is used as the origin of a new QPDA coordinate system in Figure 2b, which is represented in the form of the dashed line. In the new QPD coordinate system, the light spot is located at the coordinate origin, The new coordinate systems of QPDB and QPDC are simultaneously established according to this method.

First, the initial positions of the spots are set as the zero-point positions, i.e., every spot position is at the origin of the new QPD coordinate system. When the grating moves simultaneously in four degrees of freedom, the three light spots in the QPDA, QPDB, and QPDC assume the (x_1, y_1), (x_0, y_0), and (x_{-1}, y_{-1}) positions, respectively, which are unique coordinates for calculating the absolute position and pose.

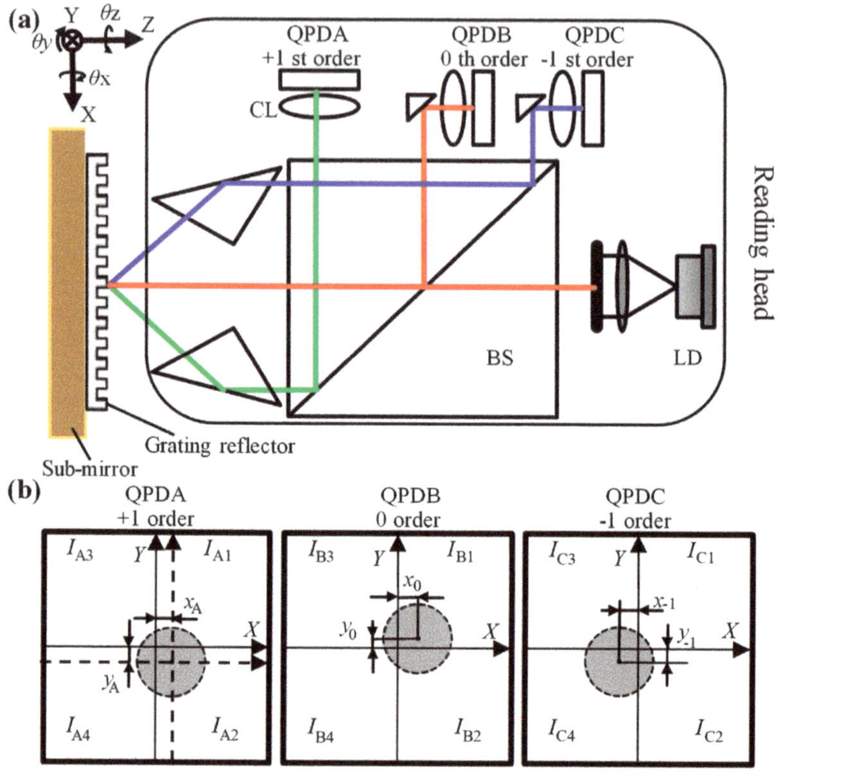

Figure 2. (**a**) The schematic of the encoder; (**b**) the spot coordinate of the QPDs.

In these coordinates, the position of the 0th-order light change is only related to θx and θy, which can be solved in Formulas (2) and (3). In addition, z and θz also cause the movement of the other spots and can be solved using Formulas (4) and (5).

$$\theta_x = \frac{k_{\theta_x} y_0}{2f} = k_{\theta xz1} \frac{k_{\theta_x} y_1}{2f} = k_{\theta xz-1} \frac{k_{\theta_x} y_{-1}}{2f}, \tag{2}$$

$$\theta_y = \frac{k_{\theta_y} x_0}{2f} = k_{\theta ydz1} \frac{k_{\theta_y} x_1}{2f} = k_{\theta ydz-1} \frac{k_{\theta_y} x_{-1}}{2f}, \tag{3}$$

$$\theta_z = k_{\theta z} \frac{(y_{-1} - y_0/k_{\theta_{xz-1}}) - (y_1 - y_0/k_{\theta_{xz1}})}{2L}, \tag{4}$$

$$z = k_z \frac{x_{-1} - x_0/k_{\theta ydz-1} + x_1 - x_0/k_{\theta ydz1}}{2}, \tag{5}$$

where k_z, $k_{\theta x}$, $k_{\theta y}$, and $k_{\theta z}$ are the parameters measured in the pre-experiment. f is the focal length of the convex lens, and L is the equivalent distance between the photodetectors A and B (or C and B). $k_{\theta xz1}$ and $k_{\theta xz-1}$ are the impact factors of θx on the +1st- and −1st-order spot positions during the movement, and so are $k_{\theta ydz1}$ and $k_{\theta ydz-1}$ of θy.

However, in the simultaneous motions of 4-DOF absolute position and pose experiment, there is an unexpected posture error inevitably generated between the grating encoder measured value ***O*** (original value, shown in Formula (6)) and the true value ***T*** (shown in Formula (7)) due to the imperfection in grating encoder alignment. This posture error can be viewed as a system error and be predicted by using a compensation matrix. The error compensation matrix is a 4 × 4 matrix (***M***), which can be obtained using Formula

(8) after the pre-experiment. With this error compensation matrix, the original value can be converted into the output of the grating encoder (θx_Encoder, θy_Encoder, θz_Encoder, z_Encoder).

$$O = \begin{bmatrix} \theta x_{1O\_main} & \theta x_{2O} & \theta x_{3O} \\ \theta y_{1O} & \theta y_{2O} & \theta y_{3O} \\ \theta z_{1O} & \theta z_{2O\_main} & \theta z_{3O} \\ z_{1O} & z_{2O} & z_{3O\_main} \end{bmatrix}, \tag{6}$$

$$T = \begin{bmatrix} \theta x_{1T\_main} & \theta x_{2T} & \theta x_{3T} \\ \theta y_{1T} & \theta y_{2T} & \theta y_{3T} \\ \theta z_{1T} & \theta z_{2T\_main} & \theta z_{3T} \\ z_{1T} & z_{2T} & z_{3T\_main} \end{bmatrix}, \tag{7}$$

$$\boldsymbol{M} \times \boldsymbol{O} = \boldsymbol{T}, \tag{8}$$

where $\boldsymbol{M}$ can be obtained in the pre-experiment with one main motion.

Due to installation errors, the 4-DOF movement is output simultaneously. The three main motions move separately, i.e., θx, θz, z, thus generating three groups of main and error motions. Then, A matrix $\boldsymbol{O}$ composed of θx_Original, θy_Original, θz_Original, and z_Original can be collected, and $\boldsymbol{M} = \boldsymbol{T} \times \boldsymbol{O}^{-1}$.

3. Experiments and Discussion

The absolute 4-DOF grating encoder test bench was built, and the schematic figure of the experiment is shown in Figure 3. The measurement truth value is mainly obtained by the dual-frequency laser interferometer and the autocollimator. A reflector is placed below the grating at 45 degrees, and a plane reflector is pasted at the grating so that the laser autocollimator can measure θx and θz.

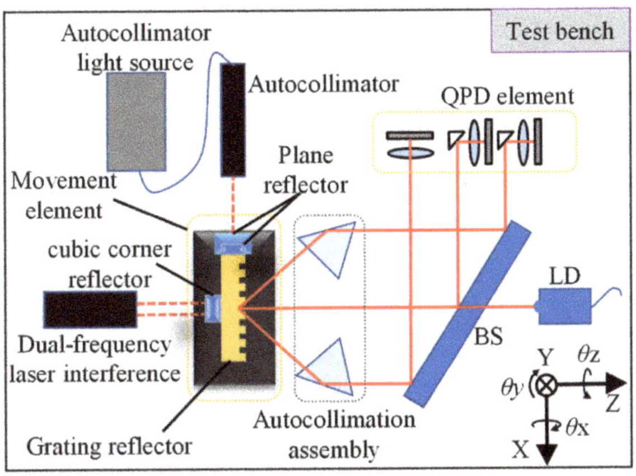

Figure 3. Schematic figure of test bench.

The actual test bench is shown in Figure 4a. The autocollimator can measure the absolute angle (θx, θz) of the grating with a measurement accuracy of 0.1″. The laser interferometer can measure the absolute displacement of the z-direction with a resolution and measurement accuracy of 0.1 nm and ±160 nm, respectively. The movement element is shown in Figure 4b, which can give the main motions of θx, θz, z, and error motion θy. The z-direction movement is controlled by the high-dynamic Z Nanopositioner (PI) (model: p-733. Z), with a resolution of 0.3 nm and a linearity of 0.03%, and the angles are controlled by a two-axis tilt table driver, with a resolution of 0.02″ and a linearity of 0.5%.

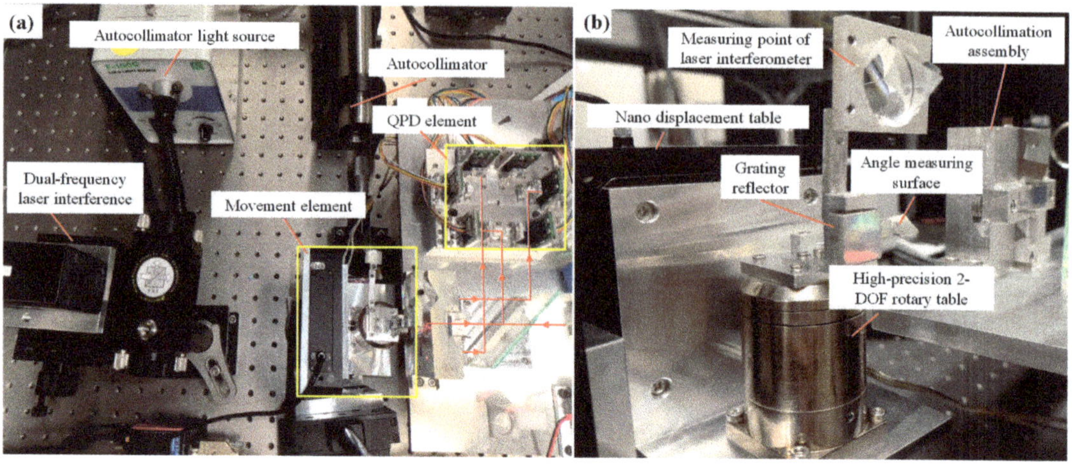

Figure 4. (**a**) The test bench; (**b**) movement element.

The value measured by the laser interferometer and autocollimator is called the true value, as described below and shown in the following figure as θx_True, θz_True, and z_True. Owing to the limitation of the autocollimator, the true value of θy was obtained during another set of repeated experiments by changing the measurement surface. The following single motion, resolution, and multiple motion experiments were carried out in this study.

The pre-experiment was performed in $\theta x(80'')$, $\theta z(80'')$, and the z-direction (100 μm) as the main motions and obtained all the constant parameters and the error compensation matrix (***M***) in Formulas (2)–(8). As is evident from Figure 2b, it is difficult to adjust the spot center to the QPD center, but when the spot is located at the center of the QPD, the measurement sensitivity and the resolution of the encoder is the highest. In this study, the distance between the spot center and the QPD center is represented by the coefficient of variation. The coefficients of variation, i.e., the ratio of standard deviation to average of $I_{\alpha 1}$, $I_{\alpha 2}$, $I_{\alpha 3}$, and $I_{\alpha 4}$ on QPDs A, B, and C are 0.1802, 0.2662, and 0.3135, respectively. Upon this, the resolutions of rotation around the x- and z-axes and the displacement along the z-axis are $0.02''$, $0.06''$, and 20 nm, respectively, as shown in Figure 5a,b. Thus, the coefficient of variation values confirm that the encoder's resolution and accuracy can be further improved, and it can work with high accuracy after being disturbed in practical use.

In Figure 5c,d, when a single freedom motion is provided by the movement element in 5 s, the residual errors are −0.60 to 0.70 arcsec in 50 arcsec θx rotation, −0.81 to 0.85 arcsec in 40 arcsec θz rotation, and −0.55 to 0.49 μm in the 100 μm z-direction displacement. This indicates that in the continuous measurement of a single degree of freedom, the encoder has good linearity and stability without a large drift.

In multiple motion experiments, the square wave rotation motions with a frequency of 0.1 Hz and the amplitude of $40''$ and $80''$ around the x- and z-axes and a z-direction motion with 100 μm at a speed of 2 μm/s are simultaneously given by the movement element. Three main movements are input concurrently, and their comparison with the true value is shown in Figure 6. Since the movement element is a square wave motion, there are data points during an abrupt change, and the original value cannot align with the true value on the timeline. Therefore, the data areas that rise or fall steadily were selected for data comparison, and the length of time was also changed in one simultaneous test. Since the movement element's accuracy is considerably high, this comparison method can characterize the measurement performance of the encoder.

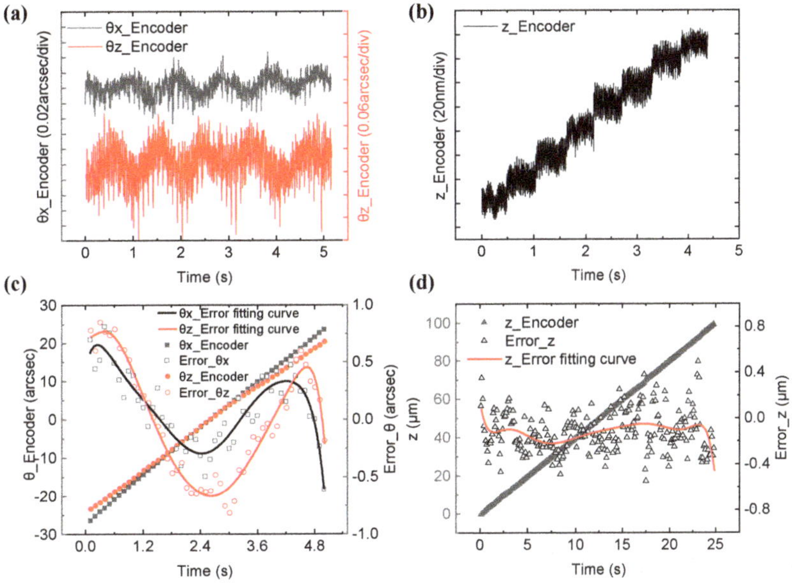

Figure 5. (**a**) Angle resolution of the encoder; (**b**) displacement resolution of the encoder; (**c**) angle error between the encoder value and the true value in a single degree; (**d**) the displacement error between the encoder value and the true value in a single degree.

Because the measurement point of the z-axis displacement is different from that of the grating encoder in periodic motion, a moving arm length linearly increases the displacement along the z-axis. Since the error matrix calculation includes the linear relationship based on this structure, the measured value of the laser interferometer represents the true value of the grating displacement herein.

During the period ($\theta x = 40''$, $\theta z = 40''$, $z = 100$ μm) shown in Figure 6d–f before the error matrix compensation, the grating encoder's original value is stable but not correct. From the output of the encoder's original signal, because of the simultaneous movement of multiple degrees of freedom, the angular measurement error around the x-axis is about $-3''$ to $7''$, while the angular measurement error around the z-axis is large, reaching the degree of about $\pm 40''$. Additionally, the angular measurement error first decreases overall and then rises and returns, mainly because the rotation around the z-axis is affected by z-axis displacement. The error of the original value is far from the true value; therefore, the original signal requires compensation by the error matrix.

Figure 6d–i show that in the output after compensation, the measurement error is greatly reduced. When the motion period ($\theta x = 40''$, $\theta z = 40''$, $z = 100$ μm) is applied to the grating, the θz error range is ($-23.58''$, $18.01''$) before compensation and ($-5.56''$, $0.98''$) after compensation. The error range in z-direction position is (-13.64, 13.94 μm) before compensation and (-2.33, 3.26 μm) after compensation. When the motion period ($\theta x = 80''$, $\theta z = 80''$, $z = 100$ μm) is applied to the grating after compensation, the θz error range is ($-6.52''$, $3.91''$), and the error range in the z-direction position is reduced to (-4.13, 3.42 μm). θx remains a relatively low error range without obvious change in these two periods. During the period ($\theta x = 40''$, $\theta z = 40''$, $z = 100$ μm) shown in Figure 6d–f, the root mean square (RMS) of the errors of the encoder value relative to the true value is shown in Table 1. The standard error around the x-axis increased by 2.26%, mainly because the fluctuations of the other three degrees of freedom were superimposed on it after compensation by the error matrix. The RMS of the errors around the z-axis was reduced by 78.99%, and the RMS of the errors of the displacement of the z-axis was reduced by 88.14%, demonstrating the effectiveness of the error matrix compensation method. As shown in Figure 6g–i, when the period is $\theta x = 80''$, $\theta z = 80''$, and $z = 100$ μm, the same compensation matrix is used, and the RMS of the errors of the encoder value is $3.02''$, $3.25''$, and 1.21 μm in θx, θz, and z-direction, respectively. When the measuring range is enlarged, the error does not increase obviously, which proves the effectiveness and reliability of the compensation method.

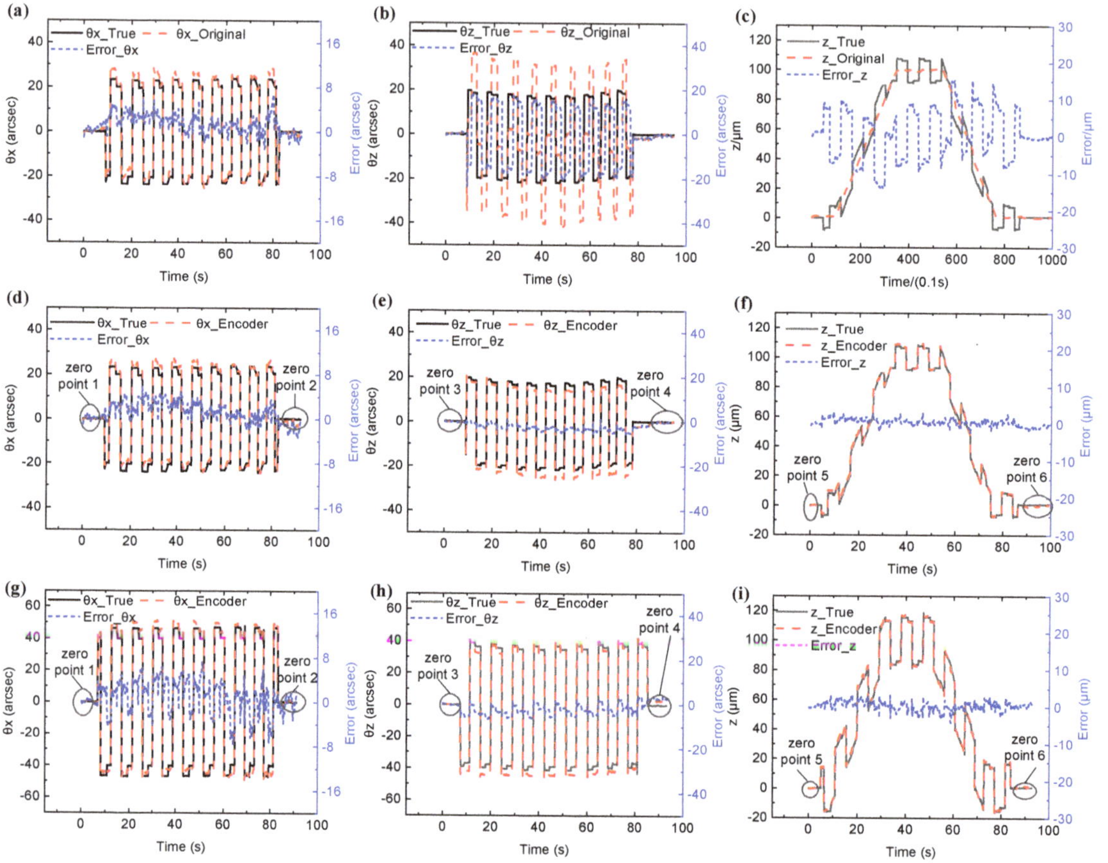

Figure 6. (**a–c**) Position and pose errors without compensation; (**d–i**) position and pose errors after compensation.

Table 1. The RMS of the error of the encoder value relative to the true value.

Error Value	Original Value	After Compensation	Error Change Percentage/%
$\theta x/('')$	2.06	2.11	+2.26
$\theta z/('')$	13.24	2.78	−78.99
$z/(\mu m)$	8.66	1.03	−88.14

Table 2 shows the average residual error of the zero-point of the absolute four-degree encoder with a different motion period. The encoder is stable in the sub-arcsecond and sub-micron when returning to the origin of position and pose in θx and z-direction. Initially, the precision of sub-angular seconds can be reached in θz; however, after a movement of long duration, the average measurement residuals of $-1.03''$ and $2.93''$ are generated, respectively. This is because in the calculation of the roll angle of the grating, the rotation movement of the light spot is simplified to a linear movement, and its initial position is not at the QPD center, resulting in the enhancement of nonlinearity and error during the measurement.

Table 2. Encoder average residual of the zero-point.

Zero-Point	1/2(″)	3/4(″)	5/6(μm)
40″40″100μm	0.10/−1.46	−0.21/−1.03	0.0002/−0.73
80″80″100μm	−0.07/−0.27	−0.14/2.93	0.10/0.62

Since the autocollimator cannot measure 3-DOF angles simultaneously, only the compensation data are shown in Figure 6. The error motion of θ_y is not caused by crosstalk but by the motion of the movement element itself due to the installation, which could be detected in the repeated experiments. The data time length in Figure 7 is different from that of Figure 6 as it does not align the time with the true value and retains the original value, which is highly unstable. After the compensation, θ_y is more stable and shows a similar change of displacement of the *z*-axis, indicating that the compensation matrix method is correct. It proves that during the simultaneous operation of the four degrees of freedom, although the DOFs affect each other, they share a relatively linear relationship within a certain range. A larger measurement range and accuracy can be achieved by optimizing the nonlinear formula and the design of the optical path onto the QPD center.

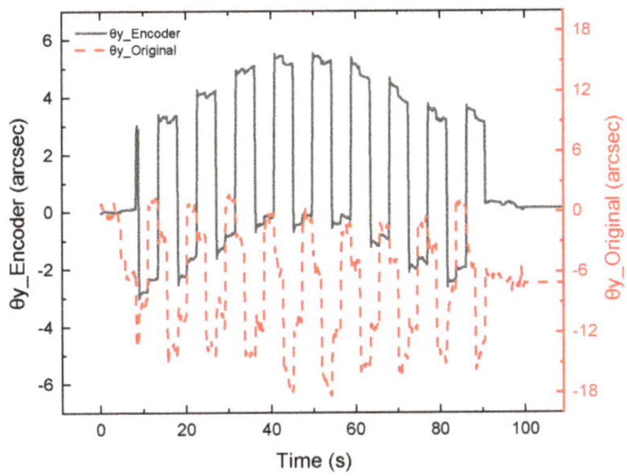

Figure 7. The θy value (40″40″100 μm) comparison between the original value and the value after compensation.

4. Conclusions

This paper proposes an ultra-precision absolute-type multi-degree-of-freedom grating encoder. Three aspects are studied as follows: Firstly, the absolute position matrix was constructed by establishing the zero-points of the diffracted light spots on QPD coordinates, and the absolute 4-DOF position and pose (θx, θy, θz, and Z) of the grating reflector in space can be decoupled. Secondly, the homogeneous error compensation matrix was proposed to greatly reduce the crosstalk error caused by installation posture error and installation distance error and improve the accuracy. Finally, a compact prototype system was designed, and the effectiveness of the proposed structure and compensation matrix was fully verified by experiments. Experiments demonstrate that this encoder can provide absolute 4-DOF position and pose monitoring with sub-arcsecond and sub-micron accuracy and high stability.

In terms of absolute measurement, the measurement of x and y can be realized at present [25–27], but the absolute measurement of the other four degrees of freedom cannot. The measurement schemes [30,31] do not discuss the possibility of absolute measurement and only refer to the increment measurement of three angles, which excludes the measurement of z-direction. The current commercial products, such as the TriAngle Products of *Trioptics* [34], can only realize the measurement of two angles. Therefore, the encoder proposed in this paper is very meaningful in the field of absolute multi-degree of freedom measurement and has great application prospects in future precision positioning.

Author Contributions: Conceptualization, S.W. and X.L.; methodology, S.W. and X.L.; software, S.W. and X.L.; validation, S.W., L.L., and X.L.; formal analysis, S.W., L.L., J.Z., N.S., and X.L.; investigation, S.W., L.L., J.Z., N.S., and X.L.; resources, X.L.; data curation, S.W. and L.L.; writing—original draft preparation, S.W., L.L., and X.L.; writing—review and editing, S.W., L.L., and X.L.; visualization, S.W.; supervision, X.L.; project administration, X.L.; funding acquisition, X.L. All authors have read and agreed to the published version of the manuscript.

Funding: This work was supported by the National Natural Science Foundation of China with no. 62275142, Basic and Applied Basic Research Foundation of Guangdong Province with no. 2021B1515120007.

Institutional Review Board Statement: Not applicable.

Informed Consent Statement: Not applicable.

Data Availability Statement: The data presented in this study are available on request from thecorresponding author.

Acknowledgments: The authors would like to thank all our laboratory members who have joined the related projects at Tsinghua University for their contributions to the achievements described in this paper.

Conflicts of Interest: The authors declare no conflict of interest.

References

1. Xuan, H.; Jing, L.D.; Chu, F.H.; Kai, W. System Analysis of Ground-based Inverse Synthetic Aperture Lidar for Geosynchronous Orbit Object Imaging. *Acta Photonica Sin.* **2018**, *47*, 12. [CrossRef]
2. Cui, X.Q.; Zhao, Y.H.; Chu, Y.Q.; Li, G.P.; Li, Q.; Zhang, L.P.; Su, H.J.; Yao, Z.Q.; Wang, Y.N.; Xing, X.Z.; et al. The Large Sky Area Multi-Object Fiber Spectroscopic Telescope (LAMOST). *Res. Astron. Astrophys.* **2012**, *12*, 1197–1242. [CrossRef]
3. Biswas, A.; Coupland, J. Synthetic aperture interferometry: Error analysis. *Appl. Opt.* **2010**, *49*, 3883–3893. [CrossRef] [PubMed]
4. Schienbein, R.; Fern, F.; Theska, R.; Supreeti, S.; Fussl, R.; Manske, E. Fundamental investigations in the design of five-axis nanopositioning machines for measurement and fabrication purposes. *Nanomanufacturing Metrol.* **2021**, *4*, 156–164. [CrossRef]
5. Tomlinson, R.; Coupland, J.M.; Petzing, J. Synthetic aperture interferometry: In-process measurement of aspheric optics. *Appl. Opt.* **2003**, *42*, 701–707. [CrossRef]
6. Jastrzebska, M. Modern Displacement Measuring Systems Used in Geotechnical Laboratories: Advantages and Disadvantages. *Sensors* **2021**, *21*, 27. [CrossRef]
7. Wang, S.C.; Xie, B.R.; Huang, S.M. Design and Analysis of Small Size Eddy Current Displacement Sensor. *Sensors* **2022**, *22*, 7444. [CrossRef]
8. Ortlepp, I.; Zollner, J.P.; Rangelow, I.W.; Manske, E. Heterodyne Standing-Wave Interferometer with Improved Phase Stability. *Nanomanufacturing Metrol.* **2021**, *4*, 190–199. [CrossRef]
9. Cai, Y.; Lou, Z.; Ling, S.; Liao, B.-s.; Fan, K.-c. Development of a Compact Three-Degree-of-Freedom Laser Measurement System with Self-Wavelength Correction for Displacement Feedback of a Nanopositioning Stage. *Appl. Sci.* **2018**, *8*, 2209. [CrossRef]
10. Hu, P.C.; Chang, D.; Tan, J.B.; Yang, R.T.; Yang, H.X.; Fu, H.J. Displacement measuring grating interferometer: A review. *Front Inform Tech El* **2019**, *20*, 631–654. [CrossRef]
11. Yu, K.; Zhu, J.; Yuan, W.; Zhou, Q.; Xue, G.; Wu, G.; Wang, X.; Li, X. Two-channel six degrees of freedom grating-encoder for precision-positioning of sub-components in synthetic-aperture optics. *Opt. Express* **2021**, *29*, 21113–21128. [CrossRef] [PubMed]
12. Yu, H.Y.; Liu, H.Z.; Ye, G.Y.; Fan, S.J.; Shi, Y.S.; Yin, L.; Chen, B.D.; Jiang, W.T. Transverse sensitivity suppression using multi-axis surface encoder with parasitic error compensation. *Appl. Phys. Lett.* **2017**, *111*, 5. [CrossRef]
13. Zhai, Z.-H.; Sun, C.-L.; Liu, Q.; Guo, L.-W.; Liu, Y.-S.; Zhang, X.; Wang, D.-T.; Meng, K.; Li, J.; Du, L.-H.; et al. Design of terahertz-wave Doppler interferometric velocimetry for detonation physics. *Appl. Phys. Lett.* **2020**, *116*, 161102. [CrossRef]
14. Yin, Y.F.; Liu, Z.W.; Jiri, G.L.T.; Yu, H.Z.; Wang, W.; Li, X.T.; Bao, H.; Li, W.H.; Hao, Q. Overview of 2D grating displacement measurement technology. *Chin. Opt.* **2020**, *13*, 1224–1238. [CrossRef]
15. Shimizu, Y.; Matsukuma, H.; Gao, W. Optical Sensors for Multi-Axis Angle and Displacement Measurement Using Grating Reflectors. *Sensors* **2019**, *19*, 5289. [CrossRef]

16. Shimizu, Y. Laser Interference Lithography for Fabrication of Planar Scale Gratings for Optical Metrology. *Nanomanufacturing Metrol.* **2021**, *4*, 3–27. [CrossRef]
17. Zhang, S.J.; Li, W.; Chen, W.; Zhang, Y.L.; Zhu, N.H. Accurate Calibration and Measurement of Optoelectronic Devices. *J. Light. Technol.* **2021**, *39*, 3687–3698. [CrossRef]
18. Yang, J.; Liu, H.; Wen, J.; Chen, L.; Shang, Y.; Chen, N.; Huang, S.; Wang, T.; Pang, F. Cylindrical vector modes based Mach-Zehnder interferometer with vortex fiber for sensing applications. *Appl. Phys. Lett.* **2019**, *115*, 051103. [CrossRef]
19. Zhu, J.; Wang, G.; Wang, S.; Li, X. A Reflective-Type Heterodyne Grating Interferometer for Three-Degree-of-Freedom Subnanometer Measurement. *IEEE Trans. Instrum. Meas.* **2022**, *71*, 1–9. [CrossRef]
20. Hong, Y.; Sato, R.; Shimizu, Y.; Matsukuma, H.; Gao, W. A New Optical Configuration for the Surface Encoder with an Expanded Z-Directional Measuring Range. *Sensors* **2022**, *22*, 3010. [CrossRef]
21. Li, X.H.; Gao, W.; Muto, H.S.; Shimizu, Y.; Ito, S.; Dian, S. A six-degree-of-freedom surface encoder for precision positioning of a planar motion stage. *Precis. Eng.-J. Int. Soc. Precis. Eng. Nanotechnol.* **2013**, *37*, 771–781. [CrossRef]
22. Li, X.H.; Shi, Y.P.; Xiao, X.; Zhou, Q.; Wu, G.H.; Lu, H.O.; Ni, K. Design and Testing of a Compact Optical Prism Module for Multi-Degree-of-Freedom Grating Interferometry Application. *Appl. Sci.* **2018**, *8*, 2495. [CrossRef]
23. Qibo, F.; Bin, Z.; Cunxing, C.; Cuifang, K.; Yusheng, Z.; Fenglin, Y. Development of a simple system for simultaneously measuring 6DOF geometric motion errors of a linear guide. *Opt. Express* **2013**, *21*, 25805–25819. [CrossRef] [PubMed]
24. Zheng, F.; Feng, Q.; Zhang, B.; Li, J. A Method for Simultaneously Measuring 6DOF Geometric Motion Errors of Linear and Rotary Axes Using Lasers. *Sensors* **2019**, *19*, 1764. [CrossRef] [PubMed]
25. Li, X.; Wang, H.; Ni, K.; Zhou, Q.; Mao, X.; Zeng, L.; Wang, X.; Xiao, X. Two-probe optical encoder for absolute positioning of precision stages by using an improved scale grating. *Opt. Express* **2016**, *24*, 21378–21391. [CrossRef] [PubMed]
26. Shi, Y.; Zhou, Q.; Li, X.; Ni, K.; Wang, X. Design and testing of a linear encoder capable of measuring absolute distance. *Sens. Actuators A Phys.* **2020**, *308*, 111935. [CrossRef]
27. Yaping, S.; Kai, N.; Xinghui, L.; Qian, Z.; Xiaohao, W. Highly accurate, absolute optical encoder using a hybrid-positioning method. *Opt. Lett.* **2019**, *44*, 5258–5261.
28. Li, R.; Xie, L.; Zhen, Y.; Xiao, H.; Wang, W.; Guo, J.; Konyakhin, I.; Nikitin, M.; Yu, X. Roll angle autocollimator measurement method based on a cylindrical cube-corner reflector with a high resolution and large range. *Opt. Express* **2022**, *30*, 7147–7161. [CrossRef]
29. Ren, W.; Cui, J.; Tan, J. A three-dimensional small angle measurement system based on autocollimation method. *Rev. Sci. Instrum.* **2022**, *93*, 055102. [CrossRef]
30. Saito, Y.; Arai, Y.; Gao, W. Detection of three-axis angles by an optical sensor. *Sens. Actuator A-Phys.* **2009**, *150*, 175–183. [CrossRef]
31. Gao, W.; Saito, Y.; Muto, H.; Arai, Y.; Shimizu, Y. A three-axis autocollimator for detection of angular error motions of a precision stage. *CIRP Ann-Manuf. Technol.* **2011**, *60*, 515–518. [CrossRef]
32. Liu, C.H.; Jywe, W.Y.; Chen, C.K.; Hsien, W.H.; Shyu, L.H.; Ji, L.W.; Liu, J.T.; Hsu, T.H.; Chen, C.D. Development of a four-degrees-of-freedom diffraction sensor. *J. Phys. Conf. Ser.* **2006**, *48*, 196–201. [CrossRef]
33. Liu, C.H.; Huang, H.L.; Lee, H.W. Five-degrees-of-freedom diffractive laser encoder. *Appl. Opt.* **2009**, *48*, 2767–2777. [CrossRef] [PubMed]
34. TriAngle-Electronic Autocollimator for Precise Optical Angle Measurement. Available online: https://trioptics.com/products/triangle-electronic-autocollimators/ (accessed on 16 November 2022).

sensors

MDPI

Article

Modeling and Parameter Identification of a 3D Measurement System Based on Redundant Laser Range Sensors for Industrial Robots

Guanbin Gao [1,2], Liulin Kuang [1,2], Fei Liu [1,2], Yashan Xing [1,2] and Qinghua Shi [3,*]

1 Faculty of Mechanical and Electrical Engineering, Kunming University of Science and Technology, Kunming 650500, China
2 Yunnan International Joint Laboratory of Intelligent Control and Application of Advanced Equipment, Kunming University of Science and Technology, Kunming 650500, China
3 Yunnan Institute, China Academy of Machinery Science and Technology Group Co., Ltd., Kunming 650031, China
* Correspondence: sqhynoffice@163.com

Abstract: The low absolute positioning accuracy of industrial robots is one of the bottlenecks preventing industrial robots from precision applications. Kinematic calibration is the main way to improve the absolute positioning accuracy of industrial robots, which greatly relies on three-dimensional (3D) measurement instruments, including laser trackers and pull rope mechanisms. These instruments are costly, and their required intervisibility space is large. In this paper, a precision 3D measurement instrument integrating multiple laser range sensors is designed, which fuses the information of multiple redundant laser range sensors to obtain the coordinates of a 3D position. An identification model of laser beam position and orientation parameters based on redundant distance information and standard spherical constraint is then developed to reduce the requirement for the assembly accuracy of laser range sensors. A hybrid identification algorithm of PSO-LM (particle swarm optimization Levenberg Marquardt) is designed to solve the high-order nonlinear problem of the identification model, where PSO is used for initial value identification, and LM is used for final value identification. Experiments of identification of position and orientation, verifications of the measuring accuracy, and the calibration of industrial robots are conducted, which show the effectiveness of the proposed 3D measurement instrument and identification methods. Moreover, the proposed instrument is small in size and can be used in narrow industrial sites.

Keywords: calibration; identification; laser range sensor; standard spherical constrain; measurement

Citation: Gao, G.; Kuang, L.; Liu, F.; Xing, Y.; Shi, Q. Modeling and Parameter Identification of a 3D Measurement System Based on Redundant Laser Range Sensors for Industrial Robots. *Sensors* **2023**, *23*, 1913. https://doi.org/10.3390/s23041913

Academic Editors: Qibo Feng, Jiakun Li and Qixin He

Received: 15 January 2023
Revised: 3 February 2023
Accepted: 6 February 2023
Published: 8 February 2023

1. Introduction

The application of industrial robots has continuously enhanced the automation and intelligence of the manufacturing industry. Compared with traditional machine tools, Industrial robots have the advantages of low cost and high flexibility and are increasingly used in grinding, milling, or 3D printing [1]. The requirements for the positioning accuracy of industrial robots are constantly upgrading in many fields [2,3], e.g., aerospace, automobile manufacturing, machining, etc. Generally, the repetitive positioning accuracy of industrial robots can reach the order of 0.01 mm, while the absolute positioning accuracy can only reach the order of 1 mm without calibration [4–6]. However, this order of absolute positioning accuracy limits industrial robots in relatively high-precision applications, such as measurement, milling, grinding, etc. The absolute positioning accuracy for an industrial robot is determined not only by its quality of components, manufacture, and assembly but by its mechanical degradation, which is mainly affected by the deformation of linkages, collision, temperature change, and other factors in practice [7–10]. Kinematic calibration is the major method to improve the absolute positioning accuracy of industrial robots, which includes four steps: establishing a kinematic model, measuring the end position of the robot, identifying the parameters of the kinematic model using the measured data, and compensating in the motion controller with the identified parameters to improve the absolute positioning accuracy of the robot [11]. The influence of thermal drift is important for high-precision industrial robots. After kinematic calibration, the error caused by non-kinematic factors such as thermal drift can be compensated to make the industrial robot obtain higher positioning accuracy [12].

The accuracy of measuring is extremely important for kinematic calibration, which is usually required to be higher than 0.2 mm in the three-dimensional (3D) space. At present, there are few instruments that can meet this requirement in practice, mainly including laser trackers, coordinate measuring machines (CMMs), pull rope calibration mechanisms, etc. [13–15]. However, CMMs with large volumes are rarely used for kinematic calibration for industrial robots since the site for kinematic calibration in practice is relatively small.

The laser tracker and the pull rope calibration mechanism are the widely used measuring instruments in robot calibration. For the laser tracker, it is necessary to install one or more reflective targets on the flange of the industrial robot to be measured. The laser emitted by the laser tracker is reflected by the targets to track and measure the end position of the industrial robot in 3D space. Gao et al. [16] measured the end position of an ER20-C10 industrial robot with an API R-20 Radian laser tracker for kinematic calibration, and the maximum positioning errors of the robot along x, y, and z decreased from 3.17 mm, 3.26 mm, and 3.30 mm to 0.39 mm, 0.68 mm, and 0.56 mm after calibration, respectively. Jiang et al. [17] calibrated an RS10N robot using a Leica laser tracker, and the maximum positioning error of the robot decreased from 4.8867 mm to 0.6421 mm. Hsiao et al. [18] completed the calibration of a PMC6VA030 industrial robot with a Faro laser tracker. The maximum positioning error and the average positioning error of the robot decreased from 6.294 mm and 2.613 mm to 1.225 mm and 0.310 mm after calibration.

Different from the laser tracker, the pull rope calibration mechanism uses one end of the rope of the mechanism wound on a photoelectric encoder. The other end is connected to the end flange of the industrial robot through a universal adapter. The displacement of the end of the robot can be measured by the length of the rope pulled out [19]. Li et al. [20] calibrated a 6-DOF robot by using a pull rope calibration mechanism with the result that the maximum positioning error of the robot decreased from 3.36 mm to 1.07 mm. Mei et al. [21] used a pull rope calibration mechanism based on SICK BCG13-E1BM0599 to measure the movement distance of a 4-DoF stacking robot, and the 3σ value of positioning error decreased from 11.73 mm to 1.79 mm after calibration, where σ is the standard deviation.

The above studies show that the calibration results with laser trackers are more accurate than with the pull rope calibration mechanism. However, in practice, the operation of the laser tracker is complex and time-consuming, the required intervisibility space is large, and the cost is also very high [5]. These factors limit its field calibration application in the workshop. Compared with the laser tracker, the measurement accuracy of the pull rope calibration mechanism is much lower, and there is information conversion from distance to position in the calibration model. As a result, its calibration accuracy is limited. Nevertheless, the cost of the pull rope calibration mechanism is also lower than the laser tracker.

Some alternative methods have been presented to meet the increasing demand for calibration in various industries. Boby et al. [22] proposed a calibration method for the industrial robot using a monocular camera as the measuring equipment, and the maximum error of the robot decreased from 8.74 mm to 5.13 mm after calibration. Yang et al. [23] proposed a kinematic calibration method based on the dynamic measurement of double ball linkages. After calibration, the average roundness error (the roundness error is the movement error of a circular path of the robot's end) of the 6-DOF robot decreased from 0.46 mm to 0.36 mm, and the range of motion errors of the robots' end in x and y directions decreased by 0.10 mm and 0.07 mm, respectively. He et al. [24] proposed a kinematic calibration method to improve the accuracy of a TKB2600 industrial robot using multiple location constraints. After calibration, the average relative offset decreased more than 50%. Icli et al. [25] presented an automated calibration method for industrial robots based on three orthogonal plunger dial indicators and four reference balls. After calibration, the maximum position error and average position error of the KUKA KR6 R700 industrial robot decreased to 0.624 mm and 0.326 mm, respectively. Guo et al. [26] presented a 3D measurement system (R-test) with three range sensors and used this equipment to calibrate the HSR-JR605 industrial robot. After calibration, the ratio of distance error to distance and the ratio of relative position error to the relative position error of the robot decreased from 14.30 μm/mm and 20.15 μm/mm to 4.4 μm/mm and 10.85 μm/mm, respectively. The R-test requires the laser beam of the laser range sensor to be strictly orthogonal. The three-dimensional measurement system based on the R-test relies on a high-precision displacement platform in the kinematic calibration of industrial robots. However, the calibration method based on vision measurement is affected greatly by camera distortion, external light, and target accuracy. Moreover, the calibration method based on the dynamic measurement of double ball linkages has a limited measuring range and can only improve the positioning accuracy in the x and y directions. The calibration method based on external constraints is usually limited by the need to constrain the end of the robot at a point or contact closely with a plane. Thus, the measured information is less, and only zero-point calibration can be performed.

To address the issues of current measurement instruments in the field of robot calibration, we developed a spatial position measurement system (SPMS) based on redundant laser range sensors for measuring the position and calibrating robots in industrial sites. The contributions and innovations of this paper are as follows.

- A small precision 3D position measuring device SPMS is proposed, which fuses the information of multiple laser range sensors. The distance information of laser range sensors is converted into position information through the standard spherical constraint to realize the 3D position measurement in a narrow space;
- An identification model of laser beam position and orientation parameters based on redundant distance information and standard spherical constraint is proposed to reduce the requirement for the assembly accuracy of laser range sensors;
- To solve the high-order nonlinear problem of the identification model, a hybrid identification algorithm is proposed, where PSO is used for initial value identification and LM is used for final value identification;
- Experiments were conducted to verify the measuring accuracy of the proposed device.

The paper is structured as follows. Section 2 provides the principle and construction of the novel measurement system. The data acquisition method based on reference sphere constraint, the establishment of the measurement coordinate system, and the identification model of the position and orientation of the laser beam are given in Section 3. A parameter identification method of position and orientation of the laser beam based on a hybrid algorithm of PSO-LM is introduced in Section 4. The experiment of identification of laser beams and verifications of the measuring accuracy and calibration of industrial robots are shown in Section 5. Section 6 concludes this paper.

2. Construction of the Measurement System

The measurement system is constructed with redundant laser range sensors to measure a reference sphere installed on the end of the industrial robot and moving along with the robot. The components and measurement principles of the system are elaborated on in this section. The sphere is only mounted on the flange for the robot calibration and not for the later experiments for parameter identification.

2.1. Components of the Measurement System

The proposed SPMS consists of laser range sensors, a reference sphere, a data acquisition and transmission circuit, and a support mechanism. In order to obtain accurate coordinates of the spherical center of the reference sphere, the laser beams in the measuring system should not interfere with each other, and the facula formed by laser beams projected on the reference sphere should be evenly distributed. Therefore, the laser displacement sensors in SPMS are installed in the regular polygon support mechanism, as shown in Figure 1.

Figure 1. The spatial position measurement system (SPMS).

The performance parameters of the range sensor are shown in Table 1. RC (resistor-capacitance circuit) filtering and high-performance ADC (analog-to-digital converter) modules are used to collect, filter and convert the analog signal output by the laser range sensors in the data acquisition and transmission circuit. The digital signal is then transformed into distance information through the MCU (microcontroller unit). The schematic diagram and photo of the circuit are shown in Figure 2.

Table 1. The performance parameters of the range sensor.

Standard Distance	Measurement Range	Linearity	Repeatability	Temperature Characteristics	Ambient Temperature [2]
100 mm	75 mm to 130 mm	±0.15% of F.S. [1]	10 μm	0.06% of F.S./°C	−10 to +50 °C

[1] F.S. denotes full scale. [2] No condensation or freezing.

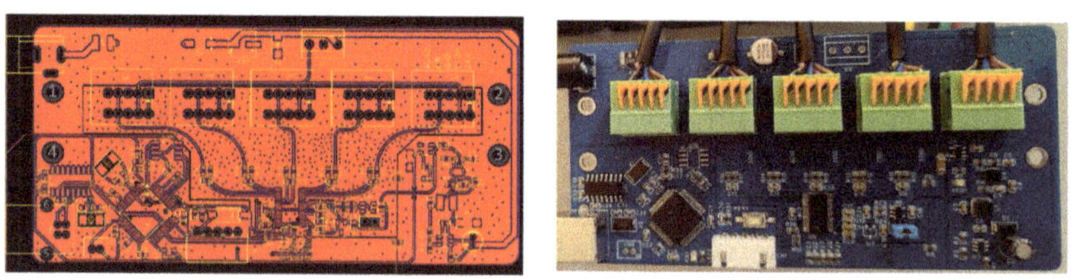

(a) (b)

Figure 2. The schematic diagram and photo of the circuit, they should be listed as: (**a**) The schematic diagram; (**b**) the photo of the circuit.

2.2. *Measurement Principle of the System*

When SPMS is working, the beam of each laser range sensor is projected on the reference sphere at the same time. The coordinates of the flare on the reference sphere are substituted into the spherical equation to obtain the position of the reference sphere center, and then the robot end position is finally measured according to the conversion relationship between the position of the reference sphere center and the position of the robot end.

The coordinates of the flare on the reference sphere can be calculated by:

$$\mathbf{F}_k = \mathbf{P}_{Lk} + \gamma_k L_k \quad (1)$$

where k is the number of laser range sensors (k = 1, 2, ... , K, K is the total number of laser range sensors in SPMS); $\mathbf{F}_k$ is the coordinates of the flare; L_k is the length of the beam that can be read from the sensor; and $\mathbf{P}_{Lk}$ and γ_k are the position and orientation vectors of the laser beam, respectively, which are determined by the installation of the sensor.

Since $\mathbf{F}_k$ is on the surface of the reference sphere, it conforms to the spherical equation, as shown in (2).

$$(x_k - x_B)^2 + (y_k - y_B)^2 + (z_k - z_B)^2 = R^2 \quad (2)$$

where x_k, y_k, and z_k are the coordinates of $\mathbf{F}_k$ in the x, y, and z direction, respectively; x_B, y_B, and z_B are the coordinates of the center of the reference sphere $\mathbf{P}_B$ in x, y, and z direction, respectively; R is the radius of the reference sphere. The coordinates of $\mathbf{P}_B$ are what we want for the reference sphere installed on the end of the robot, which can stand for the position of the robot. The coordinates of $\mathbf{P}_B$ can be written as the function of $\mathbf{P}_{Lk}$, γ_k, L_k, and R:

$$\mathbf{P}_B = f(\mathbf{P}_{Lk}, \gamma_k, L_k, R) \quad (3)$$

where L_k can be read from the sensor; R is a known constant; and $\mathbf{P}_{Lk}$ and γ_k are unknown constants that can be obtained by precision installation of the sensors or identification.

The reference sphere is fixed at the robots' end for kinematic calibration. The center position of the reference sphere represents the end position of the robot. That is, $\mathbf{P}_B$ is the end position of the industrial robot.

However, installing the sensor at a certain position and orientation precisely is very difficult, costly, and time-consuming. We proposed an identification method to obtain $\mathbf{P}_{Lk}$ and γ_k with redundant laser range information and spherical constrain.

3. Data Acquisition and Modeling for Position and Orientation of the Laser Beam

We propose the data acquisition method based on reference sphere constrain. Then, the measurement coordinate system for the SPMS is established based on Schmidt orthogonalization and normalization. Moreover, the position and orientation of the laser beam are modeled with the spherical equation.

3.1. Data Acquisition Method Based on Reference Sphere Constraint

As shown in Figure 3, p_k is the flare on the reference sphere of sensor k; $\mathbf{F}_k$ is the coordinate of p_k; O_t is the measurement coordinate system; O_w is the world system attached with the laser tracker; $\mathbf{P}_{Lk}$ = $[x_{Lk}, y_{Lk}, z_{Lk}]^T$; γ_k = $[\gamma_{kx}, \gamma_{ky}, \gamma_{kz}]^T$ and ${\gamma_{kx}}^2 + {\gamma_{ky}}^2 + {\gamma_{kz}}^2 = 1$; and $\mathbf{T}$ is the homogeneous transformation matrix between O_t and O_w; m is the number of pose changes of SPMS. Three points A_1, A_2, and A_3 are selected on the surface of the bracket of SPMS, with which O_t is established by Schmidt orthogonalization and normalization. Then, according to (2), the nonlinear least squares problem is established by orthogonal decomposition and homogeneous transformation. Finally, the position and orientation parameters of the laser beam in O_t will be obtained by solving the nonlinear least square problem.

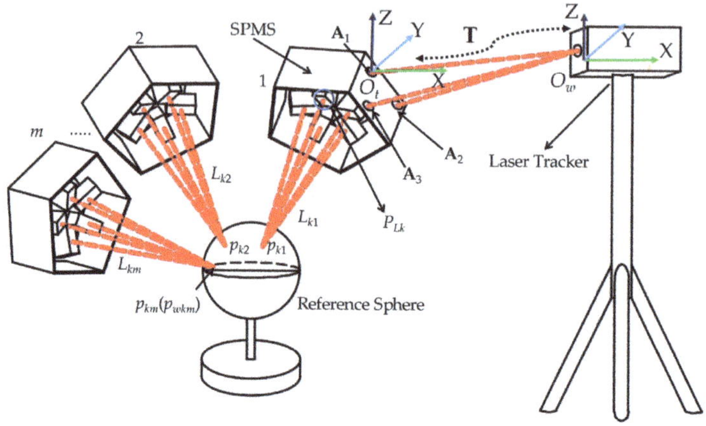

Figure 3. The data acquisition procedure of laser beams based on reference sphere constraint.

3.2. Establishment of the Measurement Coordinate System

To determine the origin and the axis direction of O_t, three points $\mathbf{A}_1$, $\mathbf{A}_2$, and $\mathbf{A}_3$, on the surface of SPMS are selected, which will be measured by the laser tracker. $\mathbf{A}_1$ is the origin of O_t. According to the calculation of $\mathbf{A}_1$, $\mathbf{A}_2$, and $\mathbf{A}_3$, we can obtain two vectors, **b** and **c**, as shown in (4)

$$\begin{gathered}\mathbf{b} = \mathbf{A}_2 - \mathbf{A}_1 \\ \mathbf{c} = \mathbf{A}_3 - \mathbf{A}_1\end{gathered} \tag{4}$$

The orthogonal vector of **b** and **c** is:

$$\mathbf{d} = \mathbf{b} \times \mathbf{c} \tag{5}$$

By Schmidt orthogonalization of vectors **b**, **c**, and **d**, the corresponding three mutually orthogonal vectors $\boldsymbol{\beta}_1$, $\boldsymbol{\beta}_2$, and $\boldsymbol{\beta}_3$ can be calculated by:

$$\begin{gathered}\boldsymbol{\beta}_1 = \mathbf{b} \\ \boldsymbol{\beta}_2 = \mathbf{c} - \frac{\langle \mathbf{c}, \boldsymbol{\beta}_1 \rangle}{\langle \boldsymbol{\beta}_1, \boldsymbol{\beta}_1 \rangle}\boldsymbol{\beta}_1 \\ \boldsymbol{\beta}_3 = \mathbf{d} - \frac{\langle \mathbf{d}, \boldsymbol{\beta}_1 \rangle}{\langle \boldsymbol{\beta}_1, \boldsymbol{\beta}_1 \rangle}\boldsymbol{\beta}_1 - \frac{\langle \mathbf{d}, \boldsymbol{\beta}_2 \rangle}{\langle \boldsymbol{\beta}_2, \boldsymbol{\beta}_2 \rangle}\boldsymbol{\beta}_2\end{gathered} \tag{6}$$

By normalization of vectors $\boldsymbol{\beta}_1$, $\boldsymbol{\beta}_2$, and $\boldsymbol{\beta}_3$, the corresponding orthogonal unit vectors $\boldsymbol{\alpha}_1$, $\boldsymbol{\alpha}_2$, and $\boldsymbol{\alpha}_3$ can be obtained. Taking $\mathbf{A}_1$ as the origin, $\boldsymbol{\alpha}_1$, $\boldsymbol{\alpha}_2$, and $\boldsymbol{\alpha}_3$ as x axis, y axis, and z axis, the measurement coordinate system O_t can be established. Since the coordinates of $\mathbf{A}_1$, $\mathbf{A}_2$, and $\mathbf{A}_3$ are measured by the laser tracker in its measurement coordinate system O_w, we can obtain the homogeneous transformation matrix **T**:

$$\mathbf{T} = \begin{bmatrix} \alpha_{11} & \alpha_{21} & \alpha_{31} & A_{1x} \\ \alpha_{12} & \alpha_{22} & \alpha_{32} & A_{1y} \\ \alpha_{13} & \alpha_{23} & \alpha_{33} & A_{1z} \\ 0 & 0 & 0 & 1 \end{bmatrix} \tag{7}$$

where α_{ij} is the component of $\boldsymbol{\alpha}_i$ ($I = 1, 2, 3$; $j = 1, 2, 3$); A_{1x}, A_{1y}, and A_{1z} are the coordinates of $\mathbf{A}_1$ in O_w.

3.3. Identification Modeling of Position and Orientation for Laser Beams

As shown in Figure 3, SPMS can be carried by a robot or other machines to move and be measured in different positions and orientations. The reference sphere is fixed, and the coordinates of the center of the reference sphere, which is fixed, are $\mathbf{P}_{Bw} = [x_{Bw}, y_{Bw}, z_{Bw}]^T$ in O_w.

The coordinate system O_t is moving with SPMS in the calibration process. The coordinates in O_t will be transformed into O_w for consistency. In O_t, the coordinates of the flare on the reference sphere of sensor k, i.e., $\mathbf{F}_k$ in O_t, can be calculated:

$$\mathbf{F}_k = \mathbf{P}_{Lk} + \gamma_k L_k \tag{8}$$

The homogeneous coordinates of $\mathbf{F}_k$ are as shown in (9), which will be used in the next homogeneous transformation.

$$\mathbf{F}_{k4\times1} = \begin{bmatrix} \mathbf{F}_k \\ 1 \end{bmatrix} \tag{9}$$

By multiplying the transformation matrix $\mathbf{T}$, the homogeneous coordinates $\mathbf{F}_{k4\times1}$ in O_w can be obtained by (10).

$$\mathbf{F}_{wk4\times1} = \mathbf{T} \cdot \mathbf{F}_{k4\times1} \tag{10}$$

By substituting $\mathbf{P}_{Lk} = [x_{Lk}, y_{Lk}, z_{Lk}]^T$ and $\gamma_k = [\gamma_{kx}, \gamma_{ky}, \gamma_{kz}]^T$ into the equation, the coordinates of $\mathbf{F}_k$ in O_w can be derived by (8)–(10).

$$\begin{bmatrix} x_{wk} \\ y_{wk} \\ z_{wk} \\ 1 \end{bmatrix} = \mathbf{T} \cdot \begin{bmatrix} x_{Lk} + \gamma_{kx} L_k \\ y_{Lk} + \gamma_{ky} L_k \\ z_{Lk} + \gamma_{kz} L_k \\ 1 \end{bmatrix} \tag{11}$$

where x_{wk}, y_{wk}, and z_{wk} are the coordinates of $\mathbf{F}_k$ in O_w, respectively. Because $\mathbf{F}_k$ is a point on the surface of the reference sphere, (12) can be obtained according to (2) and (11).

$$(x_{wk} - x_{Bw})^2 + (y_{wk} - y_{Bw})^2 + (z_{wk} - z_{Bw})^2 = R^2 \tag{12}$$

To obtain the least squares solution of x_{Lk}, y_{Lk}, z_{Lk}, γ_{kx}, γ_{ky}, and γ_{kz} of (11), (11) and (12) are rewritten as the function of d_k, which is the distance between $\mathbf{F}_k$ and $\mathbf{P}_{Bw}$:

$$d_k = f\left(x_{Lk}, y_{Lk}, z_{Lk}, \gamma_{kx}, \gamma_{ky}, \gamma_{kz}, x_{Bw}, y_{Bw}, z_{Bw}\right) \tag{13}$$

At m positions and orientations, SPMS can be measured m times, by which m groups of data, including L_{ki} and $\mathbf{A}_{1i}$, $\mathbf{A}_{2i}$, $\mathbf{A}_{3i}$ ($i = 1, 2, \ldots, m$), can be obtained. With these data, m equations can be obtained according to (13). The solution of these equations can be converted to a least square problem, as shown in (14).

$$\begin{gathered} \Delta\boldsymbol{\eta} = [\Delta\mathbf{P}_{Lk}, \Delta\boldsymbol{\gamma}_k, \Delta\mathbf{P}_{Bw}] \\ \min f(\Delta\boldsymbol{\eta}) = \sum_{i=1}^{m} \Phi(\Delta\boldsymbol{\eta}) = \sum_{i=1}^{m} \|d_{ki} - R\|^2 \end{gathered} \tag{14}$$

There are K laser range sensors in SPMS, and the position and orientation of the laser beam of each sensor have six parameters. Thus, (14) has $6K + 3$ unknown variables, including 3 for the coordinates of the center of the reference sphere. Thereby, to ensure that (14) has definite solutions, m must be greater than or equal to $6K + 3$.

4. Parameter Identification of Position and Orientation of Laser Beams

Solving (14), we can obtain the position and orientation of the laser beam of each sensor, as well as the coordinates of the reference spherical center. The Levenberg–Marquardt (LM) algorithm is a common method for solving nonlinear least squares problems with the advantages of fast convergence and strong robustness. However, (14) is a high-dimensional and strong nonlinear system of equations. When LM was used to solve (14), the results depended heavily on the initial value. Particle swarm optimization (PSO) is a swarm intelligence optimization algorithm that seeks the optimal solution through cooperation and information sharing among particles in the population. Therefore, PSO is suitable for solving multivariable and strongly nonlinear problems, but its calculation results have a certain extent of randomness. We present a PSO-LM hybrid algorithm to solve (14), where PSO is used to obtain the initial solution, and LM is used for the final solution.

4.1. Initial Parameter Identification Based on PSO

To identify the $6K + 3$ parameters in (14), the particle population size t of the PSO is set as 5000. Then, the dimensions of position vector $\mathbf{X}_{n-1}{}^{i}$ and velocity vector $\mathbf{V}_{n-1}{}^{i}$ of the ith particle are all $6K + 3$ in the $n-1$th iteration, and the iterations of $\mathbf{X}_{n-1}{}^{i}$ and $\mathbf{V}_{n-1}{}^{i}$ are:

$$\begin{aligned} \mathbf{V}_n^i &= w\mathbf{V}_{n-1}^i + c_1 r_1 (\mathbf{P}^i_{best}{}_{n-1} - \mathbf{X}_{n-1}^i) + c_2 r_2 (\mathbf{G}_{best}{}_{n-1} - \mathbf{X}_{n-1}^i) \\ \mathbf{X}_n^i &= \mathbf{X}_{n-1}^i + \mathbf{V}_n^i \end{aligned} \quad (15)$$

where $\mathbf{P}^i{}_{best\ n-1}$ is the best solution of the ith particle in the $n-1$th iteration; $\mathbf{G}_{best\ n-1}$ is the best solution of all the particles in the $n-1$th iteration; c_1 and c_2 are the learning factors; generally, $c_1 = c_2 = 2$ [27,28]; r_1 and r_2 are random numbers with a value range of [0, 1]; and w_n is the inertial factor, which can be calculated by (16).

$$w_n = w_{\max} - \frac{(w_{\max} - w_{\min})n^2}{g_n^2} \quad (16)$$

where $w_{\max}$ and $w_{\min}$ are the upper and lower bounds of the inertia factor. Generally, $w_{\max} = 0.9$, $w_{\min} = 0.4$, and g_n are the maximum number of iterations.

4.2. Final Parameter Identification Based on LM

According to the basic principle of the LM algorithm, the $n + 1$th iteration equation in solving (14) is:

$$\boldsymbol{\delta}_{n+1} = \boldsymbol{\delta}_n - (\mathbf{J}^T\mathbf{J} + \lambda_n \mathbf{I})^{-1} \mathbf{J}_n^T \boldsymbol{\varepsilon}_n \quad (17)$$

where $\boldsymbol{\delta}_n$ is a $(6K + 3) \times 1$ vector; $\mathbf{J}$ is a $(K \times m) \times (6K + 3)$ Jacobian matrix; $\mathbf{I}$ is a $(6l + 3) \times (6l + 3)$ identity matrix; $\boldsymbol{\varepsilon}_n$ is a $(K \times m) \times 1$ fitting error vector of the reference sphere; and λ_n is the damping factor. Let:

$$\mathbf{J}_{P_{Lkn}} = \begin{bmatrix} \frac{\partial d_{k1n}}{\partial x_{Lk}} & \frac{\partial d_{k1n}}{\partial y_{Lk}} & \frac{\partial d_{k1n}}{\partial z_{kL}} \\ & \cdots & \\ \frac{\partial d_{kmn}}{\partial x_{Lk}} & \frac{\partial d_{kmn}}{\partial y_{Lk}} & \frac{\partial d_{kmn}}{\partial z_{Lk}} \end{bmatrix} \quad (18)$$

$$\mathbf{J}_{\gamma kn} = \begin{bmatrix} \frac{\partial d_{k1n}}{\partial \gamma_{xk}} & \frac{\partial d_{k1n}}{\partial \gamma_{yk}} & \frac{\partial d_{k1n}}{\partial \gamma_{zk}} \\ & \cdots & \\ \frac{\partial d_{kmn}}{\partial \gamma_{xk}} & \frac{\partial d_{kmn}}{\partial \gamma_{yk}} & \frac{\partial d_{kmn}}{\partial \gamma_{zk}} \end{bmatrix} \quad (19)$$

$$\mathbf{J}_{P_{Bwkn}} = \begin{bmatrix} \frac{\partial d_{k1n}}{\partial x_{Bw}} & \frac{\partial d_{k1n}}{\partial y_{Bw}} & \frac{\partial d_{k1n}}{\partial z_{Bw}} \\ & \cdots & \\ \frac{\partial d_{kmn}}{\partial x_{Bw}} & \frac{\partial d_{kmn}}{\partial y_{Bw}} & \frac{\partial d_{kmn}}{\partial z_{Bw}} \end{bmatrix} \quad (20)$$

where d_{kmn} is the distance between $\mathbf{p}_k$ and $\mathbf{P}_{Bw}$ in the nth iteration; $\mathbf{J}_{PLkn}$, $\mathbf{J}_{\gamma kn}$, and $\mathbf{J}_{PBkn}$ are the first-order partial derivative matrices of d_{kmn} for each component of $\mathbf{P}_{Lk}$, γ_k, and $\mathbf{P}_{wBk}$; m is the number of acquisition data; and k is the number of the laser range sensor. Then, the Jacobian matrix $\mathbf{J}$ of laser beam position parameters will be:

$$\mathbf{J} = \begin{bmatrix} \mathbf{J}_{PL1n} & & \mathbf{J}_{\gamma 1n} & & \mathbf{J}_{P_{Bw1n}} \\ & \cdots & & \cdots & \cdots \\ & \mathbf{J}_{PLkn} & & \mathbf{J}_{\gamma kn} & \mathbf{J}_{P_{Bwkn}} \end{bmatrix} \quad (21)$$

The initial value of the damping factor λ_0 is:

$$\lambda_0 = \tau \cdot \max\left\{\max\left(\mathbf{J}^T\mathbf{J}\right)\right\} \tag{22}$$

where $\tau = 10^{-8}$, its value is generally obtained from experience and adjusted according to actual calculation results. The iteration value of the damping factor λ_n is determined according to the scale factor ρ_n:

$$\rho_n = \frac{\mathbf{X}_n^T(\lambda_{n-1}\mathbf{X}_n - \mathbf{J}^T\varepsilon_n)}{2} \tag{23}$$

The iteration of the damping factor λ_n is:

$$\begin{cases} \lambda_{n+1} = \lambda_n \cdot \max\left\{\frac{1}{3}, 1-(2\rho_n-1)^3\right\}, \rho_n \geq 0 \\ \lambda_{n+1} = \lambda_n\kappa_n, \kappa_{n+1} = 2\kappa_n, \rho_n < 0 \end{cases} \tag{24}$$

5. Results of Experiments and Discussion

Three experiments were conducted to verify the proposed measurement system and parameter identification method, including parameter identification of the position and orientation of laser beams, verification of measuring accuracy of SPMS, and kinematic calibration of industrial robots in this section. During the experiments, the ambient temperature is about 20 °C, and the change is small. When the laser range sensor is not preheated, its measurement accuracy is unstable [29]. Therefore, the laser range sensor was preheated for 30 min before the experiment. After the completion of preheating, the experimental data will be collected and processed.

5.1. Experiments for Identification of Laser Beams

The experimental platform for the calibration of laser beams and verification of measuring accuracy consists of SPMS, a laser tracker, an EC66 robot, and a reference sphere with a diameter of 50.8143 mm, as shown in Figure 4. The EC66 robot is only used as an actuator carrying SPMS to change position and orientation.

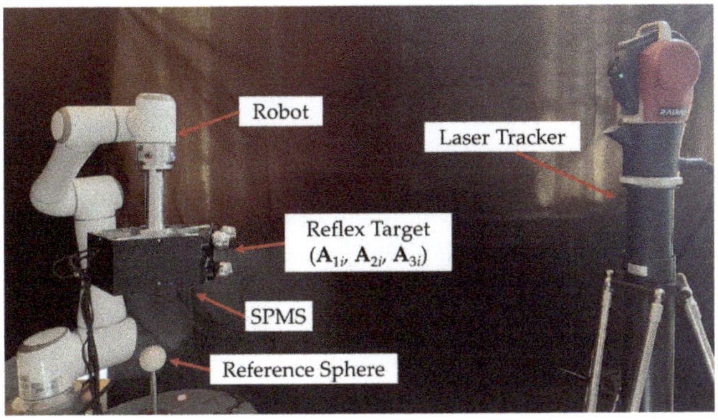

Figure 4. The platform for parameter identification and measuring accuracy.

SPMS was carried by the EC66 robot to 55 different positions and orientations, and 55 groups of L_k, $\mathbf{X}_1$, $\mathbf{X}_2$, and $\mathbf{X}_3$ were acquired. With these acquired data, the position and orientation parameters were obtained by the proposed identification method based on PSO-LM, as shown in Table 2. The spherical fitting error of each group of data is illustrated in Figure 5, which shows that nearly all the errors are smaller than 0.03 mm. The measuring accuracy, which can be indicated by the maximum measuring error required by the calibration of industrial robots, is 0.2 mm. Thus, the measuring accuracy of SPMS can meet the requirement. The measuring accuracy will be further verified in the next section. After calibration, the components of the position and orientation of $\mathbf{P}_{Lk} = [x_{Lk}, y_{Lk}, z_{Lk}]^T$ and $\boldsymbol{\gamma}_k = [\gamma_{kx}, \gamma_{ky}, \gamma_{kz}]^T$ are shown in Table 2.

Table 2. The identified position and orientation parameters of laser beams.

k	x_{Lk} (mm)	y_{Lk} (mm)	z_{Lk} (mm)	γ_{xk} (rad)	γ_{yk} (rad)	γ_{zk} (rad)
1	74.420	67.247	97.126	−0.104	0.976	0.189
2	11.612	69.074	146.755	0.210	0.975	−0.069
3	92.918	67.986	142.254	−0.190	0.981	−0.031
4	52.951	69.811	175.249	0.018	0.981	−0.194
5	25.346	70.181	97.548	0.143	0.982	0.124

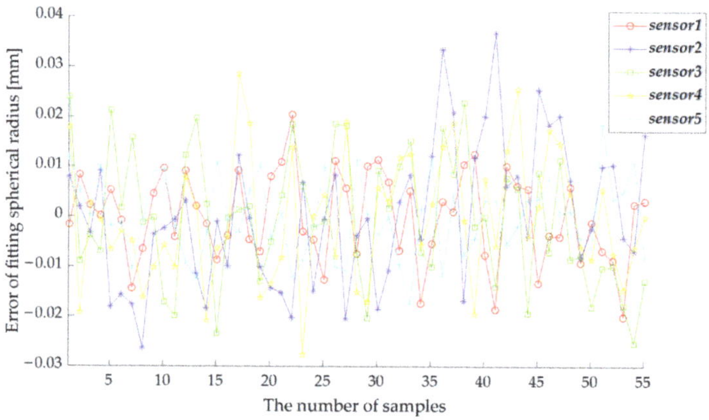

Figure 5. The spherical fitting error of each group of data.

5.2. Verification of the Measuring Accuracy

To verify the measuring accuracy of SPMS, an API R-20 Radian laser tracker is used as the reference to calculate the measuring error of SPMS, as shown in Figure 4. The reference sphere was measured 55 times using SPMS from different directions. With the measured data $\mathbf{P}_B$ and $\mathbf{P}_{Bw}$ can be obtained by (3) and (13), respectively. The difference between $\mathbf{P}_B$ and $\mathbf{P}_{Bw}$ is taken as the measuring error. The measuring errors of SPMS in 55 samples are shown in Figure 6, and the statistical information is shown in Table 3. The maximum error and average error of SPMS are 0.091 mm and 0.034 mm, respectively, which are precision enough to be applied in the calibration of industrial robots. The measurement error of SPMS mainly comes from the linearity error of laser range sensors according to the analysis and test.

5.3. Experiments of Calibration for an Industrial Robot

To further verify the effectiveness of SPMS in the calibration of industrial robots, the SPMS was used to calibrate an ER20-C10 6-DoF industrial robot. The kinematic model was established according to the modified Denavit–Hartenberg (MD-H). The position of the robot's end was measured with SPMS. With the measured data, kinematic parameters and the transformation parameter from the base of the robot to the measuring coordinate of SPMS, as well as the parameters from the flange to the reference sphere, were identified with the method in our previous publication [30]. As shown in Figure 7, the experimental platform for kinematic calibration of robots consists of SPMS, ER20-C10 robot, and a reference sphere with a diameter of 50.8143 mm.

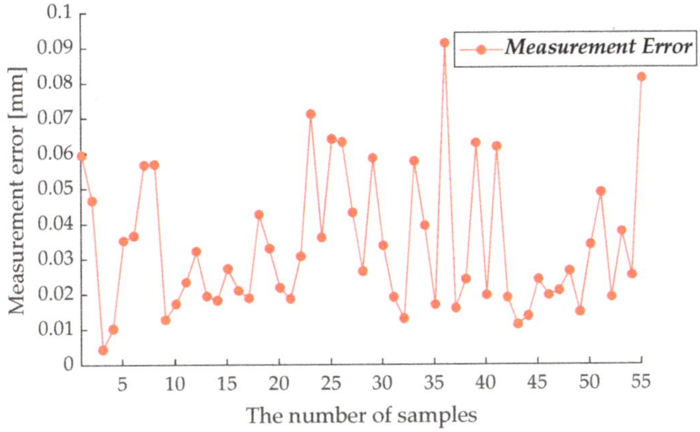

Figure 6. The measuring errors of SPMS.

Table 3. The statistics information of the measuring errors.

Index	Maximum Error	Average Error	RMS
Value (mm)	0.091	0.034	0.039

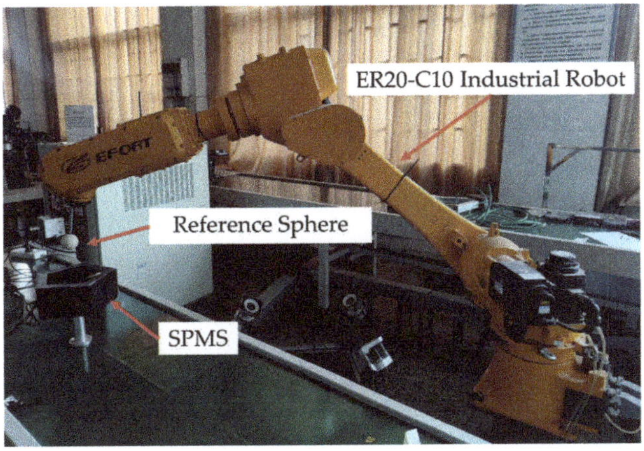

Figure 7. The experiment platform for kinematic calibration of robots.

The nominal kinematic parameters of the ER20-C10 robot before calibration are shown in Table 4.

Table 4. The kinematic parameters of the ER20-C10 robot before calibration.

No. of Joints	α_{i-1} (rad)	a_{i-1} (mm)	d_i (mm)	$\delta\theta_i$ (rad)
1	0	0.000	504.000	0
2	−pi/2	166.605	0.000	0
3	0	−782.270	0.000	0
4	−pi/2	138.826	761.350	0
5	pi/2	0.00	0.000	0

With the acquired data by SPMS, the kinematic parameters of the ER20-C10 industrial robot are identified, as shown in Table 5.

Table 5. The kinematic parameters of the ER20-C10 robot after calibration.

No. of Joints	$\alpha_{i\text{-}1}$ (rad)	$a_{i\text{-}1}$ (mm)	d_i (mm)	$\delta\theta_i$ (rad)
1	0	0.000	504.000	0
2	−pi/2	167.648	0.000	1.39×10^{-2}
3	0	−780.753	0.000	1.92×10^{-4}
4	−pi/2	139.918	759.330	3.84×10^{-4}
5	pi/2	0.00	0.000	6.63×10^{-4}

The identified kinematic parameters are compensated to the controller of the ER20-C10 industrial robot. The positioning accuracy of the industrial robot is tested according to ISO 9283:1998 [31], where a cube of 400 mm × 400 mm × 400 mm was created in the working space of the ER20-C10 industrial robot, shown in Figure 8. C_1~C_8 are the eight vertices of the cube. Five testing points P_1~P_5 on the diagonal plane C_1-C_2-C_7-C_8 of the cube are determined, among which P_1 is the center of the cube, P_2~P_5 are on the two diagonals of C_1-C_2-C_7-C_8 and being away from the vertices (10 ± 2)% of the diagonal length. After the cube was determined, reflex targets were installed at the end of the ER20-C10 robot for measurement with the laser tracker. According to ISO 9283:1998, cycle measurement 30 times should be conducted, and the average errors of P1~P5 can be obtained to evaluate the absolute positioning accuracy of the industrial robot. In the experiments, the absolute positioning (AP) was tested by the laser tracker before and after calibration, respectively. The results are shown in Table 6, which indicates that the maximum error, average error, and RMS are reduced from 1.462 mm, 1.056 mm, and 0.829 mm to 0.712 mm, 0.461 mm, and 0.388 mm, with the reduction percentage of 51.3%, 56.3%, and 53.2%, respectively. In another study by our team [30], a Radian laser tracker was used to calibrate and test an ER20-C10 industrial robot, and the maximum error, average error, and std of the robot after calibration were 0.637 mm, 0.460 mm, and 0.184 mm. Although there is still a certain gap between the calibration effect of SPMS and that of the laser tracker, the kinematic calibration of the industrial robot based on SPMS is effective and can significantly improve the absolute accuracy of the industrial robot.

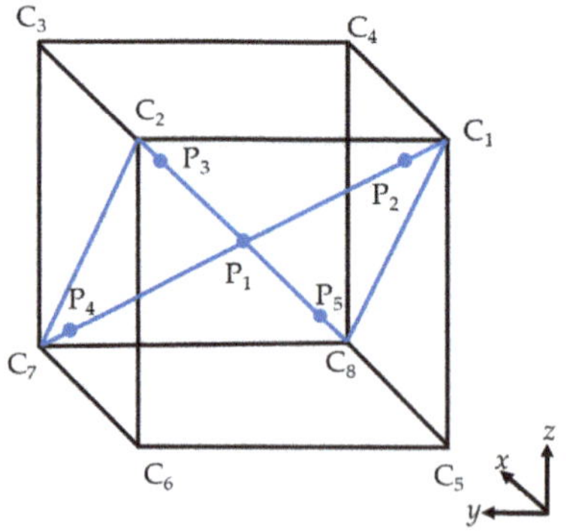

Figure 8. The test cube of the ER20-C10 industrial robot.

Table 6. The nominal kinematic parameters of the ER20-C10 robot after calibration.

Position	AP (mm) Before Calibration	AP (mm) After Calibration
P1	0.336	0.167
P2	1.462	0.712
P3	0.997	0.496
P4	1.454	0.620
P5	1.030	0.311

6. Conclusions

A spatial position measurement system (SPMS) based on redundant laser range sensors is proposed to solve industrial robots' measurement and calibration problems in narrow industrial sites. The main work of this paper is as follows:

- The conversion model is established from the distance information acquired by the laser range sensor to position information through the standard spherical constraint;
- The parameters of position and orientation of laser range sensors in the model are identified by a hybrid algorithm of PSO-LM to solve the high-dimensional and strong nonlinear problem of the model;
- Experiments were carried out to verify the conversion model, the identification method, the measuring accuracy of SPMS, and the effectiveness of industrial robot calibration. The results of the experiments show that the maximum and average error of SPMS is 0.091 mm and 0.034 mm, respectively; after calibration, the maximum error, average error, and RMS of the industrial robot are reduced from 1.462 mm, 1.056 mm, and 0.829 mm to 0.712 mm, 0.461 mm, and 0.388 mm, with the reduction percentage of 51.3%, 56.3%, and 53.2%, respectively.

Therefore, the proposed measurement device based on redundant laser range sensors is precise enough and can achieve suitable performance for industrial robot calibration.

Although many factors have been considered in the research work of this paper, there are still some problems that need to be considered or improved in future work: (1) Research on temperature compensation should be carried out to achieve more stable measurement accuracy; (2) The influence of measurement configuration on industrial robot calibration is not considered. The measurement configuration has a certain influence on the calibration results. To further improve the effect of SPMS on the kinematic calibration of industrial robots, optimization of the measurement configuration in the kinematic calibration of industrial robots based on the observability index should be studied.

Author Contributions: Conceptualization, G.G. and F.L.; methodology, F.L. and G.G.; software, L.K. and F.L.; validation, L.K. and F.L.; writing—original draft preparation, L.K. and Y.X.; writing—review and editing, G.G. and Y.X.; funding acquisition, Q.S. and G.G. All authors have read and agreed to the published version of the manuscript.

Funding: This research was funded by the National Natural Science Foundation of China under grant (52265001) and Yunnan Scientific and Technological Projects under grant (202201AS070033).

Institutional Review Board Statement: Not applicable.

Informed Consent Statement: Not applicable.

Data Availability Statement: All data will be made available on request to the correspondent author's e-mail with appropriate justification.

Conflicts of Interest: The authors declare no conflict of interest.

References

1. Mikolajczyk, T. Manufacturing Using Robot. *Adv. Mater. Res.* **2012**, *463–464*, 1643–1646.
2. Balanji, H.M.; Turgut, A.E.; Tunc, L.T. A Novel Vision-Based Calibration Framework for Industrial Robotic Manipulators. *Robot. Comput.-Integr. Manuf.* **2022**, *73*, 102248. [CrossRef]
3. Li, Z.; Li, S.; Luo, X. An Overview of Calibration Technology of Industrial Robots. *IEEE/CAA J. Autom. Sin.* **2021**, *8*, 23–36. [CrossRef]
4. Kana, S.; Gurnani, J.; Ramanathan, V.; Turlapati, S.H.; Ariffin, M.Z.; Campolo, D. Fast Kinematic Re-Calibration for Industrial Robot Arms. *Sensors* **2022**, *22*, 2295. [CrossRef]
5. Li, B.; Tian, W.; Zhang, C.; Hua, F.; Cui, G.; Li, Y. Positioning Error Compensation of an Industrial Robot Using Neural Networks and Experimental Study. *Chin. J. Aeronaut.* **2022**, *35*, 346–360. [CrossRef]
6. Sun, T.; Liu, C.; Lian, B.; Wang, P.; Song, Y. Calibration for Precision Kinematic Control of an Articulated Serial Robot. *IEEE Trans. Ind. Electron.* **2021**, *68*, 6000–6009. [CrossRef]
7. Chen, X.; Zhan, Q. The Kinematic Calibration of an Industrial Robot with an Improved Beetle Swarm Optimization Algorithm. *IEEE Robot. Autom. Lett.* **2022**, *7*, 4694–4701. [CrossRef]

8. Jiang, Y.; Yu, L.; Jia, H.; Zhao, H.; Xia, H. Absolute Positioning Accuracy Improvement in an Industrial Robot. *Sensors* **2020**, *20*, 4354. [CrossRef]
9. Luo, G.; Zou, L.; Wang, Z.; Lv, C.; Ou, J.; Huang, Y. A Novel Kinematic Parameters Calibration Method for Industrial Robot Based on Levenberg-Marquardt and Differential Evolution Hybrid Algorithm. *Robot. Comput.-Integr. Manuf.* **2021**, *71*, 102165. [CrossRef]
10. Zhao, Y.; Wen, X.; Qiao, G.; Lv, Z.; Song, A.; Kang, C. Accuracy Improvement of Serial Robot Based on Geometric Parameters Calibration. *Acta Metrol. Sin.* **2020**, *41*, 1461–1467.
11. Huang, T.; Zhao, D.; Yin, F.; Tian, W.; Chetwynd, D.G. Kinematic Calibration of a 6-Dof Hybrid Robot by Considering Multi-collinearity in the Identification Jacobian. *Mech. Mach. Theory* **2019**, *131*, 371–384. [CrossRef]
12. Vocetka, M.; Bobovský, Z.; Babjak, J.; Suder, J.; Grushko, S.; Mlotek, J.; Krys, V.; Hagara, M. Influence of Drift on Robot Repeatability and Its Compensation. *Appl. Sci.* **2021**, *11*, 10813. [CrossRef]
13. Guo, Y.; Song, B.; Tang, X.; Zhou, X.; Jiang, Z. A Calibration Method of Non-Contact R-Test for Error Measurement of Industrial Robots. *Measurement* **2021**, *173*, 108365. [CrossRef]
14. Li, Y.; Gao, G.; Na, J.; Zhang, H. Design of a Measurement System for Industrial Robots' Calibration Based on Cable Encoders. In Proceedings of the 2019 IEEE 8th Data Driven Control and Learning Systems Conference (DDCLS), Dali, China, 24–27 May 2019.
15. Lou, Z.; Zhang, J.; Gao, R.; Xu, L.; Fan, K.-C.; Wang, X. A 3D Passive Laser Tracker for Accuracy Calibration of Robots. *IEEE/ASME Trans. Mechatron.* **2022**, *27*, 5803–5811. [CrossRef]
16. Gao, G.; Gao, Y.; Liu, F.; Na, J. Calibration of Collaborative Robots Based on Position Information and Local Product of Exponentials. *J. Sens.* **2022**, *2022*, 2815164. [CrossRef]
17. Jiang, Z.; Zhou, W.; Li, H.; Mo, Y.; Ni, W.; Huang, Q. A New Kind of Accurate Calibration Method for Robotic Kinematic Parameters Based on the Extended Kalman and Particle Filter Algorithm. *IEEE Trans. Ind. Electron.* **2017**, *65*, 3337–3345. [CrossRef]
18. Hsiao, J.-C.; Shivam, K.; Lu, I.-F.; Kam, T.-Y. Positioning Accuracy Improvement of Industrial Robots Considering Configuration and Payload Effects Via a Hybrid Calibration Approach. *IEEE Access.* **2020**, *8*, 228992–229005. [CrossRef]
19. Xu, F.; Yao, E.; Feng, J.; Cao, Z. An Error Compensation Method for Incremental Pull Wire Displacement Sensor Based on Pso-Bp Neural Network. *Chin. J. Sens. Actuators* **2022**, *35*, 335–341.
20. Li, Z.; Li, S.; Luo, X. Data-Driven Industrial Robot Arm Calibration: A Machine Learning Perspective. In Proceedings of the 2021 IEEE International Conference on Networking, Sensing and Control (ICNSC), Xiamen, China, 3–5 December 2021.
21. Mei, J.; Sun, S.; Luo, Z.; Chen, L. Positioning Error Analysis and Kinematic Calibration of Robot Palletizer Based on One-Dimensional Cable Measurement System. *J. Tianjin Univ.* **2018**, *51*, 748–754.
22. Boby, R.A. Identification of Elasto-Static Parameters of an Industrial Robot Using Monocular Camera. *Robot. Comput.-Integr. Manuf.* **2022**, *74*, 102276. [CrossRef]
23. Yang, P.; Guo, Z.; Kong, Y. Plane Kinematic Calibration Method for Industrial Robot Based on Dynamic Measurement of Double Ball Bar. *Precis. Eng.* **2020**, *62*, 265–272. [CrossRef]
24. He, S.; Ma, L.; Yan, C.; Lee, C.-H.; Hu, P. Multiple Location Constraints Based Industrial Robot Kinematic Parameter Calibration and Accuracy Assessment. *Int. J. Adv. Manuf. Technol.* **2019**, *102*, 1037–1050. [CrossRef]
25. Icli, C.; Stepanenko, O.; Bonev, I. New Method and Portable Measurement Device for the Calibration of Industrial Robots. *Sensors* **2020**, *20*, 5919. [CrossRef]
26. Guo, Y.; Jiang, Z.; Song, B.; Tang, X.; Min, H.; Fu, C. A Distance Calibration Method for Kinematic Parameters of Serial Industrial Robots Considering the Accuracy of Relative Position. *Measurement* **2022**, *204*, 111842. [CrossRef]
27. MA, X.; Tan, J.; Chen, S.; Chu, Z.; Shi, L. Research on Optimal Particle Swarm Optimization for Multi-Objective Task Scheduling in Cloud Computing. *J. Electron. Meas. Instrum.* **2020**, *34*, 133–143.
28. Wang, D.; Tan, D.; Liu, L. Particle Swarm Optimization Algorithm: An Overview. *Soft Comput.* **2018**, *22*, 387–408. [CrossRef]
29. Blanco, D.; Fernández, P.; Cuesta, E.; Mateos, S.; Beltrán, N. Influence of Surface Material on the Quality of Laser Triangulation Digitized Point Clouds for Reverse Engineering Tasks. In Proceedings of the 2009 IEEE Conference on Emerging Technologies & Factory Automation, Palma de Mallorca, Spain, 22–25 September 2009.
30. Gao, G.; Liu, F.; San, H.; Wu, X.; Wang, W. Hybrid Optimal Kinematic Parameter Identification for an Industrial Robot Based on Bpnn-Pso. *Complexity* **2018**, *2018*, 4258676. [CrossRef]
31. *ISO 9283:1998*; Manipulating Industrial Robots–Performance Criteria and Related Test Methods. International Organization for Standardization: Geneva, Switzerland, 1998.

sensors

MDPI

Article

An Adaptive Hybrid Sampling Method for Free-Form Surfaces Based on Geodesic Distance

Chen Chen, Huakun Jia, Yang Lu, Xiaodong Zhang, Haohan Chen and Liandong Yu *

College of Control Science and Engineering, China University of Petroleum (East China), Qingdao 266580, China
* Correspondence: liandongyu@upc.edu.cn

Abstract: High precision geometric measurement of free-form surfaces has become the key to high-performance manufacturing in the manufacturing industry. By designing a reasonable sampling plan, the economic measurement of free-form surfaces can be realized. This paper proposes an adaptive hybrid sampling method for free-form surfaces based on geodesic distance. The free-form surfaces are divided into segments, and the sum of the geodesic distance of each surface segment is taken as the global fluctuation index of free-form surfaces. The number and location of the sampling points for each free-form surface segment are reasonably distributed. Compared with the common methods, this method can significantly reduce the reconstruction error under the same sampling points. This method overcomes the shortcomings of the current commonly used method of taking curvature as the local fluctuation index of free-form surfaces, and provides a new perspective for the adaptive sampling of free-form surfaces.

Keywords: adaptive sampling; free-form surface; non-uniform rational B-spline (NURBS); geodesic distance

Citation: Chen, C.; Jia, H.; Lu, Y.; Zhang, X.; Chen, H.; Yu, L. An Adaptive Hybrid Sampling Method for Free-Form Surfaces Based on Geodesic Distance. *Sensors* **2023**, *23*, 3224. https://doi.org/10.3390/s23063224

Academic Editors: Qibo Feng, Jiakun Li and Qixin He

Received: 3 February 2023
Revised: 14 March 2023
Accepted: 15 March 2023
Published: 17 March 2023

1. Introduction

Free-form surface parts are widely used in aerospace, automotive and other high-end equipment manufacturing fields [1]. It is one of the hotspots in the measurement field to measure their precise geometric quantities and ensure that they meet the accuracy requirements. This kind of part has a complex structure and large size, which also puts forward higher requirements for the dexterity and space of measuring equipment. The geometric measurement of a free-form surface can adopt a contact or non-contact measurement [2]. In the process of the geometric measurement of a free-form surface, the geometric error can be evaluated from two aspects: measurement accuracy and measurement efficiency. The measurement accuracy can be quantified by measuring the deviation between the actual value and the real value. When the measurement accuracy meets the measurement requirements, the measurement efficiency can be improved by optimizing the number and location of the measurement points and the measurement path [3–6].

As shown in Figure 1a, the coordinate measuring machine (CMM) equipped with a contact probe is one of the common contact measuring instruments [7]. It can obtain the point coordinates of the parts to be measured by scanning the measurement point by point. The point coordinates of the parts to be measured are reconstructed into free-form surfaces, and the geometric errors are compared with the original surfaces to accurately evaluate the machining errors generated in the manufacturing process of free-form surface parts. As shown in Figure 1b, the articulated industrial robot equipped with an optical probe is one of the common non-contact measuring instruments. The scanning measurement path is generated based on the measurement points of the parts to be measured with significant geometric characteristics, and the industrial robot drives the optical probe to move along the scanning measurement path. The scanning point cloud is reconstructed into a free surface, and the geometric error is compared with the original surface to determine the machining error of the free surface [8].

(a) (b)

Figure 1. Two measurement methods. (**a**) CMM; (**b**) Robot optical scanning system.

When measuring the geometric quantity of the parts to be measured, the number of measuring points is positively related to the measurement accuracy. The more the number of measuring points, the higher the measurement accuracy will be. When the number of measuring points reaches a certain value, the measurement accuracy will not change significantly. However, if the number of measuring points is too large, the measurement time will be significantly increased and the measurement efficiency will be reduced. Therefore, when measuring the components of the free-form surface to be measured, the number and position of the sampling points on the free-form surface should be reasonably distributed to ensure that the measurement efficiency can be improved as much as possible on the premise of meeting the measurement accuracy.

By setting a reasonable sampling plan, the corresponding measurement path is generated, and the free surface is measured using a contact measurement or non-contact measurement. Thus, the high-precision measurement of free-form surfaces can be realized, and the measurement efficiency can be significantly improved.

Blind sampling is a sampling method that does not consider the geometric characteristics of the free-form surface [9]. The number and location of sample points determine the geometric error of the reconstructed free-form surface. Uniform sampling is a simple and efficient sampling method, which is widely used in the field of free-form surface sampling. Dunbar et al. [10,11] introduced a random sampling strategy of fast disk under arbitrary dimension, which can realize blind sampling of a free-form surface simply and efficiently. Woo et al. [12] sampled the surface to be measured using the Hammersley distribution method. The research results show that blind sampling has the characteristics of simple and efficient sampling, but it cannot adaptively change the number and position of the sampling points based on the geometric characteristics of the free-form surface to be measured, thus affecting the measurement accuracy and efficiency.

Adaptive sampling generates sampling points according to the geometric characteristics of a free-form surface. In short, more sampling points are generated in the area with a large fluctuation of the free-form surface, and fewer sampling points are generated in the area with a small fluctuation of the free-form surface [13]. Ren et al. [14] proposed a method of using the curvature change matrix of adjacent points as the change index of a free-form surface. According to the proposed index, the optimal position of the newly added bidirectional curve mesh was determined, and the free-form surface was reconstructed based on the Gordon surface fitting principle. Javad et al. [15] used the optimization method and particle swarm optimization algorithm to optimize the position of the sampling points. On the basis of the initial sampling points, the optimal position of the new sampling points was determined by iterative optimization. When the geometric error between the reconstructed surface and the original surface or the number of sampling points reached the set threshold, the optimization was terminated. Gao et al. [16] took the aeroengine blade as the measurement object, and adaptively generated the sampling points based on the bending moment theory for the regions with different curvatures of the engine blade, so that the number and position of the sampling points could be accurately determined according to the fluctuation of the blade surface. In consideration of the influence of Gaussian curvature on machining error, Sang et al. [17] proposed a scanning line distribution strategy based on star pattern, which classifies peak points and anchor points, and connects the error peak points in different regions and anchor points in the same region, so that more scanning lines are generated in the region with small machining error, and fewer scanning lines are generated in the region with small machining error. Jiang et al. [18] studied a calculation method based on curve chord deviation, and adopted a two-step sampling method. First, based on the radius of the CMM probe ball as the threshold, adaptive Isoparametric sampling was carried out on the leading edge curve and trailing edge surface of the blade. Secondly, adaptive sampling points were selected on the Isoparametric based on the proposed curve chord deviation theory, and finally, the adaptive sampling of the blade was realized. Suleiman et al. [19] proposed a patch based free-form surface adaptive sampling method, which sorts patches according to the Gaussian curvature of each patch, determines the number of sampling points of each patch according to the sorting size, and selects the points of maximum curvature, minimum curvature and average curvature in the patch as the sampling points. If the number of sampling points exceeds these three types of points, the maximum curvature, minimum curvature and average curvature are taken as the sampling points, and so on until the number of sampling points reaches the threshold of the number of sampling points. Mansour [20] studied an adaptive sampling method for reducing the number of measurement points and improving the measurement efficiency with the blade as the measurement object. Based on the least square method, the minimum number of points required for the curve polynomial is found, so that the fitting error expressed by the polynomial curve and cubic curve is minimal. He et al. [21] proposed an adaptive sampling method for a free-form surface based on the machining error model. The machining error model was established based on the curvature of the free-form surface, and the relationship between Gaussian curvature and machining error was obtained. The adaptive sampling of a free-form surface was carried out based on the error model and the Hammersley principle. The number of sampling points was large in the places with large errors and small in the places with small errors. Yi et al. [22] discretized the triangular mesh of a free surface, and simplified the triangular mesh by iteratively shrinking the triangle edges. Based on the principle of minimum quadratic error, the optimal objective vertex under discrete curvature constraint is determined. By limiting the side length of the triangular mesh to control the number of sampling points, the adaptive sampling of free-form surfaces is realized. Yu et al. [23] selected the initial sampling point set, reconstructed the initial point set and solved it with the original surface to obtain the global error. The point with the largest global error is added to the initial point set to obtain the updated initial point set. The reconstructed surface and the original surface are solved to obtain the global error. The point with the maximum global error between the reconstructed surface and the original surface is added to the initial point set as a new sampling point. Cycle in turn until the global error reaches the set precision threshold or the number of sampling points reaches the set number threshold. Mian et al. [24] studied the influence of different sampling strategies on surface reconstruction accuracy, and the influence of workpiece size and machining quality on sampling methods. Gohari et al. [25] used principal component analysis to dynamically generate sampling points, thus reducing the number of sampling points, reducing the measurement cost and improving the measurement efficiency.

For the precise measurement of geometric quantities of free-form surfaces, the number of samples and the position of sampling points should be reasonably set to improve the measurement accuracy and efficiency. In this study, geodesic distance is used as the index of global fluctuation of a free-form surface. The free-form surface is divided into blocks, and the number of samples is adaptively determined according to the changes in the geometric characteristics of the free-form surface. Combined with the Isoparametric distribution, Poisson distribution, Hammersley distribution and NRook distribution, the distribution location of the sampling points is determined. Therefore, an adaptive hybrid sampling method for a free-form surface based on geodesic distance is proposed, which can effectively improve the measurement accuracy and efficiency.

The rest of this paper is organized as follows: Section 2 introduces the free-form surface modeling method; Section 3 introduces the definition of geodesic distance and its solution method in detail, including three sub-steps; in Section 4, the number of sampling points and the distribution of the sampling points are given; in Section 5, the reconstruction errors between the reconstructed surface and the original surface are solved and the results are analyzed; Section 6 summarizes the conclusions and future work prospects.

2. Free-Form Surface Modeling

Non uniform rational B-spline (NURBS) is one of the most commonly used parametric mathematical models in free-form surface geometric modeling. NURBS surfaces are widely used in the field of computer aided geometric design (CAGD), and are widely used in the geometric representation of complex components in aerospace, automobile and other fields.

The free-form surface can be represented by the control points and degrees in the u and v directions. The local adjustment of the free-form surface can be realized by adjusting the control points and weight coefficients. The NURBS surface is usually obtained by using the tensor product of two NURBS curves with two independent parameters, u and v.

The expression of a NURBS surface is as follows [26].

$$S(u,\ v) = \frac{\sum_{i=0}^{m}\sum_{j=0}^{n} w_{i,j} N_{i,p}(u) N_{j,q}(v) P_{i,j}}{\sum_{i=0}^{m}\sum_{j=0}^{n} w_{i,j} N_{i,p}(u) N_{j,q}(v)} \tag{1}$$

where $P_{i,j}$ is the control point; m is the number of control points in the u direction; n is the number of control points in the v direction; p is the degree of the parameter coordinate u; q is the degree of the parameter coordinate v; $N_{i,p}(u)$ is a basis function of order p; $N_{i,p}(v)$ is a basis function of order q; U and V are defined knot vectors, $\mathrm{U} = \{0, \cdots, 0, u_{p+1}, \cdots, u_{r-p-1}, 1, \cdots, 1\}$, $\mathrm{V} = \{0, \cdots, 0, v_{q+1}, \cdots, v_{s-q-1}, 1, \cdots, 1\}$, which specify the distribution of parameters u and v; $w_{i,j}$ is the same as the $P_{i,j}$ relevant weight coefficient. The basis function is recursively defined by the Cox–deBoor algorithm, where $N_{i,p}(u)$:

$$N_{i,1} = \begin{cases} 1 \text{ if } u_i \leqslant u \leqslant u_{i+1} \\ 0 \text{ otherwise} \end{cases} \tag{2}$$

$$N_{i,\ p}(u) = \frac{u - u_i}{u_{i+p-1} - u_i} N_{i,p-1}(u) + \frac{u_{i+p} - u}{u_{i+p} - u_{i+1}} N_{i+1,p-1}(u) \tag{3}$$

The NURBS surface is shown in Figure 2. The blue curve is the isoparm in the u direction, the red curve is the isoparm in the v direction, and the red surface and the blue curve form are the isoparm mesh of the NURBS surface.

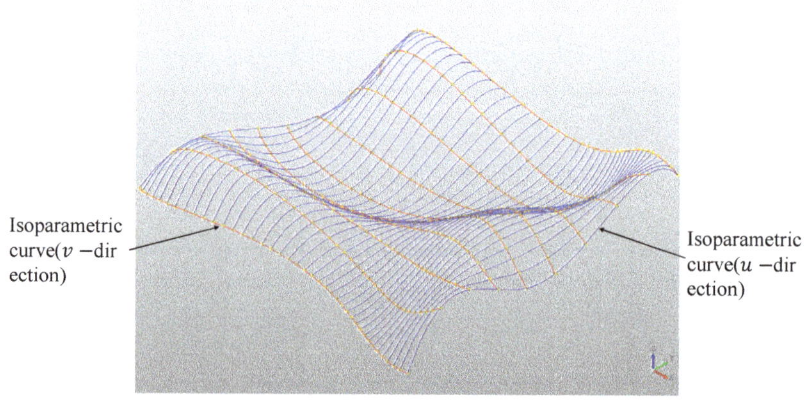

Figure 2. NURBS surface and its Isoparametric mesh.

3. Geodesic Distance

On the Riemannian manifold, the geodesic is defined as the shortest path between the points on the model surface. Geodesic distance is the distance value of geodesic. Compared with the curvature, the advantage of geodesic is that it can describe the free-form surface globally, and the curvature can only be described based on the local part of a free-form surface [27]. Therefore, the geodesic distance is used as the index of the surface fluctuation of the free-form surface in this paper, which can accurately describe the global fluctuation of the free-form surface and overcome the limitation that curvature can only describe the local fluctuation.

The geodesic distance solution method used in this paper is the thermodynamic method proposed by Crane [28]. It can be imagined that a hot needle touches a point x on the surface, which is a hot core point. With the passage of time, the heat on this point x diffuses to the rest of the surface. The heat at point y on the surface can be expressed by a thermal kernel function $k_{t,x}(y)$ to describe how the geodesic distance between any point $x,\ y$ on the Riemannian manifold $\varphi(x,\ y)$ can be recovered by the point state transformation of the thermal core, as shown in Formula (4). The geodesic distance can be recovered by solving the direction of thermal motion [29].

$$\varphi(x,y) = \lim_{t\to 0}\sqrt{-4t\log k_{t,x}(y)} \tag{4}$$

Step 1: Solve the thermal kinematics in Equation (5) by describing the propagation state of heat, and establish the temperature scalar field U. The time dispersion of the thermal propagation equation is

$$\frac{\partial u}{\partial t} = \Delta u \tag{5}$$

The thermal kinematics equation is discretized and sorted to obtain

$$(id - t\Delta)u_t = u_0 \tag{6}$$

where *id* is the identity matrix, t is the time interval, Δ is the discrete Laplacian operator, u_t is the thermal state at time t, and u_0 is the thermal state at the initial time.

Step 2: The thermal gradient direction calculated in step 1 is the same as the gradient direction of the geodesic distance. Since the gradient of the geodesic distance is a unit vector, the gradient of the geodesic distance is obtained by normalizing the thermal gradient direction.

$$X = -\nabla u/|\nabla u| \tag{7}$$

where X is the gradient of the geodesic distance.

Step 3: after the gradient of geodesic distance is obtained through step 2, the geodesic distance is solved through Formula (8):

$$\min_{\phi}\int_M |\nabla\phi - X|^2 \tag{8}$$

According to the variational method, the minimization of Formula (8) is the Poisson equation.

$$\Delta\phi = \nabla\cdot X \tag{9}$$

where ϕ is the geodesic distance between the vertex and the hot core point.

Based on the principle of the NURBS surface in part 2 of this paper, the free-form surface to be measured is shown in Figure 3, with a size of 50 mm × 50 mm. Reduce the dimension of the free-form surface from the three-dimensional space to the two-dimensional parameter domain (u,v); carry out Isoparametric sampling on the two-dimensional parameter domain (u,v); map the sampling points on the two-dimensional parameter domain to the three-dimensional space; use the collection of sampling points on the three-dimensional space to replace the free-form surface, and set the number of sampling points to (41 × 41) = 1681.

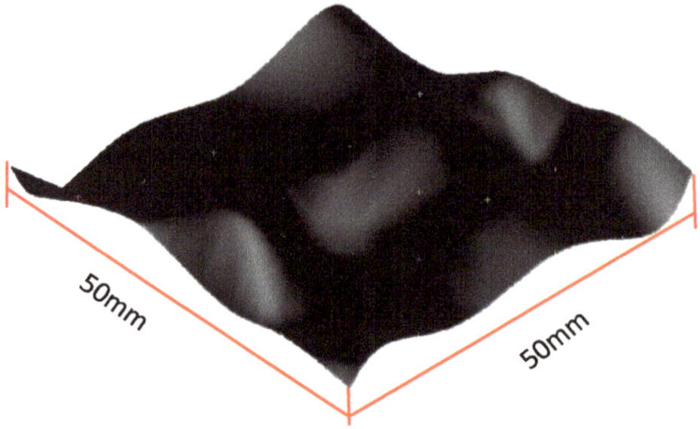

Figure 3. Free-form surface.

In order to evaluate the global fluctuation of a free-form surface, it is necessary to segment the surface. The principle of the surface segment is shown in Figures 4 and 5. As shown in Figure 4, the free-form surface to be measured is divided into four surface segments P_i in the two-dimensional parameter domain. Select the $(41 \times 41 + 1)/2 = 841$ sampling points to be used as hot core points (the red point in Figure 4 needs to be mapped to the three-dimensional space). Since the sampling points in the same row and column as the hot core point belong to the overlapping part of the two segments, the sampling points in the same row and column as the hot core point will be deleted. Solve the geodesic distance from the sampling point to the thermal core point in each surface segment P_i after deleting the duplicate sampling points; sum the geodesic distances from the sampling point to the thermal core point in each segment; obtain the sum of the geodesic distances d_i of each segment sampling point, and take it as the global fluctuation change index of the free-form surface. The sum of the geodesic distances d_i of each segment P_i is shown in Table 1.

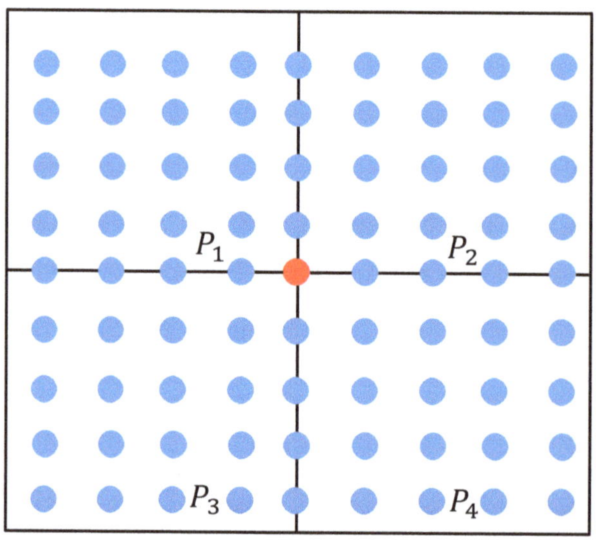

Figure 4. Schematic diagram of the geodesic distance of the first surface segment.

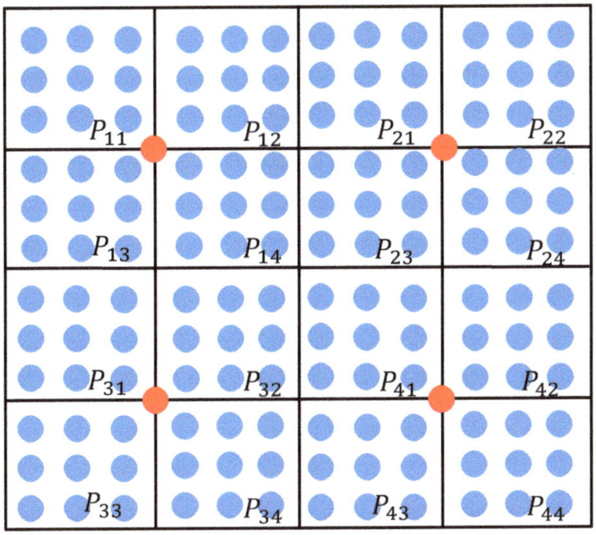

Figure 5. Schematic diagram of the segment geodesic distance of the second surface.

Table 1. Sum of the geodesic distances of the first surface segment.

Segment	Distance/mm
1	7750
2	7816
3	7578
4	7171

As shown in Figure 5, each surface segment P_i is further subdivided into four surface segments P_{ij}, and the sum of the geodesic distances d_{ij} from the sampling point to the hot core point (red point in Figure 5) in each segment P_{ij} is calculated. The results are shown in Table 2. Finally, the sum of the geodesic distances d_{ij} of each free-form surface segment is obtained, which is used as an indicator of global fluctuation. Based on this indicator, the sampling quantity of each free-form surface segment is determined.

Table 2. Sum of the geodesic distances of the second quadric surface segment.

Segment	Distance/mm
11	4207
12	4305
13	3983
14	3714
21	4174
22	4186
23	3824
24	3728
31	3546
32	3442
33	4059
34	4004
41	3366
42	3427
43	4069
44	4163

4. Sampling Strategy

In this section, firstly, the sampling quantity of each surface segment is calculated based on the sum of the geodesic distances of each surface segment solved in Section 3, and then the positions of the sampling points are generated based on the Isoparametric distribution, Poisson distribution, Hammersley distribution and NRook distribution. Finally, the geometric error evaluation method of the reconstructed surface and the original surface is determined.

4.1. Determine the Sampling Quantity

In this paper, a relative proportion method based on the sum of the geodesic distances of each segment is used to determine the P of each surface segment. The specific solution steps are as follows.

Step 1: According to the results of Section 3, calculate the average value of the sum of the geodesic distances of P_1, P_2, P_3 and P_4; obtain the ratio λ_i of the sum of the geodesic distances of P_1, P_2, P_3 and P_4 and d_i. The reference sampling number $\overline{N_i}$ of each surface segment P_i is the ratio of the total number of samples N and the number of segments φ. The sampling number of each surface segment P_i is the product of the number of reference samples $\overline{N_i}$ and λ_i. The formula is as follows:

$$\lambda_i = \frac{d_i}{\overline{d}} \tag{10}$$

$$\overline{N_i} = \frac{N}{\varphi} \tag{11}$$

$$N_i = \overline{N_i} * \lambda_i \tag{12}$$

Step 2: Each surface segment P_i continue to subdivide into surface segment P_{ij}. According to the principle of step 1, calculate the average value $\overline{d_i}$ of the sum of the geodesic distances of P_{ij}; obtain the ratio λ_{ij} of the sum of the geodesic distances of P_{i1}, P_{i2}, P_{i3} and P_{i4} and d_{ij}. The reference sampling number $\overline{N_{ij}}$ of each surface segment P_i is the ratio of the total number of samples N_i and the number of segments φ; The sampling number of each surface segment P_{ij} is the product of the number of reference samples $\overline{N_{ij}}$ and λ_{ij}. The sampling number of each surface segment P_{ij} is the calculation result of step 1. The formula is as follows:

$$\lambda_{ij} = \frac{d_{ij}}{\overline{d_i}} \tag{13}$$

$$\overline{N_{ij}} = \frac{N_i}{\varphi} \tag{14}$$

$$N_{ij} = \overline{N_{ij}} * \lambda_{ij} \tag{15}$$

According to the requirements of measurement accuracy, the total number of sampling points of the free-form surface is determined to be 1600, φ = 4. Calculate the number of sampling points of each surface segment as shown in Table 3.

4.2. Determine the Sampling Position

After determining the number of sampling points in each segment, a specific point distribution algorithm is used to distribute the sampling points. In this paper, the sampling points generated based on the Isoparametric distribution, Poisson distribution, Hammersley distribution and NRook distribution are studied.

Table 3. Number of sampling points for each surface segment.

Segment	Sample Size
11	131
12	144
13	119
14	106
21	181
22	194
23	169
24	156
31	69
32	56
33	94
34	81
41	6
42	19
43	31
44	44

(a) Isoparametric distribution

Isoparametric distribution is to map the free-form surface from the three-dimensional space to the two-dimensional parameter domain, sample the two-dimensional parameter domain according to a certain step size to obtain the sampling points in the two-dimensional parameter domain, remap the sampling points in the two-dimensional parameter domain to the three-dimensional space, and obtain the set of sampling points in the three-dimensional space. It is calculated by Formulas (16) and (17):

$$u = u_{min} + (i-1)\frac{u_{max} - u_{min}}{(N_u - 1)}; i = 1, 2, \cdots, N_u \quad (16)$$

$$v = v_{min} + (i-1)\frac{v_{max} - v_{min}}{(N_v - 1)}; i = 1, 2, \cdots, N_v \quad (17)$$

where u_{min} is the minimum value of the u direction parameter; u_{max} is the maximum value of the u direction parameter; N_u is the number of sampling points in the u direction; v_{min} is the minimum value of the v direction parameter; v_{max} is the maximum value of the v direction parameter; N_v is the number of sampling points in the v direction.

(b) Poisson distribution

Poisson distribution is a more uniform distribution mode compared to Isoparametric distribution. This distribution method can generate a random point set. The distribution method adopted is that every two points are at least on a specified minimum distance. The algorithm takes the range of the R^n sample domain, the minimum distance r between samples and the constant k as the inputs. The steps are as follows [30].

Step 1: Initialize the n dimensional background grid, which is used to store variables and spatial search data. The search cell size is $r/\sqrt{n}$, and each grid cell contains, at most, one sample. Therefore, the grid is an n dimensional integer array. The default index (-1) indicates that there is no sample, and the non-negative integer indicates the index of the sample in the single grid.

Step 2: The initial sample x_0 is randomly selected from the sample field, inserted into the background grid, and this uses index (0) to initialize the activity list (sample index list).

Step 3: When the active list is not empty, a random index is selected from the list, and evenly selects k points from the spherical ring between the surrounding radius r and the radius $2r$. For each point, check whether it is within the radius r of the existing sample (use the background grid to test only the nearby samples). If a point is far enough from the existing sample, it is taken as the next sample and added to the existing index. If no such point is found after k attempts, the index is removed from the active list.

(c) Hammersley distribution

The Hammersley distribution is one of the most prominent uniform distribution sampling algorithms at present. This method is based on the computer binary number representation method, which converts a given decimal number into a binary number, inverts the order, multiplies each number on each bit of the binary by a power series with 1/2 as the base number and the corresponding number of digits as the exponent, and cumulatively sums them. The calculation result is placed after the decimal point to form the sampling value. The sampling formula is as follows:

$$u_i = i/N \tag{18}$$

$$v_i = \sum_{j=0}^{k-1} b_{ij} 2^{-j-1} \tag{19}$$

$$k = [\log_2 N] \tag{20}$$

where N is the number of sampling points; i is the ith sampling point, and its range is $[0,\ N-1]$; b_i is the binary representation of the index; b_{ij} is the jth bit representing b_i, and the range of j is $[0,\ k-1]$; k is b number of digits of b_i.

(d) NRook distribution

NRook distribution is a uniform distribution algorithm based on the principle of a chess board. If a square is divided into $n * n$ small squares and n sampling points are placed inside, there is only one sampling point in each row and column. The algorithm steps are as follows.

Step 1: Initialize the sampling set Q and the sampling step λ according to the number of samples N. The calculation formula is as follows:

$$Q = [0,\ 1, \cdots, N-1] \tag{21}$$

$$\lambda = 1/N \tag{22}$$

Step 2: Selecting random samples n_i from the sample set Q; select the random number p_i, q_i from [0, 1], u_i is the sum of the result of multiplying the sampling step λ by n_i and a random number p_i, v_i is the sum of the result of multiplying the sampling step λ by the sampling times i and a random number q_i. The calculation formula is as follows:

$$\begin{cases} u_i = \lambda n_i + p_i \\ v_i = \lambda i + q_i \end{cases} \tag{23}$$

According to the basic principles of the four distribution modes of Isoparametric distribution, Poisson distribution, Hammersley distribution and NRook distribution, the distribution sequence of parameter $u \in [0,1]$ and parameter $v \in [0,\ 1]$ is calculated and generated. The number of samples is set to 700. The results are shown in Figure 6.

The sampling points generated by the four distribution methods, i.e., Isoparametric distribution, Poisson distribution, Hammersley distribution and NRook distribution, need to be remapped to the corresponding sections according to the surface segment interval, and the sampling points are reconstructed. This is described in detail in Section 5.

4.3. Error Comparison Method between the Reconstructed Surface and Original Surface

The reconstructed surface and the original surface after surface reconstruction need to be evaluated for the geometric error of reconstruction. The method adopted in this paper is shown in Figure 7. The Isoparametric method is used to take a series of sampling points on the original surface, and make a straight line perpendicular to the z axis of the oversampling point. The straight line has an intersection with the reconstructed surface. The x and y coordinate values of the intersection and the sampling point are the same. The difference between the z coordinate values of the sampling point and the intersection is calculated, which is the reconstructed geometric accuracy of the reconstructed surface and the original surface.

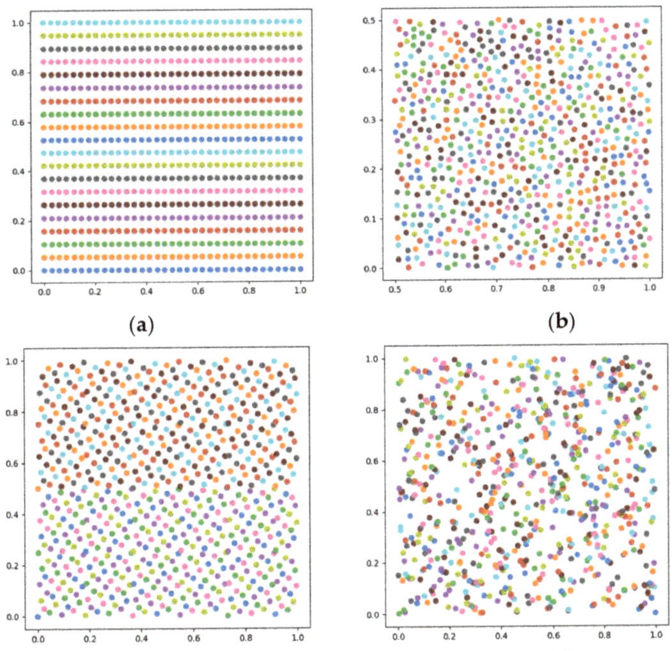

Figure 6. Sampling distribution (**a**) Isoparametric distribution; (**b**) Poisson distribution; (**c**) Hammersley distribution; (**d**) NRook distribution.

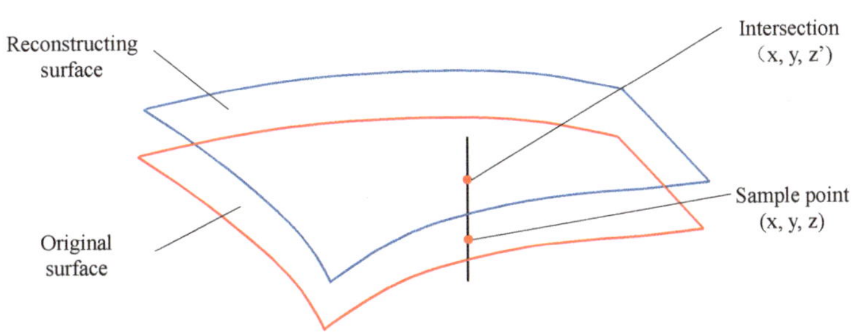

Figure 7. Schematic diagram of the reconstruction error surface.

By solving the z coordinate value difference between each sampling point on the original surface and the intersection point on the reconstructed surface, the root mean square error (RMSE) and the global maximum error (ME) are used as the reconstruction accuracy indexes of the reconstructed surface and the original surface. The formula is as follows:

$$\mathrm{MSE} = \frac{1}{N}\sum_{i=1}^{N} \|z_i - z_i'\|^2 \tag{24}$$

$$\mathrm{RMSE} = \sqrt{\frac{1}{N}\sum_{i=1}^{N} \|z_i - z_i'\|^2} \tag{25}$$

$$ME = \max|z_i - z_i'| \tag{26}$$

where N is the number of sampling points; z_i is the z coordinate value of the original surface sampling point; z_i' is the coordinate value of the reconstructed surface sampling point.

5. Experiment and Discussion

The free-form surface is reconstructed according to the different sampling strategies. The reconstructed surface is analyzed based on the reconstruction accuracy indexes, RMSE and ME, of the reconstructed surface and the original surface to evaluate the impact of the different sampling strategies on the reconstruction's accuracy.

The parameter interval obtained by the four distribution methods of Isoparametric distribution, Poisson distribution, Hammersley distribution and NRook distribution is [0, 1], and the length of the parameter u and v interval of the surface segment is 0.25. Therefore, it is necessary to map the boundary range of the divided surface to the corresponding interval. The interval range of each surface segment is shown in Figure 8.

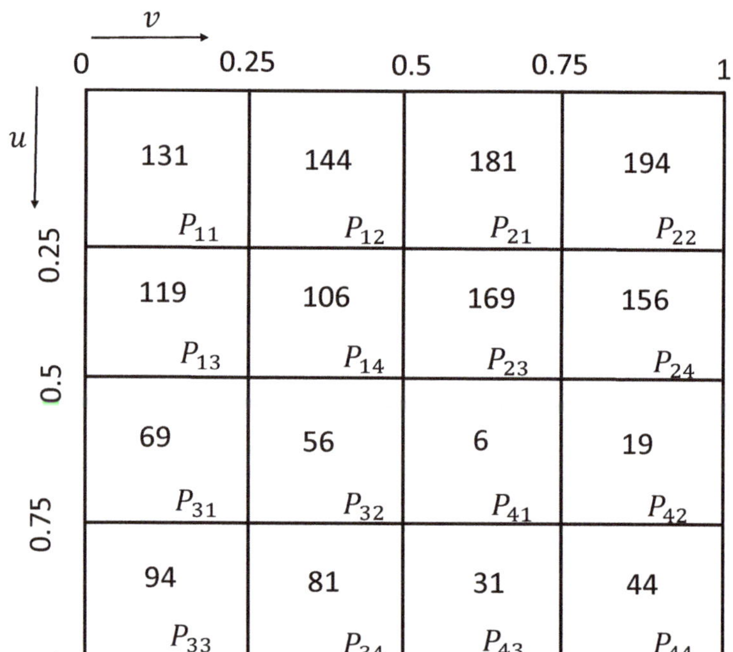

Figure 8. Interval range and sampling quantity of the surface segment.

According to the sampling number of each surface segment in Figure 8, four distribution methods, i.e., Isoparametric distribution, Poisson distribution, Hammersley distribution and NRook distribution, are used to calculate and generate sampling points, and map them to the corresponding interval range of each surface segment in Figure 8. The original free-form surface sampling point set obtained is used for the free-form surface reconstruction. The reconstructed surface is shown in Figure 9, where the green point represents the sample point, and the brown surface represents the reconstructed surface. Different sampling methods produce a different distribution of the sample points, and the reconstructed surface is also different.

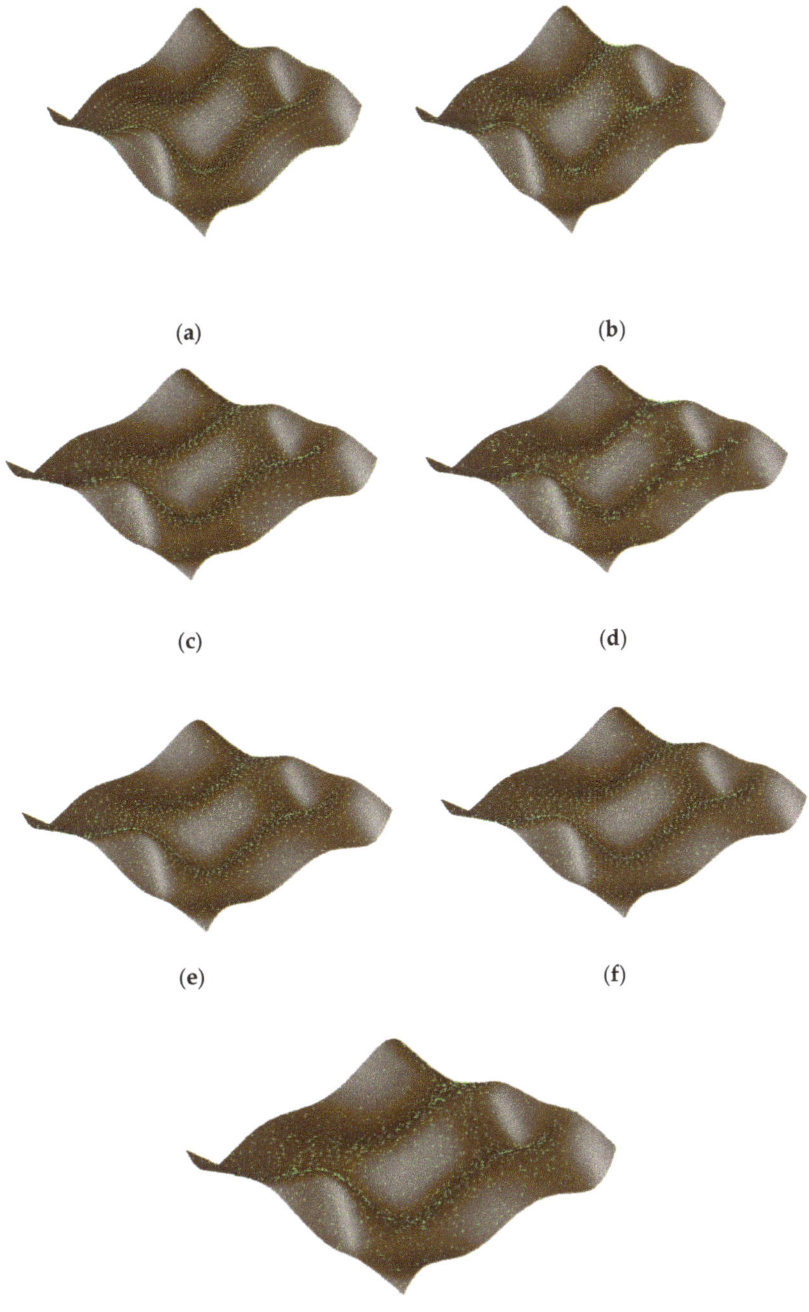

Figure 9. Sampling point distribution of the free-form surface and surface reconstruction: (**a**) Isoparametric; (**b**) Poisson; (**c**) Hammersley; (**d**) NRook; (**e**) Adaptive hybrid Poisson; (**f**) Adaptive hybrid Hammersley; (**g**) Adaptive hybrid NRook.

The reconstructed surface is compared with the original surface in terms of the reconstructed geometric error, and the ME of each point (x, y, z) on the reconstructed surface is converted into the corresponding color. The obtained error distribution results are shown in Figure 10. Figure 10a is the reconstructed geometric error without surface blocking and other parameter distribution; Figure 10b is the reconstructed geometric error of Poisson distribution; Figure 10c is the reconstructed geometric error of Hammersley distribution; and Figure 10d is the reconstructed geometric error of NRook distribution.

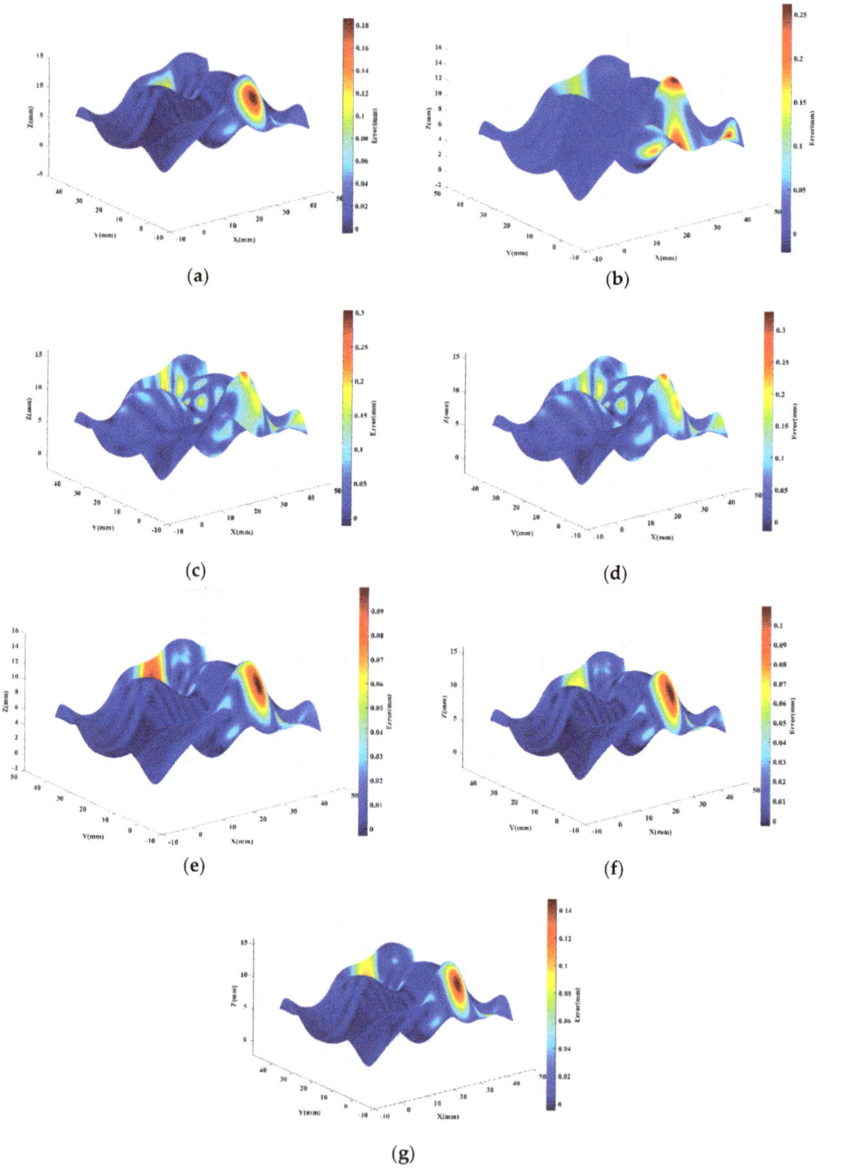

Figure 10. Reconstructed surface ME distribution map: (**a**) Isoparametric; (**b**) Poissson; (**c**) Hammersley; (**d**) NRook; (**e**) Adaptive hybrid Poisson; (**f**) Adaptive hybrid Hammersley; (**g**) Adaptive hybrid NRook.

Table 4 and Figure 11 shows the error comparison results between the reconstructed surface obtained by surface reconstruction and the original surface after the sampling points of the free-form surface are obtained by the different distribution methods. Comparing Isoparametric distribution from the aspect of reconstruction ME, the reconstruction error of the hybrid adaptive sampling method, based on geodesic distance combined with Poisson distribution, Hammersley distribution and NRook distribution, is reduced by 47.12%, 41.56% and 20.22%, respectively, compared with that of the Isoparametric distribution sampling method. Compared with the adaptive sampling method proposed in Reference [24], the reconstruction error of the hybrid adaptive sampling method based on the geodesic distance combined with Poisson distribution, Hammersley distribution and NRook distribution, is reduced by 69.83%, 66.67% and 54.49%, respectively. The reconstruction error of the adaptive hybrid sampling method, based on the geodesic distance combined with Poisson distribution, Hammersley distribution, NRook distribution and the adaptive sampling method used in Reference [14], is reduced by 62.46%, 62.15% and 52.61%, respectively, compared with Poisson distribution, Hammersley distribution and NRook distribution.

Table 4. Comparison of the reconstruction errors using different sampling methods.

Sampling Method	ME/mm	RMSE/mm
Isoparametric	0.1855	0.0266
Poisson	0.2613	0.0467
Adaptive hybrid Poisson	0.0981	0.0202
Hammersley	0.2864	0.0708
Adaptive hybrid Hammersley	0.1084	0.0189
NRook	0.3123	0.1226
Adaptive hybrid NRook	0.1480	0.0213
Adaptive sampling method [24]	0.3252	0.0486

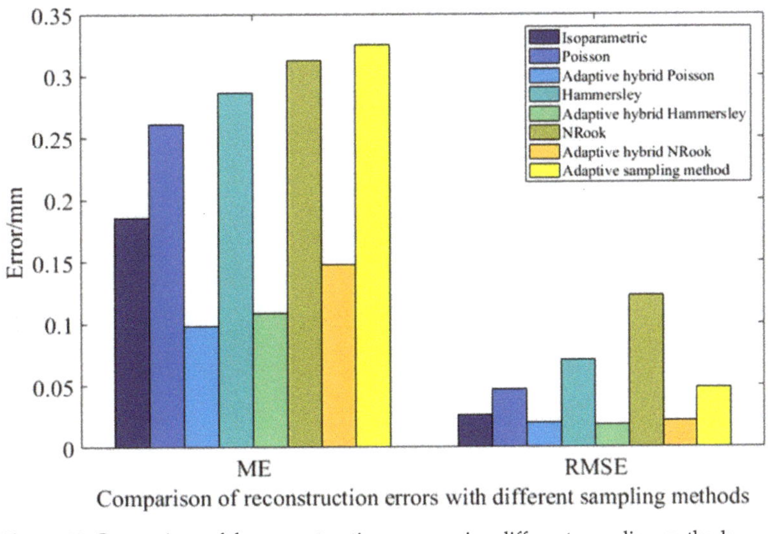

Figure 11. Comparison of the reconstruction errors using different sampling methods.

The adaptive hybrid sampling method based on geodesic distance significantly reduced the reconstruction ME compared with the general Isoparametric distribution sampling method and the adaptive sampling method in Reference [24], and the reconstruction accuracy has been greatly improved. The reconstructed ME based on the adaptive hybrid sampling method of geodesic distance is significantly improved compared with the sampling methods of Poisson distribution, Hammersley distribution and NRook distribution that are not based on the adaptive hybrid sampling method of geodesic distance, which verifies the effectiveness of the adaptive hybrid sampling method based on geodesic distance in reducing the reconstructed ME.

From the aspect of reconstruction RMSE, the adaptive hybrid sampling method based on geodesic distance combined with Poisson distribution, Hammersley distribution and NRook distribution, reduces the reconstruction RMSE by 24.06%, 28.95% and 20.00%, respectively, compared with the sampling method of Isoparametric distribution. Compared with the adaptive sampling method proposed in Reference [24], the reconstructed RMSE of the hybrid adaptive sampling method based on geodesic distance combined with Poisson distribution, Hamersley distribution and NRook distribution is reduced by 58.44%, 61.11% and 56.17%, respectively. The reconstructed RMSE of the adaptive hybrid sampling method based on geodesic distance combined with Poisson distribution, Hammersley distribution and NRook distribution is reduced by 56.75%, 73.31% and 82.63%, respectively, compared with the sampling methods of Poisson distribution, Hammersley distribution and NRook distribution.

The RMSE reconstruction of the adaptive hybrid sampling method based on the geodesic distance combined with Poisson distribution, Hammersley distribution and NRook distribution is greatly reduced compared with the sampling method of Isoparametric distribution and the adaptive sampling method in Reference [24], and the reconstruction accuracy is significantly improved. The RMSE of the adaptive hybrid sampling method based on geodesic distance combined with Poisson distribution, Hammersley distribution and NRook distribution is significantly reduced compared with the sampling method based on Poisson distribution, Hammersley distribution and NRook distribution without geodesic distance, and the reconstruction accuracy is greatly improved.

In order to further verify the effectiveness of the proposed method, this paper uses an ABB IRB1200 robot and Creaform MetraSCAN-R BLACK for experimental verification. The measurement accuracy is 25 μm, and the measurement depth is 250 mm, as shown in Figure 12.

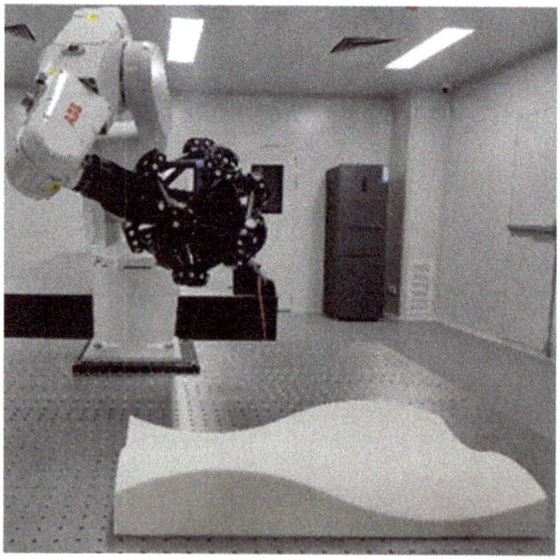

Figure 12. Measuring the surface using a MetraSCAN laser scanner.

Set the number of sample points to 100, and use the self-developed path algorithm to obtain the robot scanning measurement path, as shown in Figure 13. The free-form surface is scanned and measured. The measured results are shown in Figure 14.

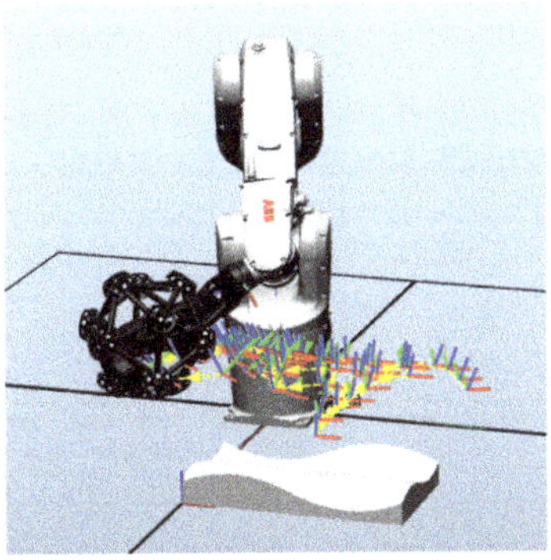

Figure 13. Robot scanning measurement path.

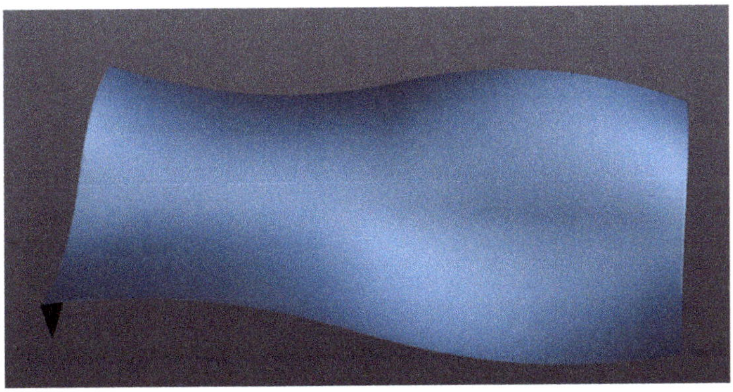

Figure 14. Point cloud reconstruction results of the free-form surface.

In order to further verify the effectiveness of the sampling algorithm, repeated measurement experiments are carried out. Taking the first measurement data as the error judgment standard, analyze the error of the two measurement experiments. The error results are shown in Figure 15. The average deviation is 0.001 mm and the standard error is 0.013 mm, which verifies the effectiveness and robustness of the sampling method proposed in this paper.

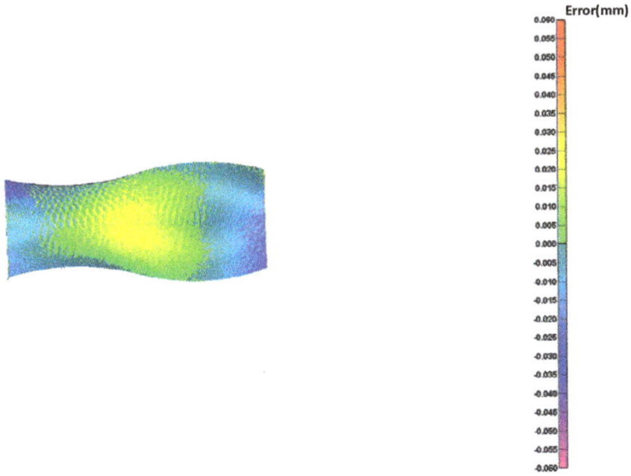

Figure 15. Error comparison results.

6. Conclusions

In this paper, an adaptive hybrid sampling method for a free-form surface based on geodesic distance is proposed, which can be used for the precise measurement of geometric parameters of free-form surfaces. This method can effectively improve measurement efficiency.

The geodesic distance is introduced as a measure of the global fluctuation of the free-form surface. The free-form surface is divided into multiple surface segments. The sampling number of each surface segment is determined according to the sum of the geodesic distances of each surface segment within the surface segment. The sampling points of each surface segment are generated based on the Poisson distribution, Hammersley distribution and NRook distribution sampling methods. Finally, the sampling points of the overall free-form surface are obtained.

The reconstruction error of the free-form surface obtained by this method was compared with the common sampling strategies. The results show that the adaptive hybrid sampling method of a free-form surface based on geodesic distance can effectively reduce the reconstruction error and significantly improve the reconstruction accuracy of free-form surfaces.

In future work, the sampling distribution method can be self-optimized by combining intelligent algorithms, to further improve the reconstruction accuracy of free-form surfaces.

Author Contributions: Methodology, C.C. and L.Y.; Software, C.C. and L.Y.; Investigation, All; Writing—original draft, C.C.; Writing—review and editing, C.C., H.J. and L.Y.; Supervision, L.Y.; Project administration, L.Y.; Funding acquisition, L.Y. All authors have read and agreed to the published version of the manuscript.

Funding: This work was supported by the National Natural Science Foundation of China (Grant Nos. 51927811/51875165/52005147), and the National Key Research and Development Program of China(2019YFE0107400/2022YFF0705700).

Institutional Review Board Statement: Not applicable.

Informed Consent Statement: Not applicable.

Data Availability Statement: Data available on request due to restrictions, e.g., privacy or ethical.

Conflicts of Interest: The authors declare no conflict of interest.

References

1. He, G.Y.; Sang, Y.C.; Wang, H.L.; Sun, G.M. A profile error evaluation method for freeform surface measured by sweep scanning on CMM. *Precis. Eng.-J. Int. Soc. Precis. Eng. Nanotechnol.* **2019**, *56*, 280–292. [CrossRef]
2. Abdullah, A.B.; Sapuan, S.M.; Samad, Z. Profile Measurement Based on Focus Variation Method for Geometrical Defect Evaluation: A Case Study of Cold Forged Propeller Blade. *Adv. Mech. Eng.* **2014**, *6*, 874691. [CrossRef]
3. Shen, Y.; Zhang, W.; Zhang, Y.; Huang, N.; Zhang, Y.; Zhu, L. Distributed Particle Swarm Optimization for the Planning of Time-optimal and Interference-free Five-axis Sweep Scanning Path. *IEEE Trans. Ind. Inform.* **2022**, *18*, 8703–8713. [CrossRef]
4. Zhang, Y.; Tang, K. Automatic Sweep Scan Path Planning for Five-Axis Free-Form Surface Inspection Based on Hybrid Swept Area Potential Field. *IEEE Trans. Autom. Sci. Eng.* **2019**, *16*, 261–277. [CrossRef]
5. Zhang, Y.; Zhou, Z.; Tang, K. Sweep scan path planning for five-axis inspection of free-form surfaces. *Robot. Comput.-Integr. Manuf.* **2018**, *49*, 335–348. [CrossRef]
6. Zhang, Y.; Zhu, L.; Zhao, P.; Hu, P.; Zhao, X. Skeleton Curve-Guided Five-Axis Sweep Scanning for Surface With Multiple Holes. *IEEE Trans. Autom. Sci. Eng.* **2021**, *19*, 2471–2486. [CrossRef]
7. Zhang, Y.; Shen, Y.; Zhang, W.; Zhang, L.; Zhu, L.; Zhang, Y. Generation of efficient and interference-free scanning path for inspecting impeller on a cylindrical CMM. *Measurement* **2022**, *198*, 111352. [CrossRef]
8. Rishi, K.; Gupta, S.K. Planning algorithms for acquiring high fidelity pointclouds using a robot for accurate and fast 3D reconstruction. *Robot. Comput.-Integr. Manuf.* **2022**, *78*, 102372. [CrossRef]
9. Huo, X.; Tran, H.D.; Shilling, K.M.; Kim, H. *Determination and Optimization of Spatial Samples for Distributed Measurements*; Sandia National Laboratories (SNL): Albuquerque, NM, USA; Livermore, CA, USA, 2010.
10. Dunbar, D.; Humphreys, G. A spatial data structure for fast Poisson-disk sample generation. *ACM Trans. Graph.* **2006**, *25*, 503–508. [CrossRef]
11. Wang, T.; Suda, R.; Assoc Comp, M. Fast Maximal Poisson-Disk Sampling by Randomized Tiling. In Proceedings of the High-Performance Graphics (HPG) Conference, Los Angeles, CA, USA, 28–30 July 2017.

12. Woo, T.C.; Liang, R.; Hsieh, C.C.; Lee, N.K. Efficient sampling for surface measurements. *J. Manuf. Syst.* **1995**, *14*, 345–354. [CrossRef]
13. Wang, J.; Jiang, X.; Blunt, L.A.; Leach, R.K.; Scott, P.J. Intelligent sampling for the measurement of structured surfaces. *Meas. Sci. Technol.* **2012**, *23*, 085006. [CrossRef]
14. Ren, M.; Kong, L.; Sun, L.; Cheung, C. A Curve Network Sampling Strategy for Measurement of Freeform Surfaces on Coordinate Measuring Machines. *IEEE Trans. Instrum. Meas.* **2017**, *66*, 3032–3043. [CrossRef]
15. Zahmati, J.; Amirabadi, H.; Mehrad, V. A hybrid measurement sampling method for accurate inspection of geometric errors on freeform surfaces. *Measurement* **2018**, *122*, 155–167. [CrossRef]
16. Gao, F.; Pan, Z.; Zhang, X.; Li, Y.; Duan, J. An adaptive sampling method for accurate measurement of aeroengine blades. *Measurement* **2021**, *173*, 108531. [CrossRef]
17. Sang, Y.; Yan, Y.; Yao, C.; He, G. A new scanning lines distribution strategy for the form error evaluation of freeform surface on CMM. *Measurement* **2021**, *188*, 109578. [CrossRef]
18. Jiang, R.S.; Wang, W.H.; Zhang, D.H.; Wang, Z.Q. A practical sampling method for profile measurement of complex blades. *Measurement* **2016**, *81*, 57–65. [CrossRef]
19. Obeidat, S.M.; Raman, S. An intelligent sampling method for inspecting free-form surfaces. *Int. J. Adv. Manuf. Technol.* **2009**, *40*, 1125–1136. [CrossRef]
20. Mansour, G. A developed algorithm for simulation of blades to reduce the measurement points and time on coordinate measuring machine (CMM). *Measurement* **2014**, *54*, 51–57. [CrossRef]
21. He, G.Y.; Sang, Y.C.; Pang, K.R.; Sun, G.M. An improved adaptive sampling strategy for freeform surface inspection on CMM. *Int. J. Adv. Manuf. Technol.* **2018**, *96*, 1521–1535. [CrossRef]
22. Yi, B.; Qiao, F.; Huang, N.; Wang, X.; Wu, S.; Biermann, D. Adaptive sampling point planning for free-form surface inspection under multi-geometric constraints. *Precis. Eng.-J. Int. Soc. Precis. Eng. Nanotechnol.* **2021**, *72*, 95–101. [CrossRef]
23. Yu, M.; Zhang, Y.; Li, Y.; Zhang, D. Adaptive sampling method for inspection planning on CMM for free-form surfaces. *Int. J. Adv. Manuf. Technol.* **2013**, *67*, 1967–1975. [CrossRef]
24. Mian, S.H.; Al-Ahmari, A.; Alkhalefah, H. Analysis and Realization of Sampling Strategy in Coordinate Metrology. *Math. Probl. Eng.* **2019**, *2019*, 9574153. [CrossRef]
25. Gohari, H.; Barari, A. A quick deviation zone fitting in coordinate metrology of NURBS surfaces using principle component analysis. *Measurement* **2016**, *92*, 352–364. [CrossRef]
26. Piegl, L.; Tiller, W. *The NURBS Book*; Springer: Berlin/Heidelberg, Germany, 1997.
27. Quan, L.; Tang, K. Polynomial local shape descriptor on interest points for 3D part-in-whole matching. *Comput.-Aided Des.* **2015**, *59*, 119–139. [CrossRef]
28. Crane, K.; Weischedel, C.; Wardetzky, M. Geodesics in Heat: A New Approach to Computing Distance Based on Heat Flow. *ACM Trans. Graph.* **2013**, *32*, 1–11. [CrossRef]
29. He, D.; Li, Y.; Li, Z.; Tang, K. Geodesic Distance Field-Based Process Planning for Five-Axis Machining of Complicated Parts. *J. Manuf. Sci. Eng.-Trans. ASME* **2021**, *143*, 061009. [CrossRef]
30. Bridson, R. Fast Poisson disk sampling in arbitrary dimensions. In Proceedings of the SIGGRAPH Sketches, San Diego, CA, USA, 5–9 August 2007.

sensors

MDPI

Article

A Targetless Method for Simultaneously Measuring Three-Degree-of-Freedom Angular Motion Errors with Digital Speckle Pattern Interferometry

Lili Shi [1], Sijin Wu [1,*], Miao Yan [1,2] and Haisha Niu [1]

1 School of Instrumentation Science and Opto-Electronics Engineering, Beijing Information Science and Technology University, Beijing 100192, China
2 Hopen Software Engineering Co., Ltd., Beijing 100192, China
* Correspondence: swu@bistu.edu.cn

Abstract: As a guide rail is the basic motion unit of precision equipment, the measurement of and compensation for its motion errors are important preconditions for precision machining and manufacturing. A targetless and simultaneous measurement method of three-degree-of-freedom (3-DOF) angular motion errors using digital speckle pattern interferometry (DSPI) is introduced in this paper. Based on the analysis of the sensitivity mechanism of DSPI to DOF errors and the formation mechanism of the phase fringes, the relationship between the angular motion errors and the distribution of the interferometric phases was established, and a new simultaneous measurement model of 3-DOF angular motion errors was further proposed. An optical setup based on a three-dimensional spatial-carrier DSPI with a right-angle symmetrical layout was used in the measurement system. Furthermore, repetitive tests, noise tests, and precision analysis were carried out to verify the performance of the system. The test results showed that the measurement resolution of the system was <1 μrad, which is capable of measuring the pitch angle, yaw angle, and roll angle at the submicron arc level simultaneously without target mirrors. The method has the advantages of no need to install cooperative targets and high measurement resolution, showing broad application prospects in many fields, including mechanical manufacturing, laser detection, aerospace, etc.

Keywords: digital speckle pattern interferometry; three degrees of freedom; angular motion errors; pitch; yaw; roll

Citation: Shi, L.; Wu, S.; Yan, M.; Niu, H. A Targetless Method for Simultaneously Measuring Three-Degree-of-Freedom Angular Motion Errors with Digital Speckle Pattern Interferometry. *Sensors* **2023**, *23*, 3393. https://doi.org/10.3390/s23073393

Academic Editor: Susana Silva

Received: 26 February 2023
Revised: 18 March 2023
Accepted: 20 March 2023
Published: 23 March 2023

1. Introduction

As the advanced manufacturing industry develops by leaps and bounds, the demand for precision and ultra-precision machine tools is growing gradually and progressively. Computer numerical control (CNC) machine tools are the core production base of the equipment manufacturing industry, and there is an increasing demand for the tools' precision requirements [1–5]. The precision detection of linear guide rails and rotary axes, which are the main precision motion units in the structure of CNC machine tools, is the key to precision machining and manufacturing [6]. Taking the linear guide rail as an example, its motion errors in three-dimensional (3D) space mainly consist of three linear errors along the axis directions (a positioning error in the direction of axis motion and two straightness errors perpendicular to the axis) and three angular errors of the motion around the axis (i.e., the pitch angle, yaw angle, and roll angle) [7,8]. The three angular errors are the key part of the six-degree-of-freedom (6-DOF) errors, and therefore, the research on the simultaneous measurement method of the three angular errors has become a topic of general interest in many fields, such as aerospace, mechanical manufacturing, and instrumentation.

At present, the measurement methods for the pitch angle, yaw angle, and roll angle can be divided into optical measurement methods and non-optical measurement methods. The latter is mainly those using electronic levels [9], capacitive sensors [10,11], etc., most of which belong to contact measurements, with a small measurement range and cable connection required during work, resulting in limited application scenarios.

The research of optical measurement methods mostly concentrates on the laser multi-DOF measurement system [12–14], which can be divided into the laser collimation method, laser interferometry method, and combination method of laser collimation and laser interferometry. Sun et al. [15] proposed a simultaneous measurement method of three-degree-of-freedom (3-DOF) based on the autocollimation. On the basis of the traditional photoelectric autocollimator, a prism instead of a plane mirror was adopted as the cooperative target to achieve a spectral dimension ampliation measurement by coating on the front surface of the prism. The pitch and yaw angles were characterized by a slope of the right-angle prism, and the roll angle was characterized by two right-angle surfaces so as to realize the simultaneous measurement of the three angles. Although this method is simple in structure, in practical applications, the beam is susceptible to the influence of multiple error sources and the measurement precision of the system may be affected by the introduction of the cooperative prism. The laser interferometry method and combination method, such as the multi-axis laser interferometer [16], etc., enjoys the advantages of high measurement resolution and large measurement range. In the working process, it is required to install optical elements [17–19] (such as plane mirror, right-angle prism, grating, etc.) on the measured objects to reflect the laser beams to the detectors, and then calculate the angle values using optical information derived from the detectors. Liu et al. [20] put forward a simultaneous measurement system of long-distance 6-DOF geometric errors based on the laser interferometry, which allows for the simultaneous measurement of 3-DOF angular errors. The proposed measurement system, which combines the geometric optics with a laser interferometer, is characterized by less measuring time and a wider measuring range than traditional laser interferometers. Cui et al. [21] came up with a simultaneous measurement system of 6-DOF errors based on the combination of laser heterodyne interferometry and laser fiber collimation. Dual-frequency laser beams that are orthogonally linear polarized were adopted as the measuring datum. With the moving unit fixed to the measured axis by two cube-corner reflectors and a beam splitter, the measured unit receives the beams through a photodetector and carries out the photoelectric conversion to obtain the three angular errors. However, all of the above-mentioned measurement methods require the installation of light sources, detectors, or cooperative target mirrors on the objects to be measured. In terms of the installation of these cooperative targets, the influence of the self-weight and installation position of the equipment need to be taken into account since they may introduce additional errors to the measurement, which, as a result, leads to the limitation of their applications in many cases.

The digital speckle pattern interferometry (DSPI) technique—as a high-precision, non-contact and target-mirror-free optical measurement method—has been applied to the research of one-degree-of-freedom and two-degree-of-freedom angular motion measurements recently. For example, a roll angle measurement method based on DSPI was proposed [22]. By studying the relationship between the change in roll angle and the distribution of the interferometric phases of DSPI, the micro roll angle was successfully measured. In addition, a large-stroke roll angle measurement method was further proposed to solve the problem of the small single measurement stroke of the roll angle [23]. Later, the single-angle measurement and double-angle simultaneous measurement of the pitch angle and yaw angle were realized based on DSPI [24]. The outstanding advantages of DSPI in measuring the DOF of rigid bodies are demonstrated in the above research methods, that is, high-precision measurement without the need for cooperative target mirrors.

Because the online measurement of the geometric errors of the moving parts of the machine tool is a very time-consuming and frequent task, and the simultaneous measurement of multi-degree-of-freedom motion errors can greatly reduce the measurement time, it is of great significance to realize the simultaneous measurement of the 3-DOF angular motion errors. However, the simultaneous measurement of the 3-DOF angle motion errors using DSPI has not been realized yet. Due to the complexity of the measurement model and problems such as angle crosstalk, the aforementioned single-angle and double-angle measurement methods based on DSPI cannot be simply extended to the simultaneous measurement method of the three angles. In this article, a 3-DOF angular motion errors geometric measurement model based on DSPI is established, and a new simultaneous measurement method of 3-DOF angular motion errors is proposed. The proposed method can not only perform dynamic measurements with high measurement resolution without the cooperation of target mirrors but also can be applied to both the linear axis and rotary axis. The theoretical analysis and test results are both provided. The theoretical measurement resolution of the three angles is also discussed.

2. Principle

2.1. Geometric Model

The motion parts in the equipment are divided into two types: the linear axis and the rotary axis. Their measurement setups are shown in Figure 1, where Figure 1a is the measurement setup of the linear axis and Figure 1b is that of the rotary axis. The three angular displacement errors around the x, y, and z directions are defined as $R_x(d)$, $R_y(d)$, and $R_z(d)$, respectively. Specifically, d refers to the linear displacement d_z of the linear axis and the angular displacement d_θ of the rotary axis. In the DSPI measurement system, the linear displacement of the linear axis along the z-axis and the angular displacement of the rotary axis around the z-axis can be regarded as the accumulation of small displacement at a high sampling rate. Therefore, this measurement method is suitable for both the linear axis and the rotary axis.

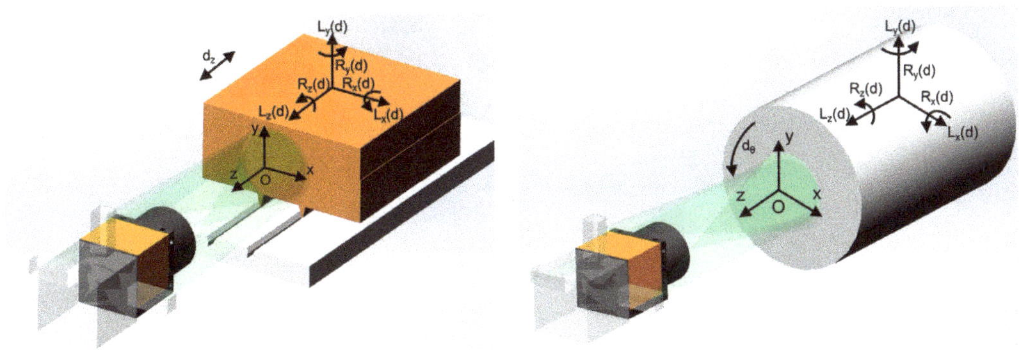

(a) (b)

Figure 1. The setups of the DOF error measurement with DSPI for (**a**) linear guide and (**b**) rotary axis.

In the two measurement setups, the DSPI device is placed directly in front of the end face of the measured axis along the z-axis. The three laser beams from the upper, left, and right sides of the DSPI device are illuminated to the end face of the measured axis at a small angle. For the DSPI device, the motions of the linear axis or the rotary axis and the resulting 6-DOF errors belong to the geometric motions of the rigid body in the six degrees of freedom. Therefore, the above measurement settings can be simplified into a geometric model as shown in Figure 2. More specifically, camera C is on the z-axis, and the three lasers with the same wavelength, namely S1, S2, and S3, are symmetrically distributed on the upper, left, and right sides of the camera. The lasers irradiate the end face of the measured object and each laser beam interferes with its own reference beam at the camera to form three independent speckle pattern interferograms. The coordinate system and coordinate values of each point are shown in Figure 2.

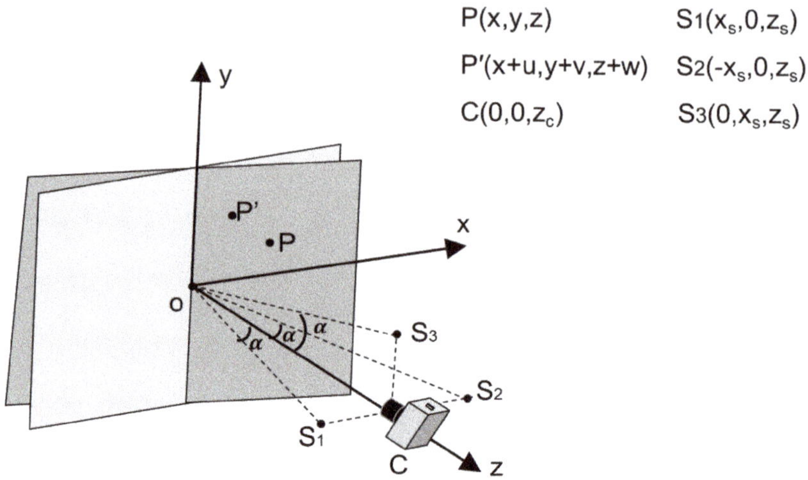

Figure 2. Geometric model of optical setups for measuring geometric motions of the rigid body by DPSI.

When there are axis motion and 6-DOF errors at the same time, the arbitrary point P on the end face of the measured axis is transformed to point P′, and the coordinates change from (x, y, z) to $(x + u, y + v, z + w)$. The three linear motions along the x, y, and z directions are defined as $L_x(d)$, $L_y(d)$, and $L_z(d)$, respectively, while the three angular motions rotating around the x, y, and z axes are $R_x(d)$, $R_y(d)$, and $R_z(d)$, respectively. As shown in Figure 2, the displacements generated in the x, y, and z directions are, respectively, represented by (u, v, w), where u is generated by $L_x(d)$, $R_y(d)$, and $R_z(d)$; v is generated by $L_y(d)$, $R_x(d)$, and $R_z(d)$; and w is generated by $L_z(d)$, $R_x(d)$, and $R_y(d)$. The expressions of u, v, and w are obtained as follows:

$$\begin{pmatrix} u \\ v \\ w \end{pmatrix} = \begin{pmatrix} L_x - x(1-\cos R_y) + \sqrt{x^2+y^2}\left[\cos\left(\arctan\frac{y}{x} + R_z\right) - \cos\left(\arctan\frac{y}{x}\right)\right] \\ L_y - y(1-\cos R_x) + \sqrt{x^2+y^2}\left[\sin\left(\arctan\frac{y}{x} + R_z\right) - \sin\left(\arctan\frac{y}{x}\right)\right] \\ L_z + \sqrt{y^2+z^2}\left[\sin\left(\arctan\frac{z}{y} + R_x\right) - \sin\left(\arctan\frac{z}{y}\right)\right] + \sqrt{x^2+z^2}\left[\sin\left(\arctan\frac{z}{x} - R_y\right) - \sin\left(\arctan\frac{z}{x}\right)\right] \end{pmatrix}. \quad (1)$$

Due to the small DOF errors, the higher-order terms of the DOF errors in Equation (1) can be ignored. Therefore, the equation is further simplified, as shown below:

$$\begin{pmatrix} u \\ v \\ w \end{pmatrix} = \begin{pmatrix} L_x - yR_z \\ L_y + xR_z \\ L_z + yR_x - xR_y \end{pmatrix}. \quad (2)$$

The relationship between the DOF errors and the geometric displacement is thereby established.

2.2. Measurement Principle

The geometric model shown in Figure 2 is composed of three independent and symmetrical digital speckle pattern interference optical paths with single-beam illumination, which are arranged symmetrically on the upper, left, and right sides of the optical axis. Each interference corresponds to an out-of-plane displacement component and an in-plane displacement component. In that case, the relationship between the distribution of the three interferometric phases and displacements is as follows [25]:

$$\begin{pmatrix} \Delta\phi_{s1} \\ \Delta\phi_{s2} \\ \Delta\phi_{s3} \end{pmatrix} = \frac{2\pi}{\lambda} \begin{pmatrix} u\sin\alpha + w(1+\cos\alpha) \\ u\sin(-\alpha) + w(1+\cos\alpha) \\ v\sin\alpha + w(1+\cos\alpha) \end{pmatrix}, \tag{3}$$

where, λ is the wavelength of the laser device and α is the illumination angle.

Considering the results shown in Equation (2), the phase increment output of the three interferences can be expressed as:

$$\begin{pmatrix} \Delta\phi_{s1} \\ \Delta\phi_{s2} \\ \Delta\phi_{s3} \end{pmatrix} = \frac{2\pi}{\lambda} \begin{pmatrix} L_x\sin\alpha + L_z(1+\cos\alpha) + yR_x(1+\cos\alpha) - xR_y(1+\cos\alpha) - yR_z\sin\alpha \\ -L_x\sin\alpha + L_z(1+\cos\alpha) + yR_x(1+\cos\alpha) - xR_y(1+\cos\alpha) + yR_z\sin\alpha \\ L_y\sin\alpha + L_z(1+\cos\alpha) + yR_x(1+\cos\alpha) - xR_y(1+\cos\alpha) + xR_z\sin\alpha \end{pmatrix}. \tag{4}$$

The phase of each interference is related to five geometric displacements, including two linear displacements and three angular displacements. When the illumination of the first and second interference optical paths is on the *xoz* plane, the interference optical path is sensitive to the remaining five geometric displacements, except for the L_y. Differently, when the illumination of the third interference optical path is on the *yoz* plane, the interference path is not sensitive to L_x. When there are one or more geometric displacements, the interferometric phases change accordingly, which explains the sensitivity of the DSPI to the DOF errors.

The derivatives of the three interferometric phases with respect to x and y are taken. Since the geometric displacement belongs to the rigid body displacements, with a spatial gradient of 0, the spatial gradients of the three interferometric phases along the x and y directions are expressed as:

$$\begin{pmatrix} \frac{\partial\Delta\phi_{s1}}{\partial x} & \frac{\partial\Delta\phi_{s1}}{\partial y} \\ \frac{\partial\Delta\phi_{s2}}{\partial x} & \frac{\partial\Delta\phi_{s2}}{\partial y} \\ \frac{\partial\Delta\phi_{s3}}{\partial x} & \frac{\partial\Delta\phi_{s3}}{\partial y} \end{pmatrix} = \frac{2\pi}{\lambda} \begin{pmatrix} -R_y(1+\cos\alpha) & R_x(1+\cos\alpha) - R_z\sin\alpha \\ -R_y(1+\cos\alpha) & R_x(1+\cos\alpha) + R_z\sin\alpha \\ -R_y(1+\cos\alpha) + R_z\sin\alpha & R_x(1+\cos\alpha) \end{pmatrix}. \tag{5}$$

It is revealed in Equation (5) that the spatial gradient of each digital speckle pattern interferometric phase is only related to one or two angular displacements, indicating that the generation of spatial fringes in the phase map is caused by the corresponding angular displacements only. When angular displacement variation occurs, the spatial gradient of the phases at a certain time is the same in the whole field, which suggests that the phase fringes are uniformly changed and the fringe spacing determined by the angular displacement between the two samplings is equal.

The expression of the three angular displacement errors can be obtained by solving part of the information in Equation (5). Considering that the spatial distribution of the phases varies uniformly in the whole field, the difference operation can be used to replace the differential operation, and the three angular displacement errors can be expressed as:

$$\begin{pmatrix} R_x(d) \\ R_y(d) \\ R_z(d) \end{pmatrix} = \frac{\lambda}{2\pi} \begin{pmatrix} \frac{\Delta\phi_{s3}(x_1,y_1,d) - \Delta\phi_{s3}(x_2,y_2,d)}{(1+\cos\alpha)(y_1-y_2)} \\ \frac{\Delta\phi_{s1}(x_3,y_3,d) - \Delta\phi_{s1}(x_4,y_4,d)}{(1+\cos\alpha)(x_3-x_4)} \\ \frac{\Delta\phi_{s2}(x_5,y_5,d) - \Delta\phi_{s2}(x_6,y_6,d) - \Delta\phi_{s1}(x_5,y_5,d) + \Delta\phi_{s1}(x_6,y_6,d)}{2\sin\alpha(y_5-y_6)} \end{pmatrix}, \tag{6}$$

where, $(x_n, y_n, d), (n \in N_+)$ stands for the spatial coordinates of the point, and d represents the displacement of the measured axis.

According to the above mathematical model, it can be seen that the 3-DOF angular motion errors are linearly related to the digital speckle pattern interferometric phases, and the measurement resolution of the 3-DOF angular motion mainly depends on the phase measurement resolution. Usually, in the measurement using DSPI, the phase measurement resolution can reach $\pi/10$, and be up to $\pi/25$ under optimal conditions. In general, with the phase resolution of $\pi/10$, the laser wavelength of 532 nm, and the illumination angle of 20°, the measurement resolution of the pitch, yaw, and roll angles can reach 0.14 μrad, 0.14 μrad, and 0.39 μrad, respectively, in theory when two symmetrical points about the origin with a distance of 100 mm are selected for analysis.

3. Results and Discussion

3.1. Experimental Setup

The measurement system proposed in this paper is shown in Figure 3. Three lasers with a central wavelength of 532 nm were utilized as the light source. A disc with a thickness of 1 mm and a radius of 32 mm was taken as the measured object, which was fixed to a six-axis piezoelectric oscillating table (Harbin Core Tomorrow Science & Technology Co., Ltd., Harbin, China, H63. XYZTR1S). An area of 13.4 mm × 13.4 mm in the center of the measured object was selected as the measurement area. By driving the oscillating table, tiny yaw, pitch, and roll motions were generated. The overall optical setup was composed of three interference parts, which had independent light sources and shared an imaging device and a detector. The illumination direction of the interference part containing Laser 3 was on the *yoz* plane, while the illumination directions of the other two interference parts were on the *xoz* plane. All illumination angles were set to about 26°. Taking the interference part where Laser 1 was located as an example, the laser beam was split into a reference beam and an object beam after passing through beam splitter 1 (BS1). The object beam was irradiated to the measured object after passing through mirror 1 (M1) and beam expander 1 (BE1), and then the diffusely reflected beam was collected by the imaging lens and captured by the camera through an aperture. The reference beam hit on the surface of the camera at a specific angle near the center of the aperture to interfere with the object beam after entering the optical fiber through the coupling lens 1 (CL1), yielding speckle pattern interferograms. When the geometric motion occurred on the measured disc, the optical path of the object beam changed while the optical path of the reference beam remained stable, thus changing the interferometric phase.

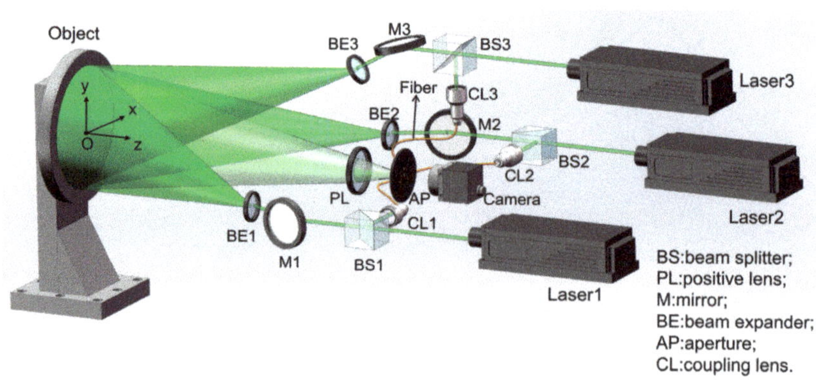

Figure 3. Optical setup of the simultaneous measurement system of 3-DOF angular motion errors using DSPI.

The three interferograms corresponding to the three interference parts were superimposed. By performing Fourier transform on the interferograms, the interferometric information was separated into three pairs of frequency components on the spectrum due to the difference in spatial-carrier frequencies of the three interference parts. Three halos, marked as A, B, and C in the spectrum, as shown in Figure 4a, were selected and used to generate three phase maps that are shown in Figure 4b–d by the means of inverse Fourier transform. These phase maps were derived from the three interference parts that contained Laser 1, Laser 2, and Laser 3, respectively.

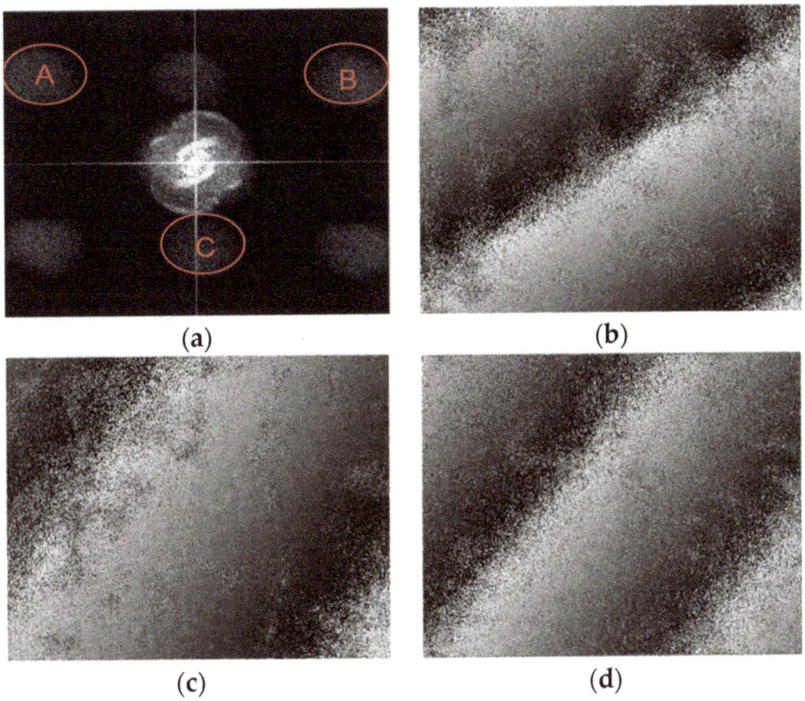

Figure 4. Frequency spectrum and phase maps: (**a**) spectrum obtained after Fourier transform; (**b**) wrapped phase map corresponding to frequency component A; (**c**) wrapped phase map corresponding to frequency component B; (**d**) wrapped phase map corresponding to frequency component C.

After the phases shown in Figure 4 were unwrapped, several points at the edge of the phase maps were used to solve the pitch, yaw, and roll angles by performing the calculation shown in Equation (6). The results showed that the pitch, yaw, and roll angles were 10.11 μrad, 10.12 μrad, and 10.23 μrad, respectively, which were basically identical to the actual output values of the piezoelectric oscillating table (all of the three angles were nominally 10 μrad), verifying the feasibility of the method preliminarily.

3.2. Simultaneous Measurements of Pitch, Yaw, and Roll Angles

The pitch angle, yaw angle, and roll angle were simultaneously loaded by driving the six-axis piezoelectric oscillating table. The angle-loading ranges of the pitch angle, yaw angle, and roll angle set in the test were within 10–60 μrad, respectively. The three angle channels of the piezoelectric oscillating table were controlled to output the same angle value simultaneously, with a loading step of 5 μrad. Finally, a total of 11 data points were obtained, and the measurement process was repeated three times to obtain three sets of repeated test data. The measurement results and average deviation are shown in Figure 5.

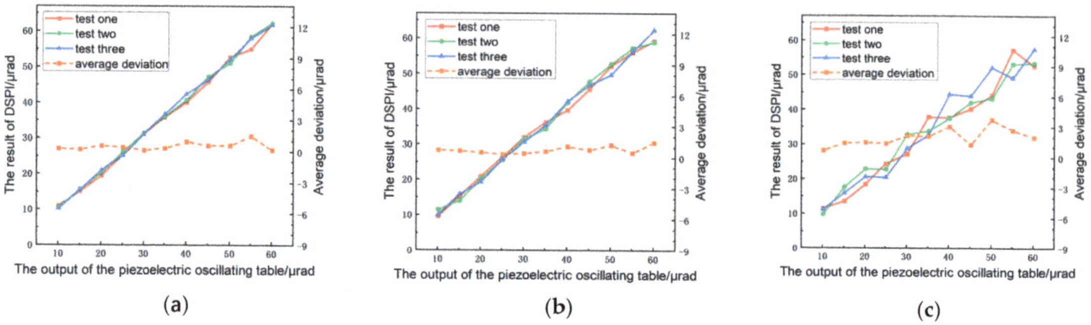

Figure 5. Simultaneous measurement results: (**a**) the measurement values of pitch, (**b**) the measurement values of yaw, (**c**) the measurement values of roll.

The coincidence of the three groups of tests can be obviously reflected by the values shown in Figure 5. In particular, the coincidence of the measurement curves of the pitch and yaw angles was very good, while the coincidence of the roll angle data was slightly poor, but the fluctuation was within the acceptable limits. Under different measurement steps, the average deviation of pitch ranged from 0.14 μrad to 1.45 μrad, with an average relative error of 3.54% compared with the piezoelectric oscillating table. The average deviation of yaw was within the range of 0.33–1.45 μrad, with an average relative error of 3.71%. The average deviation of roll was within the range of 0.68–3.72 μrad, with an average relative error of 8.32%.

The close-loop piezoelectric oscillating table used in the experiment has good linearity and repeatability accuracy, reaching 0.25% F.S. and 0.2% F.S., respectively, with a maximum stroke of 600 μrad per axis. The closeness of the three-angle measurement results to the outputs of the piezoelectric oscillating table indicates that the proposed method has high accuracy in measurement, and proves that this method can reliably realize the simultaneous measurement of the 3-DOF angular motion errors.

3.3. Repeatability Test

Aiming to verify the measurement reliability of the proposed measurement system, the pitch angle, yaw angle, and roll angle were measured repeatedly 100 times with a loading angle of 20 μrad by the piezoelectric oscillating table. The measurement results are shown in Figure 6. It can be seen that the measured pitch angle fluctuated within the range of 19.11–23.06 μrad, with an average of 21.39 μrad and a standard deviation of 0.81 μrad. The measured yaw angle fluctuated between 19.03 μrad and 22.50 μrad with an average of 21.16 μrad and a standard deviation of 0.85 μrad. The measured roll angle fluctuated between 16.03 μrad and 21.91 μrad with an average of 18.33 μrad and a standard deviation of 1.82 μrad. The low standard deviation of the measurement values of the three angles verified the high repeatability of the overall measurement. The factors affecting the fluctuation of the measurement results of the three angles mainly include phase noise, environmental disturbance, etc.

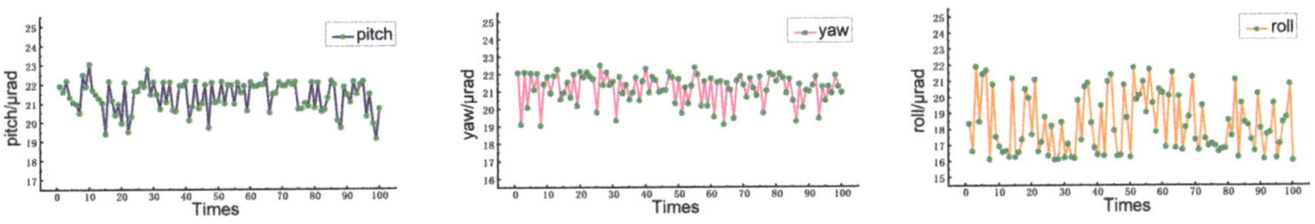

(a) (b) (c)

Figure 6. Repeatability experiment results: (**a**) results from pitch loading, (**b**) results from yaw loading, (**c**) results from roll loading.

3.4. Noise Evaluation

In order to verify the influence of random errors caused by noise interference on the measurement system, measurements were carried out 50 times, with the original measurement system remaining unchanged and the measured object remaining stationary. The data set obtained by DSPI is exhibited in Figure 7.

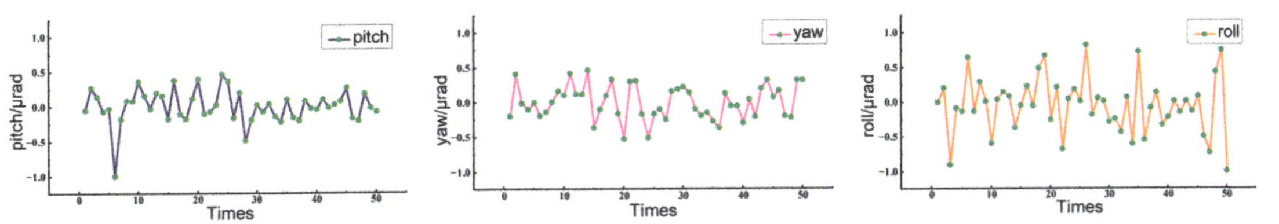

(a) (b) (c)

Figure 7. Noise evaluation experiment: (**a**) measurement results of pitch, (**b**) measurement results of yaw, (**c**) measurement results of roll.

The data in Figure 7 show that when the object was stationary, phase noise was detected at all three angles, of which the standard deviations were 0.238 μrad, 0.242 μrad, and 0.403 μrad, respectively. If twice the standard deviation was taken as the noise value, the three-angle noise values were 0.476 μrad, 0.484 μrad, and 0.806 μrad, respectively, which were all less than 1 μrad and close to the theoretical measurement resolution. It is verified that the DSPI measurement system has a high resolution and can be used for the simultaneous measurement of the pitch angle, yaw angle, and roll angle at the submicron arc level. During the test, the measurement resolution can be further improved and optimized by adopting a shorter wavelength laser source, and especially, expanding the measurement area. In this test, the measurement area was 13.4 mm × 13.4 mm, so the distance between the two selected points was short. If the measurement area is larger and two points at a greater distance can be selected for calculating the angular motion errors, then the measurement resolution will be improved.

The experimental results show that the measurement performance of the roll angle is weaker than that of the pitch angle and yaw angle. This is consistent with theoretical analysis. As can be seen from Equation (6), both the pitch and yaw angles can be solved from one phase map, while the roll angle requires two phase maps to solve. This shows that the measurement accuracy of the roll angle is more susceptible to phase errors. Moreover, roll angle measurement has a higher sensitivity coefficient. In these experiments, the illumination angles were 26°, so the sensitivity coefficient of roll measurement was about twice those of pitch angle and yaw angle measurements. These explain why roll angles were slightly less measured. However, the measurement of the roll angle as well as the measurements of pitch and yaw angles is a high-precision measurement method.

4. Conclusions

A simultaneous measurement method of 3-DOF angular motion errors based on DSPI is proposed. The method does not require any cooperative target and does not require any pretreatment of the measured object, which is a true non-contact, pollution-free measurement method. It not only eliminates the inconvenience and measurement errors caused by the installation of the target mirrors, but also provides a feasible solution for applications where the target mirror cannot be installed. Because this method belongs to the optical interference method, its measurement resolution and measurement accuracy are very high. Moreover, compared with the traditional optical interference methods of 3-DOF angular motion error measurement, in addition to eliminating the need for a cooperative target, the optical setup of this method is simple, easy to integrate, and highly reliable. This method mainly relies on the diffused light from the rough surface of the measured object for interference. Therefore, in the field of engineering, it can be applied to most measured objects except for objects with a specular surface.

In this article, the measuring principle of the proposed method is described in detail, and the performance of the measurement method is also demonstrated through theoretical analysis and experimentation. It can be concluded from the theoretical analysis and experimental results that this measurement method has an edge in terms of precision, requiring no target mirrors, and is applicable to both the linear axis and the rotary axis. Therefore, it has a certain application prospect in rigid body angle measurement and other fields, laying a foundation for the research of the simultaneous measurement of 6-DOF errors.

Author Contributions: Conceptualization, S.W.; methodology, L.S. and S.W.; investigation, L.S.; validation, L.S. and M.Y.; data curation, L.S. and M.Y.; writing—original draft preparation, L.S.; writing—review and editing, S.W. and H.N.; project administration, S.W.; funding acquisition, S.W. and H.N. All authors have read and agreed to the published version of the manuscript.

Funding: This work was funded by the National Natural Science Foundation of China, No. 52075045, the Natural Science Foundation of Beijing Municipality, China, No. 4212047, and Research Project of Beijing Education Committee, No. 202111232020.

Institutional Review Board Statement: Not applicable.

Informed Consent Statement: Not applicable.

Data Availability Statement: Data are available from the corresponding authors on reasonable request.

Conflicts of Interest: The authors declare no conflict of interest.

References

1. Song, R.; Niu, J.; Chen, D. Current Development and Countermeasure of Ultra-Precision Machine Tools. *Key Eng. Mater.* **2010**, *455*, 632–636. [CrossRef]
2. Yang, Z.; Chen, C.; Chen, F.; Li, G. Progress in the Research of Reliability Technology of Machine Tools. *Chin. J. Mech. Eng.* **2013**, *49*, 130–139. [CrossRef]
3. Liu, C.H.; Hsieh, W.H.; Chang, Z.Y.; Tzou, G.Y.; Hanson, S.G.; Hwang, Y.L. Recent Advances in Precision Machinery and Manufacturing Technology. *Adv. Mech. Eng.* **2014**, *6*, 508592. [CrossRef]
4. Schneider, F.; Das, J.; Kirsch, B.; Linke, B.; Aurich, J.C. Sustainability in Ultra Precision and Micro Machining: A Review. *Int. J. Precis. Eng. Manuf.-Green Tech.* **2019**, *6*, 601–610. [CrossRef]
5. Yip, W.S.; To, S.; Zhou, H. Current status, challenges and opportunities of sustainable ultra-precision manufacturing. *J. Intell. Manuf.* **2022**, *33*, 2193–2205. [CrossRef]
6. Li, X.; Yang, X.; Gao, L.; Su, Z.; Wei, X.; Lv, Z.; Liang, J.; Li, H.; Fang, F. Rapid Measurement and Identification Method for the Geometric Errors of CNC Machine Tools. *Appl. Sci.* **2019**, *9*, 2701. [CrossRef]
7. MA, Y.; LI, Y. Motion Error of Rolling Guide Based on Uncertainty of Geometric Error. *Chin. J. Mech. Eng.* **2019**, *55*, 11–18. [CrossRef]
8. Wei, X.; Su, Z.; Yang, X.; Lv, Z.; Yang, Z.; Zhang, H.; Li, X.; Fang, F. A Novel Method for the Measurement of Geometric Errors in the Linear Motion of CNC Machine Tools. *Appl. Sci.* **2019**, *9*, 3357. [CrossRef]

9. Wang, C.; Chen, G.; Xu, G. Electronic Gradienter Based on ICL 7135. *Instrum. Exp. Tech.* **2008**, *10*, 8135.
10. Park, S.R.; Jang, J.S.; Yang, S.H. Roll error measurement system for linear moving stages using four capacitance sensors. *Int. J. Adv. Manuf. Technol.* **2016**, *84*, 2579–2587. [CrossRef]
11. Fan, X.; Yu, Z.; Peng, K.; Chen, Z.; Liu, X. A Compact and High-Precision Capacitive Absolute Angular Displacement Sensor. *IEEE Sens. J.* **2020**, *20*, 11173–11182. [CrossRef]
12. Feng, Q.; Zhang, B.; Cui, C.; Kuang, C.; Zhai, Y.; You, F. Development of a simple system for simultaneously measuring 6DOF geometric motion errors of a linear guide. *Optics Express* **2013**, *21*, 25805–25819.
13. Hsieh, H.L.; Pan, S.W. Development of a grating-based interferometer for six-degree-of-freedom displacement and angle measurements. *Optics Express* **2015**, *23*, 2451–2465. [CrossRef] [PubMed]
14. Cai, Y.; Sang, Q.; Lou, Z.F.; Fan, K.C. Error Analysis and Compensation of a Laser Measurement System for Simultaneously Measuring Five-Degree-of-Freedom Error Motions of Linear Stages. *Sensors* **2019**, *19*, 3833. [CrossRef] [PubMed]
15. Sun, C. Research on Multi-Degree-of-Freedom Measurement Method for Precision Engineering. Ph.D. Thesis, University of Changchun Institute of Optics, Fine Mechanics and Physics Chinese Academy of Sciences, Changchun, China, 2021.
16. Ikram, M.; Hussain, G. Michelson interferometer for precision angle measurement. *Appl. Opt.* **1999**, *38*, 113–120. [CrossRef] [PubMed]
17. Gao, W.; Saito, Y.; Muto, H.; Arai, Y.; Shimizu, Y. A three-axis autocollimator for detection of angular error motions of a precision stage. *CIRP Ann. Manuf. Technol.* **2011**, *60*, 515–518. [CrossRef]
18. Yin, Y.; Cai, S.; Qiao, Y. Design, fabrication, and verification of a three-dimensional autocollimator. *Appl. Opt.* **2016**, *55*, 9986–9991. [CrossRef]
19. Li, W.; Mu, Q.; Wang, S. Roll angle measurement with a large range based on the photoelectronic autocollimator. *Optoelectron. Lett.* **2016**, *12*, 74–76. [CrossRef]
20. Liu, C.S.; Pu, Y.F.; Chen, Y.T.; Luo, Y.T. Design of a Measurement System for Simultaneously Measuring Six-Degree-Of-Freedom Geometric Errors of a Long Linear Stage. *Sensors* **2018**, *18*, 3875. [CrossRef]
21. Cui, C.; Feng, Q.; Zhang, B.; Zhao, Y. System for simultaneously measuring 6DOF geometric motion errors using a polarization maintaining fiber-coupled dual-frequency laser. *Optics Express* **2016**, *24*, 6735–6748. [CrossRef]
22. Wu, S.; Yang, J.; Li, W.; Wu, F.; Dong, M. Precision roll angle measurement based on digital speckle pattern interferometry. *Meas. Sci. Technol.* **2019**, *30*, 045005. [CrossRef]
23. Yang, J.; Wu, S.; Li, W.; Wu, F.; Liu, J.; Yang, L. Precise measurement of large roll angle using digital speckle pattern interferometry. In Proceedings of the 10th International Symposium on Precision Engineering Measurements and Instrumentation (ISPEMI 2018), Kunming, China, 8–10 August 2018.
24. Wu, F.; Wu, S.; Li, W.; Li, X. Simultaneous and precision measurement of yaw and pitch using digital speckle pattern interferometry. *Opt. Eng.* **2021**, *60*, 034103. [CrossRef]
25. Fang, Y.; Wu, S.; Yang, L. Synchronous Measurement of Three-Dimensional Deformations Using Tri-Channel Spatial-Carrier Digital Speckle Pattern Interferometry. *Appl. Mech. Mater.* **2017**, *868*, 316–322. [CrossRef]

sensors

MDPI

Article

Accurate Calibration of a Large Field of View Camera with Coplanar Constraint for Large-Scale Specular Three-Dimensional Profile Measurement

Rongsheng Lu [1,2,*], Zhizhuo Wang [1] and Zhiting Zou [1]

1 School of Instrument Science and Opto-Electronics Engineering, Hefei University of Technology, Hefei 230009, China
2 Anhui Province Key Laboratory of Measuring Theory and Precision Instrument, Hefei University of Technology, Hefei 230009, China
* Correspondence: rslu@hfut.edu.cn

Abstract: In the vision-based inspection of specular or shiny surfaces, we often compute the camera pose with respect to a reference plane by analyzing images of calibration grids, reflected in such a surface. To obtain high precision in camera calibration, the calibration target should be large enough to cover the whole field of view (FOV). For a camera with a large FOV, using a small target can only obtain a locally optimal solution. However, using a large target causes many difficulties in making, carrying, and employing the large target. To solve this problem, an improved calibration method based on coplanar constraint is proposed for a camera with a large FOV. Firstly, with an auxiliary plane mirror provided, the positions of the calibration grid and the tilt angles of the plane mirror are changed several times to capture several mirrored calibration images. Secondly, the initial parameters of the camera are calculated based on each group of mirrored calibration images. Finally, adding with the coplanar constraint between each group of calibration grid, the external parameters between the camera and the reference plane are optimized via the Levenberg-Marquardt algorithm (LM). The experimental results show that the proposed camera calibration method has good robustness and accuracy.

Keywords: coplanar constraint; camera calibration; large FOV; specular measurement

Citation: Lu, R.; Wang, Z.; Zou, Z. Accurate Calibration of a Large Field of View Camera with Coplanar Constraint for Large-Scale Specular Three-Dimensional Profile Measurement. *Sensors* **2023**, *23*, 3464. https://doi.org/10.3390/s23073464

Academic Editors: Qibo Feng, Jiakun Li and Qixin He

Received: 11 February 2023
Revised: 22 March 2023
Accepted: 22 March 2023
Published: 25 March 2023

1. Introduction

In recent years, the vision measurement system has been widely used in industrial production due to its high precision, non-contact, real-time capabilities, etc. [1,2]. At the same time, for some special objects, such as car windshield [3], painted body shell [4], polishing mold, stainless steel products, and other smooth surface objects, the demand for three-dimensional measurement is greater and greater. Meanwhile, the traditional three-dimensional reconstruction method [5–7] is not ideal for the reconstruction of the bright surface. The two-dimensional feature information of the image obtained by the camera mainly comes from the surrounding environment of the shiny surface, rather than the surface itself. For the high reflection characteristics of the shiny surface, the reference pattern is usually placed around it, and the reference pattern modulated by the surface helps realize three-dimensional reconstruction of itself [8–12]. In this case, the calibration accuracy of the reference plane and camera directly affects the subsequent three-dimensional reconstruction accuracy of the shiny surface. Meanwhile, to measure more area of surface, a camera with a large FOV is needed. However, for calibration in large FOV, the targets with large areas and high precision are not only difficult to make, but they are also inconvenient to carry and use.

For the calibration of the catadioptric system, many scholars have proposed methods of using an auxiliary plane mirror to estimate the external parameters between the camera and the reference object [13]. Kumar et al. [14] proposed using the orthogonal constraint between the direction vector of the connection from the corresponding point of the object to the mirror image, as well as the column vector of the rotation matrix to list linear equations for solving, and each set of equations requires at least five calibration images. However, the calculated position parameter has a large error with the true value, which is harmful to the subsequent parameter optimization. Takahashi et al. [15] obtain the unique solution of three P3P problems (perspective-three-point problem) from three mirror images based on the orthogonal constraint. However, if the reference object is smaller than a certain size, a wrong solution will be obtained. The method proposed by Hesch et al. [16] also obtains the solutions of three P3P problems from three mirror images, but it can only select an optimal solution from 64 candidate solutions after re-projection error evaluation. Xin et al. [17] directly estimate the camera rotation matrix by the SVD decomposition of the sum of the rotation matrices. Additionally, they calculate the translation vector by solving overdetermined linear equations. While it is more sensitive to noise, the algorithm stability is poor. Bergamasco [18] proposed a method to locate coplanar circles from images by means of a non-cooperative evolutionary game and refined the estimation of camera parameters by observing a set of coplanar circles. However, the accuracy of this method is low.

For the calibration of the camera with a large FOV, scholars consider combing several two-dimensional small targets into a large three-dimensional target. While in the methods proposed in the paper [19,20], all intrinsic parameters of the camera cannot be obtained because the polynomial projection mode is used. Meanwhile, in the methods proposed in the paper [21,22], the relative positions between the small targets are subject to certain restrictions, which makes it difficult to be applied in real applications. Occlusion-resistant markers, such as Charuco [23] or RUNETag [24], are also robust options, but they present fewer points for calibration.

To solve this problem, we use a LCD monitor as a reference plane to produce the calibration grid. It not only solves the problem of difficulty in manufacturing, carrying, and using large-sized objects, but it also can be used as a carrier for projecting encoded patterns when measuring bright surfaces due to its ability to produce free patterns. Bergamasco [25,26] also used a monitor that displays dense calibration grids for camera calibration, but it requires multiple frames, and when dense grid points are spread over the display, the curvature of the display surface will greatly affect the accuracy and robustness of calibration. Therefore, this article calibrates using a smaller calibration grid on the monitor and covers the camera's field of view by moving the position of the calibration grid, which to some extent reduces the impact of display surface curvature, and ultimately it achieves high accuracy and robustness.

Firstly, by moving the calibration grid on the reference plane and changing the tilt angle of the plane mirror on the optical platform to obtain multiple sets of mirrored calibration images, the internal and external parameters of the camera are computed by Zhang's [27] calibration method. Secondly, the orthogonality constraint calibration method and P3P algorithm proposed in [15,16] are used to obtain the external parameters from the reference plane to the camera. Finally, the LM [28] algorithm is used to obtain the optimal solution of the external parameters with the coplanar constraint of multiple calibration grid positions. At the same time, using the method of reconstructing the smooth mirror shape from a single image proposed in [12], three-dimensional measurement experiments are carried out to indirectly verify the accuracy of the calibration method proposed.

2. Geometry of Camera Pose Estimation

2.1. Plane Mirror Reflection Model

As is shown in Figure 1, in the camera coordinate system C, the plane mirror can be described by the plane parameters $\Pi = \{n, d\}$. The unit vector n denotes the normal vector of the mirror plane, d represents the distance between the origin of C and the plane [17], and R_{s2c} and T_{s2c} are the rotation matrix and the translation vector between the reference plane coordinate system and the camera coordinate system. $\boldsymbol{P}$ is a feature point on the reference plane.

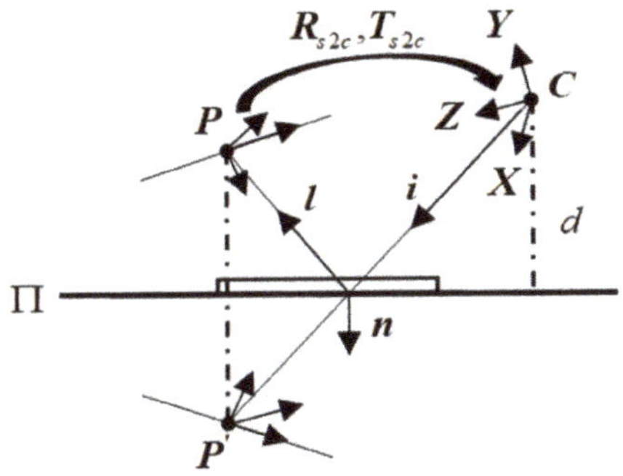

Figure 1. Calibration principle for reference plane. The camera $\boldsymbol{C}$ observes a point $\boldsymbol{P}$ on the reference plane via the plane mirror Π. We denote by $\boldsymbol{i}$ the incident ray and by $\boldsymbol{l}$ the reflected ray, $\boldsymbol{R}_{s2c}$ and $\boldsymbol{T}_{s2c}$ denote the pose parameters between the reference plane and the camera, $\boldsymbol{n}$ denotes the normal of the mirror, and d is the distance between $\boldsymbol{C}$ and Π.

Based on the reflection property of the mirror, the relationship between this point and its mirror point is given by:

$$\begin{bmatrix} \boldsymbol{P'} \\ 1 \end{bmatrix} = \boldsymbol{M}_1 \cdot \begin{bmatrix} \boldsymbol{P} \\ 1 \end{bmatrix}, \boldsymbol{M}_1 = \begin{bmatrix} \boldsymbol{I} - 2 \cdot \boldsymbol{n} \cdot \boldsymbol{n}^T & 2 \cdot d \cdot \boldsymbol{n} \\ \boldsymbol{O} & 1 \end{bmatrix} \tag{1}$$

This denotes the symmetric transformation induced by Π. Note that $\boldsymbol{M}_1 = \boldsymbol{M}_1{}^{-1}$, and $(\boldsymbol{I} - 2 \cdot \boldsymbol{n} \cdot \boldsymbol{n}^T)$ is a Householder matrix. Let $\boldsymbol{M}_2$ describe the rigid transformation that transforms points from the reference to the camera frame:

$$\boldsymbol{M}_2 = \begin{bmatrix} \boldsymbol{R}_{s2c} & \boldsymbol{T}_{s2c} \\ \boldsymbol{O} & 1 \end{bmatrix} \tag{2}$$

2.2. Mirror-Based Camera Projection Model

The perspective projection model is a camera imaging model widely used in computer vision [23]. The mapping relation between any three-dimensional point $\boldsymbol{P}_w$ in the space and its corresponding pixel point $\boldsymbol{v} = [x\ y\ 1]^T$ in the image can be described as:

$$\boldsymbol{v} = \boldsymbol{s} \cdot \boldsymbol{A} \cdot [\boldsymbol{R}\ \boldsymbol{T}] \cdot \boldsymbol{P}_w \tag{3}$$

where s is a nonzero scale factor, $\boldsymbol{A}$ is the intrinsic parameters matrix of the camera, and $\boldsymbol{R}$ and $\boldsymbol{T}$ are the rotation matrix and the translation vector between the camera coordinate system and the world coordinate system. Taking the mirror reflection into account, concatenate the camera model with the mirror reflection, the mirror-based camera projection model becomes:

$$\boldsymbol{v} = s \cdot \boldsymbol{A} \cdot \boldsymbol{M}_1 \cdot \boldsymbol{M}_2 \cdot \boldsymbol{P}_w \tag{4}$$

$\boldsymbol{R}$ and $\boldsymbol{T}$ can be written as:

$$\begin{cases} \boldsymbol{R} = (\boldsymbol{I} - 2 \cdot \boldsymbol{n} \cdot \boldsymbol{n}^T) \cdot \boldsymbol{R}_{s2c} \\ \boldsymbol{T} = (\boldsymbol{I} - 2 \cdot \boldsymbol{n} \cdot \boldsymbol{n}^T) \cdot \boldsymbol{T}_{s2c} + 2 \cdot d \cdot \boldsymbol{n} \end{cases} \tag{5}$$

According to the Equation (5), we need at least three specular reflection images to calculate $\boldsymbol{R}_{s2c}$ and $\boldsymbol{T}_{s2c}$.

2.3. Computation of External Parameters

By changing the tilt angle of the plane mirror, we can obtain mirrored images at different positions and compute external parameters by the P3P algorithm [16]. Let $j, j' \in \{1,2,3\}$, and $\boldsymbol{R}_j$ represents the rotation matrix of the mirrored image at the j position of the plane mirror. Assume unit vector $\boldsymbol{m}_{jj'}$ is perpendicular to $\boldsymbol{n}_j$ and $\boldsymbol{n}_{j'}$, so we can obtain:

$$\boldsymbol{R}_j \cdot \boldsymbol{R}_{j'}{}^T \cdot \boldsymbol{m}_{jj'} = (\boldsymbol{I} - 2 \cdot \boldsymbol{n}_j \cdot \boldsymbol{n}_j{}^T) \times (\boldsymbol{I} - 2 \cdot \boldsymbol{n}_{j'} \cdot \boldsymbol{n}_{j'}{}^T) \cdot \boldsymbol{m}_{jj'} = \boldsymbol{m}_{jj'} \tag{6}$$

$\boldsymbol{R}_j \cdot \boldsymbol{R}_{j'}{}^T$ is a special orthogonal matrix, which has two complex conjugate eigenvalues, and one eigenvalue equals 1. So $\boldsymbol{m}_{jj'}$ is the eigenvector of $\boldsymbol{R}_j \cdot \boldsymbol{R}_{j'}{}^T$ corresponding to the eigenvalue of 1. According to the cross-product properties of the eigenvector, the unit normal vectors corresponding to the three positions of the plane mirror can be calculated.

$$\boldsymbol{n}_1 = \frac{\boldsymbol{m}_{13} \times \boldsymbol{m}_{12}}{\|\boldsymbol{m}_{13} \times \boldsymbol{m}_{12}\|},\ \boldsymbol{n}_2 = \frac{\boldsymbol{m}_{21} \times \boldsymbol{m}_{23}}{\|\boldsymbol{m}_{21} \times \boldsymbol{m}_{23}\|},\ \boldsymbol{n}_3 = \frac{\boldsymbol{m}_{13} \times \boldsymbol{m}_{23}}{\|\boldsymbol{m}_{13} \times \boldsymbol{m}_{23}\|}. \tag{7}$$

According to the Equation (5), $\boldsymbol{R}_{s2c}$ can be calculated. In the case of an ideal condition without noise, the three rotation matrices calculated by three $\boldsymbol{R}_j$ should be equal. While they are not equal in fact due to the noise. Therefore, the average of the rotation matrices should be calculated [20].

$$\overline{\boldsymbol{R}} = [(\hat{\boldsymbol{R}}^T \cdot \hat{\boldsymbol{R}})^{1/2}]^{-1} \cdot \hat{\boldsymbol{R}},\ \text{where } \hat{\boldsymbol{R}} = \frac{1}{3} \cdot \sum_{j=1}^{3} \boldsymbol{R}_{s2cj} \tag{8}$$

The rest of the parameters $[\boldsymbol{T}, d_1, d_2, d_3]^T$ can be solved by linear equations constructed by the Equation (5). So far, all of the initial values of the pose parameters have been calculated.

$$\begin{bmatrix} (\boldsymbol{I} - 2 \cdot \boldsymbol{n}_1 \cdot \boldsymbol{n}_1{}^T) & 2 \cdot \boldsymbol{n}_1 & o & o \\ (\boldsymbol{I} - 2 \cdot \boldsymbol{n}_2 \cdot \boldsymbol{n}_2{}^T) & o & 2 \cdot \boldsymbol{n}_2 & o \\ (\boldsymbol{I} - 2 \cdot \boldsymbol{n}_3 \cdot \boldsymbol{n}_3{}^T) & o & o & 2 \cdot \boldsymbol{n}_3 \end{bmatrix} \cdot \begin{bmatrix} \boldsymbol{T} \\ d_1 \\ d_2 \\ d_3 \end{bmatrix} = \begin{bmatrix} \boldsymbol{T}_1 \\ \boldsymbol{T}_2 \\ \boldsymbol{T}_3 \end{bmatrix} \tag{9}$$

2.4. Optimization with Coplanar Constraint

Linear solutions are usually sensitive to noise, we can minimize the reprojection error of back-projection by adjusting $\boldsymbol{R}_{s2c}$, $\boldsymbol{T}_{s2c}$, $\boldsymbol{n}$ and d with coplanar constraint. As is shown in Figure 2, we move the calibration grid on the LCD monitor for W times and rotate the plane mirror corresponding to each grid position for M times. The grid has N characteristic corners. Let $\boldsymbol{R}_{ji}$ represent the rotation matrix of the mirrored image of the j grid at the i plane mirror position. In the same way, $\boldsymbol{T}_{ji}$ is translation vector, $\boldsymbol{n}_{ji}$ represents the normal vector of the mirror, d_{ji} represents the distance between the origin of the camera coordinate system and the plane mirror, $\boldsymbol{R}_{s2cj}$ represents rotation matrix from the j checkerboard coordinate system to the camera coordinate system, and $\boldsymbol{T}_{s2cj}$ represents the translation vector. $\boldsymbol{P}_k$ represents the k feature point of the grid in the reference plane coordinate system. $\boldsymbol{q}_{jik}$ represents the projection point of the k feature point of the j grid at the i planar mirror position. $\widetilde{\boldsymbol{q}}_{jik}$ represents the back-projection point. The back-projection process can be written as:

$$\widetilde{\boldsymbol{q}}_{jik} = \lambda_{ji} \cdot \boldsymbol{A} \cdot (\boldsymbol{R}_{ji} \cdot \boldsymbol{P}_k + \boldsymbol{T}_{ji}) \tag{10}$$

where λ_{ji} is a nonzero scale factor, $\boldsymbol{A}$ represents the intrinsic matrix of the camera, and $\boldsymbol{R}_{ji} = (\boldsymbol{I} - 2 \cdot \boldsymbol{n}_{ji} \cdot \boldsymbol{n}_{ji}{}^T) \cdot \boldsymbol{R}_{s2cj}$, $\boldsymbol{T}_{ji} = (\boldsymbol{I} - 2 \cdot \boldsymbol{n}_{ji} \cdot \boldsymbol{n}_{ji}{}^T) \cdot \boldsymbol{T}_{s2cj} + 2 \cdot d_{ji} \cdot \boldsymbol{n}_{ji}$.

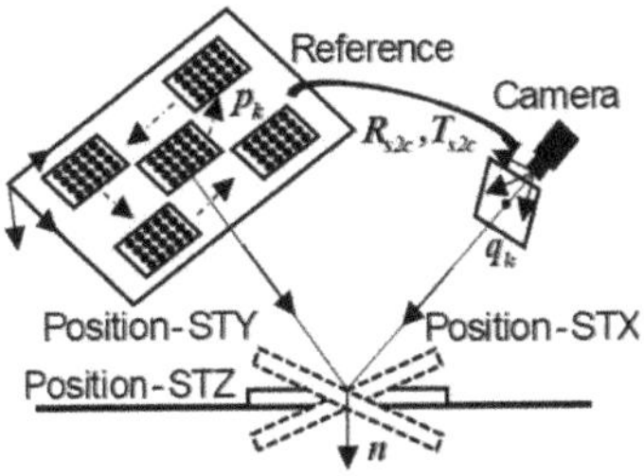

Figure 2. Structure of measurement system. The calibration grid moves on the reference plane for M times. A feature point $\boldsymbol{P_k}$ of the grid is reflected in the image point q_k via the mirror at position STX, STY, STZ. We can obtain three calibration images at each grid location. Then, $W \times 3$ calibration images can calculate the intrinsic matrix $\boldsymbol{A}$, as well as the pose parameters $\boldsymbol{R_{ji}}$ and $\boldsymbol{T_{ji}}$. Finally, $\boldsymbol{R_{s2c}}$, $\boldsymbol{T_{s2c}}$, $\boldsymbol{n}$, and d can be calculated by Equations (8) and (9).

Combined with the Equation (10), the reprojection error function of the back-projection can be expressed as:

$$Errpro = \sum_{j=1}^{W}\sum_{i=1}^{M}\sum_{k=1}^{N} \|\boldsymbol{q_{jik}} - \tilde{\boldsymbol{q}}_{jik}(\boldsymbol{R_{s2cj}}, \boldsymbol{T_{s2cj}}, \boldsymbol{n_{ji}}, d_{ji}, \boldsymbol{P_k})\|^2 \tag{11}$$

Let $\boldsymbol{P_{jk}}$ represent the k feature point of the j checkboard in the camera coordinate system.

$$\boldsymbol{P_{jk}} = (\boldsymbol{R_{s2cj}} \cdot \boldsymbol{P_k} + \boldsymbol{T_{s2cj}}) \tag{12}$$

Since the reference plane can be regarded as a standard plane, the coplanar constraint of the W grids should be added. Let $Perr$ represent the fitting effect evaluation value of plane fitting function: $[fitresult, Perr] = createFit(dx,\ dy,\ dz)$. The input of the function is $\boldsymbol{P_{jk}}$. The smaller the $Perr$ value is, the better the coplanar effect will perform. In addition, $\boldsymbol{R_{s2cj}}, j \in \{1, \ldots W\}$ are equal in theory. Let $\boldsymbol{R_{av}}$ represent the average rotation matrix [24]. The error $Rerr$ between $\boldsymbol{R_{s2cj}}$ and $\boldsymbol{R_{av}}$ can be written as:

$$Rerr = \sum_{j=1}^{W} \|\boldsymbol{R_{s2cj}} - \boldsymbol{R_{av}}\|^2 \tag{13}$$

The smaller the $Rerr$ value is, the better the coplanar effect will perform. Likely, the five plane mirror positions with zero tilt angle on the optical platform also have coplanar characteristics. Therefore, the corresponding normal vectors $\boldsymbol{n_{j1}}$ are theoretically equal. The average normal vector $\boldsymbol{n_{av}}$ can also be calculated.

$$Nerr = \sum_{j=1}^{W} \|\boldsymbol{n_{j1}} - \boldsymbol{n_{av}}\|^2 \tag{14}$$

In the ideal condition, $Perr = 0,\ Rerr = 0,\ Nerr = 0$. Therefore, the cost function can be regarded as two major components: the reprojection error term $Errpro$ and the coplanar constraint term $(Perr,\ Rerr,\ Nerr)$. We can establish the cost function in the case of equality constraints:

$$\begin{cases} F = \min \sum_{j=1}^{W}\sum_{i=1}^{M}\sum_{k=1}^{N} \|\boldsymbol{q_{jik}} - \tilde{\boldsymbol{q}}(\boldsymbol{R_{s2cj}}, \boldsymbol{T_{s2cj}}, \boldsymbol{n_{ji}}, d_{ji}, \boldsymbol{P_k})\|^2 + Errcop \\ Errcop = Perr + Rerr + Nerr \end{cases} \tag{15}$$

where $\boldsymbol{R_{s2cj}}$, $\boldsymbol{T_{s2cj}}$, $\boldsymbol{n_{ji}}$ and d_{ji} are parameters to be optimized. The calculation of the specific LM algorithm can be realized by the tool function $lsqnonlin()$ in Matlab.

2.5. Three-Dimensional Measurement Principle of a Single Camera

In the monocular measurement system, we observe the images of the grid pattern, reflected in the unknown surface when the pose of the camera is known, and establish the reflection correspondence between the three-dimensional reference points and the two-dimensional image points. The depth of the reflection points on the surface is parameterized, and the surface shape is fitted by a polynomial. Therefore, the measurement of the surface shape is converted into an optimization problem: minimizing the error between the reference points and the corresponding points through the surface back projection [12]. The principle of the measurement system is shown in Figure 3. $\boldsymbol{O}$ is the origin of the camera coordinate frame, $\boldsymbol{m}$ is a feature point on the reference plane, $\boldsymbol{p}$ is a reflection point of the surface, and $\boldsymbol{v}$ is a projection point on the normalized image plane. $\boldsymbol{p}$ and $\boldsymbol{v}$ are called reflection correspondences. $\boldsymbol{l}$ is the reflected ray at $\boldsymbol{p}$, and $\boldsymbol{i}$ is the incident ray. $\boldsymbol{R}_{s2c}$ and $\boldsymbol{T}_{s2c}$ are the rotation matrix and translation vector from reference plane coordinate frame to camera coordinate frame. Obviously, $\boldsymbol{v}$ is on the incident ray $\boldsymbol{i}$. The relationship between $\boldsymbol{p}$ and $\boldsymbol{v}$ is given by

$$p = s \cdot v \tag{16}$$

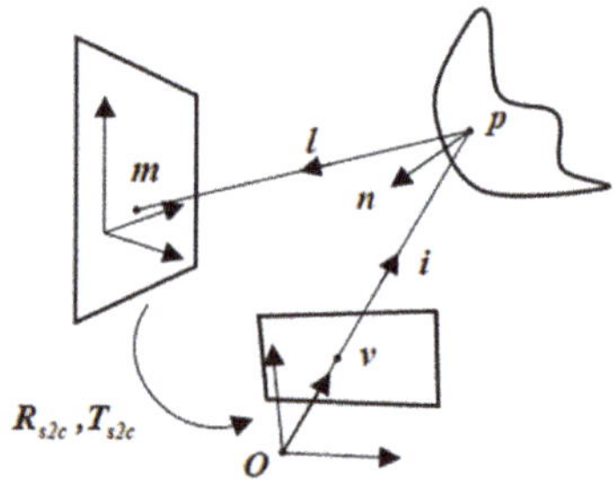

Figure 3. Principle of mirror surface measurement. A pinhole camera centered at $\boldsymbol{O}$ is observing a mirror surface point $\boldsymbol{p}$ that reflects a reference point $\boldsymbol{m}$ to an image point $\boldsymbol{v}$. We refer to $\boldsymbol{m}$ and $\boldsymbol{v}$ as reflection correspondences. The reflected ray $\boldsymbol{l}$ is determined by $\boldsymbol{m}$ and $\boldsymbol{p}$. We denote by $\boldsymbol{i}$ the incident ray for image point $\boldsymbol{v}$ and by $\boldsymbol{n}$ the normal at $\boldsymbol{p}$. $\boldsymbol{R}_{s2c}$ and $\boldsymbol{T}_{s2c}$ denote the pose parameters between the reference plane and the camera.

s is the depth of the corresponding reflected point $\boldsymbol{p}$. Correspondingly, the normal $\boldsymbol{n}$ to the surface at $\boldsymbol{p}$ can be written as:

$$n = \left(\frac{\partial p}{\partial x}, \frac{\partial p}{\partial y}, \frac{\partial p}{\partial z}\right)^T \tag{17}$$

Suppose the coordinates of the normalized image points $\{v_1, v_1, \ldots, v_m\}$ and points on the reference plane $\{\boldsymbol{m}_1, \boldsymbol{m}_2, \ldots, \boldsymbol{m}_m\}$ are known. The principle of back projection is shown in Figure 4. The three-dimensional reflection point on the mirror corresponds to the normalized image plane coordinates $(x_i, y_i)^T$ that can be expressed as $\boldsymbol{p}_i = s_i(x_i, y_i, 1)^T$. The unit vector of the incident ray is $\boldsymbol{i}_i = (x_i, y_i, 1)^T / \|(x_i, y_i, 1)^T\|$, the unit vector of the reflected ray is $\boldsymbol{l}_i = \boldsymbol{i}_i - 2 \cdot \langle \widetilde{\boldsymbol{n}}_i, \boldsymbol{i}_i \rangle \cdot \widetilde{\boldsymbol{n}}_i$, and $\widetilde{\boldsymbol{n}}_i = \boldsymbol{n}_i / \|\boldsymbol{n}_i\|$. Let $\boldsymbol{R}_{s2c} = (\boldsymbol{r}_1\ \boldsymbol{r}_2\ \boldsymbol{r}_3)$, $\boldsymbol{r}_3$ represents the coordinates of the unit vector in the Z-axis direction of the reference plane coordinate frame in the camera coordinate frame. $\boldsymbol{T}_{s2c}$ indicates the coordinates of the origin of the reference plane coordinate frame in the camera coordinate frame. The reference plane can be represented by the vector $\boldsymbol{q} = (\boldsymbol{r}_3{}^T, -\boldsymbol{r}_3{}^T \cdot \boldsymbol{T}_{s2c})^T$, such that $\left\langle \boldsymbol{q}, (\hat{\boldsymbol{m}}_i^T, 1)^T \right\rangle = 0$ for any point on the reference plane. Back-projection can be achieved by computing the point $\hat{\boldsymbol{m}}$, the intersection of the reflected ray with the reference plane.

$$\hat{\boldsymbol{m}}_i = \boldsymbol{p}_i - (\langle \boldsymbol{r}_3, \boldsymbol{p}_i \rangle - \boldsymbol{r}_3{}^T \cdot \boldsymbol{T}_{s2c}) / \langle \boldsymbol{r}_3, \boldsymbol{l}_i \rangle \cdot \boldsymbol{l}_i \tag{18}$$

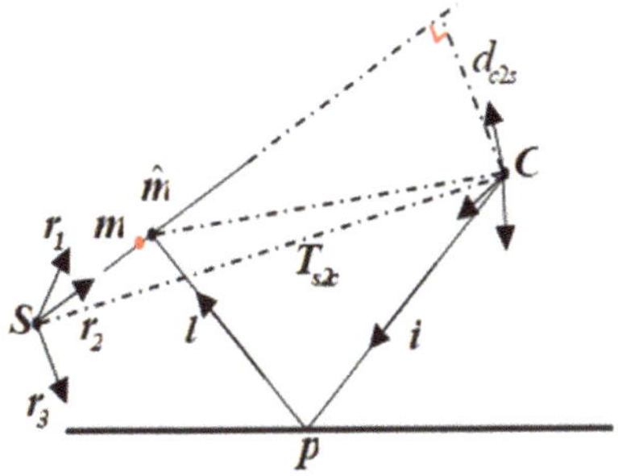

Figure 4. Principle of back projection. The rotation matrix $\boldsymbol{R}_{s2c}$ can be written as $(r_1\ r_2\ r_3)$. r_3 denotes the unit vector in the Z-axis of the reference plane. $\boldsymbol{T}_{s2c}$ denotes the distance between **S** and **C**. The reflected ray *l* intersects the reference plane at the point $\hat{m}$. The point $\hat{m}$ satisfies $-r_3{}^T \cdot \hat{m} = d_{s2c}$. We denote by d_{c2s} the distance between **C** and the reference plane.

In Equation (18), $\hat{m}_i$ is a function of depth s. We can build an optimization model to minimize the error between the back projection point and the real point on the reference plane. That means solving a nonlinear least-squares problem to estimate the depth of the mirror.

$$\min_{s} \sum_{i=1}^{m} \|\hat{m}_i(s) - m_i\|^2 \tag{19}$$

For minimizing problems in (19), we can also iteratively calculate s with the LM algorithm. The initial surface can be regarded as a plane.

3. Experimental Verification

3.1. Calibration Experiment

To verify the accuracy and universality of the calibration method proposed in this paper, a monocular vision system measurement experiment was designed (Figure 5). The whole measurement system consists of an optical platform, standard plane mirror, LCD monitor, and large FOV camera. The focal length of the camera is 8 mm; the image resolution is 1280 pixel × 1024 pixel, and the pixel size is 4 μm; the FOV of the camera is 820 mm × 670 mm, which is much bigger than grid image. When the measurement distance is about 1000 mm, the field of view of the camera is 820 mm × 670 mm. The LCD is 19 inches in size and has a pixel size of 0.2451 mm. In order to approach a large field of view measurement scene, we use a 90 × 120 mm checkerboard image as a calibration target, which is much smaller than the camera's field of view range.

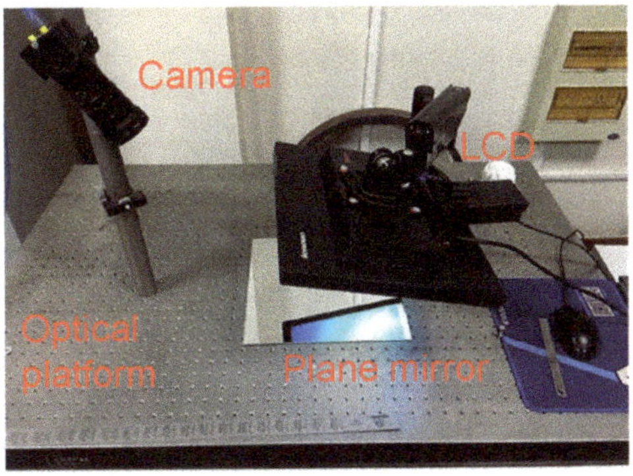

Figure 5. Experiment setup. The whole measurement system consists of an optical platform, standard plane mirror, LCD monitor, and large FOV camera.

The LCD faces the standard plane mirror on the optical platform. The grid image on the LCD is captured by the camera through the plane mirror. In the experiment, the grid image is moved on the LCD. Each grid position corresponds to three positions of a plane mirror, which are the position STZ on the optical platform, the position STX around the X-axis, and the position STY around the Y-axis. In this way, it not only ensures that the three positions of the plane mirror intersect with each other to satisfy the orthogonality constraint, but also ensures that there is an obvious height difference to satisfy the conditions of Zhang's calibration method.

Figure 6 is a set of mirrored images of the grid taken by the camera for calibration. The grid image was moved five times, and the five positions of the grid basically filled the whole LCD screen to cover the whole FOV of the camera. In the five pose conversion parameters from the reference coordinate system to the camera coordinate system, the rotation matrices are equal, and the translation vectors change with the motion of the grid in theory. In the same way, the plane mirrors at the STZ position corresponding to the five grid images are also coplanar, so the corresponding mirror normal vectors are equal. This is the coplanar constraint described in Section 2.4.

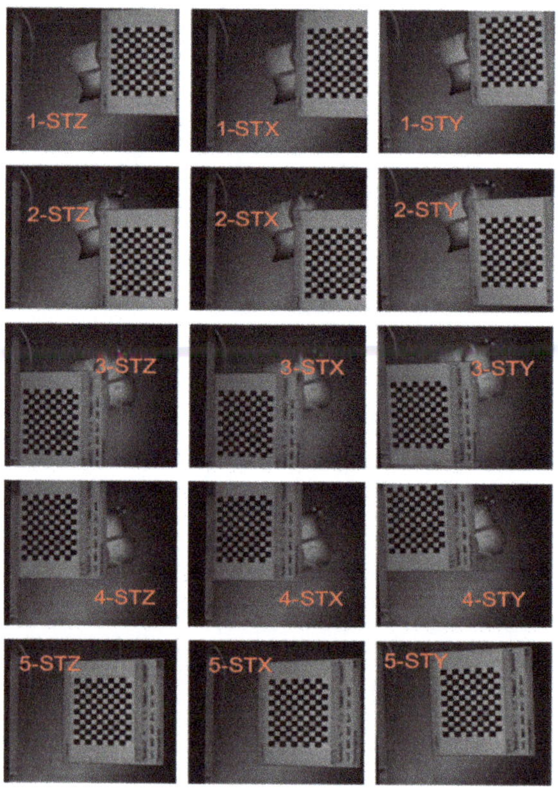

Figure 6. Checkerboard mirror image taken by the camera for calibration. The calibration grid moves on the reference plane five times. We change the position of the plane mirror to position STX, STY, and STZ for each grid. Then, we can obtain $5 \times 3 = 15$ calibration images.

Figure 7 describes the mirrored grid positions and the real grid positions with and without coplanar constraints. The mirrored grid positions corresponding to the STZ positions of the five plane mirrors are coplanar. Fitting the plane of five mirrored grid positions, the average distance error RMSE is 0.14 mm, which is consistent with Figure 7a,c. However, the coplanarity of the five grid positions restored is obviously different.

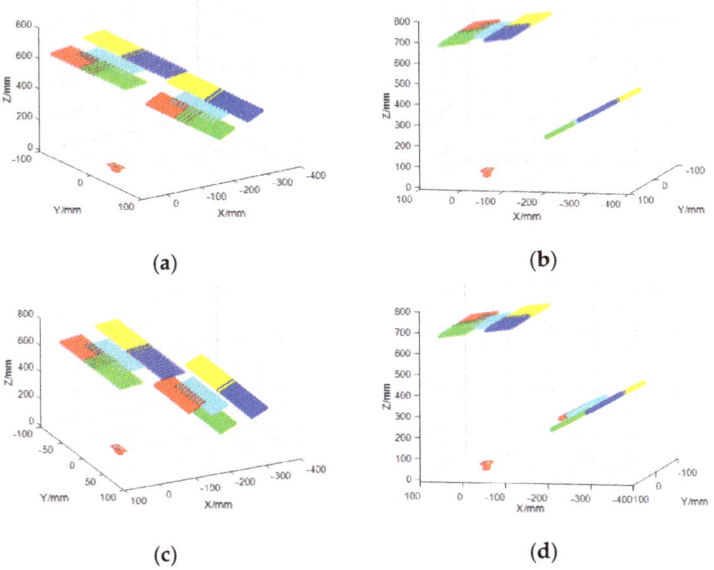

Figure 7. Calibration results with and without coplanar constraints. • points denote the real feature points on the reference plane. ∗ points denote the mirrored feature points. (**a**) Calibration results with coplanar constraints in view 1. (**b**) Calibration results with coplanar constraints in view 2. (**c**) Calibration results without coplanar constraints in view 1. (**d**) Calibration results without coplanar constraints in view 2. Yellow: Grid location 1-STZ. Blue: Grid location 2-STZ. Green: Grid location 3-STZ. Red: Grid location 4-STZ. Cyan: Grid location 5-STZ.

As is shown in Figure 7c,d, the positions of each chessboard are not only poor in coplanarity, but they also have a large offset in the relative positions, which can not comply with the law of mirror reflection.

Figure 8a is the coplanarity of the five grids performs well with coplanar constraint, RMSE = 0.11 mm. However, the five grids without coplanar constraint have poor coplanarity, RMSE = 6.45 mm. Figure 8b is the reprojection error of the two methods after back projection. The average reproject error of the method proposed in this paper is 0.1641 pixels, and in paper [16], it is 0.1419 pixels.

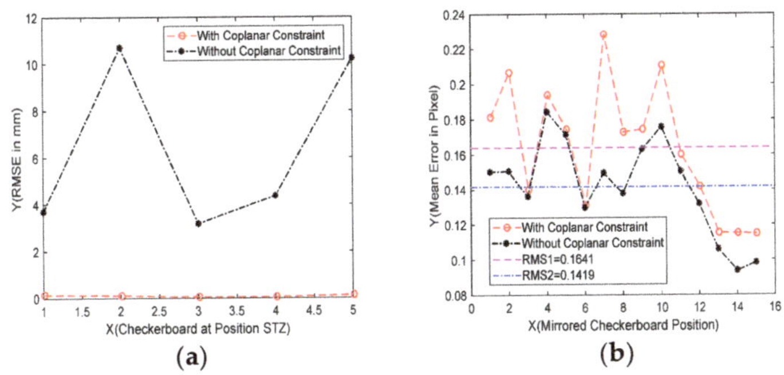

Figure 8. Comparison of error with and without constraints. (**a**) Coplanarity error of the two methods. With constraints: RMSE = 0.11 mm. Without constraints: RMSE = 6.45 mm. (**b**) Reprojection error of the two methods. With constraints: RMS = 0.1641 pixel. Without constraints: RMS = 0.1419 pixel.

The two methods are similar in terms of calibration accuracy, and the reprojection error without coplanar constraint is smaller. However, for the reference plane, the calibration result of this method is locally optimal. With coplanar constraints, the reprojection optimization model can unify the positions of five checkerboards and optimize the calibration results as a whole. Therefore, the calibration method in this paper sacrifices part of the calibration accuracy to improve the reliability of the algorithm. This calibration result is more suitable for practical measurement.

3.2. Measurement of the Step Surface

After the calibration of the reference plane, we can carry out a three-dimensional measurement experiment according to Section 2.5. As is shown in Figure 9a,b, a standard plane mirror is placed on the optical platform, and the mirror feature point calculation is performed at the STZ position. Then place the standard gauge block between the optical table and the planar mirror, so the mirror position is 8.74 mm higher than before, and the mirror feature points are calculated at the higher mirror position. Fit the mirror surface with feature points by *createFit*(), and then use the point-to-plane distance formula to calculate the distance from each feature point to the fitting plane, and then take the average value. Compare it with the actual distance of 8.74 mm to indirectly verify the accuracy of the calibration method proposed in this paper. The mirror feature points of the first mirror position are shown in Figure 9c. The plane fitting model is as follows:

$$f(x,y) = p_{00} + p_{10} \cdot x + p_{01} \cdot y \tag{20}$$

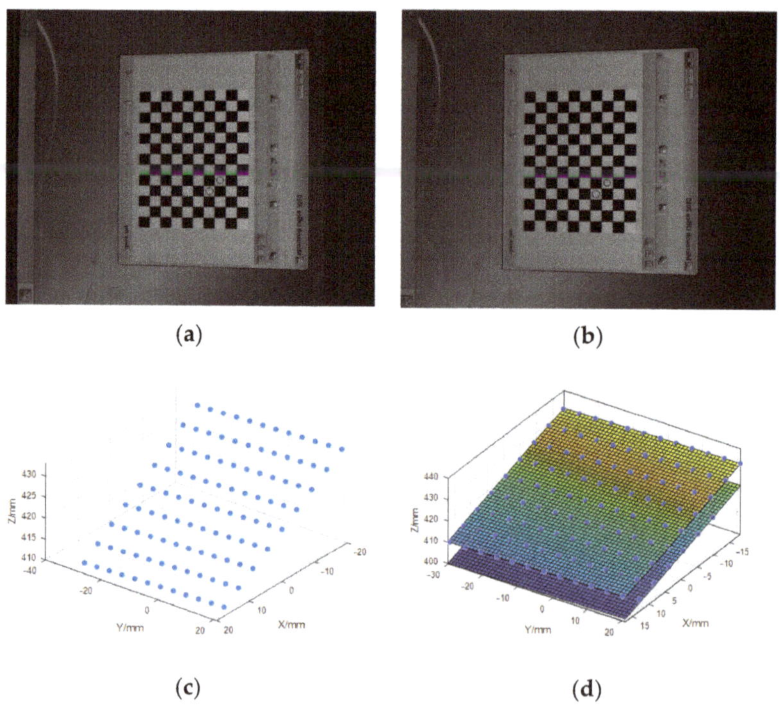

Figure 9. Restoration of plane mirror feature points at position 5. (**a**) Mirrored checkerboard image before placing the standard gauge block. (**b**) Mirrored checkerboard image after placing the standard gauge block. (**c**) Mirror feature points at position STZ. (**d**) Plane fitting of two mirror feature points. The distance between the two planes is 8.68 mm.

We can obtain the coefficients of the plane: $p_{00} = 421.4000$, $p_{10} = -0.6167$, $p_{01} = 0.0267$, and the RMSE = 0.02 mm. To have an intuitive display effect, the first and second mirror positions are shown together in Figure 9d. The average distance of the two mirror positions is 8.68 mm. The difference with the actual distance of 8.74 mm is 0.06 mm, and the relative error is 0.69%.

3.3. Measurement of the Spherical Mirror

In addition, we also measure the spherical mirror surface. The principle of the experiment is the same as that of the mirror. Firstly, measure five sets of spherical characteristic points, with 108 points in each group as measurement data. Then, place the spherical mirror on a coordinate measuring machine (model: MC850) with the highest resolution of 1 um for sampling.

The number of detection points is 202, which is used as reference data. Since the coordinate system of the coordinate measuring machine is not unified with the camera coordinate system, it is necessary to use Cloud-Compare software to unify the measurement data and reference data with the method of iterative closest point (ICP). The ICP registration of the measured feature points and the reference feature points is shown in Figure 10b.

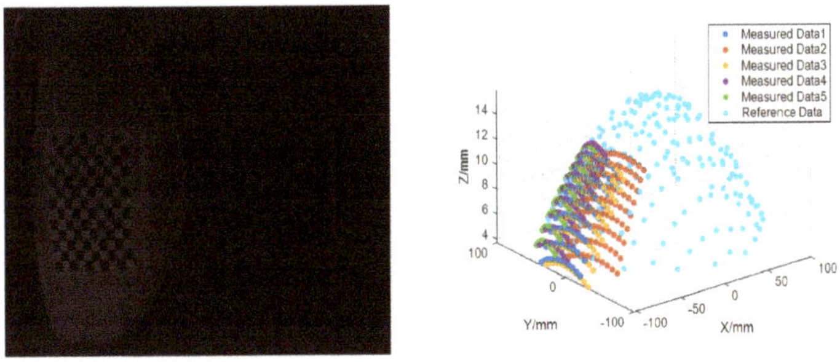

(a) (b)

Figure 10. Restoration of spherical mirror feature points. (**a**) Mirrored checkerboard image of the spherical mirror at position 5. (**b**) Display of feature points and reference data.

The spherical equation is fitted to the reference data through Cloud-Compare software. As shown in Figure 11a, the spherical equation is:

$$z = -459.621 + \sqrt{(475.617^2 - (x - 0.506226)^2 - (y - 0.264729)^2)} \tag{21}$$

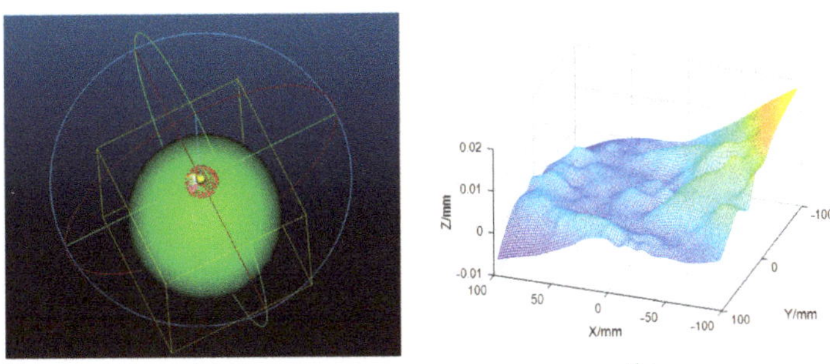

(a) (b)

Figure 11. Spherical equation fitting of reference points. (**a**) Spherical equation fitting. The fitting result: radius of the spherical mirror is 475.62 mm. RMSE = 0.01 mm, the manufacturing error can be ignored. (**b**) Surface fitting error.

Additionally, RMSE = 0.01 mm. The fitting error distribution is shown in Figure 11b. We can obtain the spherical mirror radius from the Equation (21) (Table 1).

Table 1. The fitting results of measurement data and the measurement error results compared with the reference data.

Point-Data	Radius (mm)	RMSE (mm)	Error (%)
Data 1	477.94	0.02	0.49
Data 2	478.32	0.02	0.57
Data 3	472.48	0.02	0.66
Data 4	478.54	0.01	0.61
Data 5	472.37	0.02	0.68

In the experiment, we use a cubic polynomial to initialize the spherical mirror surface because we treat the mirror surface as unknown. Supposing we directly use the spherical equation to iteratively optimize the mirror surface, the measurement accuracy will perform better.

4. Conclusions

This paper proposes a calibration method based on coplanar constraints for a camera with a large FOV. The whole experiment process is divided into two parts. The first is the calibration of a large FOV camera and the reference plane. By adjusting the tilt angle of the planar mirror and moving the grid image on the LCD monitor, the camera acquires multiple sets of calibration images and then obtains the optimal solution of the external parameters between the camera and the LCD monitor with the coplanar constraint. The other is shiny surface reconstruction. When the pose of the reference plane is known, we can establish the dense reflection correspondence between normalized image plane two-dimensional feature points, reference plane three-dimensional feature points, and bright surface reflection points, and we can iteratively calculate the reflection point depth information. In terms of calibration accuracy, the calibration accuracy of the method proposed in this paper is similar to that of [16]. At the same time, in the step surface and spherical surface measurement experiments, the results also indirectly prove the accuracy of the proposed method. The universality of the method has important research significance for further application to the multi-camera measurement system in the future.

Author Contributions: Conceptualization, R.L.; Methodology, Z.Z.; Software, Z.W.; Validation, Z.Z.; Investigation, Z.Z.; Writing—original draft, Z.Z.; Writing—review & editing, Z.W.; Supervision, R.L.; Project administration, R.L.; Funding acquisition, R.L. All authors have read and agreed to the published version of the manuscript.

Funding: This work was supported by the National Natural Science Foundation of China (NSFC) (Grant No. 51875164), as well as the National Key Research and Development Program of China (No. 2018YFB2003801).

Institutional Review Board Statement: Not applicable.

Informed Consent Statement: Not applicable.

Data Availability Statement: Data will be made available on request.

Conflicts of Interest: The authors declare no conflict of interest.

References

1. Morandi, P.; Brémand, F.; Doumalin, P.; Germaneau, A.; Dupré, J. New Optical Scanning Tomography using a rotating slicing for time-resolved measurements of 3D full field displacements in structures. *Opt. Lasers Eng.* **2014**, *58*, 85–92. [CrossRef]
2. Yu, L.; Pan, B. High-speed stereo-digital image correlation using a single color high-speed camera. *Appl. Opt.* **2018**, *57*, 31. [CrossRef] [PubMed]

3. Xu, J.; Xi, N.; Zhang, C.; Shi, Q. Windshield shape inspection using structured light patterns from two diffuse planar light sources. In Proceedings of the 2009 IEEE/RSJ International Conference on Intelligent Robots and Systems, St. Louis, MO, USA, 10–15 October 2009; IEEE: Piscataway, NJ, USA, 2009; pp. 11–15. [CrossRef]
4. Balzer, J.; Hfer, S.; Beyerer, J. Multiview specular stereo reconstruction of large mirror surfaces. In Proceedings of the IEEE Conference on Computer Vision and Pattern Recognition, Colorado Springs, CO, USA, 20–25 June 2011; IEEE: Piscataway, NJ, USA, 2011; pp. 2537–2544.
5. Zuo, C.; Feng, S.; Huang, L.; Tao, T.; Yin, W.; Chen, Q. Phase shifting algorithms for fringe projection profilometry: A review. *Opt. Lasers Eng.* **2018**, *109*, 23–59. [CrossRef]
6. Song, Z. High-speed 3D shape measurement with structured light methods: A review. *Opt. Lasers Eng.* **2018**, *106*, 119–131.
7. Liu, Y.; Fu, Y.; Cai, X.; Zhong, K.; Guan, B. A novel high dynamic range 3D measurement method based on adaptive fringe projection technique—ScienceDirect. *Opt. Lasers Eng.* **2020**, *128*, 106004. [CrossRef]
8. Shengpeng, F.U. Imaging Simulation Method for Specular Surface Measurement. *J. Mech. Eng.* **2015**, *51*, 17–24.
9. Halstead, M.A.; Barsky, B.A.; Klein, S.A.; Mandell, R.B. Reconstructing curved surfaces from specular reflection patterns using spline surface fitting of normals. In Proceedings of the Conference on Computer Graphics and Interactive Techniques, New Orleans, LA, USA, 1 August 1996.
10. Tarini, M.; Lensch, H.P.; Goesele, M.; Seidel, H.-P. 3D acquisition of mirroring objects using striped patterns. *Graph. Model.* **2005**, *67*, 233–259. [CrossRef]
11. Savarese, S.; Chen, M.; Perona, P. Local Shape from Mirror Reflections. *Int. J. Comput. Vis.* **2005**, *64*, 31–67. [CrossRef]
12. Liu, M.; Hartley, R.; Salzmann, M. Mirror Surface Reconstruction from a Single Image. *IEEE Trans. Pattern Anal. Mach. Intell.* **2015**, *37*, 760–773. [CrossRef] [PubMed]
13. Sturm, P.; Bonfort, T. *How to Compute the Pose of an Object without a Direct View?* Springer: Berlin/Heidelberg, Germany, 2006.
14. Kumar, R.K.; Ilie, A.; Frahm, J.M.; Pollefeys, M. Simple calibration of non-overlapping cameras with a mirror. In Proceedings of the IEEE Conference on Computer Vision and Pattern Recognition, Anchorage, AK, USA, 23–28 June 2008; IEEE: Piscataway, NJ, USA, 2008; pp. 1–7.
15. Takahashi, K.; Nobuhara, S.; Matsuyama, T. Mirror-based Camera Pose Estimation Using an Orthogonality Constraint. *IPSJ Trans. Comput. Vis. Appl.* **2016**, *8*, 11–19. [CrossRef]
16. Hesch, J.A.; Mourikis, A.I.; Roumeliotis, S.I. Mirror-Based Extrinsic Camera Calibration, Algorithmic Foundation of Robotics VIII. *Springer Tracts Adv. Robot.* **2009**, *57*, 285–299. [CrossRef]
17. Li, X.; Long, G.; Guo, P.; Liu, J.; Zhang, X.; Yu, Q. Accurate mirror-based camera pose estimation with explicit geometric meanings. *Sci. China Technol. Sci.* **2014**, *57*, 2504–2513. [CrossRef]
18. Bergamasco, F.; Cosmo, L.; Albarelli, A.; Torsello, A. Camera Calibration from Coplanar Circles. In Proceedings of the International Conference on Pattern Recognition IEEE Computer Society, Stockholm, Sweden, 6 December 2014; pp. 2137–2142.
19. Li, W.; Chu, J.; Meng, H.; Wang, J.; Li, X.; Xing, X. Calibration method with separation patterns of a single-camera. *Proc. SPIE* **2006**, *6269*, 303–304.
20. Yang, N.; Huo, J.; Yang, M.; Wang, W.X. A calibration method of camera with large field-of-view based on spliced small targets. *Guangdianzi Jiguang/J. Optoelectron. Laser* **2013**, *24*, 1569–1575.
21. Sun, J.; Liu, Z.; Zhang, G. Camera Calibration Based on Flexible 3D Target. *Acta Opt. Sin.* **2009**, *29*, 3433–3439.
22. Liu, Z.; Li, F.; Li, X.; Zhang, G. A novel and accurate calibration method for cameras with large field of view using combined small targets. *Measurement* **2015**, *64*, 1–16. [CrossRef]
23. An, G.H.; Lee, S.; Seo, M.-W.; Yun, K.; Cheong, W.-S.; Kang, S.-J. Charuco Board-Based Omnidirectional Camera Calibration Method. *Electronics* **2018**, *7*, 421. [CrossRef]
24. Bergamasco, F.; Albarelli, A.; Cosmo, L.; Rodolà, E.; Torsello, A. An Accurate and Robust Artificial Marker based on Cyclic Codes. *IEEE Trans. Pattern Anal. Mach. Intell.* **2016**, *38*, 2359. [CrossRef] [PubMed]
25. Bergamasco, F.; Cosmo, L.; Gasparetto, A.; Albarelli, A.; Torsello, A. Parameter-Free Lens Distortion Calibration of Central Cameras. In Proceedings of the 2017 IEEE International Conference on Computer Vision (ICCV), Venice, Italy, 22–29 October 2017; IEEE: Piscataway, NJ, USA, 2017.
26. Bergamasco, F.; Albarelli, A.; Cosmo, L.; Torsello, A.; Rodola, E.; Cremers, D. Adopting an unconstrained ray model in light-field cameras for 3D shape reconstruction. In Proceedings of the 2015 IEEE Conference on Computer Vision and Pattern Recognition (CVPR), Boston, MA, USA, 7–12 June 2015.
27. Zhang, Z. A Flexible New Technique for Camera Calibration. *IEEE Trans. Pattern Anal. Mach. Intell.* **2000**, *22*, 1330–1334. [CrossRef]
28. Moré, J.J. The Levenberg-Marquardt algorithm: Implementation and theory. In *Numerical Analysis*; Watson, G.A., Ed.; Springer: Berlin/Heidelberg, Germany, 1978; p. 630.

sensors

MDPI

Article

A Novel Optical Instrument for Measuring Mass Concentration and Particle Size in Real Time

Jingxiu Zhang [1,*], Zhiwei Zhang [2], Longfei Hou [3] and Weihu Zhou [4,5,*]

1 School of Aeronautic Science and Engineering, Beihang University, Beijing 100191, China
2 School of Precision Instrument and Opto-Electronics Engineering, Tianjin University, Tianjin 300072, China
3 College of Automation Engineering, Nanjing University of Aeronautics and Astronautics, Nanjing 210016, China
4 University of Chinese Academy of Sciences, Beijing 100049, China
5 Institute of Microelectronics of the Chinese Academy of Sciences, Beijing 100029, China
* Correspondence: jingxiu.zhang@buaa.edu.cn (J.Z.); zhouweihu@ime.ac.cn (W.Z.)

Abstract: Particle mass and particulate size are two important parameters used to characterize the aerosol. Currently, there are a few methods for measuring particle mass concentration and particle size. However, the existing methods have their own shortcomings. In this article, we describe a novel laser scattering instrument that measures mass concentration and particle size in real time over a wide concentration range. This instrument combines laser scattering and time-of-flight aerodynamics in one optical device. There are two innovations in this paper: (1) Two APD detectors are used to receive scattered light. One receives forward-scattered light and the other receives side-scattered light. The advantage is that the sensitivity of the detector is increased greatly, and the ratio of forward and side scattering is used to further obtain the size and shape information of the particles. (2) In order to measure the high concentrations of aerosol, a high-speed ADC and FPGA is combined to achieve an anti-overlap algorithm objective. It has been verified by experiments that the anti-overlapping algorithm can effectively improve the applicability of the aerodynamic particle size spectrometer under high concentration conditions.

Keywords: particle size; aerodynamic particle size; anti-overlap algorithm; particulate matter

Citation: Zhang, J.; Zhang, Z.; Hou, L.; Zhou, W. A Novel Optical Instrument for Measuring Mass Concentration and Particle Size in Real Time. *Sensors* **2023**, *23*, 3616. https://doi.org/10.3390/s23073616

Academic Editor: Nikolay Kazanskiy

Received: 20 February 2023
Revised: 26 March 2023
Accepted: 27 March 2023
Published: 30 March 2023

1. Introduction

Industry's rapid development in recent years has greatly contributed to China's economic progress, but it has also caused numerous air quality problems. Haze weather frequently occurs as industrial pollution intensifies. Among all air pollutants, particulate matter is the most serious and has the greatest effect on human health. Air pollution control has become a public concern to ensure a reasonable development of industry [1]. Hence, the detection of air quality is particularly important. The monitoring and analysis of the concentration of particulate matter in the atmosphere becomes a prerequisite for environmental governance.

Particle size and mass are two important parameters used to characterize an aerosol. The atmospheric particulate concentration is a basic parameter for characterizing the spatial distribution of the particulate matter, which is generally expressed by mass concentration and quantity concentration [2]. According to the aerodynamic diameter, atmospheric particulate matter can be divided into four levels: (1) total suspended particulate matter (TSP) with a diameter of less than 100 μm; (2) respirable particulate matter with a diameter of less than 10 microns (PM10); (3) a diameter less than 2.5 μm fine particles (PM2.5), which can be suspended in the atmosphere for a long time, bringing an important impact on air quality and visibility; (4) ultrafine particles with a diameter less than 1 micron (PM1), which can easily enter various tissues of the human body by respiratory system. PM2.5 and PM1 have a small particle size and strong chemical activity, and are prone to carry toxic and harmful substances (such as heavy metals, viruses, microorganisms, etc.). They stay in the atmosphere for a long time and move a long distance. Therefore, PM1 and PM2.5 have greater impacts on human health and atmospheric environmental quality [3–5].

The standards for ambient air quality (PM1, PM2.5, and PM10), exposure assessment for inhalable particulates, and FDA guidelines for pharmaceutical aerosol characterization are based on mass and aerodynamic size of the particles. Most of this aerosol characterization will benefit from the real-time measurement of mass-weighted aerodynamic size distributions, which will significantly reduce the time required to characterize aerosols and provide higher resolution particle size data. The demand for concentration statistics over different particle sizes is increasing with environmental requirements.

At present, there are some common methods for measuring particle mass concentration and particle size: the β-ray absorption method [6,7], the QCM(Quartz crystal microbalance) [8,9], the charge transfer method [10], the optical scattering method (laser scattering method) [11–13], and the time-of-flight aerodynamic method [14–16]. The β-ray absorption method has good mass sensitivity [6,17]; however, the disadvantage of the β-ray absorption method is that it is time consuming and expensive [17]. Additionally, the sampling paper tape inside must be replaced manually. The β-ray absorption method and QCM method have good mass sensitivity, but are unable to measure particle size without a size selective inlet. The charge transfer method is the most used for nanoparticles.

The optical scattering method is widely used for measuring particle size distribution in real time [17,18]. Optical measurements of particle velocities are widely used to study particle dynamics and gas flows. The measurement of particle size is used in a variety of fields, including pollution and contamination monitoring, respirable particle mass monitoring, and spray nozzle performance monitoring. Optical scattering has the following advantages: (1) accurate particle counting when the concentration is low; (2) good signal-to-noise ratios when the particles are larger (e.g., >100 μm); and (3) low cost. There are several disadvantages to the optical method: (1) If particle density is unknown, optical size does not equal geometric size since it depends on particle shape and refractive index; this error is exacerbated when particle size distribution is converted to mass concentration. (2) Particle concentrations will be underestimated because multiple particles are present at the same time in the measure volume, causing coincidence errors. Due to these reasons, optical scattering methods are mostly used in clean environments [17].

Time-of-flight velocimetry, also called transit time, two-spot, or two-focus velocimetry, is the most common technique for measuring velocity. Using this method, two beams of light radiation (laser radiation) are directed through a volume of particles. Two pulse signals will be generated when a particle passes through both beams. In order to measure particle velocity over a known distance, a timing signal is initiated contemporaneously with the first pulse and terminated concurrently with the second pulse. The advantages of time-of-flight aerodynamic method is that it is less dependent on the particle refractive index and density than optical scattering method. Good agreement between the time-of-flight aerodynamics and direct mass measurements has been reported [19]. However, a commercial instrument that uses time-of-flight aerodynamics (for example the APS3321 from TSI) cannot measure high concentrations [20], mainly due to the defects of its signal processing circuit and its data processing algorithm [21].

According to these methods and instrument characteristics, Table 1 briefly compares the advantages and disadvantages of the methods for measuring particle concentration and particle size.

Table 1. Methods for measuring particulate concentration and size.

Method	Advantage	Disadvantage
β-ray absorption	Good mass sensitivity.	Time consuming and expensive. Sampling paper tape replaced manually. Data output period is long (1 data/1 h). Measuring different particles sizes need cutting head.
QCM (Quartz crystal microbalance)	Good mass sensitivity.	Cannot measure particle size. Highly affected by water vapor. Measuring different particles sizes needs cutting head.
Charge transfer	System is simple. Mainly used for engine exhaust nanoparticle size detection.	Influenced by factors such as particle size changes, composition changes, and water vapor.
Optical scattering (Laser scattering)	Low cost. Accurate in low concentration. Suitable for large particle.	Inaccurate mass concentration. Concentrations will be underestimated because of overlapping particles.
Time of flight (Only have APS3321)	Suitable for 0.5–20 μm. High measurement accuracy. Measure mass concentration at the same time. High resolution, good stability. Unaffected by water vapor.	Be interfered with by overlapping particles in high concentrations due to defects of its signal processing circuit. Very expensive. High technical complexity.

In this paper, we design a novel laser scattering instrument that measures mass concentration and particle size in real time over a wide concentration range. The novelty of this instrument is that it combines laser scattering and time-of-flight aerodynamic in one optical device. We use two APDs in order to increase the sensitivity and obtain more information of the particles. In order to measure higher particle concentrations, we use digital acquisition technology for implementing anti-overlapping algorithms, which solve the problem of overlapping particles interfering with each other in high concentrations.

2. Principle of Measurement

2.1. Instrument Description

The novel instrument is shown in Figure 1. Using a pump with a damping chamber at a total flow rate of 5 L/min, clean air and aerosol (air with particulates) are drawn into the optical chamber through the sheath nozzle in a continuous stream. The air is filtered to remove particulates and becomes clean air through a HEPA filter. The clean air is then drawn back into the optical chamber around the inlet nozzle as sheath flow to reduce particles and protect the optics from particle contamination. The remaining 1 L/min of air with particulate matter through the inlet of the sheath nozzle enters the optical chamber. As particles pass through the measurement volume, they are illuminated by a parallel laser beam with a wavelength of 635 nm. The avalanche photo detector (APD) captures side-scattered light in the scattering angle of 30° to 120° using a spherical mirror. A lens focuses the forward-scattered light onto the second APD. Signals from the APD are converted into digital form by high-speed ADCs and processed by FPGAs. In order to maintain pressurized drop balance in the two flow paths, an orifice is used to maintain the aerosol-to-sheath flow ratio. The HEPA filters are used to filter dust from the sheath flow before affecting the flow rate. This damping chamber is used to reduce the air jitter that is caused by the pump, thereby maintaining a steady flow rate at the nozzle.

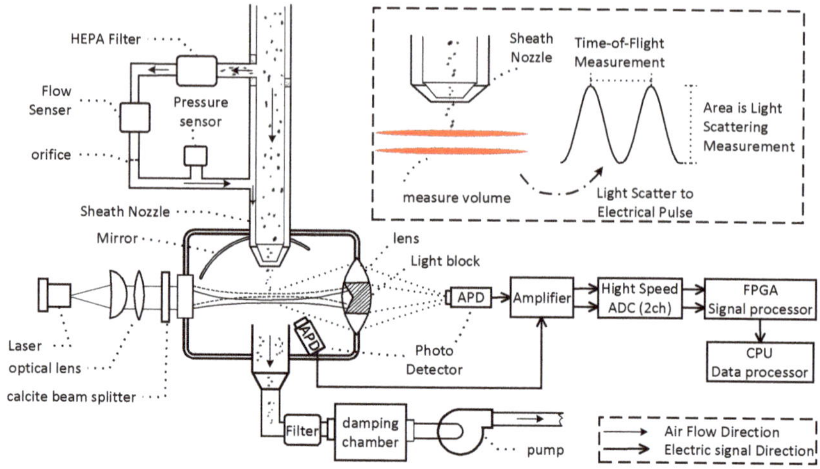

Figure 1. Schematic diagram of the novel instrument. The particle beam, laser beam, and the axis of the mirror are orthogonal to each other.

The laser is shaped by an optical lens, then split into two beams by calcite spaced 100 μm apart. Each beam is 1 mm wide and 40 μm thick. Figure 2 shows the light intensity over the measuring volume, double-peaked from top to bottom. A reduced pressure is created and maintained by the vacuum pump, so that the clean gas at the nozzle is ejected at the same speed. With the same pressure, different particles will move through the measuring volume from top to bottom at different rates. Larger or heavier particles scatter more lights and move over the measuring volume more slowly. This affects the relationship among particle size, time of flight (TOF) and intensity, as seen in Figure 3. It can be used to estimate the aerodynamic particle size of particles ranging from PM0.3 to PM20 based on this property.

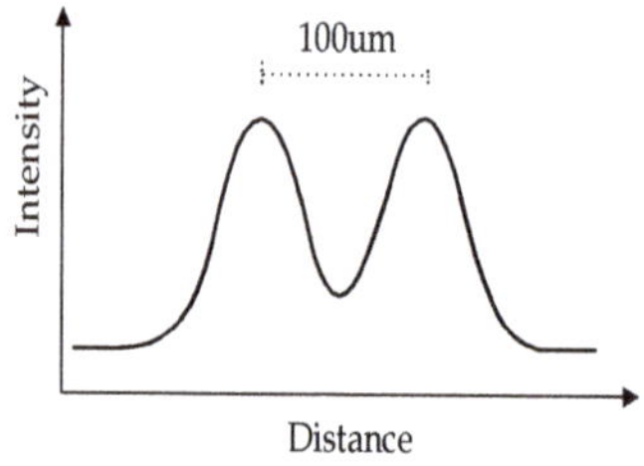

Figure 2. Light intensity profile over measurement volume.

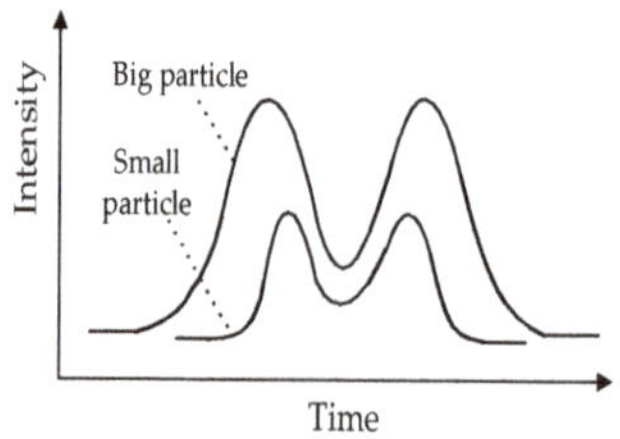

Figure 3. Relationship among particle size, TOF, and intensity.

2.2. Intensity Ratio of Backward Scattering to Forward Scattering

The laser scattering method principle calculates particle size distribution from the scattered light intensity distribution (scattering pattern). When the laser irradiates the particles, if the particle size exceeds the wavelength of the laser, the particles will scatter the light in the same direction as the laser light (forward scattering). If the particle size is approximately equal to or smaller than the wavelength of the light, the scattered light increases in perpendicular directions (lateral) and in directions backward (backward scattering) [22–26], show as Figure 4.

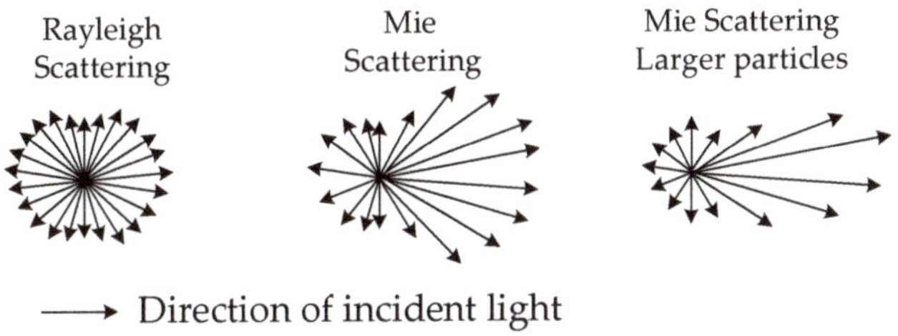

Figure 4. Relationship between size and scattered light intensity distribution.

The device has two photo detectors, one of which is used to collect sideward and backward-scattered light (scattering angles between 30° and 120°), the other to collect forward-scattered light. It offers two benefits: (1) The signal-to-noise ratio is increased by combining the signal values from the two detectors. (2) As the particle diameter decreases, the ratio of backward scattering plus sideward scattering to forward scattering increases. As a result, this ratio can be utilized to increase the precision of small particle measurements.

2.3. Time of Flight

The Bernoulli's equation is a basic equation in fluid dynamics [27]. According to Bernoulli's equation (Equation (1)), when the pressure difference between the inside and outside of the nozzle of the instrument is 15 kPa, the air spray velocity of the sheath nozzle will be 150 m/s.

$$P_1 + \frac{1}{2}\rho v_1^2 = P_2 + \frac{1}{2}\rho v_2^2 \tag{1}$$

The velocity of particles relative to the air in the nozzle can reach approximately one third the speed of sound (Table 2) [28].

Table 2. Particle properties in nozzle.

Particle Diameter (μm)	Relative Velocity (cm/s)	Particle Reynolds Number
0.5	40.0	0.01
1.0	1750.0	1.16
3.0	6490.0	12.90
10.0	10,600.0	69.60
15.0	11,500.0	114.00
20.0	12,300.0	163.00

The Reynolds number is the ratio of inertial forces to viscous forces within a fluid The larger the particle's Reynolds number, the slower the particles move in the fluid [27,29].

As shown in Figure 1, the distance between two parallel beams is 100 μm; based on Table 1, PM0.5-PM20 particles have a flight time (Time-of-Flight) ranging approximately from 700 to 3500 nanoseconds.

2.4. Digital Signal Processing and Anti-Overlap Algorithm

This paper proposes the use of high-speed ADC for photoelectric signal acquisition, and FPGA for signal processing to replace previous analog circuits [21]. The advantage is that digital signal processing techniques is capable of processing complex waveforms. With the help of ADC and FPGA digital circuitry, it is possible to process waveforms shown in Figure 5, which are impossible for analog circuits [20,30].

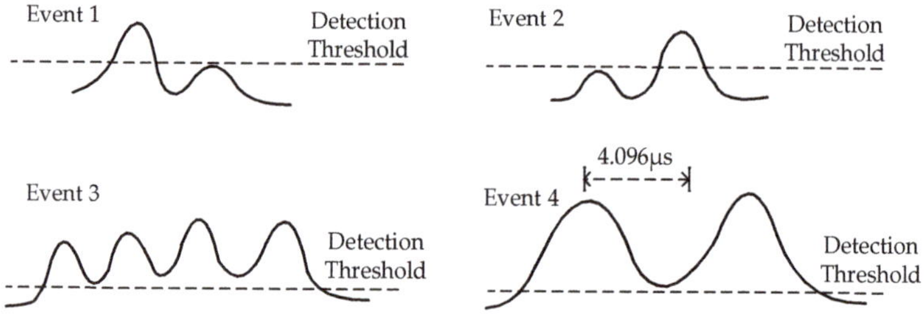

Figure 5. Complicate events which analog circuits cannot handle. In events 1 and 2, when the signal from a small particle cannot remain above the threshold, only one crest is detected, and no time-of-flight measurements are taken. In the case of event 3, although the signal remains above the threshold, three or more crests are detected as a result of coincidence. Such events are logged, but concentration and flight time are not recorded. Event 4 is outside the timer's maximum range, and in this case, the signal re-mains above the threshold until it moves outside the timer's range, and only one crest is observed. Event 4 is typically caused by large or recirculating particles, and in this case, the event will be logged, but no time-of-flight is recorded.

In analog circuits, the threshold is set to a fixed value. The background signal output by the APD fluctuates widely due to temperature, gain, etc. If the threshold is lowered, the instrument will not operate correctly. The digital signal processing techniques proposed in this paper can be used to dynamically detect the APD's background value so that the threshold value also fluctuates with the fluctuation of the background signal, causing waveforms such as events 1 and 2 to be processed accurately. For detailed processing flow, please refer to Figure 6, in which V is the voltage value of the APD output; V1 is the lowest value detected in the current loop; and the threshold V2 is equal to the manually set value S plus V1. Event 4 is also easily handled using the processing flow shown in Figure 6.

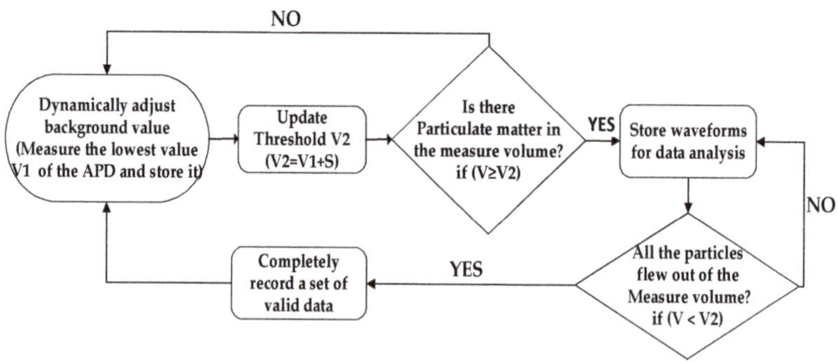

Figure 6. Processing flow of signal acquisition and storage.

It is easy to process the waveforms of event 3 using digital processing techniques. Event 3 is called particle overlap. There are many ways to handle event 3. This paper proposed an anti-overlap algorithm, which can be achieved using the data processing flow shown in Figure 7.

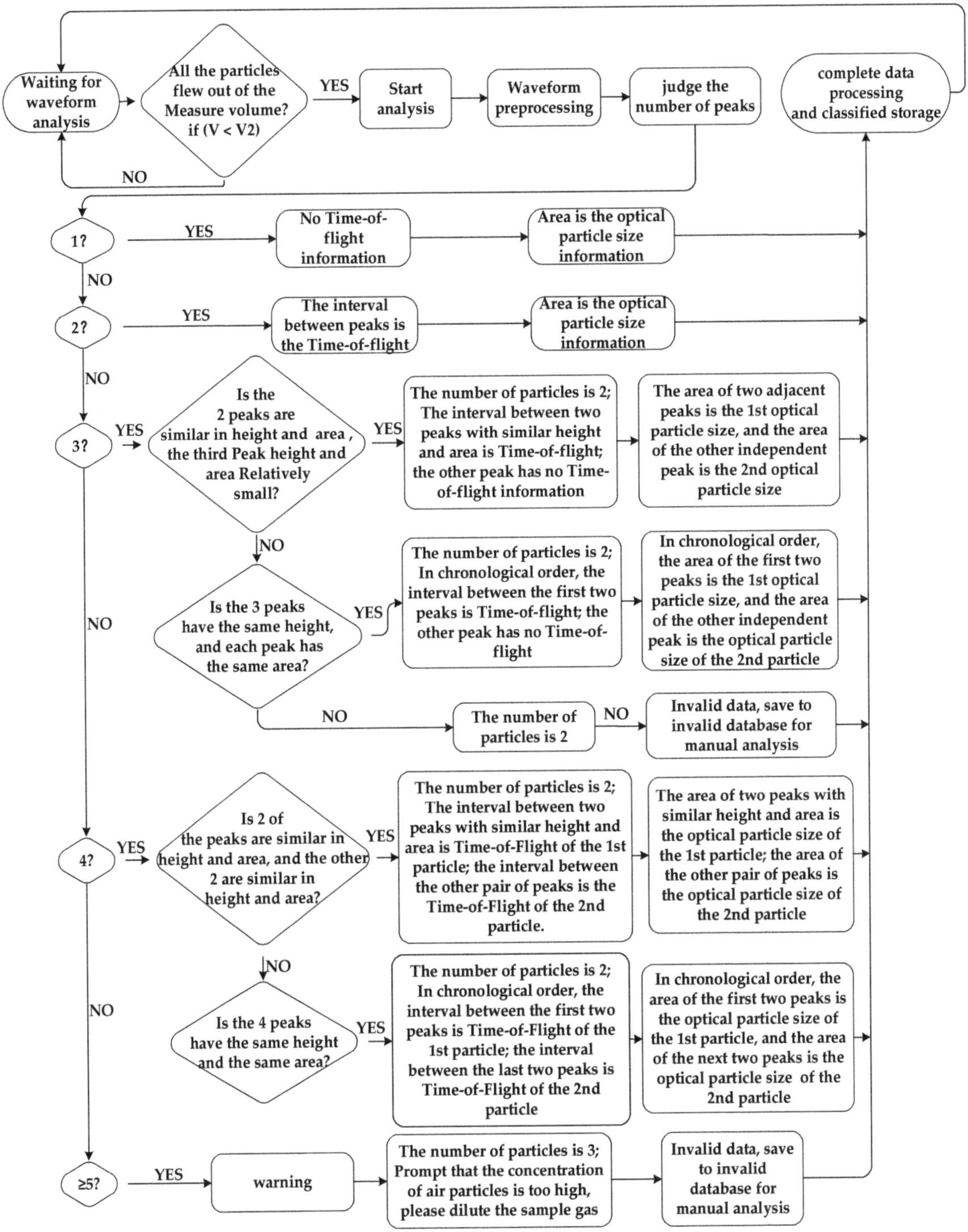

Figure 7. Processing flow for anti-overlap algorithm.

The system works in an uninterrupted and real-time mode. As soon as a particle flies out of the measuring volume, the data is analyzed. To ensure that no data are lost, the algorithm for particulate matter needs to be completed before the next particulate matter flies out of the measurement volume. The time for particles to fly through the measurement volume is approximately 600 ns–4000 ns, which means that the algorithm needs to be completed within 600 ns. Therefore, an FPGA must be used to implement the algorithm.

3. Results and Discussion

Some experiments have been conducted to test the instrument's performance using TSI Company's 3410U and CMAG 3475 aerosol generators. These generators generate particles of varying sizes for experimentation, and the relationship between particle concentration and scattered light intensity is observed [31]. The 3410U is an aerosol generator, which produces aerosols of different sizes from 0.5 μm to 100 μm. CMAG 3475 aerosol generator with Sinclair–Lamer condensation technology produces aerosols of varying sizes (0.5 mm to 8 mm) by controlling temperature and airflow. CMAG 3475 produces particulates in quantities per unit volume.

The specific experiments are as follows:

(1) A calibration experiment for TOF of Aerodynamic Particles was conducted in which 3410U was used to generate 11 different standards of particulate matter. Measure the time-of-flight of 11 different diameters of standard particles and use these data to calibrate the instrument. By calibrating the time-of-flight of 10 different diameters of standard particles, the instrument can measure the time-of-flight of similar particles in the future.

(2) Experiment for "scattered light intensity of p". After TOF calibration in experiment 1, several different diameters of standard particles are measured again. The instrument performance is analyzed by analyzing the standard deviation of the measured TOF data and light scattering intensity data. The results of the analysis can be used to determine how well the instrument is functioning, and whether or not any adjustments need to be made. The smaller the standard deviation, the better the instrument's performance.

(3) Experiment for "work in high concentrations". In this experiment, the CMAG 3475 was used to generate standard particulate matter. This was performed to test the accuracy of the instrument in measuring the number concentration at different concentrations by controlling the change in the number concentration per unit volume. The CMAG 3475 was chosen because it is capable of producing a range of different concentrations of particulate matter while also providing a constant number concentration per unit volume. This allowed for a more accurate test of the instrument's accuracy in measuring the number concentration of particulate matter at different concentrations.

(4) Experiment for ambient aerosols. In this experiment, a parallel comparison experiment with the particle analyzer of the β-ray principle was performed at the same location. The purpose of the experiment is to evaluate the accuracy of the analyzer by comparing the results of the particles measured by β-ray.

3.1. Calibration Experiment for TOF of Aerodynamic Particle

In theory, there exists a precise relationship between aerodynamic size and velocity. However, the actual system has some (albeit minor) variations. A series of tests using single spheres of uniform sizes is highly recommended to calibrate the system. Based on these tests, the empirical relationship between TOF measurements and aerodynamic particle size will be established. Figure 8 is a calibration curve using 11 different standards of particulate matter.

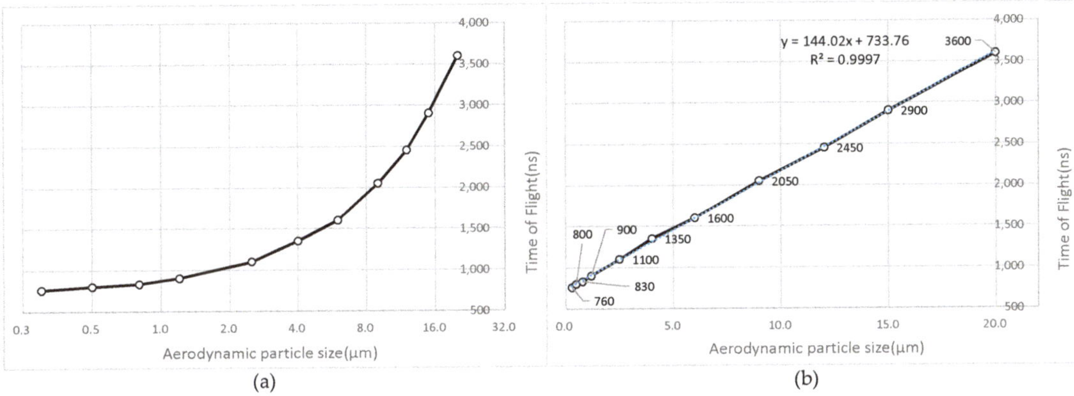

Figure 8. TOF of aerodynamic particle size of PM0.3-PM20. (**a**) A logarithmic base of 2 is used for the abscissa to facilitate viewing the values. (**b**) In order to view linearity comfortably, the abscissa is normal.

Time-of-flight data for 11 different standard particle diameters were obtained, and Figure 8 illustrates the results. The time-of-flight of particles 0.3–20 μm is approximately 700 ns to 4000 ns. In Figure 8b, it is evident that the particle size is linearly proportional to time-of-flight. Figure 9 shows the data collected by ADC when calibrating the instrument with PM1.2.

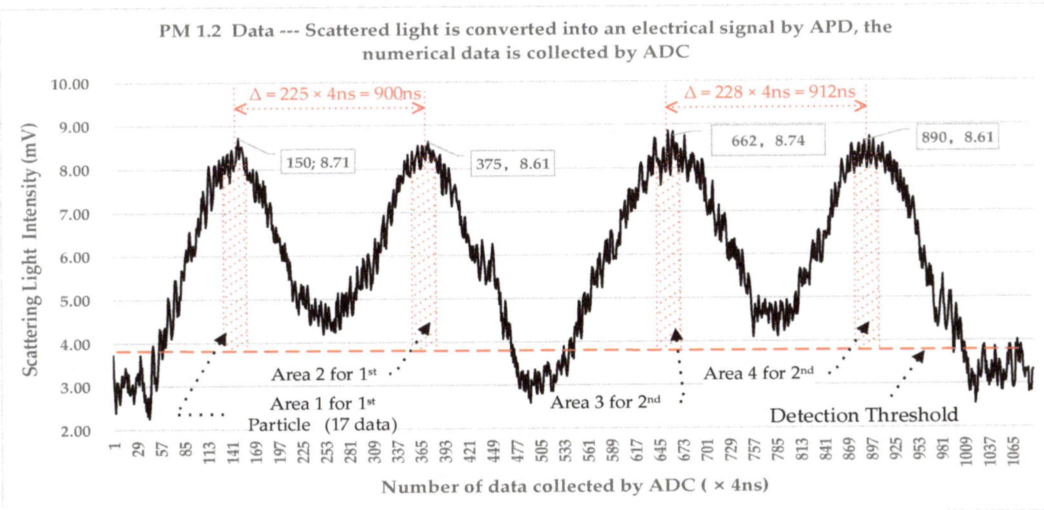

Figure 9. TOF and Scattering Light Intensity of two PM1.2 particles; TOF = Δ = Numbers × 4 ns.

As shown in Figure 9, the TOF of PM1.2 is around 900 ns. In order to normalize the data, for each particle with TOF, the scattering light intensity value is taken from the highest point on each peak to take its adjacent 16 points, and there are 34 points in total for the 2 peaks. The scattering light particle size of the first PM1.2 is area 1 plus area 2, and the scattering light particle size of the second PM1.2 is area 3 plus area 4.

In order to obtain time-of-flight data for 0.3 μm standard particles, one APD is utilized to receive forward-scattered light and the other APD for side-scattered light; these two APD's signals are then summed for the same moment of data to increase the signal-to-noise ratio. By using two APDs, the system can measure the time of flight of particles with much higher accuracy and resolution than if only one APD was used. The two APDs also allow for measurements of both forward and side-scattered light, which gives a better overall picture of the particles' behavior.

3.2. Scattered Light Intensity of Particle

After time-of-flight calibration, which means that the time of flight corresponds to the particle diameter, samples of standard particles (PM0.8, PM2.2, PM4) were tested with two objectives: one for analyzing the standard deviation of TOF, and the other for analyzing the light intensity of the forward and side-scattered light of the particles. For different diameters of particulate matter, the ratio of forward scattering to side scattering is obtained. To analyze instrument performance, 1000 FWHM data were collected for each particle sample (PM0.8, PM2.2, and PM4.0). Its data heatmap is shown in Figure 10.

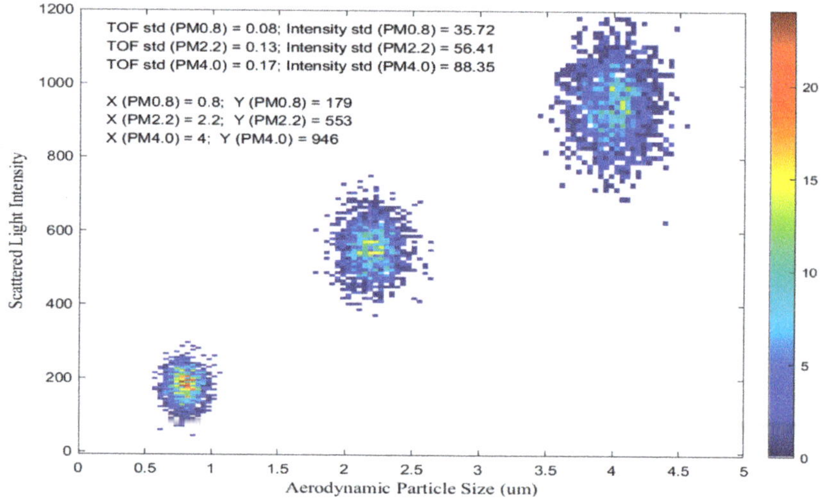

Figure 10. Scattered Light Intensity of PM0.8\PM2.2\PM4.

Analyzing the data in Figure 10 and Table 3, it is found that the standard deviation of TOF increases with particle size. In addition, the standard deviation of scattered light intensity increases with particle size. The standard deviation of TOF for the same particles is much smaller than the standard deviation of light scattering. This indicates that the TOF measurement is much more precise and accurate than light scattering measurement when it comes to determining the size of particles. This is because the TOF measurement measures the time it takes for a particle to travel a certain distance, so any change in particle size would lead to a change in the time of flight. However, the light scattering measurement is less accurate because it measures the amount of light reflected by the particles, and the amount of light reflected does not necessarily depend on the size of the particle. Therefore, the TOF measurement is more reliable for measuring particle size.

According to the data in Table 3, scattering light intensity increases as particle size increases. The ratio of forward scattering light to backward scattering light increases as particle size increases. This is because, in accordance with Mie scattering theory, the larger the particle's diameter, the stronger the forward scattering light relative to the side-scattering light. This is due to the fact that larger particles have a larger cross-sectional area, which increases the amount of light that is scattered in the forward direction. At the same time, the size of the particles also increases the probability of light being scattered in the backward direction, resulting in a higher ratio of forward to backward scattering light.

Table 3. Standard deviation of TOF and Intensity of PM0.8\PM2.2\PM4.

Particle Matter	Std of TOF	Total Intensity of Scattered Light			Ratio (F/B)	Std of Light Intensity	Std of Light Scattering Particle
		Sum	Backward	Forward			
0.8	0.08	179	110	69	0.62	35.72	0.16
2.2	0.13	553	312	241	0.77	56.41	0.22
4.0	0.17	946	496	450	0.91	88.35	0.37

3.3. Work in High Concentrations

A particle generator CMAG 3475 is used to generate PM2.2 in different numbers of concentrations from 0.1 to 10,000 particles/cm^3. At these different concentrations, our instrument was compared with the TSI 3321. The concentration data curves obtained by these two analyzers are shown in Figure 11.

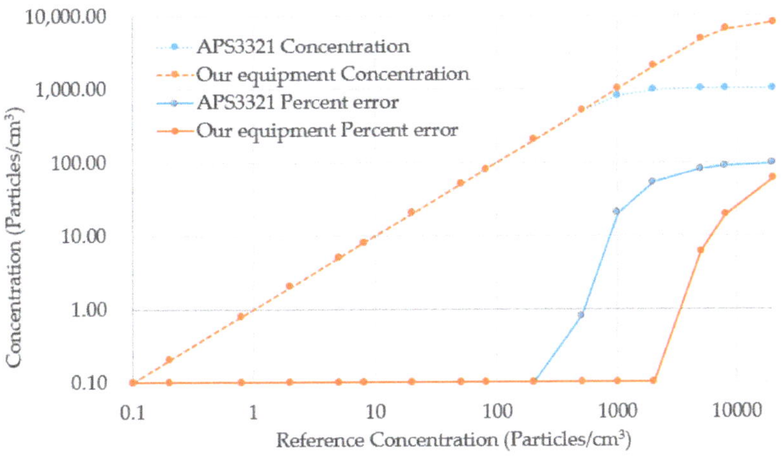

Figure 11. Concentration linearity and percent error of our equipment vs. APS 3321.

When the concentration is below 800 particles/cm^3, both instruments exhibit very good linearity and accuracy. The linearity and accuracy of the APS3321 deteriorates when the concentration of generated particulate matter exceeds 1000 particles/cm^3. The linearity and accuracy of our equipment is excellent up to a concentration of 8000 units/cm^3. Our instrument measures maximum particle concentrations of 8000 particles/cm^3, which is much higher than the 1000 particles/cm^3 measured by APS3321 [32]. Due to the overlap of particles, the analog circuitry used in the APS3321 is unable to process complex signals caused by high concentrations. In contrast, our instrument uses an ADC to acquire the signal and anti-overlap algorithms implemented in a FPGA to increase linearity and accuracy at high concentrations. This means that our instrument is capable of accurately measuring higher particle concentrations than the APS3321, which is limited by its analog circuitry and the complexity of signals caused by high concentrations. Additionally, our instrument uses an ADC and anti-overlap algorithms to further improve accuracy and linearity. However, there are some drawbacks to this approach. One is that it can be more expensive to produce an instrument with digital circuitry. Another is that the digital approach can be more complex and difficult to troubleshoot than an analog approach.

3.4. Experiment for Ambient Aerosols

In this experiment, we compare the data obtained by the β-ray equipment in the standard air station. Through analyzing the data, it can be seen that the PM2.5 and PM10 data obtained by our instrument (TOF) are consistent with the data of β-ray (Thermo-Fisher Model 5014i) made by Thermo-Fisher Environmental Instrument, USA.

Using the method proposed by Thomas M. Peters [16], the number concentration measured by the aerodynamic particle size spectrometer was converted into a mass concentration for comparison with the data of β-ray method. For each TOF channel, the differential mass concentration (dM_{Dae}) was calculated as follows:

$$dM_{Dae} = dN_{Dae}\frac{\pi}{6} D_{ve}^3 \rho_P \tag{2}$$

where Dae is aerodynamic diameter, N is number of particles, D_{ve} is volumetric equivalent diameter, and ρ_P is density of the particle. In this paper we set ρ_P as 1.8 g /cm^3 for fine PM2.5 and ρ_P as 2.7 g/cm^3 for coarse PM10 [16]. Figures 12 and 13 are actual measurement data of PM2.5 and PM10 using TOF and β-ray equipment, respectively.

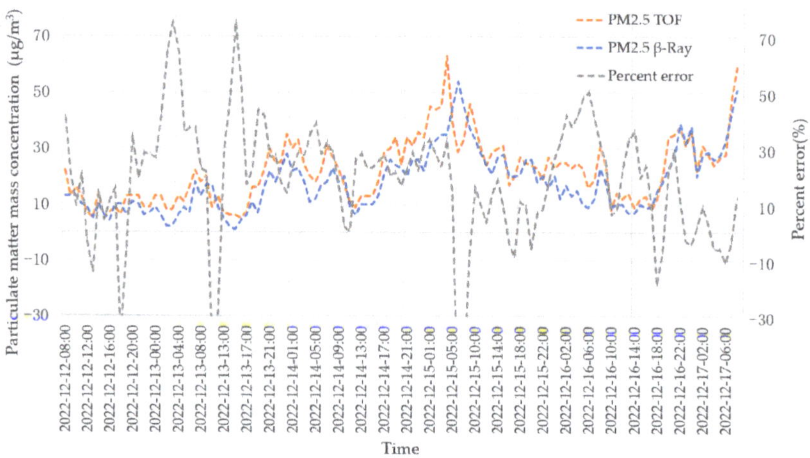

Figure 12. Actual measured concentration of PM2.5.

The distance between the two instruments is about 10 m. On 15 December 2022, from 3:00 to 5:00, the β-ray lost 2 h of data because the tape needed to be replaced manually. During the whole measurement period, the concentration value recorded by TOF is consistent with the measured value of β-ray, the mass concentration data of TOF is based on the assumption of the density of the analyzed particles, and there are some deviations from the average concentration data of β-ray. TOF equipment outputs data every 5 min, which is the average value within 5 min. β-ray outputs data once per hour, which is the average value within an hour. When the concentration of particulate matter in the aerosol fluctuates greatly within 1 h, the peak value of TOF data is higher, and it can reflect the real concentration at that time more correctly. This is because TOF equipment can measure the concentration of particles every 5 min, providing more accurate and real-time data than the β-ray equipment, which provides data once per hour. It is acceptable to have some deviations when compared to a β-ray instrument. Even in heavily polluted conditions, TOF instruments can accurately reflect the true picture of pollutants.

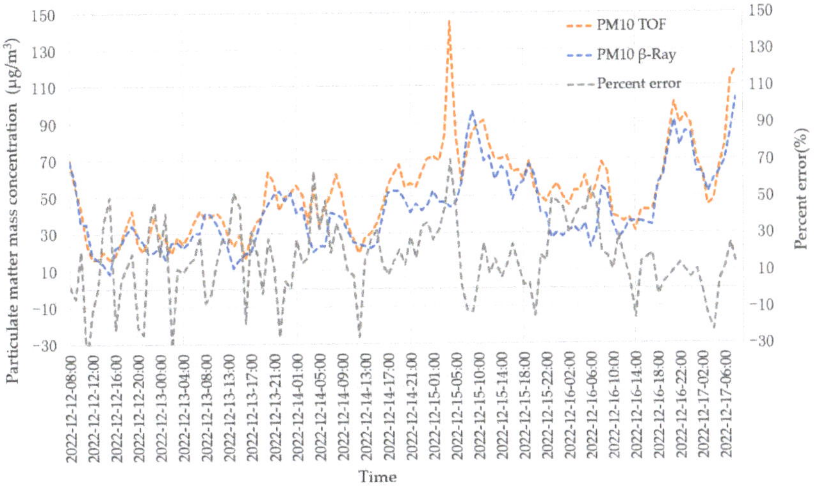

Figure 13. Actual measured concentration of PM10.

3.5. Performance Comparison

After the above comparative experiments and actual measurements, the feasibility of the method for measuring particle size proposed in this paper is confirmed. The method described in this article has many advantages over other types of equipment. The following Table 4 compares our equipment with other equipment mentioned in this article.

Table 4. Performance comparison of various instruments and equipment.

Instruments	Method	Measure (Particles Size)			Measure (Mass)	
		Range	Numbers	Resolution	Range	Resolution
Thermo Model 5014i [1]	β-ray	\	\	\	Indirect ~10 mg/m^3	0.1 μg/m^3
QCM200 [2]	QCM	\	\	\	Direct ~10 mg/m^3	0.2 ng/m^3
TSI 3091 [3]	Charge transfer	5.6~560 nm	~10^7 p/cm^3	4 nm @56 nm	Indirect	10 ng/m^3
TSI 8533 or 8534 [4]	laser scattering	0.1~15 μm	~50,000 p/cm^3	1.57 μm @2.2 μm	Indirect ~150 mg/m^3	1 μg/m^3
GRIMM model 1.107 [5]	laser scattering	0.25~32 μm	~2000 p/cm^3	0.5 μm @2.2 μm	Indirect	0.4 μg/m^3
TSI3321 [6]	TOF and laser scattering	0.5~20 μm	~1000 p/cm^3	0.15 μm @2.2 μm	Indirect	0.1 μg/m^3
Our equipment [7]	TOF and laser scattering	0.3~20 μm	~8000 p/cm^3	0.13 μm @2.2 μm	Indirect	0.1 μg/m^3

[1] Model 5014i [33]; [2] QCM200 [9]; [3] TSI3091 [34]; [4] Xiaoliang Wang's Instrument [17]; [5] Hans Grimm's Instrument [18]; [6] TSI3321 [14–16,35]; [7] TOF range can be extended to 0.3~40 μm.

Compared to other methods, the TOF method has many advantages, such as that the TOF method can measure particles of varying sizes and masses simultaneously. It is also unaffected by water vapor, making it ideal for measuring particles in humid conditions. Additionally, the TOF method offers high measurement accuracy and real-time performance, and does not require any consumables.

Compared with TSI3321, our equipment can measure forward scattering light and backward scattering light at the same time. The addition of forward scattering light and backward scattering light increases the measurement sensitivity, and the ratio of forward to backward scattering light can further enrich the information of particles. Using digital acquisition technology and digital signal processing technology, various complex algorithms can be designed for high-concentration overlapping events to achieve a wide measurement dynamic range, and high concentrations can be measured. Additionally, digital signal acquisition provides for the establishment of a particle model library, which can be used to develop new applications. This combination of technologies allows for the acquisition of more accurate and detailed data on the particles in a sample.

4. Conclusions

Through verification, the method proposed in this paper can increase the minimum resolution particle size of TOF from 0.5 μm to 0.3 μm by employing 2 APDs to receive forward-scattered light and side-scattered light, and the ratio of forward scattering light to side scattering light can be used to further obtain the size and shape information of the particles.

The experiment for high concentrations shows that the anti-overlap algorithm proposed in this paper can effectively improve the applicability of the aerodynamic particle size spectrometer for high-concentration conditions. Based on the anti-overlap algorithm, our instrument can work at concentrations up to 8000 particles/cm^3, much higher than the 1000 particles/cm^3 of APS3321 [32].

The experiment for ambient aerosols shows the 5-day data of the comparison between the TOF instrument and the β-ray instrument to measure the ambient air. It can be seen that the concentration data of the two instruments are consistent. Since the mass concentration data of TOF is based on the assumption of the density of the analyzed particles, there are some deviations from average concentration data of β-ray.

It is concluded that the two APD methods used in this paper can improve the sensitivity of the instrument, and the anti-overlap algorithm based on the digital method can increase the upper limit of the instrument's detection concentration. It laid the foundation for the development of a new generation of aerodynamic particle size spectrometer.

Author Contributions: Conceptualization, W.Z.; Methodology, J.Z. and Z.Z.; Software, L.H. All authors have read and agreed to the published version of the manuscript.

Funding: This research was funded by [Beijing Municipal Science & Technology Commission] grant number [20220481111].

Institutional Review Board Statement: Not applicable.

Informed Consent Statement: Not applicable.

Data Availability Statement: Data is contained within the article.

Conflicts of Interest: The authors declare no conflict of interest.

References

1. Angelino, M.; Fernández-Yáñez, P.; Xia, H.; Page, G. Large-eddy simulation with modeled wall stress for complex aerodynamics and stall prediction. *AIAA J.* **2021**, *59*, 1225–1237. [CrossRef]
2. Zhang, J.; Zhang, Z.; Hou, L. Detection of Particle Concentration and Particle Size Based on Aerodynamic Particle Size Spectrometer. *Adv. Multimed.* **2022**, *2022*, 4152186. [CrossRef]
3. Chen, R.; Kan, H.; Chen, B.; Huang, W.; Bai, Z.; Song, G.; Pan, G. Association of particulate air pollution with daily mortality: The China Air Pollution and Health Effects Study. *Am. J. Epidemiol.* **2012**, *175*, 1173–1181. [CrossRef] [PubMed]
4. Cheng, Z.; Jiang, J.; Fajardo, O.; Wang, S.; Hao, J. Characteristics and health impacts of particulate matter pollution in China (2001–2011). *Atmos. Environ.* **2013**, *65*, 186–194. [CrossRef]
5. Pope Iii, C.A.; Burnett, R.T.; Thun, M.J.; Calle, E.E.; Krewski, D.; Ito, K.; Thurston, G.D. Lung cancer, cardiopulmonary mortality, and long-term exposure to fine particulate air pollution. *JAMA* **2002**, *287*, 1132–1141. [CrossRef]

6. Shin, S.E.; Jung, C.H.; Kim, Y.P. Analysis of the measurement difference for the PM10 concentrations between Beta-ray absorption and gravimetric methods at Gosan. *Aerosol Air Qual. Res.* **2011**, *11*, 846–853. [CrossRef]
7. Su, X.; Sutarlie, L.; Loh, X.J. Sensors and analytical technologies for air quality: Particulate matters and bioaerosols. *Chem.-Asian J.* **2020**, *15*, 4241–4255. [CrossRef]
8. Kurosawa, S.; Park, J.-W.; Aizawa, H.; Wakida, S.-I.; Tao, H.; Ishihara, K. Quartz crystal microbalance immunosensors for environmental monitoring. *Biosens. Bioelectron.* **2006**, *22*, 473–481. [CrossRef]
9. Ngo, N.D.; Lee, J.; Kim, M.-W.; Jang, J. Measurement of PM 2.5 mass concentration using an electrostatic particle concentrator-based quartz crystal microbalance. *IEEE Access* **2019**, *7*, 170640–170647. [CrossRef]
10. Brown, R. Tutorial review: Simultaneous measurement of particle size and particle charge. *J. Aerosol Sci.* **1997**, *28*, 1373–1391. [CrossRef]
11. Czitrovsky, A.; Oszetzky, D.; Nagy, A.; Gál, P.; Szymanski, W.; Lupkovics, G. Laser monitoring of the air pollution by aerosols. In Proceedings of the ICO20: Optical Devices and Instruments, Changchun, China, 9 December 2005; pp. 57–62.
12. Heyder, J. Deposition of inhaled particles in the human respiratory tract and consequences for regional targeting in respiratory drug delivery. *Proc. Am. Thorac. Soc.* **2004**, *1*, 315–320. [CrossRef] [PubMed]
13. Bohren, C.F.; Huffman, D.R. *Absorption and Scattering of Light by Small Particles*; John Wiley & Sons: Hoboken, NJ, USA, 2008.
14. Wood, S.H.; Prather, K.A. Time-of-flight mass spectrometry methods for real time analysis of individual aerosol particles. *TrAC Trends Anal. Chem.* **1998**, *17*, 346–356. [CrossRef]
15. Salt, K.; Noble, C.A.; Prather, K.A. Aerodynamic particle sizing versus light scattering intensity measurement as methods for real-time particle sizing coupled with time-of-flight mass spectrometry. *Anal. Chem.* **1996**, *68*, 230–234. [CrossRef]
16. Peters, T.M. Use of the aerodynamic particle sizer to measure ambient PM10–2.5: The coarse fraction of PM10. *J. Air Waste Manag. Assoc.* **2006**, *56*, 411–416. [CrossRef]
17. Wang, X.; Chancellor, G.; Evenstad, J.; Farnsworth, J.E.; Hase, A.; Olson, G.M.; Sreenath, A.; Agarwal, J.K. A novel optical instrument for estimating size segregated aerosol mass concentration in real time. *Aerosol Sci. Technol.* **2009**, *43*, 939–950. [CrossRef]
18. Grimm, H.; Eatough, D.J. Aerosol measurement: The use of optical light scattering for the determination of particulate size distribution, and particulate mass, including the semi-volatile fraction. *J. Air Waste Manag. Assoc.* **2009**, *59*, 101–107. [CrossRef] [PubMed]
19. Sioutas, C. Evaluation of the measurement performance of the scanning mobility particle sizer and aerodynamic particle sizer. *Aerosol Sci. Technol.* **1999**, *30*, 84–92. [CrossRef]
20. Peters, T.M.; Leith, D. Concentration measurement and counting efficiency of the aerodynamic particle sizer 3321. *J. Aerosol Sci.* **2003**, *34*, 627–634. [CrossRef]
21. Hairston, P.P.; Dorman, F.D.; Sem, G.J.; Agarwal, J.K. Apparatus for Measuring Particle Sizes and Velocities. U.S. Patent 5,561,515, 1 October 1996.
22. Uemura, Y.; Fujimura, M.; Hashimoto, T.; Kawai, H. Application of Light Scattering from Dielectric Cylinder Based upon Mie and Rayleigh—Gans—Born Theories to Polymer Systems. I. Scattering from a Glass Fiber. *Polym. J.* **1978**, *10*, 341–351. [CrossRef]
23. Yguerabide, J.; Yguerabide, E.E. Light-scattering submicroscopic particles as highly fluorescent analogs and their use as tracer labels in clinical and biological applications: II. Experimental characterization. *Anal. Biochem.* **1998**, *262*, 157–176. [CrossRef]
24. Fan, X.; Zheng, W.; Singh, D.J. Light scattering and surface plasmons on small spherical particles. *Light: Sci. Appl.* **2014**, *3*, e179. [CrossRef]
25. Hong, S.-H.; Winter, J. Size dependence of optical properties and internal structure of plasma grown carbonaceous nanoparticles studied by in situ Rayleigh-Mie scattering ellipsometry. *J. Appl. Phys.* **2006**, *100*, 064303. [CrossRef]
26. Ross, D.J.; Sigel, R. Mie scattering by soft core-shell particles and its applications to ellipsometric light scattering. *Phys. Rev. E* **2012**, *85*, 056710. [CrossRef]
27. Anderson, J. *EBOOK: Fundamentals of Aerodynamics (SI Units)*, 6th ed.; McGraw Hill: New York, NY, USA, 2011.
28. Baron, P.A. Calibration and use of the aerodynamic particle sizer (APS 3300). *Aerosol Sci. Technol.* **1986**, *5*, 55–67. [CrossRef]
29. Dioguardi, F.; Mele, D.; Dellino, P. A new one-equation model of fluid drag for irregularly shaped particles valid over a wide range of Reynolds number. *J. Geophys. Res. Solid Earth* **2018**, *123*, 144–156. [CrossRef]
30. Stein, S.W.; Gabrio, B.J.; Oberreit, D.; Hairston, P.; Myrdal, P.B.; Beck, T.J. An evaluation of mass-weighted size distribution measurements with the model 3320 aerodynamic particle sizer. *Aerosol Sci. Technol.* **2002**, *36*, 845–854. [CrossRef]
31. Wang, T.; Jiang, C.; Zou, J.; Zhou, H.; Lin, X.; Chen, H.; Puccioni, G.P.; Wang, G.; Lippi, G.L. Second-order correlation function supported optical sensing for particle detection. *IEEE Sens. J.* **2021**, *21*, 19948–19958. [CrossRef]
32. TSI. Aerodynamic Particle Sizer Spectrometer, (2015). TSI Specif. 2022. Available online: https://tsi.com/products/particle-sizers/particle-size-spectrometers/aerodynamic-particle-sizer-(aps)-spectrometer-3321/ (accessed on 1 July 2022).
33. Intra, P.; Yawootti, A.; Sampattagul, S. Field evaluation of an electrostatic pm 2.5 mass monitor. *Songklanakarin J. Sci. Technol.* **2018**, *40*, 347–353.
34. Zimmerman, N.; Pollitt, K.J.G.; Jeong, C.-H.; Wang, J.M.; Jung, T.; Cooper, J.M.; Wallace, J.S.; Evans, G.J. Comparison of three nanoparticle sizing instruments: The influence of particle morphology. *Atmos. Environ.* **2014**, *86*, 140–147. [CrossRef]

35. Mitchell, J.P.; Nagel, M.W. Time-of-flight aerodynamic particle size analyzers: Their use and limitations for the evaluation of medical aerosols. *J. Aerosol Med.* **1999**, *12*, 217–240. [CrossRef]

sensors

MDPI

Communication

A Compact and High-Precision Three-Degree-of-Freedom Grating Encoder Based on a Quadrangular Frustum Pyramid Prism

Shengtong Wang [1], Baiqi Liao [1], Ningning Shi [1] and Xinghui Li [1,2,*]

1 Tsinghua Shenzhen International Graduate School, Tsinghua University, Shenzhen 518055, China; wct21@mails.tsinghua.edu.cn (S.W.); lbq20@mails.tsinghua.edu.cn (B.L.)
2 Tsinghua-Berkerley Shenzhen Institute, Tsinghua University, Shenzhen 518055, China
* Correspondence: li.xinghui@sz.tsinghua.edu.cn

Abstract: A compact and high-precision three-degrees-of-freedom (DOF; X, Y, and Z directions) grating encoder based on the quadrangular frustum pyramid (QFP) prisms is proposed in this paper to solve the insufficient installation space problem of the reading head of the multi-DOF in high-precision displacement measurement applications. The encoder is based on the grating diffraction and interference principle, and a three-DOF measurement platform is built through the self-collimation function of the miniaturized QFP prism. The overall size of the reading head is 12.3 × 7.7 × 3 cm^3 and has the potential for further miniaturization. The test results show that three-DOF measurements can be realized simultaneously in the range of X-250, Y-200, and Z-100 μm due to the limitations of the measurement grating size. The measurement accuracy of the main displacement is below 500 nm on average; the minimum and maximum errors are 0.0708% and 2.8422%, respectively. This design will help further popularize the research and applications of multi-DOF grating encoders in high-precision measurements.

Keywords: displacement measurement; grating encoder; multi-degrees-of-freedom; compactness

Citation: Wang, S.; Liao, B.; Shi, N.; Li, X. A Compact and High-Precision Three-Degree-of-Freedom Grating Encoder Based on a Quadrangular Frustum Pyramid Prism. *Sensors* **2023**, *23*, 4022. https://doi.org/10.3390/s23084022

Academic Editor: Nélia J. Alberto

Received: 1 March 2023
Revised: 12 April 2023
Accepted: 13 April 2023
Published: 15 April 2023

1. Introduction

Precision displacement measurements play a remarkable role in national ultraprecision machining, national defense, and other fields [1–3]. At present, ultraprecision displacement measurements predominantly involve electrical and optical measurement methods. Electrical sensors typically include inductive, eddy current, and capacitive sensors, among which eddy current and capacitive displacement sensors are most commonly used for precision measurements. Eddy current is mainly aimed at the mm-level measurement range and can generally only provide micron accuracy [4–7]. Capacitive displacement sensors can achieve nanometer accuracy in a single degree-of-freedom (DOF). However, the measurement range is generally compressed to the level of hundreds of microns [8]. This compression is mainly due to the small linear range of the capacitance value [9]. Capacitive displacement sensors can also be used in three DOF, but only achieve micron-level accuracy [10]. A high-precision measurement of capacitive time grating is also currently observed, and the accuracy can reach submicron or even higher [11,12] in a stable electromagnetic environment. Current optical methods for precision displacement measurement include a laser interferometer and grating encoder, which both achieve nanometer or even higher accuracy in mm-level ranges with a complex system. Therefore, the two schemes have always complemented each other in the field of large-range precision measurements [13,14]. However, the laser interferometer is limited by the laser wavelength, which is the measurement standard [15,16]; therefore, in actual use, the wavelength is easily affected by the temperature and humidity in the environment, thus affecting accuracy [17–19]. Laser interferometers are generally used for measurements in a controlled environment in the laboratory, which can achieve high accuracy and stability [20–22]. The measurement standard of the grating encoder is the grating pitch [23,24]. The standard of this physical structure is substantially stable, and its measurement accuracy mainly depends on the accuracy and uniformity of the grating pitch [25–28]. In addition, absolute measurements are possible [29] with nanometer-level accuracy [30], and angle measurements can be realized under the appropriate optical path [31,32]. Therefore, a grating encoder is a widely used measurement method in actual industrial production and other applications.

However, in practical applications, the high-quality grating reading head will also increase the difficulty in controlling equipment motion due to the limited installation or operating space of production equipment; thus, high integration and stability of the grating encoder are required in most situations. Commercial products, such as Heidenhain, can only currently achieve multi-DOF by assembling together the sensor components [33]. Therefore, considering industrial applications, multi-DOF and integrated miniaturization are the largest obstacles to practical promotion.

Scholars have also conducted a series of research to achieve multi-DOF and compactness. Multi-DOF mostly comprises two to three DOF [34–37] and four to six DOF [32,38–41]. However, only a few scholars, mainly led by enterprises, such as the LIP6031D series products proposed by Heidenhain in July 2021, have researched miniaturization [42,43]. These products realize two-DOF for single reading head reads and assemble these sensors with low number DOF to form a sensor assembly with up to five DOF. The difficulty of miniaturization lies in the necessary batch collimation of diffracted light for the three-DOF grating encoder. Integrating this module and miniaturizing its volume is difficult. Scholars at home and abroad generally use prisms [40], convex lenses [44], or diffraction gratings [45] for batch collimation of diffracted light modules. Integrating the ordinary prism is complicated, and the convex lens needs focal length matching. This condition limits the miniaturization of the system and the low diffraction efficiency of the grating, which is not conducive to the detection of light intensity. Thus, the three solutions are not conducive to the development of miniaturization and integration.

This paper generally proposes a miniaturized three-DOF grating encoder based on quadrangular frustum pyramid (QFP) prisms to solve the urgent need for multi-DOF integration miniaturization and provides a reference for future six-DOF miniaturization. This encoder also addresses the difficulty of multi-DOF grating encoder application promotion and promotes research progress in multi-DOF miniaturization.

2. Principles and Method

The main principle of the three-DOF compaction scheme proposed in this paper is to design the encoder using two QFP prisms. The schematic of the QFP prism is shown in Figure 1. The draft angle α of the collimated part is 12.924° when the collimation operation is performed corresponding to the diffraction angle of the 1-micron periodic grating. However, the experimental draft angle is approximately 12.9° because of the defects in the processing technology.

This paper develops and designs a three-DOF encoder scheme using the QFP, and the schematic is shown in Figure 2. The main principle is as follows: the polarization state of the laser diode is adjusted when a laser with a wavelength of 660 nm is emitted from the laser diode (LD). Therefore, the energy of the P-light and S-light, which are separated by the polarizer beam splitter (PBS1), is the same. At this time, the S-light is first reflected, passed through a quarter-wave plate (QWP1), and then converted into circularly polarized light and irradiated on the reference grating. The four beams diffracted by the reference grating are X_{+1}, X_{-1}, Y_{+1}, and Y_{-1}. After collimation by QFP1, the four beams demonstrate parallel emission, pass through QWP1 again, and then transform into P-light to pass through PBS1. Similarly, the P-light from PBS1 follows the same rule. Finally, four parallel beams, namely X_{+1}', X_{-1}', Y_{+1}', and Y_{-1}', are obtained from the measurement grating, converted to S-light, and then reflected on PBS1 after passing through QWP2 again. At this time, two sets of four diffracted beams from various gratings enter the optical path subdivision module together under the performance of PBS1. These beams interfere with each other under the joint action of the QWP and PBS. Photoelectric signals are generated with phase information of 0°, 90°, 180°, and 270°. The four groups of signals can eliminate the influence of the energy fluctuation of the homodyne signal on the signal calculation and enhance the stability and accuracy of the signal. The displacement in the three directions of X, Y, and Z can finally be obtained in accordance with the signal calculation.

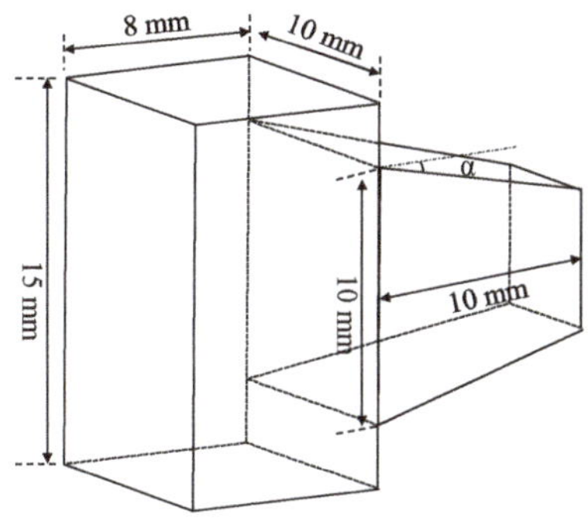

Figure 1. Schematic of the Quadrangular Frustum Pyramid Prism structure.

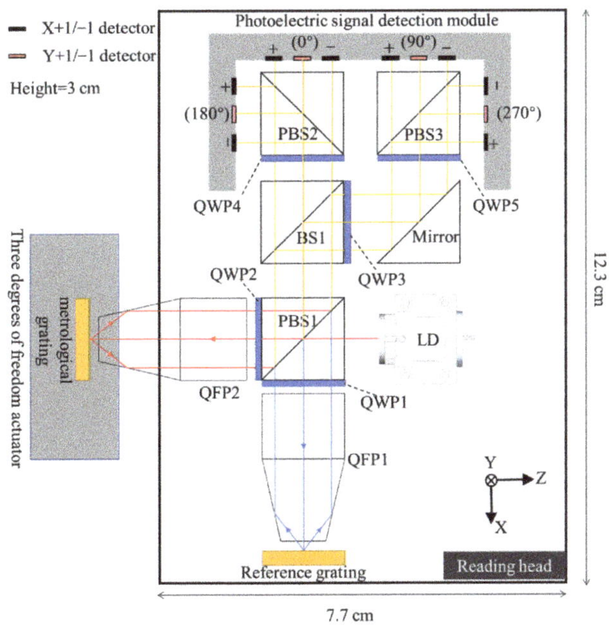

Figure 2. Schematic of the three-DOF reading head: (Beam Splitter (BS)).

The phase change of the signal is used for measurement in three-DOF encoders. The Doppler frequency shift effect causes phase changes in the diffracted light direction when the grating measurement moves along the X and Y directions. The principles in the two directions are the same. Therefore, this paper takes the X direction as an example. When the grating measurement is displaced along the X direction, the phase of the diffracted lights $U_{S_{X\pm1}}$ in the X direction changes, as shown in Equation (1), while the reference beams $U_{r_{X\pm1}}$ remain unchanged.

$$\begin{cases} U_{r_{X+1}} = U_{r_{X-1}} = U_0 \\ U_{S_{X\pm1}} = U_0 e^i(\Omega_{X\pm1} + \Phi) \end{cases}, \tag{1}$$

where the phase changes are Ω and Φ, corresponding to the displacements in the X and Z directions, respectively.

$$\begin{cases} \Omega_{X\pm1} = \pm 2\pi \frac{\Delta x}{g} \\ \Phi = 2\pi \frac{\Delta z(1+\cos\theta)}{\lambda} \end{cases}, \tag{2}$$

where ΔX and ΔZ are the displacements in the X and Z directions, respectively, g is the grating pitch, θ is the diffraction angle, and λ is the wavelength of the incident light.

After light synthesis,

$$U_{X\pm1} = Us_{X\pm1} + Ur_{X\pm1}. \tag{3}$$

The light intensity is

$$I_{X\pm1} = U_{X\pm1} \cdot \overline{U_{X\pm1}} = 2U_0^2[1 + cos(\Omega_{X\pm1} + \Phi), \tag{4}$$

The light intensity information of the corresponding four phases is as follows:

$$\begin{cases} I_{X\pm1(0^\circ)} = b_1 + a_1\cos(\Omega_{\pm1} + \Phi) \\ I_{X\pm1(90^\circ)} = b_2 + a_2\cos(\Omega_{\pm1} + \Phi + pi/2) \\ I_{X\pm1(180^\circ)} = b_3 + a_3\cos(\Omega_{\pm1} + \Phi + pi) \\ I_{X\pm1(270^\circ)} = b_4 + a_4\cos(\Omega_{\pm1} + \Phi + 3 \times pi/2) \end{cases}, \tag{5}$$

where a_1–a_4 is the amplitude of the ideal interference signal, and b_1–b_4 is the amplitude of the DC bias signal.

The influence of DC bias fluctuations can be removed, and the amplitude changes of the interference signal can be preserved.

$$\begin{cases} S_{X+1} = \frac{I_{X+1}(0^\circ) - I_{X+1}(180^\circ)}{I_{X+1}(0^\circ) + I_{X+1}(180^\circ)}, \\ S'_{X+1} = \frac{I_{X+1}(90^\circ) - I_{X+1}(270^\circ)}{I_{X+1}(90^\circ) + I_{X+1}(270^\circ)}, \\ S_{X-1} = \frac{I_{X-1}(0^\circ) - I_{X-1}(180^\circ)}{I_{X-1}(0^\circ) + I_{X-1}(180^\circ)}, \\ S'_{X-1} = \frac{I_{X-1}(90^\circ) - I_{X-1}(270^\circ)}{I_{X-1}(90^\circ) + I_{X-1}(270^\circ)}, \end{cases} \tag{6}$$

According to the trigonometric function, the calculation formula for the corresponding displacement can be obtained as follows:

$$\begin{cases} \Delta X = \frac{g}{4\pi}\left\{\arctan\left(\frac{S_{X+1}}{S'_{X+1}}\right) - \arctan\left(\frac{S_{X-1}}{S'_{X-1}}\right)\right\}, \\ \Delta Z = \frac{\lambda}{4\pi(1+\cos\theta)}\left\{\arctan\left(\frac{S_{X+1}}{S'_{X+1}}\right) + \arctan\left(\frac{S_{X-1}}{S'_{X-1}}\right)\right\}, \end{cases} \tag{7}$$

The calculation method in the Y direction is the same as that in the X direction and will not be repeated herein.

In actual data processing, direct data processing according to formulas (5)–(7) cannot obtain accurate displacement results because the actual incremental signal is not an ideal cosine signal. First, the results are affected by noise interference. Second, the optical component errors (e.g., the splitting ratio of the beam splitting prism is not ideal at 1:1), system installation errors (e.g., adjusting the angle between the fast axis of the wave plate and the X-axis direction to the optimum 45° is difficult), and errors in photoelectric conversion also contribute to the aforementioned phenomenon. The signal can be expressed as follows:

$$S_{X+1} = \frac{I_{X+1(180^o)} - I_{X+1(0^o)}}{I_{X+1(180^o)} + I_{X+1(0^o)}} \approx A_1 \sin(\Omega + \Phi) + B_1, \tag{8}$$

$$S_{X+1}{}' = \frac{I_{X+1(90^o)} - I_{X+1(270^o)}}{I_{X+1(90^o)} + I_{X+1(270^o)}} \approx A_3 \cos(\Omega + \varphi - \Phi) + B_3, \tag{9}$$

$$S_{X-1} = \frac{I_{X-1(180^o)} - I_{X-1(0^o)}}{I_{X-1(180^o)} + I_{X-1(0^o)}} \approx A_2 \sin(-\Omega + \Phi) + B_2, \tag{10}$$

$$S_{X-1}{}' = \frac{I_{X-1(90^o)} - I_{X-1(270^o)}}{I_{X-1(90^o)} + I_{X-1(270^o)}} \approx A_4 \cos(-\Omega + \Phi - \sigma) + B_4, \tag{11}$$

where A_1–A_4 is the fluctuation amplitude of the interference signal, σ is the phase deviation from the ideal signal, and B_1–B_4 is the maintained DC component.

Therefore, the signal processing process must be performed to minimize the relative error of the calculated results to the true value. First, the high-frequency noise signal will be removed by the low-pass filtering method without changing the phase. Second, amplitude regularization will generally correct the maximum absolute value of the signal to maintain consistency. Furthermore, the phase compensation can adjust the phase information of each signal according to the ideal phase set. Finally, arctangent counting will provide a displacement result based on the grating pitch.

3. Experiments and Discussion

This paper designed the experiments in three DOF based on the above principles, and the object pictures of the test bench are shown in Figure 3. The pixelated part is a fixed device that has nothing to do with this paper.

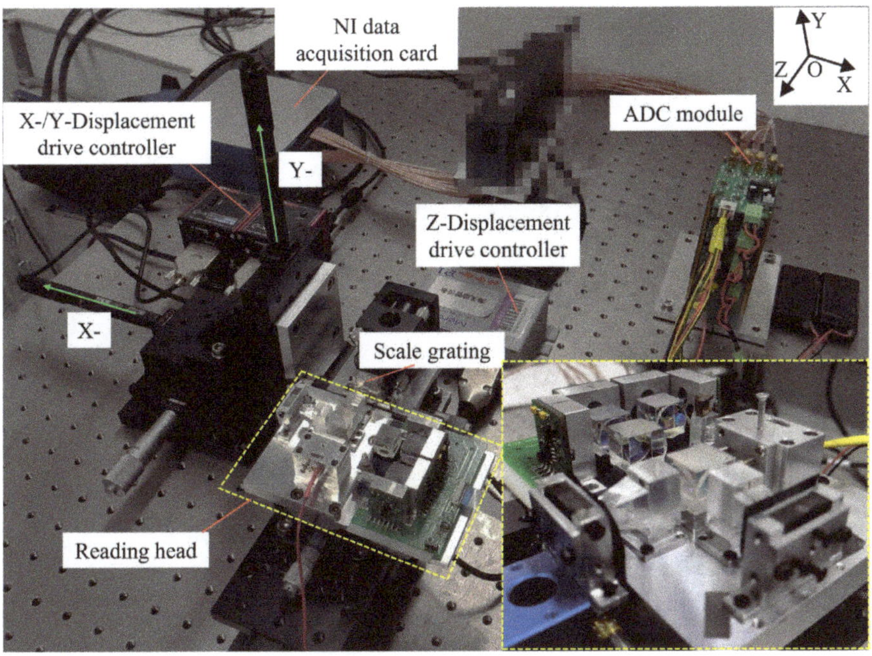

Figure 3. Three-DOF grating encoder test platform.

The test platform comprises the following three main parts: data acquisition, displacement, and reading head. The data acquisition part mainly includes an NI acquisition card, an ADC module, and a computer. The displacement part is the Advanced Positioning Technology (APT) drive controller of THORLABS and the corresponding displacement drive (X and Y directions). The Z direction is controlled and displaced by the micron displacement stage of the Physik Instrumente (PI). The main parameters are shown in Table 1. The lower right corner of Figure 3 shows an enlarged picture of the reading head, which was built in accordance with the schematic of Figure 2.

Table 1. Driver brand name and performance parameter.

Instrument	Brand and Model Name	Repeatability/μm	Resolution/μm
X-displacement driver	THORLABS/Z825B	0.2000	0.0500
Y-displacement driver	THORLABS/Z825B	0.2000	0.0500
Z-displacement driver	PI/M-112	0.2500	0.0500

Figure 4 shows the displacement driver designed for a three-DOF reading head, which can output displacements in three directions simultaneously or one output alone. However, a cosine error is observed in the actual test. Therefore, the displacement in one-direction measurement will still output the displacement in other directions in the test and will be measured using the three-DOF encoder.

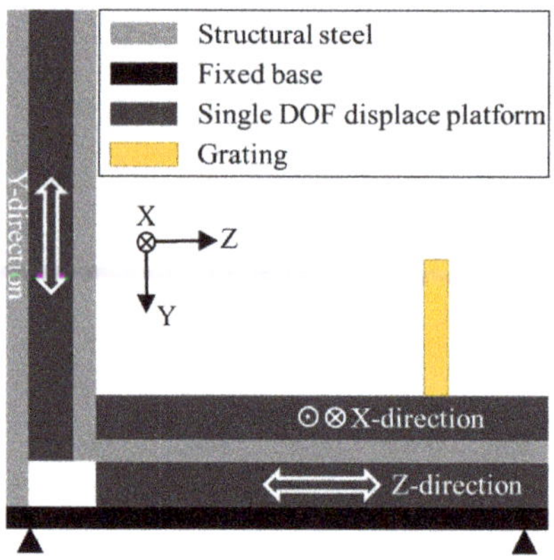

Figure 4. Displacement driver designed for the three-DOF reading head.

The three-DOF measurement experiment was conducted on the basis of this experimental platform. First, the pre-experiment was performed. The displacement value input by the stage controller is taken as a standard reference. Then, the following corrections were performed in accordance with the pre-experiment. The grating pitch in the X/Y direction of the grating and the wavelength of the light wave are corrected after the pre-experiment, and the corrected values are 1.1340, 1.0196, and 0.6879 μm. This paper uses a light spot with a diameter of 0.5 mm for experiments to further miniaturize the encoder. Therefore, the theoretical limit measurement range in the Z direction is 250 μm, while the range in the X and Y directions is mainly determined using the area of the measurement grating. However, the encoder will fail to perform the measurement when the displacement in the X and Y directions is excessively large and the crosstalk displacement in the Z direction exceeds the theoretical limit of 250 μm due to the cosine error. Therefore, the ranges selected for the experiment are X-250, Y-200, and Z-100 μm.

Table 2 shows the measurement results of the three DOF in the main direction of the X, Y, and Z directions. The driver velocities are 500 μm/s and 100 μm/s in the X and

Y directions, respectively. The drive velocities are 1 μm/s for Z1 and Z2 and 4 μm/s for Z3. The last column of Table 2 shows the percentage error of the main displacement measurements from the input values. The table reveals the following: the minimum displacement error percentage in the X, Y, and Z directions can reach 0.0708%, 0.1389%, and 0.3339%, respectively, and the minimum errors are below 200, 300, and 400 nm, respectively. In the six groups of experiments, statistics are established on the accuracy error of the main displacement, and the average error is 455 nm. In addition to the main motion direction, measurements in the two other directions during all the tests are also stable. This finding proves the stability and accuracy of measurements in three DOF.

Table 2. Measurement results of three DOF in X, Y, and Z directions. The bold indicates main displacements.

Motion Axis	Input Displacement/μm	Measured Value			
		X/μm	Y/μm	Z/μm	Motion Error/%
X1	250	**−251.3361**	2.8269	−0.6948	0.5344
X2	250	**−248.8365**	5.3853	−0.5163	−0.4654
X3	250	**−249.8230**	4.4111	−0.8086	−0.0708
Y1	200	−4.0984	**200.7770**	−4.6850	0.3885
Y2	200	−2.4667	**198.9402**	−3.5440	−0.5299
Y3	200	−3.6587	**200.2778**	−5.2337	0.1389
Z1	100	−1.1824	1.1926	**102.8422**	2.8422
Z2/Reverse movement	100	0.9627	−1.0207	**−99.0848**	−0.9152
Z3	100	−0.0091	0.0088	**100.3339**	0.3339

Figure 5 shows that the measurement data (plotted in black) is first linearly fitted, and the error (plotted in red) is the difference between the measured value and the fitted curve. The fitted curve equation shown below assumes that the driver velocity is constant.

$$X/Y/Z = V \times T, \tag{12}$$

where the X/Y/Z are the fit values, V is the driver velocity, and T is the time of movement.

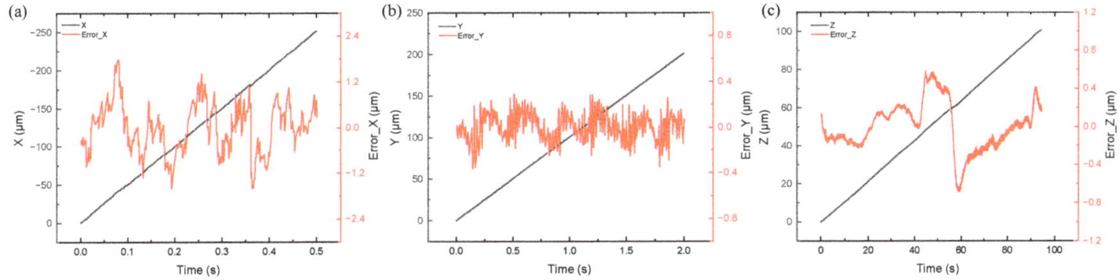

Figure 5. Displacement measurement results and errors of three-DOF: (**a**) X direction; (**b**) Y direction; (**c**) Z direction.

Figure 5a shows that the predominant fluctuations in displacement measurements in the X direction are within ±2 μm mainly due to the unideal speed of the driver operation during motion. Periodic fluctuations are observed in the straight-line error of the drive screw. In addition, a problem with speed fluctuations is found in the Y and Z directions, which is considered a common problem for screw-driven drivers. The displacement velocity is the largest in all three directions, thus also demonstrating the largest displacement fluctuation. Figure 5b illustrates that the fluctuation error is less than 400 nm in the Y direction. Figure 5c shows that the fluctuation error is less than 800 nm in the Z direction.

In other words, the displacement fluctuation of the PI displacement driver is larger than that of the THORLABS displacement driver. This finding is consistent with the repeatability parameter trend shown in Table 1.

An equal displacement test of 100 μm was conducted in the opposite direction of the X and Y directions to test the simultaneous measurement capability of three DOF, and the movements of the Y and X directions were turned on successively. The experimental results are X (−100.7862 μm), Y (−98.0961 μm), and Z (−1.2153 μm), and the measurement error percentages in the X and Y directions are 0.7862% and −1.9039%, respectively. The measurement error in the Y direction increases by approximately two times compared with the single movement in the Y direction due to the simultaneous movement of the two directions. A crosstalk error should be observed between the two movements of the X and Y directions.

The final test conclusion indicates that the structure can achieve submicron accuracy at 250 μm (X direction), 200 μm (Y direction), and 100 μm (Z direction), and the overall structure is small with a size of $12.3 \times 7.7 \times 3\ cm^3$. The measurement range and accuracy can further increase with this three-DOF measurement method after the improvement of the grating processing technology in the future.

4. Conclusions

A compact and high-precision three-DOF grating displacement encoder is established in this paper based on the QFP prism, and the photoelectric signal is stabilized and improved through the optical path subdivision module. Three DOF can be measured simultaneously, the accuracy of the main displacement is below 500 nm on average, and the minimum error is 7.08‱. The grating processing quality and area will be further improved in the future. Therefore, the performance of the grating encoder will be further enhanced to achieve a range of millimeters and an accuracy of 1‱.

Author Contributions: Conceptualization, S.W., B.L. and X.L.; methodology, S.W. and X.L.; software, S.W. and X.L.; validation, S.W. and X.L.; formal analysis, S.W., B.L., N.S. and X.L.; investigation, S.W. and X.L.; resources, X.L.; data curation, S.W. and B.L.; writing—original draft preparation, S.W. and X.L.; writing—review and editing, S.W. and X.L.; visualization, S.W.; supervision, X.L.; project administration, X.L.; funding acquisition, X.L. All authors have read and agreed to the published version of the manuscript.

Funding: This work was supported by the National Natural Science Foundation of China with no. 62275142 and the Basic and Applied Basic Research Foundation of Guangdong Province with no. 2021B1515120007.

Institutional Review Board Statement: Not applicable.

Informed Consent Statement: Not applicable.

Data Availability Statement: The data presented in this study are available on request from the Corresponding author.

Acknowledgments: The authors would like to thank all our laboratory members who have joined the related projects at Tsinghua University for their contributions to the achievements described in this paper.

Conflicts of Interest: The authors declare no conflict of interest.

References

1. Gao, W.; Kim, S.W.; Bosse, H.; Haitjema, H.; Chena, Y.L.; Lu, X.D.; Knapp, W.; Weckenmann, A.; Estler, W.T.; Kunzmann, H. Measurement technologies for precision positioning. *CIRP Ann.-Manuf. Technol.* **2015**, *64*, 773–796. [CrossRef]
2. Schmidt, R.H. Ultra-precision engineering in lithographic exposure equipment for the semiconductor industry. *Philos. Trans. A Math. Phys. Eng. Sci.* **2012**, *370*, 3950–3972. [CrossRef] [PubMed]
3. Donges, A.; Noll, R. *Laser Measurement Technology*; Springer Series in Optical Sciences; Springer: Berlin/Heidelberg, Germany, 2015; Volume 188.

4. Wang, S.C.; Xie, B.R.; Huang, S.M. Design and Analysis of Small Size Eddy Current Displacement Sensor. *Sensors* **2022**, *22*, 7444. [CrossRef] [PubMed]
5. Xue, X.; Dong, Y.; Wu, X. Motion Induced Eddy Current Sensor for Non-Intrusive Vibration Measurement. *IEEE Sens. J.* **2020**, *20*, 735–744. [CrossRef]
6. Nabavi, M.R.; Nihtianov, S.N. Design Strategies for Eddy-Current Displacement Sensor Systems: Review and Recommendations. *IEEE Sens. J.* **2012**, *12*, 3346–3355. [CrossRef]
7. Mizuno, T.; Enoki, S.; Hayashi, T.; Asahina, T.; Shinagawa, H. Extending the Linearity Range of Eddy-Current Displacement Sensor With Magnetoplated Wire. *IEEE Trans. Magn.* **2007**, *43*, 543–548. [CrossRef]
8. Krupa, A.; Lackowski, M.; Jaworek, A. Capacitance sensor for measuring void fraction in small channels. *Measurement* **2021**, *175*, 109046. [CrossRef]
9. Beverte, I.; Cabulis, U.; Gaidukovs, S. Polytetrafluoroethylene Films in Rigid Polyurethane Foams' Dielectric Permittivity Measurements with a One-Side Access Capacitive Sensor. *Polymers* **2021**, *13*, 1173. [CrossRef]
10. Li, X.; Wang, R.; Du, H.; Lu, Y. Three-dimensional micro-displacement measurement method based on capacitance-grating sensor. *Measurement* **2022**, *187*, 110179. [CrossRef]
11. Peng, K.; Yu, Z.; Liu, X.; Chen, Z.; Pu, H. Features of Capacitive Displacement Sensing That Provide High-Accuracy Measurements with Reduced Manufacturing Precision. *IEEE Trans. Ind. Electron.* **2017**, *64*, 7377–7386. [CrossRef]
12. Liu, X.; Peng, K.; Chen, Z.; Pu, H.; Yu, Z. A New Capacitive Displacement Sensor With Nanometer Accuracy and Long Range. *IEEE Sens. J.* **2016**, *16*, 2306–2316. [CrossRef]
13. Qibo, F.; Bin, Z.; Cunxing, C.; Cuifang, K.; Yusheng, Z.; Fenglin, Y. Development of a simple system for simultaneously measuring 6DOF geometric motion errors of a linear guide. *Opt. Express* **2013**, *21*, 25805–25819. [CrossRef] [PubMed]
14. Yu, H.; Chen, X.; Liu, C.; Cai, G.; Wang, W. A survey on the grating based optical position encoder. *Opt. Laser Technol.* **2021**, *143*, 107352. [CrossRef]
15. Ortlepp, I.; Zöllner, J.-P.; Rangelow, I.W.; Manske, E. Heterodyne Standing-Wave Interferometer with Improved Phase Stability. *Nanomanuf. Metrol.* **2021**, *4*, 190–199. [CrossRef]
16. Aasi, J.; Abbott, B.P.; Abbott, R.; Abbott, T.; Abernathy, M.R.; Ackley, K.; Adams, C.; Adams, T.; Addesso, P.; Adhikari, R.X.; et al. Advanced LIGO. *Class. Quantum Grav.* **2015**, *32*, 41. [CrossRef]
17. Liu, W.; Yu, Z.; Duan, F.; Hu, H.; Fu, X.; Bao, R. Robust five-degree-of-freedom measurement system with self-compensation and air turbulence protection. *Opt. Express* **2023**, *31*, 4652–4666. [CrossRef]
18. Zheng, F.; Feng, Q.; Zhang, B.; Li, J. A Method for Simultaneously Measuring 6DOF Geometric Motion Errors of Linear and Rotary Axes Using Lasers. *Sensors* **2019**, *19*, 1764. [CrossRef]
19. Liang, A.J.; He, P.Y.; Hsieh, H.L. Symmetrical Double Diffraction Laser Encoder. In Proceedings of the Conference on Optical Micro- and Nanometrology VII, Strasbourg, France, 25–26 April 2018.
20. Zhang, Y.; Guzman, F. Fiber-based two-wavelength heterodyne laser interferometer. *Opt. Express* **2022**, *30*, 37993–38008. [CrossRef]
21. Zhao, Y.; Zhang, B.; Feng, Q. Measurement system and model for simultaneously measuring 6DOF geometric errors. *Opt. Express* **2017**, *25*, 20993–21007. [CrossRef]
22. Cui, C.; Feng, Q.; Zhang, B.; Zhao, Y. System for simultaneously measuring 6DOF geometric motion errors using a polarization maintaining fiber-coupled dual-frequency laser. *Opt. Express* **2016**, *24*, 6735–6748. [CrossRef]
23. Hu, P.C.; Chang, D.; Tan, J.B.; Yang, R.T.; Yang, H.X.; Fu, H.J. Displacement measuring grating interferometer: A review. *Front. Inf. Technol. Electron. Eng.* **2019**, *20*, 631–654. [CrossRef]
24. Morlanes, T.; de la Pena, J.L.; Sanchez-Brea, L.M.; Alonso, J.; Crespo, D.; Saez-Landete, J.B.; Bernabeu, E. Optoelectronic device for the measurement of the absolute linear position in the micrometric displacement range. In Proceedings of the Conference on Photonic Materials, Devices and Applications, Seville, Spain, 9–11 May 2005; pp. 862–870.
25. Shimizu, Y. Laser Interference Lithography for Fabrication of Planar Scale Gratings for Optical Metrology. *Nanomanufacturing Metrol.* **2021**, *4*, 3–27. [CrossRef]
26. Liu, Y.; Li, X.; Huang, J.; Wang, Z.; Zhao, X.; Zhao, B.; Jiang, L. High-Uniformity Submicron Gratings with Tunable Periods Fabricated through Femtosecond Laser-Assisted Molding Technology for Deformation Detection. *ACS Appl. Mater. Interfaces* **2022**, *14*, 16911–16919. [CrossRef]
27. Liang, J.; Wang, C.; Lu, H.; Wang, X.; Ni, K.; Zhou, Q. Method for fabricating large-area gratings with a uniform duty cycle without a spatial beam modulator. *Opt. Express* **2021**, *29*, 27791–27806. [CrossRef] [PubMed]
28. Deng, X.; Tan, W.; Tang, Z.; Lin, Z.; Cheng, X.; Li, T. Scanning and Splicing Atom Lithography for Self-traceable Nanograting Fabrication. *Nanomanuf. Metrol.* **2022**, *5*, 179–187. [CrossRef]
29. Li, X.; Wang, H.; Ni, K.; Zhou, Q.; Mao, X.; Zeng, L.; Wang, X.; Xiao, X. Two-probe optical encoder for absolute positioning of precision stages by using an improved scale grating. *Opt. Express* **2016**, *24*, 21378–21391. [CrossRef]
30. Shi, Y.; Ni, K.; Li, X.; Zhou, Q.; Wang, X. Highly accurate, absolute optical encoder using a hybrid-positioning method. *Opt. Lett.* **2019**, *44*, 5258–5261. [CrossRef]
31. Ren, W.; Cui, J.; Tan, J. A Novel Enhanced Roll-Angle Measurement System Based on a Transmission Grating Autocollimator. *IEEE Access* **2019**, *7*, 120929–120936. [CrossRef]
32. Liu, C.H.; Huang, H.L.; Lee, H.W. Five-degrees-of-freedom diffractive laser encoder. *Appl. Opt.* **2009**, *48*, 2767–2777. [CrossRef]

33. Heidenhain. MULTI-DOF_Measurement-Technology_JA. Available online: https://semiconductor.heidenhain.com/fileadmin/semiconductor/MULTI-DOF_Measurement-Technology_JA.pdf (accessed on 21 February 2023).
34. Yang, F.; Zhang, M.; Zhu, Y.; Ye, W.; Wang, L.; Xia, Y. Two Degree-of-Freedom Fiber-Coupled Heterodyne Grating Interferometer with Milli-Radian Operating Range of Rotation. *Sensors* **2019**, *19*, 3219. [CrossRef]
35. Saito, Y.; Arai, Y.; Gao, W. Detection of three-axis angles by an optical sensor. *Sens. Actuators A-Phys.* **2009**, *150*, 175–183. [CrossRef]
36. Zhu, J.; Wang, G.; Xue, G.; Zhou, Q.; Li, X.; Zhu, J.; Jiang, J.; Han, S.; Zeng, L. Heterodyne three-degree-of-freedom grating interferometer for ultra-precision positioning of lithography machine. In Proceedings of the 2021 International Conference on Optical Instruments and Technology: Optoelectronic Measurement Technology and Systems, Online, 8–10 April 2022.
37. Ye, W.; Cheng, R.; Zhang, M.; Zhu, Y.; Wang, L.; Hu, J.; Li, X. Grating Interferometer with Redundant Design for Performing Wide-Range Displacement Measurements. *Sensors* **2022**, *22*, 3738. [CrossRef] [PubMed]
38. Wang, S.; Luo, L.; Zhu, J.; Shi, N.; Li, X. An Ultra-Precision Absolute-Type Multi-Degree-of-Freedom Grating Encoder. *Sensors* **2022**, *22*, 9047. [CrossRef] [PubMed]
39. Li, X.H.; Gao, W.; Muto, H.S.; Shimizu, Y.; Ito, S.; Dian, S. A six-degree-of-freedom surface encoder for precision positioning of a planar motion stage. *Precis. Eng.-J. Int. Soc. Precis. Eng. Nanotechnol.* **2013**, *37*, 771–781. [CrossRef]
40. Yu, K.; Zhu, J.; Yuan, W.; Zhou, Q.; Xue, G.; Wu, G.; Wang, X.; Li, X. Two-channel six degrees of freedom grating-encoder for precision-positioning of sub-components in synthetic-aperture optics. *Opt. Express* **2021**, *29*, 21113–21128. [CrossRef]
41. Matsukuma, H.; Ishizuka, R.; Furuta, M.; Xinghui, L.; Shimizu, Y.; Wei, G. Reduction in Cross-Talk Errors in a Six-Degree-of-Freedom Surface Encoder. *Nanomanuf. Metrol.* **2019**, *2*, 111–123. [CrossRef]
42. Li, X.H.; Shi, Y.P.; Xiao, X.; Zhou, Q.; Wu, G.H.; Lu, H.O.; Ni, K. Design and Testing of a Compact Optical Prism Module for Multi-Degree-of-Freedom Grating Interferometry Application. *Appl. Sci.* **2018**, *8*, 2495. [CrossRef]
43. Cai, Y.; Lou, Z.; Ling, S.; Liao, B.-S.; Fan, K.-C. Development of a Compact Three-Degree-of-Freedom Laser Measurement System with Self-Wavelength Correction for Displacement Feedback of a Nanopositioning Stage. *Appl. Sci.* **2018**, *8*, 2209. [CrossRef]
44. Gao, W.; Saito, Y.; Muto, H.; Arai, Y.; Shimizu, Y. A three-axis autocollimator for detection of angular error motions of a precision stage. *CIRP Ann.-Manuf. Technol.* **2011**, *60*, 515–518. [CrossRef]
45. Kimura, A.; Gao, W.; Kim, W.; Hosono, K.; Shimizu, Y.; Shi, L.; Zeng, L.J. A sub-nanometric three-axis surface encoder with short-period planar gratings for stage motion measurement. *Precis. Eng.* **2012**, *36*, 576–585. [CrossRef]

sensors

MDPI

Article

A Laser Plane Attitude Evaluation Method for Rail Profile Measurement Sensors

Le Wang, Hao Wang, Qiang Han, Yue Fang, Shengchun Wang *, Ning Wang, Guoqing Li and Shengwei Ren

Infrastructure Inspection Research Institute, China Academy of Railway Sciences Corporation Limited, Beijing 100081, China; 15121630@bjtu.edu.cn (L.W.)
* Correspondence: wangshengchun@rails.cn

Abstract: The non-coplanar lasers on both sides of the rail during full-section rail profile measurement based on line-structured light vision will cause the measured profile to be distorted, resulting in measurement errors. Currently, in the field of rail profile measurement, there are no effective methods for evaluating laser plane attitude, and it is impossible to determine the degree of laser coplanarity quantitatively and accurately. This study proposes an evaluation method based on fitting planes in response to this problem. Real-time fitting of laser planes with three planar targets of different heights provides information about the laser plane attitude on both sides of the rails. On this basis, laser coplanarity evaluation criteria were developed to determine whether the laser planes on both sides of the rails are coplanar. Using the method in this study, the laser plane attitude can be quantified and accurately assessed on both sides, effectively resolving the problem with traditional methods that can only assess the laser plane attitude qualitatively and roughly, thereby providing a solid foundation for calibration and error correction of the measurement system.

Keywords: laser plane; attitude evaluation; coplanarity evaluation; rail profile

Citation: Wang, L.; Wang, H.; Han, Q.; Fang, Y.; Wang, S.; Wang, N.; Li, G.; Ren, S. A Laser Plane Attitude Evaluation Method for Rail Profile Measurement Sensors. *Sensors* **2023**, *23*, 4586. https://doi.org/10.3390/s23104586

Academic Editor: Simone Borri

Received: 10 March 2023
Revised: 17 April 2023
Accepted: 21 April 2023
Published: 9 May 2023

1. Introduction

Rails play an important role in the maintenance and repair of railway lines. As a result of regularly inspecting the rail profile, as well as evaluating the state parameters of these rails, such as vertical wear and side wear [1–3], it is possible to gain a better understanding of not only the state of the rails but also how to grind the rails, as this is a crucial part of railway operations and maintenance [4,5]. The measurement of rail profiles using line-structured light vision is based on the principle of triangulation and features high speed, high precision, and noncontact. As a mainstream method of dynamic detection of rail profiles globally, it can detect the parameters of in-service rails, such as vertical wear and side wear [6–8]. In most cases, a line-structured light sensor is placed on each side of a rail in order to obtain the profile data of the left and right half-sections of the rail. The full-section profile of the rail is then produced by splicing these two half-sections together [9–11]. In this process, if the laser beams of the line-structured light sensors are not on the same plane on either side of the rail, the measurement profile will be distorted to a certain degree, resulting in errors in rail profiling. It is therefore necessary to accurately assess the laser plane attitude on both sides of the rail to ensure that the laser planes on both sides are coplanar in order to obtain high-precision full-section profile data. When the lasers on both sides are installed, they are incident on the calibration plate, forming a line of intersection between the two lasers. The calibration plate is marked with a scale line. In order to determine whether the laser planes are coplanar as required, it is necessary to visually observe the degree of coincidence between the intersection lines of the laser planes on both sides and the scale line of the calibration plate. Obviously, this method is limited to a qualitative analysis and cannot provide a quantitative or accurate assessment of the plane attitude of the lasers on both sides.

Currently, little attention is being paid to the attitude of the laser planes on both sides when measuring rail profiles. In most existing studies, the focus is on adjusting the lasers in order to make them coplanar. As an example, Zhan et al. proposed a mechanism and method for the adjustment of laser planes. A mechanical adjustment mechanism could be used to translate and rotate the line-structured laser [12]. Through a high-precision manual translation stage, Chen et al. were able to scan the profiles of large workpieces [13]. In addition, some scholars have been able to align the laser planes on both sides by examining calibration techniques. In one study, Wu et al. were able to calibrate profile measurement components globally through the use of special mechanical parts, such as rotating arms [14]. In their study, Zhang et al. documented the positions of the laser profiler and markers by photographing them and converting the adjustment of coplanarity into the position adjustment of the laser profiler [15]. Ju et al. calibrated the laser planes by using the contour line calibration method [16]. Wang et al. also identified a method for correcting the error in rail profile measurement caused by non-coplanar lasers. Using projection transformation, a laser non-coplanarity correction model was proposed, as well as a reference coordinate system based on the longitudinal direction of rails. Half-section profile data were projected onto an auxiliary plane perpendicular to the longitudinal direction of the rail, and the projection profile was used to correct the measurement results [17]. Currently, few reports have been published on the use of the laser plane attitude evaluation method for both sides of the rail during the full-section measurement of the rail profile. In summary, existing methods cannot quantitatively and accurately evaluate the attitude of laser planes on both sides.

The author previously studied the distribution characteristics of laser non-coplanar error in rail profile measurement sensors and proposed a correction method for laser non-coplanar error and a calibration method for rail longitudinal parameters. However, during the installation process of rail profile measurement devices, the laser coplanar adjustment operation is mainly guided by observing the degree of collinearity of the laser line with the naked eye, which has a certain degree of blindness. In order to avoid this blindness and guide laser coplanar adjustment operations, this paper proposes a quantitative evaluation method for laser plane posture. When the measuring device is installed, three planar targets of different heights are used to obtain the attitude information of both laser planes in real time. Such information is used to establish a laser coplanarity evaluation criterion, which will be used to guide the alignment of the two line-structured lasers. As opposed to naked-eye evaluations with low precision, poor real-time performance, and subjectivity, the proposed method uses computer vision evaluations for laser coplanarity adjustment, with high precision, excellent real-time performance, and visualization, which can reduce calibration errors. In light of this, it is of great importance to improve the accuracy of the measurement of rail profiles in full sections.

2. Basic Principle

Figure 1 presents a schematic diagram of the full-section rail profile measurement based on line-structured light vision. Both sides of the rail are equipped with a line-structured light sensor, and the lasers of both sensors are positioned in the same plane in order to obtain the left and right half-sections of the rail. In order to obtain the full-section profile of the rail, the half-section profiles on both sides are spliced together according to the calibration parameters [18]. Using the scanning motion, the rail profile for the entire railway line can be measured in the full section. The degree of coplanarity of the laser planes on both sides of the rail is an important factor in determining the accuracy of a full-section rail profile measurement system based on line-structured light vision. Ideally, the laser planes on both sides should be coplanar in order to ensure the accuracy of rail profile measurements. As shown in Figure 2, the origin is the center of the top surface of the rail, the laser plane is the XOZ plane, and the horizontal direction is the X axis of the world coordinate system, O-XYZ. The non-coplanarity of the laser planes on both sides of the rail can therefore be expressed as the rotation of the laser plane around the X and

Z axes, respectively. In Figure 3, the lasers on both sides are not coplanar, in which the left laser plane is perpendicular to the longitudinal direction of the rail, but the right laser plane rotates around the Z axis, at an angle. In this case, the right laser plane does not remain perpendicular to the longitudinal direction of the rail, nor is it coplanar to the left laser plane. While the left camera still captures the profile data of the cross-section perpendicular to the rail longitudinal direction, the right camera captures the profile data not perpendicular to the rail longitudinal direction, referred to as the oblique-section profile data. In comparison with the cross-section profile of the rail, the oblique-section profile is stretched in a certain direction, and this stretching direction is directly related to the angle between the laser planes and the longitudinal direction of the rail. As a result of this stretching, the rail profile is distorted, causing deviations in the positioning of the feature points, leading to an increase in the measurement error of the rail wear. More generally, the measured profiles on both sides of the rail are not the cross-section profile of the rail when the laser planes are not coplanar or perpendicular to the longitudinal direction of the rail. In addition, the measurements will be distorted, resulting in greater errors in determining rail profile.

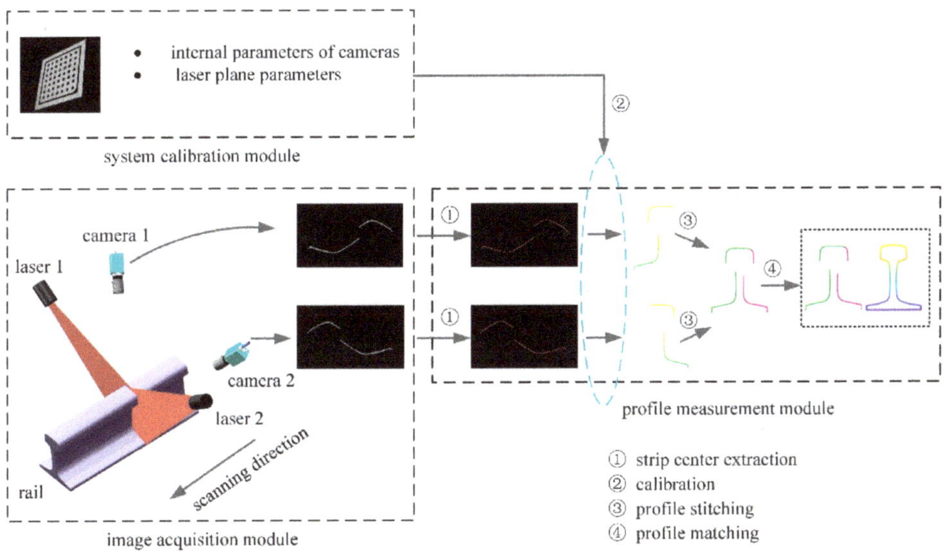

Figure 1. Schematic diagram of the full-section rail profile measurement based on line-structured light vision.

At present, the laser coplanarity of the rail profile measurement system is determined by observing the degree of overlap of two laser lines with the human eye. As shown in Figure 4, the surface of the aluminum alloy ruler is engraved with a long scale line. When installing the laser scan sensors on both sides, the laser planes on both sides are projected onto the same scale line of the aluminum alloy ruler, as shown in Figure 5. The intersection line of the laser planes on both sides and the scale line of the aluminum alloy ruler are observed with the naked eye to determine whether the laser planes on both sides meet the coplanar installation requirements based on the degree of overlap between the intersection line and the scale line. Obviously, this method evaluates the attitude of a two-dimensional plane through one-dimensional lines, which can only be qualitatively evaluated and cannot be quantitatively evaluated and has a certain degree of blindness. Due to the low level of visualization, it cannot effectively guide the coplanar installation operation of two laser scan sensors. The main purpose of this article is to propose a quantitative evaluation method for laser plane attitude, which visualizes and quantifies the adjustment process of the laser plane, thereby guiding the coplanar installation operation of two laser scan sensors.

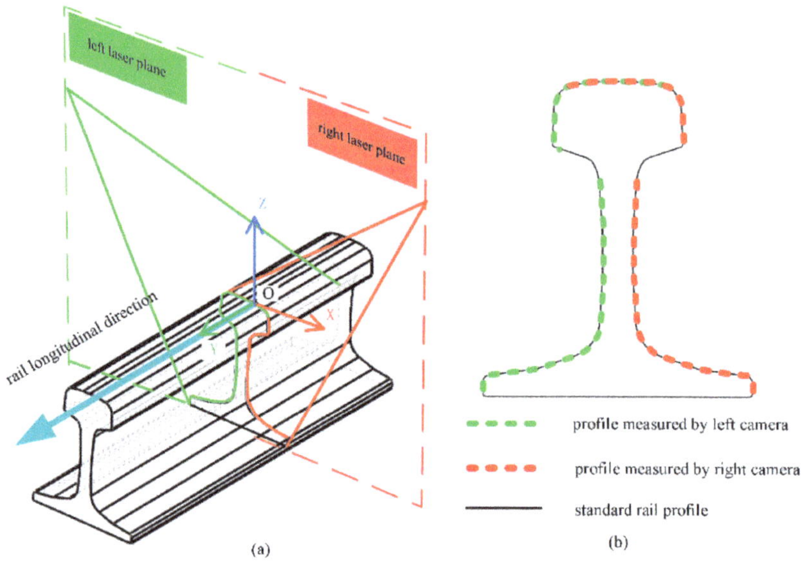

Figure 2. Schematic diagram of coplanar lasers on both sides: (**a**) coplanar lasers and (**b**) profile measurement results.

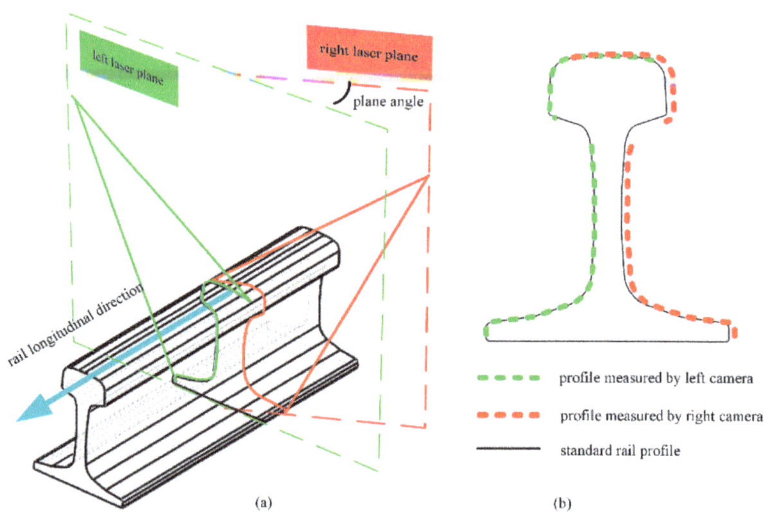

Figure 3. Schematic diagram of non-coplanar lasers on both sides: (**a**) non-coplanar lasers and (**b**) profile measurement results.

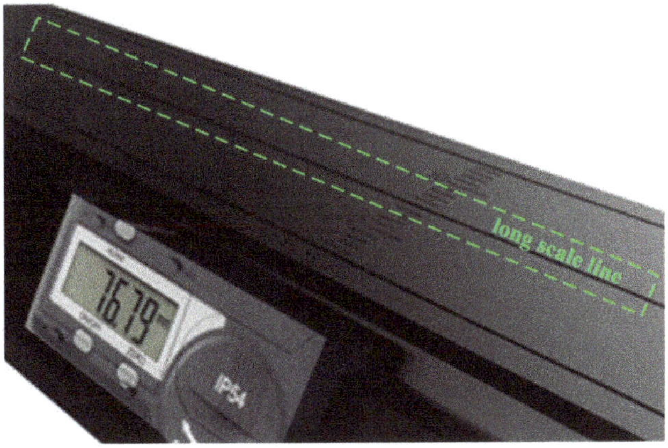

Figure 4. Aluminum alloy ruler.

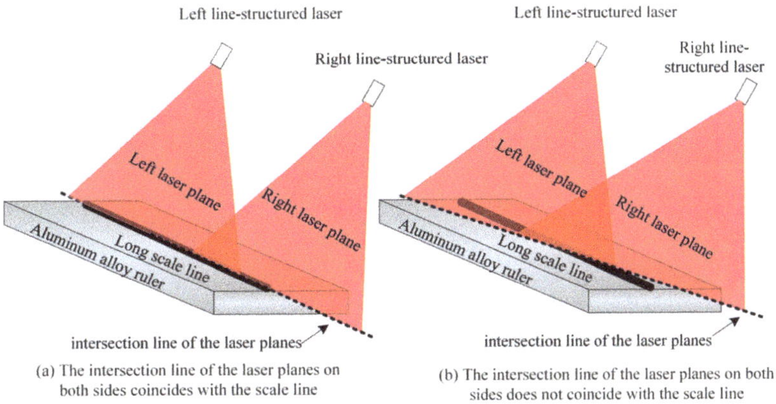

Figure 5. Schematic diagram of the traditional laser plane evaluation method for rail profile measurement system.

3. Materials and Methods

As can be seen from the analysis above, the existing method is based on visual observation of whether the lasers on both sides are coplanar; however, there are many uncertainties involved in the evaluation process, which cannot guarantee the accuracy of calibration. Therefore, this study proposes a method for evaluating laser coplanarity based on fitting planes, which can serve as a guide for the adjustment of laser coplanarity on both sides. Figure 6 illustrates the visualized laser plane adjustment device for the full-section rail profile measurement system. It contains an additional target compared to the existing full-section rail profile measurement device. The special target consists of a convex calibration block and three planar calibration plates. There are three upper surfaces on the convex calibration block, and the upper surface in the center is higher than the upper surfaces on both sides. As shown in Figure 7, three calibration plates are placed on the three upper surfaces, which are numbered 1, 2, and 3, from left to right. Consequently, the corresponding target coordinate systems, *tcs*1, *tcs*2, and *tcs*3, are established, with the center of each calibration plate as the origin and the target plane as the XOY plane.

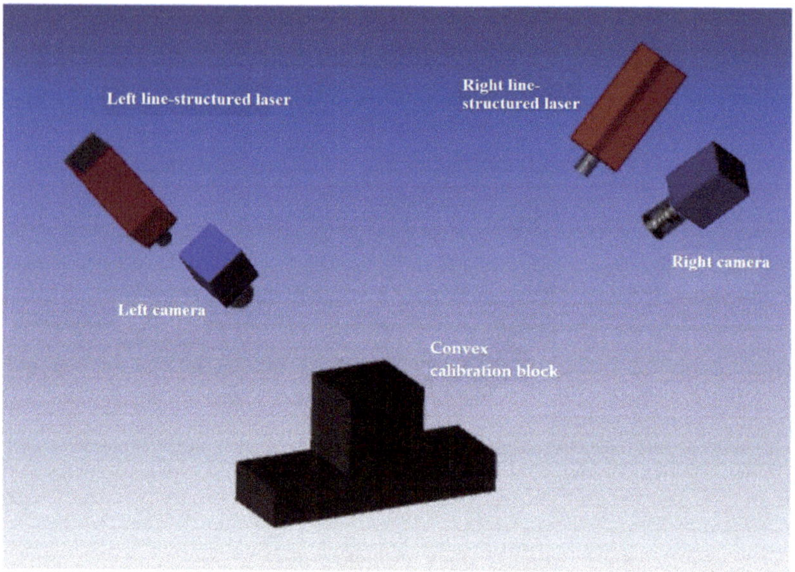

Figure 6. The visualized laser plane adjustment device for the full-section rail profile measurement system.

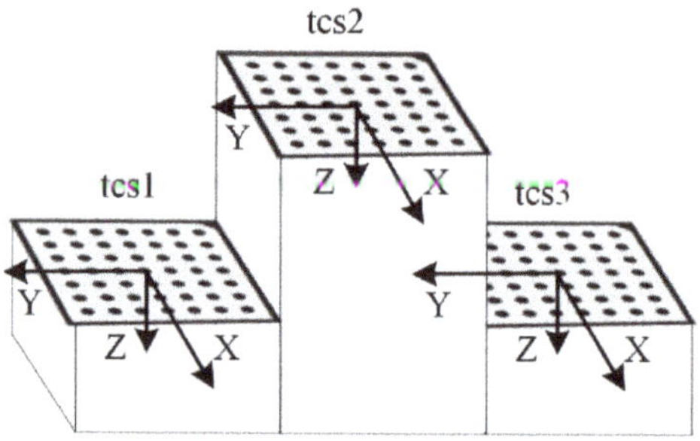

Figure 7. The coordinate system, which is composed of the convex calibration block and three planar targets.

An overview of the laser plane attitude evaluation for the full-section rail profile measurement system is presented in Figure 8. This system consists of a system calibration module, an image acquisition module, and a coplanarity evaluation module. The specific realization process for each module is described in more detail below.

The system calibration module is responsible for obtaining the internal and external parameters of the cameras. The calibration method previously described [19] is employed in this study to simultaneously collect images of the planar target calibration plates in different attitudes through the left and right cameras in order to obtain the cameras' internal parameters.

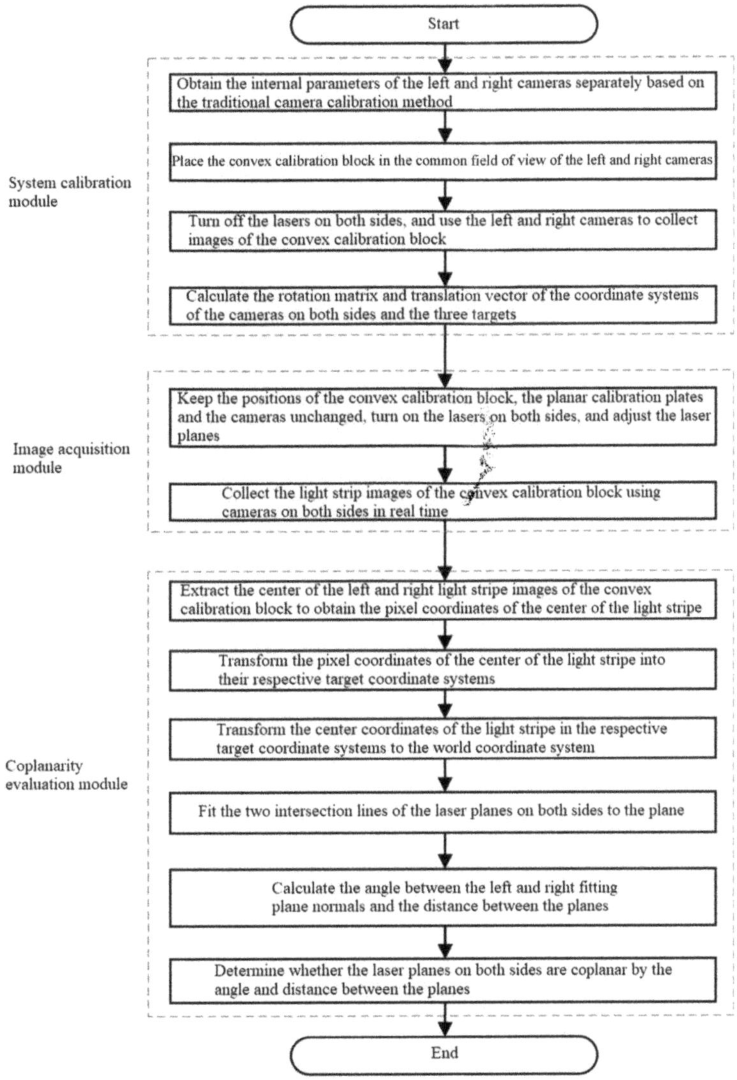

Figure 8. Flowchart of laser coplanarity evaluation of the full-section rail profile measurement system.

The convex calibration block is located in the common field of view of the left and right cameras, ensuring that the left laser plane intersects the target planes of $tcs1$ and $tcs2$, and the right laser plane intersects the target planes of $tcs2$ and $tcs3$. After the lasers on both sides are turned off, the left and right cameras are used to capture images of the convex calibration block. Due to occlusion, as shown in Figure 9, the left camera can capture the entire calibration plates No. 1 and No. 2, but only a portion of calibration plate No. 3, and the right camera can capture the entire calibration plates No. 2 and No. 3, but only a portion of calibration plate No. 1. The coordinate systems of the left and right cameras are expressed as $ccs1$ and $ccs2$, respectively, and R_{tcs1}^{ccs1}, t_{tcs1}^{ccs1} represent the rotation matrix and translation vector of the coordinate systems $ccs1$ and $tcs1$. Since the internal parameters of the cameras are known, based on the camera calibration method as previously described, the rotation matrix, R_{tcs1}^{ccs1}, R_{tcs2}^{ccs1}, and translation vector, t_{tcs1}^{ccs1}, t_{tcs2}^{ccs1}, of the coordinate systems from the left camera *ccs*1 to No. 1 and No. 2 targets, as well as the rotation matrix, R_{tcs2}^{ccs2}, R_{tcs3}^{ccs2}, and translation vector, t_{tcs2}^{ccs2}, t_{tcs3}^{ccs2}, of the coordinate systems from the right camera *ccs*2 to No. 2 and No. 3 targets are calculated based on the images of the convex calibration block.

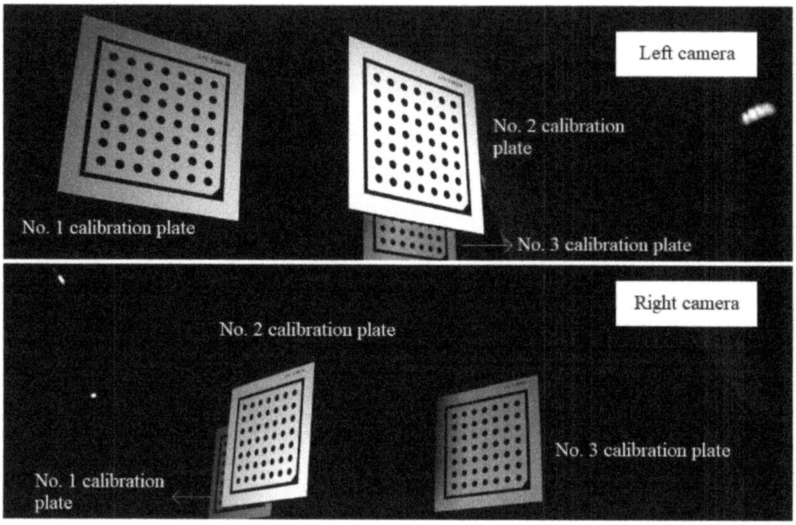

Figure 9. The images of the convex calibration block captured by the left and right cameras.

The image acquisition module is used to acquire the real-time light stripe images of the convex calibration block during the process of adjusting the laser planes. During the calibration process, the positions of the convex calibration block, the planar calibration plates, and the cameras remain unchanged, while the lasers on both sides are turned on and adjusted as required. With a suitable exposure time, the cameras on both sides are used to collect the light strip images of the convex calibration block in real time. The light strip image sequence of the convex calibration block is denoted as follows:

$$I = \{I_{ij}/i = 1,2, j = 1,2,3\ldots n\} \tag{1}$$

where $i = 1$ is the light stripe image of the convex calibration block collected by the left camera, $i = 2$ is the light stripe image of the convex calibration block collected by the right camera, and n is the number of their respective light stripe images. The intersection line of the laser planes and the calibration plates form a light stripe, as shown in Figure 10. The light stripe image of the convex calibration block collected by the left camera shows the intersection lines between the left laser plane and the planar calibration plates No. 1 and No. 2, which are designated as *l*1 and *l*2, respectively. The light stripe image of the convex calibration block collected by the right camera shows the intersection lines between the right laser plane and the planar calibration plates No. 2 and No. 3, which are denoted as *r*2 and *r*3, respectively.

Based on the system calibration parameters and the convex calibration block light stripe image, the coplanarity evaluation module calculates the parameters of the left and right laser planes in order to determine whether the lasers on both sides are coplanar. The coplanarity evaluation module operates in six steps, according to the data processing flow.

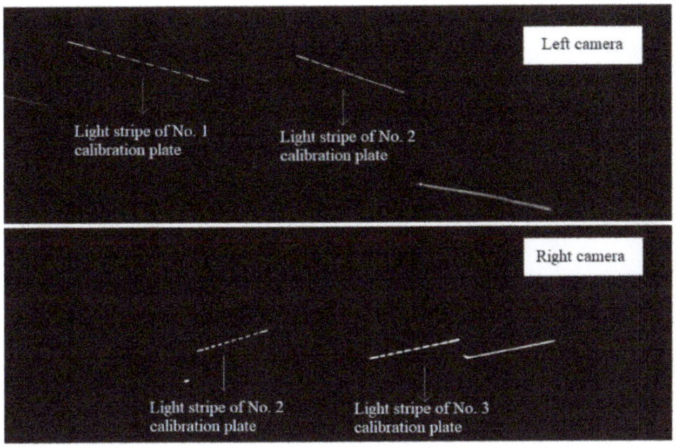

Figure 10. The left and right light stripe images of the convex calibration block.

Step 1: The pixel coordinates of the light stripe centers are obtained by extracting the centers of the left and right light stripe images of the convex calibration block. In Figure 10, the light strip images of the convex calibration block collected by the cameras on both sides are shown, and they indicate the intersection lines *l*1, *l*2, *r*2, and *r*3. As follows, the light stripe centers are extracted using traditional algorithms (such as maximum value methods, grayscale center-of-gravity methods, Steger methods, template-matching methods, etc.), and the pixel coordinates of the left and right light stripe centers are obtained.

$$P_i = \left\{ (u_i, v_i)^T | 0 \le u \le width - 1, 0 \le v \le height - 1, i = 1,2 \right\} \tag{2}$$

where the width is the image width, and the height is the image height; P_1 is any point on the center of the left light stripe image; and P_2 is any point on the center of the right light stripe image.

Step 2: The pixel coordinates of the left and right light stripe image centers of the convex calibration block are converted into the corresponding target coordinate system. Different calibration plates correspond to different external parameters. The coordinate transformation process is described below, using the light strip image of the convex calibration block captured by the left camera in Figure 10 as an example. First, as shown in Figure 10, the light stripe *l*1 of the calibration plate No. 1 and the light stripe *l*2 of the calibration plate No. 2 are located. Then, for the pixel coordinate of the light stripe center of the calibration plate No. 1, $(u_i, v_i)^T$, the pixel coordinate of the center of the light stripe *l*1, $(u_i, v_i)^T$, is transformed into the coordinate system of the No. 1 target according to the internal and the external parameters (R_{tcs1}^{ccs1} and t_{tcs1}^{ccs1}) of the camera. Similarly, the pixel coordinate of the center of the light stripe *l*2, (u_i, v_i), is transformed into the coordinate system of the No. 2 target according to the internal and external parameters (R_{tcs2}^{ccs1} and t_{tcs2}^{ccs1}) of the camera. A similar coordinate transformation is applied to the light stripe image of the convex calibration block captured by the right camera. As a result, we have the coordinate of the intersection line *l*1 in the coordinate system of target No. 1, *tcs*1 ($P_{tcs1}^{l1} = \left(x_{tcs1}^{l1}, y_{tcs1}^{l1}, z_{tcs1}^{l1}\right)^T$); the coordinate of the intersection line *l*2 in the coordinate system of target No. 2, *tcs*2 ($P_{tcs2}^{l2} = \left(x_{tcs2}^{l2}, y_{tcs2}^{l2}, z_{tcs2}^{l2}\right)^T$); the coordinate of the intersection line *r*2 in the coordinate system of target No. 2, *tcs*2 ($P_{tcs2}^{r2} = \left(x_{tcs2}^{r2}, y_{tcs2}^{r2}, z_{tcs2}^{r2}\right)^T$); and the coordinate of the intersection line *r*3 in the coordinate system of target No. 3, *tcs*3 ($P_{tcs3}^{r3} = \left(x_{tcs3}^{r3}, y_{tcs3}^{r3}, z_{tcs3}^{r3}\right)^T$).

Step 3: The coordinate system of target No. 2 is regarded as the world coordinate system, *wcs*, and the coordinates of the light stripe centers in the respective target coordinate systems are transformed into the world coordinate system. The transformation relationship between the coordinate systems $tcs2$, $tcs1$, and $tcs3$ is calculated. The rotation matrix, R_{tcs1}^{tcs2}, and translation vector, t_{tcs1}^{ccs2}, from the target coordinate system $tcs2$ to the target coordinate system $tcs1$ are calculated by Equation (3):

$$\begin{cases} R_{tcs1}^{tcs2} = \left(R_{tcs2}^{ccs1}\right)^{-1} \cdot R_{tcs1}^{ccs1} \\ t_{tcs1}^{tcs2} = \left(R_{tcs2}^{ccs1}\right)^{-1} \cdot \left(t_{tcs1}^{ccs1} - t_{tcs2}^{ccs1}\right) \end{cases} \tag{3}$$

The rotation matrix and translation vector from the target coordinate system $tcs2$ to the target coordinate system $tcs3$ are calculated by Equation (4):

$$\begin{cases} R_{tcs3}^{tcs2} = \left(R_{tcs2}^{ccs2}\right)^{-1} \cdot R_{tcs3}^{ccs2} \\ t_{tcs3}^{tcs2} = \left(R_{tcs2}^{ccs2}\right)^{-1} \cdot \left(t_{tcs3}^{ccs2} - t_{tcs2}^{ccs2}\right) \end{cases} \tag{4}$$

The world coordinates of the light stripe center on the intersection lines $l2$ and $r2$ are the same as the coordinates in the target coordinate system, that is, $P_{wcs}^{l2} = P_{tcs2}^{l2}$ and $P_{wcs}^{r2} = P_{tcs2}^{r2}$. The coordinates of the intersection lines $l1$ and $r3$ in the world coordinate system, wcs, namely $P_{wcs}^{l1} = \left(x_{wcs}^{l1}, y_{wcs}^{l1}, z_{wcs}^{l1}\right)^T$ and $P_{wcs}^{r3} = \left(x_{wcs}^{r3}, y_{wcs}^{r3}, z_{wcs}^{r3}\right)^T$, are obtained by Equations (5) and (6), respectively. Similarly, we have the coordinate of the intersection line $l2$ in the world coordinate system ($P_{wcs}^{l2} = \left(x_{wcs}^{l2}, y_{wcs}^{l2}, z_{wcs}^{l2}\right)^T$), the coordinate of the intersection line $r2$ in the world coordinate system ($P_{wcs}^{r2} = \left(x_{wcs}^{r2}, y_{wcs}^{r2}, z_{wcs}^{r2}\right)^T$), and the coordinate of the intersection line $r3$ in the world coordinate system ($P_{wcs}^{r3} = \left(x_{wcs}^{r3}, y_{wcs}^{r3}, z_{wcs}^{r3}\right)^T$).

$$\begin{bmatrix} P_{wcs}^{l1} \\ 1 \end{bmatrix} = \begin{bmatrix} x_{wcs}^{l1} \\ y_{wcs}^{l1} \\ z_{wcs}^{l1} \\ 1 \end{bmatrix} = \begin{bmatrix} R_{tcs1}^{tcs2} & t_{tcs1}^{tcs2} \\ 0 \quad 0 \quad 0 & 1 \end{bmatrix} \begin{bmatrix} P_{tcs1}^{l1} \\ 1 \end{bmatrix} = \begin{bmatrix} R_{tcs1}^{tcs2} & t_{tcs1}^{tcs2} \\ 0 \quad 0 \quad 0 & 1 \end{bmatrix} \begin{bmatrix} x_{tcs1}^{l1} \\ y_{tcs1}^{l1} \\ z_{tcs1}^{l1} \\ 1 \end{bmatrix} \tag{5}$$

$$\begin{bmatrix} P_{wcs}^{r3} \\ 1 \end{bmatrix} = \begin{bmatrix} x_{wcs}^{r3} \\ y_{wcs}^{r3} \\ z_{wcs}^{r3} \\ 1 \end{bmatrix} = \begin{bmatrix} R_{tcs3}^{tcs2} & t_{tcs3}^{tcs2} \\ 0 \quad 0 \quad 0 & 1 \end{bmatrix} \begin{bmatrix} P_{tcs3}^{r3} \\ 1 \end{bmatrix} = \begin{bmatrix} R_{tcs3}^{tcs2} & t_{tcs3}^{tcs2} \\ 0 \quad 0 \quad 0 & 1 \end{bmatrix} \begin{bmatrix} x_{tcs3}^{r3} \\ y_{tcs3}^{r3} \\ z_{tcs3}^{r3} \\ 1 \end{bmatrix} \tag{6}$$

Step 4: In the world coordinate system, the two intersection lines ($l1$ and $l2$) of the left laser plane and the two intersection lines ($r2$ and $r3$) of the right laser plane are sequentially fitted to the plane, and the parameters of the left and right laser planes are obtained. The fitting process is described below, taking the intersection lines $l1$ and $l2$ as an example.

In $P_{wcs}^{i} = \left(x_{wcs}^{i}, y_{wcs}^{i}, z_{wcs}^{i}\right)^T, i = 1, 2, 3 \ldots k$ is any point on the intersection lines $l1$ and $l2$ of the left laser plane and the calibration plates No. 1 and No. 2, where $k = m + n$, and m and n are the number of points on the intersection lines $l1$ and $l2$, respectively. The following matrix is thus constructed:

$$M = \begin{bmatrix} x_{wcs}^{1} - \overline{x}, x_{wcs}^{1} - \overline{y}, x_{wcs}^{1} - \overline{z} \\ \vdots \\ x_{wcs}^{k} - \overline{x}, y_{wcs}^{k} - \overline{y}, x_{wcs}^{1} - \overline{z} \end{bmatrix} \tag{7}$$

where $\overline{x} = \frac{1}{k}\sum\limits_{i=1}^{k} x_{wcs}^{i}, \overline{y} = \frac{1}{k}\sum\limits_{i=1}^{k} y_{wcs}^{i}, \overline{z} = \frac{1}{k}\sum\limits_{i=1}^{k} z_{wcs}^{i}$. The point $\overline{y}(\overline{x}, \overline{y}, \overline{z})^T$ is designated as the center of gravity of the plane. If $S = M^T \cdot M$, where S has three eigenvalues, then the eigenvector corresponding to the smallest eigenvalue is the normal of the fitting plane, $\boldsymbol{n}^l$. The left laser plane is constructed with $(\overline{x}, \overline{y}, \overline{z})^T$ as a point on the plane and the vector $\boldsymbol{n}^l$ as the normal. Similarly, the intersection lines $r2$ and $r3$ of the right laser plane are fitted to the plane to obtain the normal of the right laser plane, $\boldsymbol{n}^r$, and the right laser plane is constructed.

Step 5: The angle between the normals of the fitting planes on the left and right sides (α) and the distance between the planes (d) are calculated. The angle, α, can be calculated by Equation (8):

$$\alpha = \arccos\left(\frac{\boldsymbol{n}^l \cdot \boldsymbol{n}^r}{|\boldsymbol{n}^l||\boldsymbol{n}^r|}\right) \tag{8}$$

In order to calculate the distance between the two planes, the corresponding plane coordinate systems $pcs1$ and $pcs2$ are established with the center of gravity of the fitting planes on the left and right sides as the origin and the normal direction as the Z axis. The rotation of the coordinate system around the Z axis will not affect the direction of the plane normal in the plane coordinate system. For simplicity, the rotation around the Z axis is set to 0 here. The rotation matrix and translation vector from the plane coordinate system $pcs1$ to the world coordinate system (wcs) are denoted as R_{wcs}^{pcs1} and t_{wcs}^{pcs1}, and the rotation matrix and translation vector from the world coordinate system (wcs) to the plane coordinate system $pcs2$ are expressed as R_{pcs2}^{wcs} and t_{pcs2}^{wcs}. As a result, the corresponding homogeneous transformation matrices are constructed, as shown below.

$$H_{wcs}^{pcs1} = \begin{bmatrix} R_{wcs}^{pcs1} & t_{wcs}^{pcs1} \\ 0 \;\; 0 \;\; 0 & 1 \end{bmatrix} \tag{9}$$

$$H_{pcs2}^{wcs} = \begin{bmatrix} R_{pcs2}^{wcs} & t_{pcs2}^{wcs} \\ 0 \;\; 0 \;\; 0 & 1 \end{bmatrix} \tag{10}$$

Then the homogeneous transformation matrix of the coordinate system $pcs1$ and $pcs2$ can be expressed as follows:

$$H_{pcs2}^{pcs1} = H_{wcs}^{pcs1} \cdot H_{pcs2}^{wcs} \tag{11}$$

Therefore, for any point in the coordinate system $pcs2$, $P_{pcs2} = (x_{pcs2}, y_{pcs2}, z_{pcs2})^T$, it can be transformed into the coordinate system $pcs1$ by Equation (12), where $P_{pcs1} = (x_{pcs1}, y_{pcs1}, z_{pcs1})^T$ is the corresponding coordinate of the point in the coordinate system $pcs1$.

$$\begin{bmatrix} P_{pcs1} \\ 1 \end{bmatrix} = \begin{bmatrix} x_{pcs1} \\ y_{pcs1} \\ z_{pcs1} \\ 1 \end{bmatrix} = H_{pcs2}^{pcs1} \cdot P_{pcs2} = H_{pcs2}^{pcs1} \begin{bmatrix} x_{pcs2} \\ y_{pcs2} \\ z_{pcs2} \\ 1 \end{bmatrix} \tag{12}$$

N points on the right laser plane are arbitrarily taken, as shown in Equation (13), and transformed into the coordinate system $pcs1$ through Equation (9). Then, we have Equation (14).

$$P_{pcs2}^i = \left(x_{pcs2}^i, y_{pcs2}^i, z_{pcs2}^i\right)^T, i = 1,2,3\ldots N \tag{13}$$

$$P_{pcs1}^i = \left(x_{pcs1}^i, y_{pcs1}^i, z_{pcs1}^i\right)^T, i = 1,2,3\ldots N \tag{14}$$

Since the fitting plane on the left coincides with the XOY plane of the coordinate system $pcs1$, $\left|z_{pcs1}^i\right|, i = 1,2,3\ldots N$ is the distance from these N points to plane 1, and the distance (d) from plane 2 to plane 1 can be expressed as follows:

$$d = \frac{1}{N}\sum_{i=1}^{i=N}\left|z_{pcs1}^i\right| \tag{15}$$

Step 6: Both the distance between the planes and the angle between the normals determine whether the lasers on both sides are coplanar. If the angle between the two planes is 0, it means that the two planes are parallel or coincident. If the distance from any point on one of the planes to the other plane is 0, the two planes coincide. To eliminate the influence of errors, two parameters are used to determine whether the lasers on both sides are coplanar, namely the angle (α) and the distance (d). As long as the angle and distance satisfy Equation (16), the coincidence degree of the left and right laser planes is high, and the laser planes on both sides of the rail are appropriate.

$$d \leq T_d \ \& \ \alpha \leq T_\alpha \tag{16}$$

where T_d, T_α are the thresholds of the distance (d) and the angle (α), which can be determined in accordance with the accuracy requirements.

When the laser planes are adjusted, the left and right cameras collect real-time light stripe images of the convex calibration block and obtain the sequence of light stripes, as shown in Equation (2). Then, the coplanarity evaluation module processes the light strip image sequence of the convex calibration block in real time, calculates the laser plane parameters on both sides of the rail, draws the laser planes on both sides in real time in the display window, and displays the angle and distance between the normals. By utilizing Equation (16), the module determines if the laser planes on both sides of the rail are appropriate, allowing the full-section rail profile measurement system to visualize the alignment of both laser planes and effectively determine if the lasers are coplanar on both sides. The effect is shown in Figure 11.

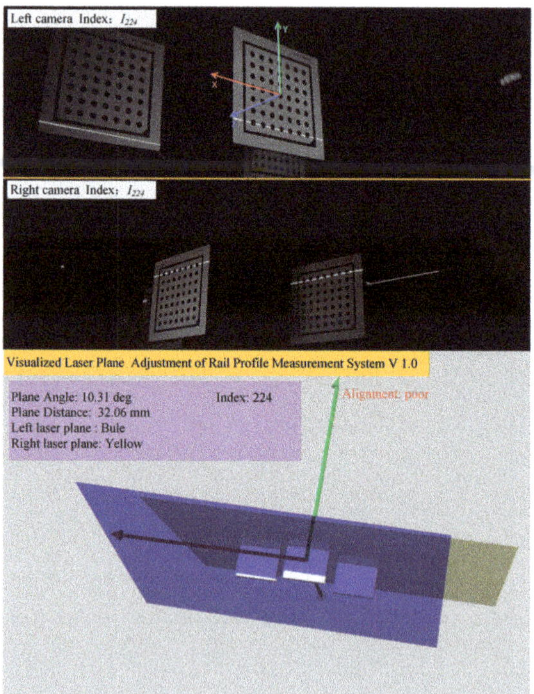

Figure 11. Schematic diagram of the visualized adjustment of the laser planes on both sides of the rail.

4. Results and Discussion

4.1. Experimental Design

To confirm the advantages of the proposed laser coplanarity evaluation method, such as a strong real-time performance, high accuracy, and visualization, a laser coplanarity evaluation experiment was designed, and the experimental device shown in Figure 12 was constructed. Among them, the camera is the Ranger 3 high-speed camera produced by the German SICK company (Waldkirch, Germany), with a resolution of 2560 × 832 pixels, and the line laser is produced by the Canadian Osela company, with a wavelength of 660 nm and 450 nm, respectively. The laptop used for test is ThinkPad T480, Intel®, Core™, i7-8550U CPU@1.80GHz, 1.99 GHz, 32.0 GB memory, 64-bit operating system.

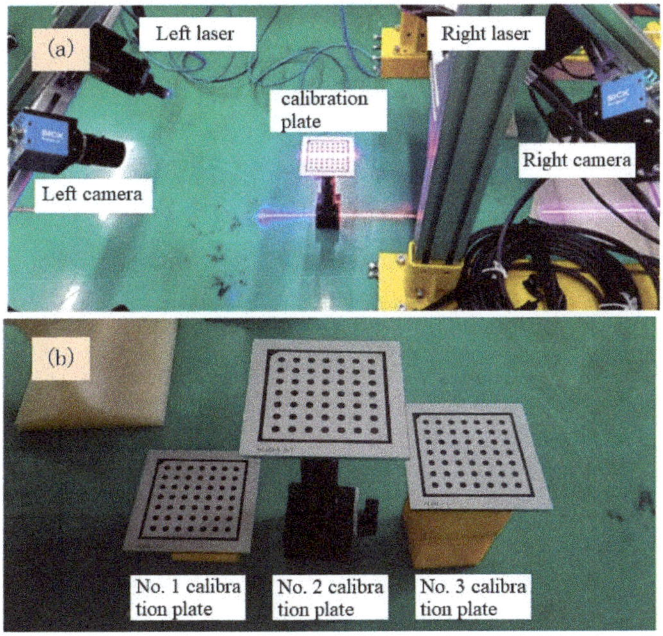

Figure 12. Experimental device for laser coplanarity evaluation: (**a**) calibration of the internal parameters of the cameras and (**b**) simplified convex calibration block and calibration plates.

First, the internal parameters of the left and right cameras are calibrated based on the planar target, and then the simplified convex calibration block is placed in the common field of view of the two cameras, ensuring that the left laser plane intersects the calibration plates No. 1 and No. 2, and the right laser plane intersects the calibration plates No. 2 and No. 3. Then, the lasers on both sides are turned off, and the images of the calibration plates are collected through the cameras on both sides. The result is shown in Figure 9. After that, the lasers on both sides are turned on, and the light strip images of the three calibration plates are collected in real time through the cameras on both sides. The result is shown in Figure 10. The parameters of the laser planes on both sides are obtained in real time from the cross-sectional laser images of the three calibration plates, and the two laser planes are drawn in real time in the program window. The lasers on both sides are adjusted according to the two laser planes displayed in real time in the program window until they satisfy the coplanarity standard. The following three groups of experiments were carried out:

The first group of experiments is a real-time verification experiment. On the basis of the abovementioned experimental device, the laser planes on both sides are continuously adjusted, and two cameras collect 500 target images each in real time. At the same time, the coplanar degree of the two laser planes is calculated by the proposed method, and the real-time performance of the proposed method is evaluated by the time of program brushing the new window.

The second group of experiments is the accuracy verification experiment. In this experiment, the accuracy of the proposed method is evaluated by the laser plane fitting error, laser plane distance measurement error, and laser plane angle measurement error. First, in the process of laser plane adjustment on both sides, 500 consecutive target images are collected by two cameras, corresponding to 500 × 2 = 1000 laser plane positions; therefore, the accuracy of the proposed method is evaluated with these 1000 laser plane fitting errors. As shown in Figure 13, we kept the camera and two targets of different heights stationary and placed the line laser on a precision displacement table and a rotating table. Then, the controlled line laser moved from 0 mm translational motion to 10 mm, with a step length of 1 mm. The light stripe images of two targets were collected at each position. With the position of 0 mm as the reference, the proposed method was used to calculate the distance between the laser plane at the other 10 different positions and the laser plane at the reference position. Finally, controlling the line laser, we rotated from 0 deg to 2 deg, with a step size of 0.200 deg. Similarly, two target light stripe images were collected at each position, and the angle between the remaining ten different laser planes and the reference position laser plane was calculated using the proposed method, using the 0 deg position as a reference. In order to compare the measurement accuracy of the rail profile after laser coplanar adjustment, a comparative experiment was designed using a standard worn rail as the measurement object. The standard worn rail are shown in Figure 14, with a vertical wear of 11.00 mm. When the laser planes of two laser scan sensors are not coplanar (in order to highlight the effect, the angle between the laser planes is about 3 deg), collect the full-section profile of the steel rail 20 times and calculate its vertical wear. Under the guidance of the proposed method, two laser scan sensors were adjusted to meet the installation requirements of coplanarity. The full profile of the steel rail was collected 20 times again, and its vertical wear was calculated.

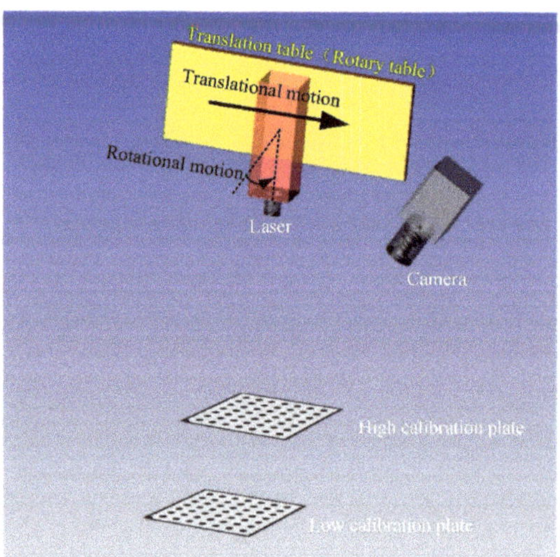

Figure 13. Schematic diagram of accuracy verification experiment.

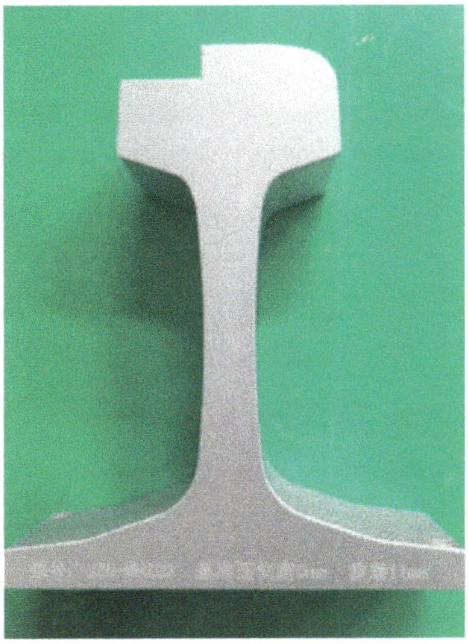

Figure 14. Standard worn rail with a vertical wear of 11.00 mm. (NO. JZG-MHZ02, horizontal wear 0 mm, vertical wear 11 mm).

The third group of experiments is the repeatability verification experiment. Keep the measuring device stationary, continuously collect 500 pairs of target images through two cameras, and also evaluate the repeatability of the proposed method with the fitting error of the laser plane.

4.2. Experimental Result

4.2.1. Real-Time Verification Experiment Results

Figure 15 exhibits the laser planes at three typical positions during the adjustment process. It can be seen that the program window displays the laser planes on both sides in real time and calculates the current coplanarity of the laser planes on both sides in real time, according to the standard. Therefore, as opposed to traditional methods that rely on visually observing laser beams, the proposed method allows for the visualization of the three-dimensional laser plane in real time, thus avoiding blindness caused by observing two-dimensional laser beams. Figure 16 shows the corresponding refresh window time of the two cameras. The average refresh time is 0.03 s, and the frame rate is about 33 frames/second, which can fully meet the real-time requirements.

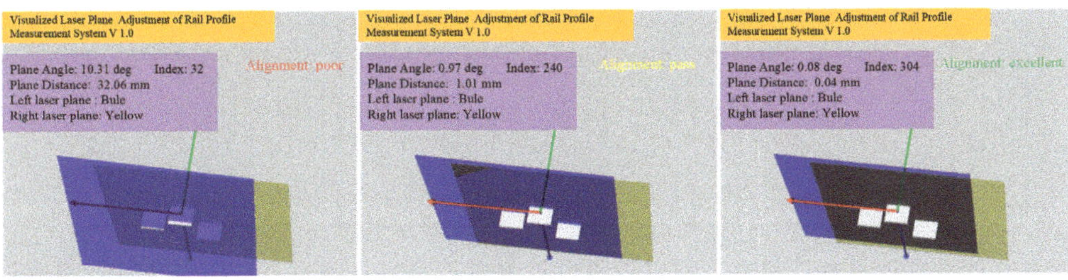

Figure 15. The laser planes at three typical positions.

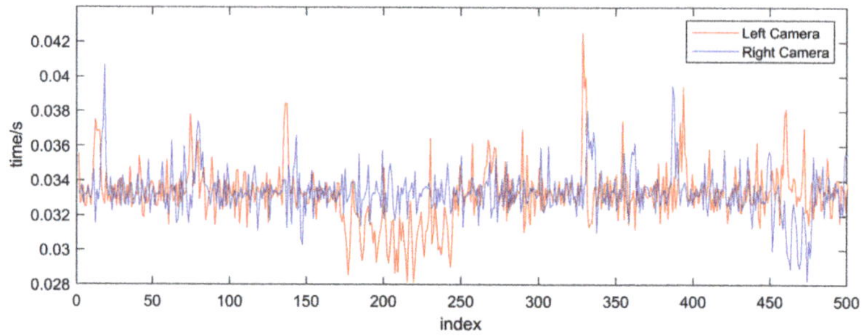

Figure 16. The time for the program to refresh the window.

4.2.2. Accuracy Verification Experiment Results

The system calibration module calculates the transformation matrix between the camera coordinate system and the target coordinate system through a single planar target image, with known camera internal parameters. Figure 17 shows the reprojection error diagram. The average reprojection errors of the No. 1 calibration plate and the No. 2 calibration plate images obtained by the left camera are 0.037 pixel and 0.025 pixel, respectively. The average reprojection errors of the images of the No. 2 calibration plate and the No. 3 calibration plate obtained by the right camera are 0.026 pixel and 0.027 pixel, respectively, and the projection errors are within a reasonable range.

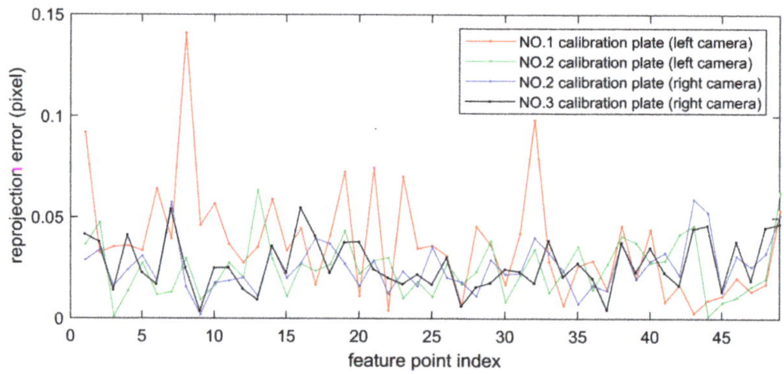

Figure 17. Reprojection errors of three calibration plates.

The laser plane fitting results and statistical values are given in Figure 18 and Table 1, respectively. It can be seen that, during the random change of the laser plane attitude, the average values of the laser plane fitting errors on both sides obtained by the proposed method are 0.062 mm and 0.054 mm, respectively, which are smaller than the laser plane calibration error of the rail profile measurement sensor.

Table 1. Statistical results of the laser plane fitting error (mm).

Camera	Average Value	Standard Deviation	Maximum
Left camera	0.062	0.017	0.119
Right camera	0.054	0.017	0.121

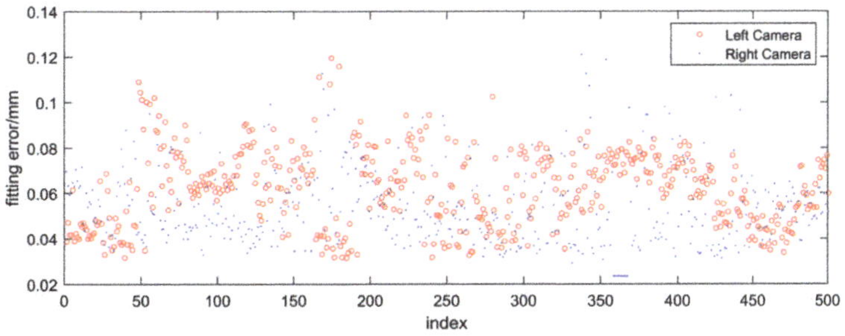

Figure 18. The results of laser plane fitting error.

The results of laser plane distance and angle measurement are shown in Table 2, where MV represents the measured value, AV represents the actual value, and ME represents the measurement error. It can be seen that when the laser plane is moving in a translational motion, the maximum value of the distance measurement error of the laser plane is 0.13 mm, and the average error is 0.09 mm. When the laser plane is rotating, the maximum value of the angle measurement error of the laser plane is 0.019 deg, and the average value is 0.009 deg. The measurement errors of plane distance and plane angle are within an acceptable range.

Table 2. Laser plane distance and angle measurement results.

Location Index		1	2	3	4	5	6	7	8	9	10
Distance (mm)	AV	1.00	2.00	3.00	4.00	5.00	6.00	7.00	8.00	9.00	10.00
	MV	1.11	2.09	3.06	3.87	5.06	5.94	7.14	7.93	9.12	10.06
	ME	0.11	0.09	0.06	0.13	0.06	0.06	0.14	0.07	0.12	0.06
Angle (deg)	AV	0.200	0.400	0.600	0.800	1.000	1.200	1.400	1.600	1.800	2.000
	MV	0.206	0.400	0.581	0.807	1.008	1.191	1.415	1.616	1.804	1.991
	ME	0.006	0.000	0.019	0.007	0.008	0.009	0.015	0.016	0.004	0.009

The results of rail wear before and after coplanar adjustment are shown in Figure 19, and Table 3 provides the corresponding error statistics. It can be seen that, after the coplanarity adjustment, the average measurement error of the rail's vertical wear decreased from 0.279 mm to 0.037 mm. Therefore, the proposed method can guide the laser coplanar adjustment process and improve the accuracy of rail profile measurement.

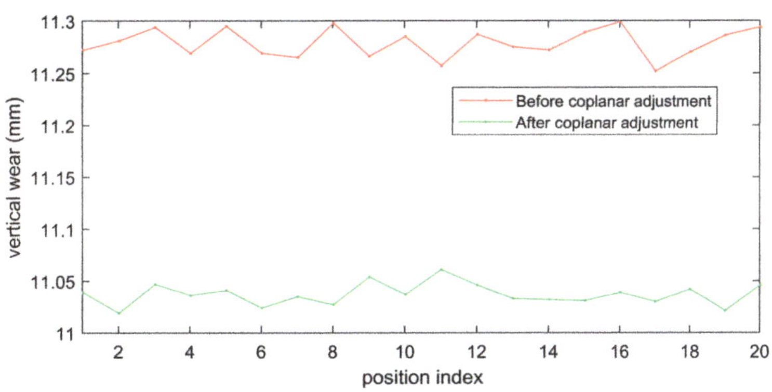

Figure 19. Measurement results of rail wear before and after laser coplanar adjustment.

Table 3. Measurement results of rail wear before and after coplanar adjustment (mm).

Statistical Value	Mean Error	Standard Deviation
Before coplanar adjustment	0.279	0.014
After coplanar adjustment	0.037	0.011

Therefore, the proposed laser coplanar evaluation method has high accuracy. Based on the accurate acquisition of laser plane parameters on both sides, the degree of laser plane on both sides can be quantitatively evaluated, which is superior to the traditional qualitative evaluation method based on visual inspection.

4.2.3. Repeatability Validation Test Results

The repeatability experimental results and statistical values are shown in Figure 20 and Table 4, respectively. It can be seen that when the laser plane is not adjusted, the average values of the fitting errors of the two laser planes obtained by the proposed method are 0.022 mm and 0.021 mm, respectively, and the standard deviations are 0.013 mm and 0.012 mm, respectively, indicating that the method has high repeatability.

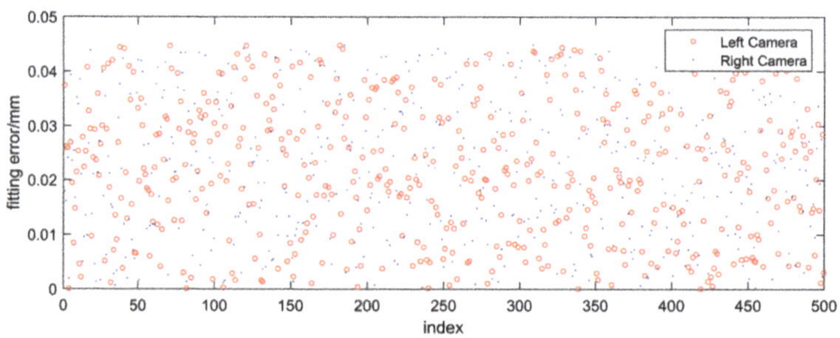

Figure 20. The results of the repeatability verification experiment.

Table 4. Statistical results of the repeatability verification experiment (mm).

Camera	Average Value	Standard Deviation	Maximum
Left camera	0.022	0.013	0.045
Right camera	0.021	0.012	0.044

4.3. Discussions

Based on past experience, when the target area accounts for more than a quarter of the entire image, it can ensure that the target attitude evaluation has high accuracy. Conversely, when the target area accounts for a small proportion in the image, a certain degree of measurement accuracy will be lost. Due to the limitations of the on-site installation environment of the rail profile measurement sensor, larger targets cannot be used, and the target area accounts for significantly less than a quarter, which limits the accuracy of laser planar attitude assessment. Therefore, in response to the accuracy issue caused by the small proportion of target areas, the author intends to perform further research work around target optimization and attitude evaluation algorithm optimization in the next step.

Although using this method can ensure laser coplanar installation on both sides of the rail, the next question is how to ensure the laser plane perpendicular to the longitudinal direction of the rail. To address the issue, it is necessary to obtain the longitudinal direction vector of the rail in the camera coordinate system (or world coordinate system). We proposed a solution to calibrate the rail longitudinal direction in our previous research [8]. By combining the longitudinal vector parameters of the rail and the laser plane parameters, it is possible to further quantitatively evaluate whether the installation of laser scan sensors meets the requirements.

5. Conclusions

Considering the difficulty of evaluating laser plane attitude in the measurement of rail profiles, this paper presents a method for evaluating laser plane attitude using fitting planes, analyzes the impact of non-coplanar lasers on plane profile measurement results, elaborates on how the laser plane attitude calculation is calculated, and constructs the criterion for evaluating laser coplanarity. The experimental results indicate that the proposed method has the advantages of high accuracy, real-time performance, and excellent visualization. This method improves the laser coplanarity adjustment process from the original naked-eye evaluation with low precision, poor real-time performance and subjectivity to computer-based evaluation with high precision, and strong real-time performance and visualization. By doing so, it reduces the calibration error of the traditional calibration method and provides a theoretical basis for improving the accuracy and reliability of rail profile measurement systems. In on-site applications, factors such as changes in ambient light and target posture affect the on-site evaluation efficiency and calibration accuracy of the proposed method. Therefore, it is necessary to improve the evaluation algorithm for railway on-site application scenarios to further improve the efficiency and accuracy of on-site evaluation.

Author Contributions: Conceptualization, L.W. and S.W.; methodology, L.W., G.L. and S.R.; validation, N.W., Y.F. and L.W.; formal analysis, S.W.; investigation, S.W.; resources, H.W.; data curation, S.W.; writing—original draft preparation, L.W.; writing—review and editing, S.W.; visualization, L.W.; supervision, H.W.; project administration, Q.H.; funding acquisition, Q.H. All authors have read and agreed to the published version of the manuscript.

Funding: The work described in this paper was partially supported by Science and technology research and development plan of China National Railway Group Co., Ltd., Grant NO. P2021G014; and Research and Development Plan of China Academy of Railway Sciences Co., Ltd., Grant NO. 2021YJ309.

Institutional Review Board Statement: Not applicable.

Informed Consent Statement: Not applicable.

Data Availability Statement: Data available on request from the authors.

Conflicts of Interest: The authors declare no conflict of interest.

References

1. Ye, J.; Stewart, E.; Clive, R. Use of a 3D model to improve the performance of laser-based railway track inspection. *Proc. Inst. Mech. Eng. Part F J. Rail Rapid Transit* **2018**, *233*, 337–355. [CrossRef]
2. Molleda, J.; Usamentiaga, R.; Millara, A.F.; Garcia, D.F.; Manso, P.; Suarez, C.M.; Garcia, I. A profile measurement system for rail quality assessment during manufacturing. *IEEE Trans. Ind. Appl.* **2016**, *52*, 2684–2692. [CrossRef]
3. Karaduman, G.; Karakose, M.; Aydin, I.; Akin, E. Contactless rail profile measurement and rail fault diagnosis approach using featured pixel counting. *Intell. Autom. Soft Comput.* **2020**, *26*, 455–463. [CrossRef]
4. Li, Y.; Zhong, X.; Ma, Z.; Liu, H. The outlier and integrity detection of rail profile based on profile registration. *IEEE Trans. Intell. Transp. Syst.* **2020**, *21*, 1074–1085. [CrossRef]
5. Fuchs, J.; Müller, G.; Sazgetdinov, K.; Wipfler, E.; Nerlich, I. A methodology for alignment of measured rail profiles in turnouts as a basis for reliable vehicle/track interaction simulations. *Veh. Syst. Dyn.* **2023**, *61*, 821–837. [CrossRef]
6. Wang, S.; Wang, H.; Zhou, Y.; Liu, J.; Dai, P.; Du, X.; Wahab, M.A. Automatic laser profile recognition and fast tracking for structured light measurement using deep learning and template matching. *Measurement* **2020**, *169*, 108362. [CrossRef]
7. Wang, C.; Li, Y.; Ma, Z.; Zeng, J.; Jin, T.; Liu, H. Distortion rectifying for dynamically measuring rail profile based on self-calibration of multiline structured Light. *IEEE Trans. Instrum. Meas.* **2018**, *67*, 678–689. [CrossRef]
8. Wang, L.; Zhou, Q.; Fang, Y.; Wang, S.; Wang, H.; Li, G.; Ren, S.; Dai, P.; Han, Q.; Wang, F. Rail longitudinal calibration method for profile measurement system. *Acta Opt. Sin.* **2021**, *41*, 1012004.
9. Sun, J.; Liu, Z.; Zhao, Y.; Liu, Q.; Zhang, G. Motion deviation rectifying method of dynamically measuring rail wear based on multi-line structured-light vision. *Opt. Laser Technol.* **2013**, *50*, 25–32. [CrossRef]
10. Zhao, X.; Wang, S.; Wang, H.; Xia, C.; Du, X.; Ren, S. Fast tracking algorithm of rail profile under vehicle dynamice conditions. *China Railw. Sci.* **2020**, *41*, 145–155.

11. Feng, K.; Yu, L.; Zhan, D.; Zhang, D. Research on fast and robust matching algorithm in inspection of full cross-section rail profile. *J. China Railw. Soc.* **2019**, *41*, 162–167.
12. Zhan, D.; Wu, X.; Xiang, W.; Xiang, W.; Cao, W.; Tang, L.; Wang, K.; Deng, H. Laser Adjusting Mechanism and Method for 2D Laser Measuring Instrument. China Patent CN111982059A, 20 August 2020.
13. Chen, S.; Wang, Z.; Xiao, G.; Yan, T. A three-dimensional scanning method for large workpiece by monocular structure light. *LaserInfrared* **2018**, *48*, 1358–1362.
14. Wu, B.; Feng, Y.; Yao, P. A Calibration Device for Two Dimensional Laser Displacement Sensor. China Patent CN110793459A, 30 October 2019.
15. Zhang, D.; Li, Q.; Chen, D.J. Coplanar Adjustment System and Method of Laser Profilometer Array. China Patent CN110057314B, 1 April 2019.
16. Ju, B.; Zhu, H.T. Study on line structure light-plane calibration in rail matching. *LaserInfrared* **2017**, *47*, 957–962.
17. Wang, L.; Wang, H.; Wang, F.; Fang, Y.; Han, Q.; Wang, S.; Zhou, Q.; Li, G.; Ren, S.; Wang, N. Correction Method for Laser Noncoplanar Error for Rail Profile Measurement System. *Chin. J. Lasers* **2022**, *49*, 0604002.
18. Sugiyama, H.; Yada, M.; Yamamoto, H.; Kurihara, J.; Ohbayashi, H.; Shimokawa, Y.; Mizuno, M.; Tanimoto, M. Wheel and rail profile wear on small radius curved tracks and its effect on derailment coefficients: Measurement and simulation. *Int. J. Railw. Technol.* **2013**, *2*, 85–98. [CrossRef]
19. Zhang, Z. A flexible new technique for camera calibration. *IEEE Trans. Pattern Anal. Mach. Intell.* **2000**, *22*, 1330–1334. [CrossRef]

sensors

Communication

A High-Precision Real-Time Pose Measurement Method for the Primary Lens of Large Aperture Space Telescope Based on Laser Ranging

Heng Shi *, Junfeng Du, Lihua Wang, Jiang Bian, Guohan Gao, Dun Liu, Bin Fan and Hu Yang

Institute of Optics and Electronics, Chinese Academy of Sciences, Chengdu 610209, China
* Correspondence: shiheng@ioe.ac.cn

Abstract: The aperture of space telescopes increases with their required resolution, and the transmission optical systems with long focal length and diffractive primary lens are becoming increasingly popular. In space, the changes in the pose of the primary lens relative to the rear lens group have a significant impact on the imaging performance of the telescope system. The measurement of the pose of the primary lens in real-time and with high-precision is one of the important techniques for a space telescope. In this paper, a high-precision real-time pose measurement method for the primary lens of a space telescope in orbit based on laser ranging is proposed, and a verification system is established. The pose change of the telescope's primary lens can be easily calculated through six high-precision laser distance changes. The measurement system can be installed freely, which solves the problems of complex system structure and low measurement accuracy in traditional pose measurement techniques. Analysis and experiments show that this method can accurately obtain the pose of the primary lens in real-time. The rotation error of the measurement system is 2×10^{-5} degrees (0.072 arcsecs), and the translation error is 0.2 μm. This study will provide a scientific basis for high-quality imaging of a space telescope.

Keywords: pose measurement; laser ranging; space telescope; high-precision

Citation: Shi, H.; Du, J.; Wang, L.; Bian, J.; Gao, G.; Liu, D.; Fan, B.; Yang, H. A High-Precision Real-Time Pose Measurement Method for the Primary Lens of Large Aperture Space Telescope Based on Laser Ranging. *Sensors* **2023**, *23*, 4833. https://doi.org/10.3390/s23104833

Academic Editors: Michele Norgia, Qibo Feng, Jiakun Li and Qixin He

Received: 23 March 2023
Revised: 8 May 2023
Accepted: 11 May 2023
Published: 17 May 2023

1. Introduction

A membrane optical telescope is a new transmission-type imaging space telescope based on special functional materials [1,2]. Theoretically, it overturns the traditional imaging systems, which rely on the primary lens with a curved surface. Technically, it breaks through the bottleneck of increasing the aperture, such as telescope weight, tolerance control of primary lens surface, and envelope size of primary lens [3,4]. Similar projects, such as the Membrane Optic Imager Real-Time Exploitation (MOIRE) program supported by US Defense Advanced Research Projects Agency (DARPA) [5–8], have been called the 21st-century space disruptive imaging technology.

In space, the relative pose (positions and orientations) of the optical elements, especially between the primary lens and the rear lens group, will change due to the changes in the thermal environment and the structure, which will seriously affect the imaging quality of the telescope [9–11]. Furthermore, it is found that the tilt and decentration of the primary lens will make the telescope imaging quality decrease obviously [12,13]. Therefore, the pose measurement of the telescope primary lens has become difficult to overcome urgently in membrane optical imaging technology [14,15].

Traditional pose measurement techniques usually adopt multiple sets of measuring equipment and sensors, which are complicated in structure and require high installation accuracy [16–18]. It is difficult to meet the pose measurement requirements of the primary lens of a large aperture transmission space telescope, and the calculation process is relatively cumbersome [19–21]. Due to excessive system devices and installation errors, more errors will be introduced [22,23].

This paper presents a high-precision real-time measurement method for the pose of the primary lens of a telescope based on laser ranging. First, the high-precision distance information can be obtained using the laser ranging method, and then the high-precision pose information of the primary lens can be obtained by calculation. Finally, the pose of the primary lens can be calculated by using six distances obtained by laser ranging, and the measurement system can be installed freely, which solves the problems of complex system structure and low measurement accuracy in traditional pose measurement techniques. The system has the advantages of a simple structure, convenient installation, and high measurement accuracy.

2. Telescope System Design

We take a long focal length diffraction imaging optical system as an example; the schematic diagram of the telescope optical system is shown in Figure 1. The diameter of the primary lens is 1 m, the primary lens is a plane, and there are annular microstructures prepared on the surface so that the light can converge on the relay mirror. The relay mirror is 300 mm in diameter, and the relay mirror and the rear lens group are fixed in a frame. The relay mirror is used to fold the optical path and shorten the total length of the system. The rear lens group is responsible for correcting aberrations and providing the final image. The primary lens is extended to the right place by trusses, and the distance between the primary lens and the relay mirror is about 3 m. In order to ensure the imaging quality of the telescope, the Modulation Transfer Function (MTF) @62.5 lp/mm should be up to 0.1. To meet the MTF requirement, the system needs to have good optical quality in terms of resolution, contrast, and uniformity. The MTF is a measure of the ability of the system to transfer contrast from the object to the image, and it depends on many factors, such as the aperture size, aberrations, and diffraction effects, which is a challenging requirement for a system with such a long focal length. Defocus tolerance is important for maintaining the focus of the system over time and under different environmental conditions. The focal depth is proportional to the square of the aperture diameter and the wavelength and inversely proportional to the focal length. Therefore, the defocus of the system should be less than 1/4 focal depth during the whole life cycle [24]. In summary, the long focal length diffraction imaging optical system requires careful design and manufacturing to achieve high MTF and low defocus tolerance. The use of annular microstructures, relay mirrors, and rear lens groups are effective ways to improve the optical performance of the system.

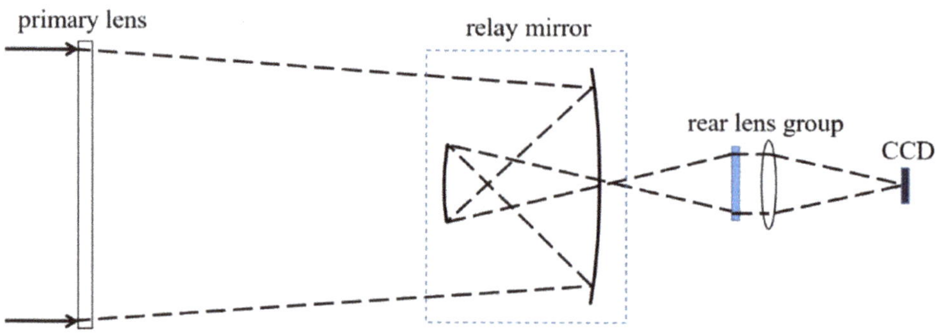

Figure 1. Schematic diagram of the telescope optical system.

As we all know, MTF is a key index in the evaluation criteria of an optical system. The index assigned by the telescope system to the MTF drop caused by the change of the pose of the primary lens is 0.98; that is, the MTF drop caused by the change of the pose of the primary lens is less than 2%. Optical analyses show that the decentration of the primary lens is most sensitive to the telescope optical system. Therefore, a tolerance of 1% is assigned to decentration for MTF drop, 0.5% for distance, and 0.5% for tilt. Through

Monte Carlo simulation, it can be known that in order to ensure that decentration caused MTF drop within 1%, decentration should be controlled within ±0.02 mm. Similarly, the distance should be controlled within ±0.1 mm, and the tilt should be controlled within ±0.02°. According to Shannon's sampling theorem, their corresponding measurement accuracy requirements should be 1/5 of the control requirements. In order to achieve the requirement that MTF drops not more than 2%, the control requirements and measurement accuracy requirements of the pose of the primary lens are provided in Table 1.

Table 1. Influence of each degree of freedom of pose on imaging and measurement accuracy requirements.

Pose Tolerance of Primary Lens	MTF Influence	Tolerance	Measurement Requirements
Distance (translation along the z-axis)	0.5%	±0.1 mm	20 μm
Decentration (translation along the x/y axis)	1%	±0.02 mm	4 μm
Tilt (rotation around x/y axis)	0.5%	±0.02°	0.004°

3. Principle of Measurement Method

3.1. Establishment of Optical Model of Measurement System

The method is derived from the principle of the Stewart Gough platform, where the platform and base are connected by six legs at three points on the platform and six points on the base to form a hexapod geometry platform, as shown in Figure 2. The Stewart Gough platform is a mechanical system that uses actuators to drive changes in the length of legs inducing changes in the pose of the platform. For the pose measurement system, the platform legs are the laser optical paths, which constitute the laser optical paths between the primary lens and the relay mirror. In traditional Stewart Gough platforms, forward kinematics is used to map expected changes in platform pose to changes in leg length. For the pose measurement system of the primary lens of the telescope, the laser ranging lengths are known, and the pose of the primary lens must be determined. This is achieved by using an inverse kinematics technique, which generated a sensitivity matrix used to calculate primary lens pose changes from the laser ranging length changes.

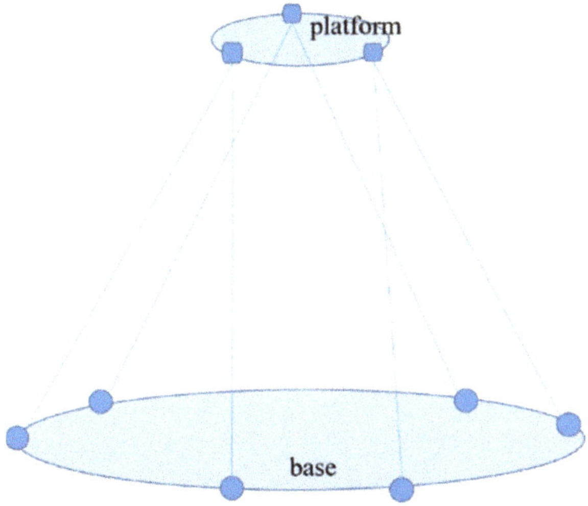

Figure 2. Hexapod geometry Stewart Gough platform.

In this method, six laser collimators are fixed on the edge of the relay mirror, and three retroreflectors are fixed on the edge of the primary lens. Each retroreflector provides reflected light for the corresponding two collimators so that six laser distances can be obtained. According to the changes in the six measured distances and the initial coordinates of the collimators and retroreflectors, the pose of the primary lens can be calculated. The schematic diagram of the measurement principle is shown in Figure 3, where C1~C6 are the laser collimators and A1~A3 are the retroreflectors.

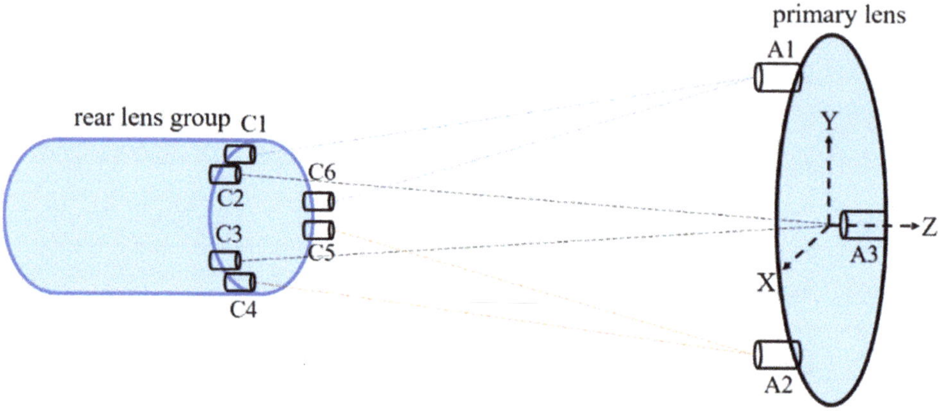

Figure 3. The schematic diagram of the pose measurement principle.

3.2. Pose Calculation Algorithm

The sensitivity matrix is the basis of the pose calculation technique. The sensitivity matrix defines the sensitivity of the pose changes of the primary lens relative to the relay mirror to changes in the laser length. The change in the laser length is defined as the sensitivity matrix multiplied by the changes in the pose

$$\overrightarrow{\Delta l} = S \cdot \overrightarrow{p} \tag{1}$$

where S is the sensitivity matrix, $\overrightarrow{p}$ is the pose vector which includes three translations and three rotations, and $\overrightarrow{\Delta l}$ is the laser length change. This is a simple linear algebraic equation, which is the basis of the Stewart Gough platform pose calculation.

The sensitivity matrix is formed using the coordinates data measured by a laser tracker. The laser path direction vectors must be established first. The coordinates of retroreflectors a_i, and the coordinates of collimators c_i are measured. The direction vectors e_i of the laser are created

$$e_i = \frac{c_i - a_i}{|c_i - a_i|}, (i = 1, \ldots, 6) \tag{2}$$

The sensitivity matrix S is created by cross-multiplying the coordinates of the laser collimators c_i with the direction vectors. The origin of coordinates is located in the center of the primary lens.

$$S = \begin{bmatrix} c_1 \times e_1 & e_1 \\ \vdots & \vdots \\ c_6 \times e_6 & e_6 \end{bmatrix} \tag{3}$$

The sensitivity matrix is a 6×6 matrix that defines the sensitivity of the system to the six degrees of freedom of rotation around and translation along the x-, y-, and z-axis of the primary lens.

After the sensitivity matrix is established, we measure six distances from the retroreflectors to the laser collimators $L0_i$ in the initial state. Then, the six distances from the retroreflectors to the laser collimators in real-time L_i are measured. And the changes in distance are calculated

$$\Delta l_i = L_i - L0_i, (i = 1, \ldots, 6) \quad (4)$$

In forward kinematics, the sensitivity matrix is multiplied by the desired pose to determine the changes in path length. The inverse of the sensitivity matrix is needed to calculate the changes in the pose from the changes in laser path length. The changes in the pose of the primary lens relative to the relay mirror $\vec{p}$ are calculated

$$\vec{p} = S^{-1} \cdot \vec{\Delta l} \quad (5)$$

where S^{-1} is the inverse of the sensitivity matrix, $\vec{\Delta l}$ the changes of laser leg length, which can be obtained by two laser rangefinders.

4. Experiments

4.1. Laser Rangefinders

Laser rangefinders are the key equipment in experiments; the stability of their measured data largely determines the accuracy of experimental results. In addition, they have a long working distance range of up to 50 m, allowing for measurements to be taken over large distances. The laser rangefinders used in the experiment are two sets of IDS3005 laser rangefinders from ATTOCUBE Company (Munich, Germany), and the measurement principle is shown in Figure 4a. One rangefinder weighs less than 2 kg, which makes it a significant advantage for use in space telescopes. With a resolution of 1 pm and a repetition accuracy of 2 nm, one rangefinder has three laser channels and can measure three distances of data simultaneously. When two rangefinders are connected to a computer by a switch, six laser channels can be measured at the same time, which constitutes six laser paths of the measurement system.

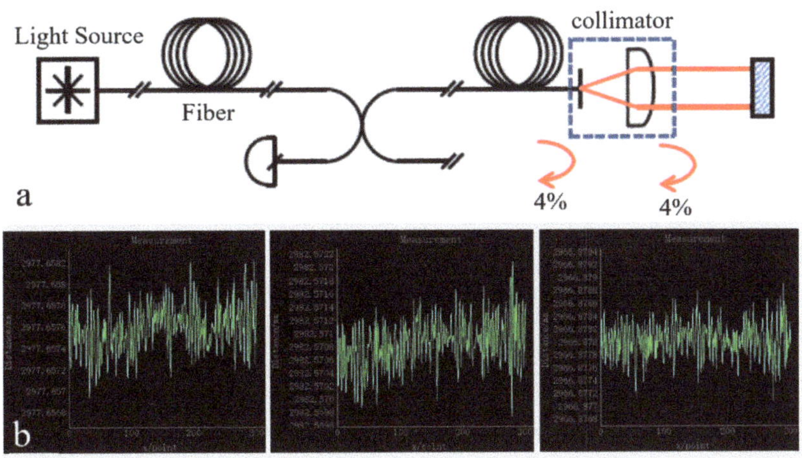

Figure 4. Principle and accuracy test for the laser rangefinder. (**a**) Principle of the laser rangefinder, which is based on Fabry-Perot interference. (**b**) Accuracy test for the laser rangefinder, the laser ranging accuracy is ±0.5 μm.

IDS3005 laser rangefinder adopts the principle of Fabry-Perot interference [25], which has many advantages: The size of the laser collimators can be very small; the beam spacing can be arranged freely; directly measuring the metal surface; directly measuring the Angle by three beams; allowing a large target Angle deviation; little environmental impact; easy installation and debugging; used in vacuum, low temperature, radiation environment. These advantages make the measurement fairly simple. Each channel consists of a telecom fiber that is coupled to a collimator that transmits a beam of laser to the retroreflector. The backward reflection of the fiber tip is the reference arm of the rangefinder, and the measuring arm is the path through the collimator to the retroreflector and back again. The measured optical path difference (OPD) corresponds to the distance between the fiber tip and the retroreflector.

The measurement accuracy of the laser rangefinder was tested. The test was carried out in a standard laboratory environment, where the retroreflectors were fixed at about 3 m away from the collimators of the laser rangefinder, three channels of one laser rangefinder were measured in 5 min, and the measured distance data were recorded by a data recording software, as shown in Figure 4b. According to the measured data, it was determined that the laser rangefinder is accurate to within ± 0.5 μm in such experimental conditions, with little variation between two separate rangefinders. This is consistent with the measurement accuracy data given by the manufacturer. Therefore, the error is acceptable and can meet the experimental requirements. Therefore, we can use these instruments for the pose measurement experiment.

Then, in order to determine the pose measurement accuracy under such laser ranging accuracy, the relationship between the laser ranging accuracy and the pose measurement accuracy was calculated by 10,000 Monte Carlo simulations, and the results are shown in Table 2. This experimental system is highly accurate and precise in terms of rotation around the x and y-axes, rotation around the z-axis, and translation along the x-, y-, and z-axes. For example, in the experimental system, when the laser ranging accuracy is ± 0.5 μm, the accuracy of rotation around the x and y axes (tilt) Rx/Ry is 0.03″, the accuracy of rotation around the z-axis (roll) Rz is 0.07″, the accuracy of translation along the x and y axes (decentration) is 0.35 μm. The accuracy of translation along the z-axis (optical axis direction of the telescope) is 0.15 μm, respectively, indicating highly precise control of the optical axis direction of the telescope. Overall, this experimental system is highly precise and would be ideal for conducting experiments that require highly accurate measurements of position and orientation.

Table 2. Analysis accuracy (σ).

Degree of Freedom	Accuracy
Rx,y	0.03″
Rz	0.07″
x,y	0.35 μm
z	0.15 μm

4.2. Ground Test

In order to verify the pose measurement method and its accuracy, an experimental system was built, as shown in Figure 5. For convenience, a frame was used instead of the primary lens of the telescope and its center was aligned with the center of the relay mirror. At the same time, the frame was fixed to a high-precision hexapod platform to provide an accurate reference pose. Three retroreflectors were attached to the edge of the frame. In addition, six laser collimators were fixed to the edge of the relay mirror and connected to the laser rangefinders by six fibers. Each retroreflector and collimator weigh only about 20 g, which makes the overall measurement system very light and flexible.

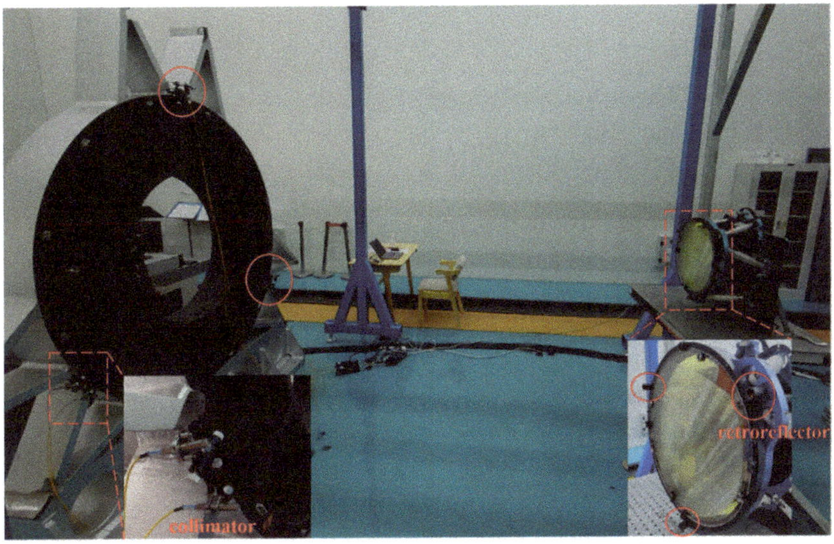

Figure 5. Test system. Three retroreflectors (red circles on the right of Figure 5) were attached to the edge of the frame, and six laser collimators (red circles on the left of Figure 5) were fixed to the edge of the relay mirror to verify the pose measurement method and its accuracy.

After the experimental equipment was installed, the frame was first adjusted to the initial position by the hexapod platform, and the six laser collimators were aligned with the corresponding retroreflectors. Six initial distances were recorded at the initial position, and then the hexapod platform was controlled to drive the frame; the change values of the six new distances were recorded, and the pose can be calculated by the algorithm described above. The pose obtained by this measurement method can be compared with the reference pose of the hexapod platform to verify the accuracy of this measurement system.

To determine the accuracy of rotation and translation of the pose measurement method, the hexapod platform was driven to rotate around the x-axis for step movement every 1×10^{-4} degree and move along the x-axis for step movement every 0.4 μm. The pose data were measured once a second in 5 min. Figure 6 shows the test results of 1×10^{-4} degrees and 0.4 μm step test of the measurement system, respectively. The error of rotation is 2×10^{-5} degrees (0.072 arcsec), and the error of translation is 0.2 μm. The spikes on the curve are caused by the fact that the hexapod platform has not been stabilized during the motion. The test results of rotation around the y/z-axis and translation along the y/z-axis are almost the same. The experimental results show that the accuracy of the measurement system for tilt reaches 1×10^{-5} degree level, and the measuring accuracy for decentration reaches 100 nanometers level, both of which are much better than the measurement requirements. The measurement and calculation were carried out on an ordinary laptop computer, and the single calculation time is 5~10 milliseconds, which achieves the purpose of high-precision real-time measurement.

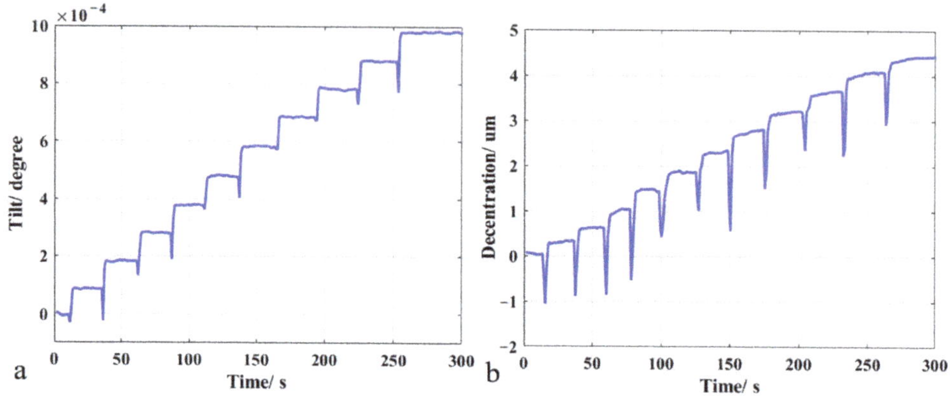

Figure 6. Test results of 1×10^{-4} degrees and 0.4 μm step test of the measurement system. (**a**) The tilt error of the measurement system is 2×10^{-5} degrees, (**b**) The decentration error of the measurement system is 0.2 μm.

4.3. Error Analysis

During the experiment, there were some errors introduced by the environment and other factors. The analysis shows that the errors are about tens of nanometers. These error sources can be divided into three categories, namely, random errors, systematic errors and environmental errors.

The random errors include detector noise, optical noise, reading error, etc. For the measurement system, the typical value of these noises is about a few nanometers and can be further reduced by averaging the distance data. Systematic errors include some nonlinear periodic errors, which may be caused by optical crosstalk in the light source and electrical crosstalk between reference channels. These errors should be controlled by the design of the rangefinder system, which has little influence on the measurement results. Finally, the environmental errors are mainly caused by ambient temperature and humidity changes and air disturbance, which affect the stability of ranging data. Fortunately, we can reduce these errors caused by environmental factors through the environmental information compensation module of the laser rangefinder and finally control the errors caused by environmental factors within tens of nanometers.

Overall, the errors introduced by the environment and other factors can be mitigated through careful design of the rangefinder system and the use of compensation modules to reduce the impact of environmental factors. By minimizing these errors, the accuracy and reliability of the laser rangefinder can be greatly improved, making it an essential tool for a wide range of applications, from surveying and mapping to industrial positioning and robotics.

5. Conclusions

In this paper, a reasonable solution for pose measurement of a long focal length optical telescope in orbit is proposed. Considering the simplicity and accuracy, a measurement system is provided to realize high-precision real-time measurement. The step test results show that the rotation error of the measurement system is 2×10^{-5} degrees (0.072 arcsecs), the translation error is 0.2 μm, and the single calculation time is 5~10 milliseconds, achieving the purpose of high-precision real-time measurement. This method has the advantages of simple structure, flexible installation, lightweight, non-contact, high-precision and real-time, which lays a technical foundation for the pose measurement of the primary lens of the space telescope. This measurement method also can be applied to other optical systems and other related systems.

Author Contributions: Conceptualization, H.S.; methodology, H.S., J.D. and J.B.; formal analysis, H.S. and L.W.; investigation, H.S., G.G. and D.L.; original draft preparation, H.S.; review and editing, H.S., J.D., L.W., J.B., G.G., D.L., B.F. and H.Y.; funding acquisition, D.L., B.F. and H.Y. All authors have read and agreed to the published version of the manuscript.

Funding: This research was funded by the National Natural Science Foundation of China (61905254 and 62075220) and the National Key Research and Development Program of China (No. 2016YFB0500200).

Institutional Review Board Statement: Not applicable.

Informed Consent Statement: Not applicable.

Data Availability Statement: Data underlying the results presented in this paper are not publicly available at this time but may be obtained from the authors upon reasonable request.

Acknowledgments: We thank the National Natural Science Foundation of China and the National Key Research and Development Program of China. The authors would like to thank the reviewers and the editors for their valuable comments and constructive suggestions that helped to improve the paper significantly.

Conflicts of Interest: The authors declare no conflict of interest.

References

1. Arbabi, E.; Arbabi, A.; Kamali, S.M.; Horie, Y.; Faraon, A. High efficiency double-wavelength dielectric metasurface lenses with dichroic birefringent meta-atoms. *Opt. Express* **2016**, *24*, 18468–18476. [CrossRef] [PubMed]
2. Genevet, P.; Capasso, F.; Aieta, F.; Khorasaninejad, M.; Devlin, R. Recent advances in planar optics: From plasmonic to dielectric metasurfaces. *Optica* **2017**, *4*, 139–152. [CrossRef]
3. Zhang, Y.; Jiao, J.; Wang, B.; Jin, J.; Su, Y. Transmissive diffractive membrane optic for large aperture lightweight optical telescope. *Proc. SPIE* **2015**, *9622*, 96220G.
4. Liu, D.; Wang, L.; Yang, W.; Wu, S.; Fan, B.; Wu, F. Stray light characteristics of the diffractive telescope system. *Opt. Eng.* **2018**, *57*, 025105. [CrossRef]
5. MacEwen, A.H.; Breckinridge, J.B. Large diffractive/refractive apertures for space and airborne telescopes. *Proc. SPIE* **2013**, *8739*, 873904.
6. Britten, A.J.; Dixit, N.S.; DeBruyckere, M.; Steadfast, D.; Hackett, J.; Farmer, B.; Poe, G.; Patrick, B.; Atcheson, D.P.; Domber, L.J.; et al. Large-aperture fast multilevel Fresnel zone lenses in glass and ultrathin polymer films for visible and near-infrared imaging applications. *Appl. Opt.* **2014**, *53*, 2312–2316. [CrossRef]
7. Atcheson, P.; Stewart, C.; Domber, J.; Whiteaker, K.; Cole, J.; Spuhler, P.; Seltzer, A.; Britten, A.J.; Dixit, N.S.; Farmer, B.; et al. MOIRE-Initial Demonstration of a Transmissive Diffractive Membrane Optic for Large Lightweight Optical Telescopes. *Proc. SPIE* **2012**, *8442*, 844221.
8. Rahlves, M.; Rezem, M.; Boroz, K.; Schlangen, S.; Reithmeier, E.; Roth, B. Flexible, fast, and low-cost production process for polymer based diffractive optics. *Opt. Express* **2015**, *23*, 3614–3622. [CrossRef]
9. Zhao, B.; Shi, W.; Zhang, J.; Zhang, M.; Qi, X.; Li, J.; Li, F.; Tan, J. Six Degrees of Freedom Displacement Measurement System for Wafer Stage Composed of Hall Sensors. *Sensors* **2018**, *18*, 2030. [CrossRef]
10. Fan, K.C.; Chen, M.J.; Huang, W.M. A Six-Degree-Of-Freedom Measurement System for The Motion Accuracy of Linear Stages. *Int. J. Mach. Tools Manufact.* **1998**, *38*, 155–164. [CrossRef]
11. Hu, Y.; Miyashita, L.; Watanabe, Y.; Ishikawa, M. Robust 6-DOF motion sensing for an arbitrary rigid body by multi-view laser Doppler measurements. *Opt. Express* **2017**, *25*, 30371–30387. [CrossRef] [PubMed]
12. Henselmans, R.; Nijkerk, D.; Lemmen, M.; Rijnveld, N.; Kamphues, F. Design, analysis, and testing of the optical tube assemblies for the ESO VLT four laser guide star facility. *Proc. SPIE* **2012**, *8447*, 84474N.
13. Kim, D.W.; Esparza, M.; Quach, H.; Rodriguez, S.; Kang, H.; Feng, Y.; Choi, H. Optical Technology for Future Telescopes. *Proc. SPIE* **2021**, *11761*, 1176103.
14. Salvadé, Y.; Schuhler, N.; Lévêque, S.; Floch, L.S. High-accuracy absolute distance measurement using frequency comb referenced multiwavelength source. *Appl. Opt.* **2008**, *47*, 2715–2720. [CrossRef] [PubMed]
15. Jia, X.; Liu, Z.; Tao, L.; Deng, Z. Frequency-scanning interferometry using a time-varying Kalman filter for dynamic tracking measurements. *Opt. Express* **2017**, *25*, 25782–25796. [CrossRef]
16. Yan, L.; Xie, J.; Chen, B.; Lou, Y.; Zhang, S. Absolute distance measurement using laser interferometric wavelength leverage with a dynamic-sideband-locked synthetic wavelength generation. *Opt. Express* **2021**, *29*, 8344–8357. [CrossRef] [PubMed]
17. Han, S.; Kim, Y.; Kim, S. Parallel determination of absolute distances to multiple targets by time-of-flight measurement using femtosecond light pulses. *Opt. Express* **2015**, *23*, 25874–25882. [CrossRef]
18. Gao, G.; Wang, L.; Shi, H.; Liu, D.; Fan, B.; Guan, C. Facile large-area uniform photolithography of membrane diffractive lens based on vacuum assisted self contact method. *Sci. Rep.* **2020**, *10*, 9005. [CrossRef]

19. Dale, J.; Hughes, B.; Lancaster, J.A.; Lewis, J.A.; Reichold, J.A.; Warden, S.M. Multi-channel absolute distance measurement system with sub ppm-accuracy and 20m range using frequency scanning interferometry and gas absorption cells. *Opt. Express* **2014**, *22*, 24869–24893. [CrossRef]
20. Zhu, K.; Guo, B.; Lu, Y.; Zhang, S.; Tan, Y. Single-spot two-dimensional displacement measurement based on self-mixing interferometry. *Optica* **2017**, *4*, 729–735. [CrossRef]
21. Lu, C.; Liu, G.; Liu, B.; Chen, F.; Gan, Y. Absolute distance measurement system with micron-grade measurement uncertainty and 24m range using frequency scanning interferometry with compensation of environmental vibration. *Opt. Express* **2016**, *24*, 30215–30224. [CrossRef] [PubMed]
22. Liu, C.S.; Hsu, H.C.; Lin, Y.X. Design of a six-degree-of-freedom geometric errors measurement system for a rotary axis of a machine tool. *Opt. Lasers Eng.* **2020**, *127*, 105949. [CrossRef]
23. Friedrich, C.; Kauschinger, B.; Ihlenfeldt, S. Spatial force measurement using a rigid hexapod-based end-effector with structure-integrated force sensors in a hexapod machine tool. *Measurement* **2019**, *145*, 350–360. [CrossRef]
24. Gao, G.; Shi, H.; Wang, L.; Liu, D.; Wang, J.; Du, J.; Bian, J.; Fan, B.; Yang, H. Large Aperture High Diffraction Efficiency Off-axis Fresnel Lens Fabrication and Analysis. *Opt. Express* **2022**, *30*, 28932–28940. [CrossRef]
25. Thurner, K.; Quacquarelli, F.P.; Braun, P.P.; Savio, C.D.; Karrai, K. Fiber-based distance sensing interferometry. *Appl. Opt.* **2015**, *54*, 3051–3063. [CrossRef]

sensors

MDPI

Article

Sensing Method for Wet Spraying Process of Tunnel Wall Based on the Laser LiDAR in Complex Environment

Degang Xu *, Qing Song *, Shiyu Fang and Yanrui Guo

School of Automation, Central South University, Changsha 410083, China
* Correspondence: dgxu@csu.edu.cn (D.X.); songqing@csu.edu.cn (Q.S.); Tel.: +86-0731-8887-9190 (D.X.); +86-187-1104-3216 (Q.S.)

Abstract: In tunnel lining construction, the traditional manual wet spraying operation is labor-intensive and can be challenging to ensure consistent quality. To address this, this study proposes a LiDAR-based method for sensing the thickness of tunnel wet spray, which aims to improve efficiency and quality. The proposed method utilizes an adaptive point cloud standardization processing algorithm to address differing point cloud postures and missing data, and the segmented Lamé curve is employed to fit the tunnel design axis using the Gauss–Newton iteration method. This establishes a mathematical model of the tunnel section and enables the analysis and perception of the thickness of the tunnel to be wet sprayed through comparison with the actual inner contour line and the design line of the tunnel. Experimental results show that the proposed method is effective in sensing the thickness of tunnel wet spray, with important implications for promoting intelligent wet spraying operations, improving wet spraying quality, and reducing labor costs in tunnel lining construction.

Keywords: 3D LiDAR point cloud; intelligent detection methods; normalization; shotcrete

Citation: Xu, D.; Song, Q.; Fang, S.; Guo, Y. Sensing Method for Wet Spraying Process of Tunnel Wall Based on the Laser LiDAR in Complex Environment. *Sensors* **2023**, 23, 5167. https://doi.org/10.3390/s23115167

Academic Editor: Vittorio Passaro

Received: 13 April 2023
Revised: 15 May 2023
Accepted: 22 May 2023
Published: 29 May 2023

1. Introduction

With the rapid development of 3D laser scanning technology, the application of laser technology is rapidly expanding and offering decreased costs and increased accuracy. This technology has various applications, including road detection [1,2], object recognition [3,4], surface reconstruction [5,6] and tunnel detection [7,8].

In China, the total length of highways is reported to be 4,846,500 km, including 17,236.1 km of highway tunnels as of 2018 [9]. It is estimated that by 2030, the total number of tunnels in operation in China will reach 17,000, exceeding 30,000 km in length [10]. Therefore, the development of intelligent tunnels in construction and monitoring is becoming increasingly urgent. The use of 3D laser scanning technology in tunnel construction and monitoring can improve construction efficiency, ensure safety, and reduce labor costs.

Laser technology has become an indispensable tool in the intelligent development of tunnels [11,12]. The use of laser scanners for acquiring 3D data of excavation surfaces in tunnels was first proposed by Lemy et al. [13]. They determined the displacement of the excavation surface by comparing the point clouds obtained at different times. This study demonstrated the potential of LiDAR in data collection during tunnel excavation. In subsequent years, Fekete et al. [7,14] used 3D laser scanning for quality control in drill and blast tunnels. LiDAR scanning allowed for precise monitoring of excavation and support installation during construction. More recent research has further explored the potential of LiDAR in tunnel construction activities, such as rock mass classification [15], drill bit wear detection [16], and automatic surveying of tunnel sections [17]. However, research on tunnel shotcrete is still lacking. Shotcrete is commonly used in the construction of railway and highway tunnels [18]. Tunnel shotcrete spraying is a process of rapidly applying shotcrete to the rock or concrete surface to prevent tunnel collapse during excavation [19,20]. Currently, shotcrete spraying is a manual process, resulting in unstable construction quality and slow construction progress. There is a need for further research on the intelligent construction of shotcrete in tunnels.

LiDAR-based 3D object detection is essential for automating tunnel shotcrete spraying, as it directly relates to understanding the tunnel environment. A previous study proposed a novel neural network structure based on LiDAR for detecting the area of tunnel shotcrete spraying [21], demonstrating the real-time monitoring of the tunnel profile and shotcrete area. Ranjbarnia [22] studied the effects of various parameters such as sprayed concrete thickness, soil geomechanical properties, tunnel depth, and fault plane dip angle using the 3D finite difference analysis algorithm and centrifuge physical model, but mainly focused on crossing faults in urban tunnels, which is not applicable for the construction analysis of tunnels in progress. Oreste [23] proposes a new calculation procedure based on the combined use of two calculation methods, convergence confinement and hyperstatic reaction methods, to analyze the factors of shotcrete and determine the trend of the lining safety factor over time. Unlike previous studies, our research focuses on detecting the thickness of shotcrete to enable intelligent detection for large spatial arch spraying processes in complex tunnel scenarios. To achieve this, we developed a LiDAR-based intelligent detection method. Our contribution lies in proposing a novel approach to address the specific challenge of shotcrete thickness detection.

In summary, the main contributions of this paper are as follows:

- This paper proposes a new method for the intelligent detection of wet shotcrete thickness on the tunnel arch surface during large spatial arch spraying processes in complex tunnel scenarios, which has not been previously explored. This method addresses the need for automated and accurate shotcrete spraying to improve the construction quality and progress.
- An innovative adaptive tunnel standardization processing algorithm is introduced, which mathematically describes the inner contour of the tunnel. This algorithm can accurately detect the tunnel arch surface, which is a prerequisite for detecting the thickness of shotcrete, and can adapt to different tunnel shapes and sizes.
- The proposed algorithm has demonstrated robust and reliable performance in detecting tunnel shotcrete thickness during tunnel construction in China. This method contributes significantly to the field of tunnel construction by improving construction quality and efficiency while reducing costs and risks associated with manual inspection.

The structure of this paper is as follows: Section 2 presents the problem and describes the process of data acquisition. In Section 3, an innovative adaptive method for normalizing point cloud data is proposed. Section 4 discusses a fitting method used to obtain a mathematical model of the inner contour of a large-scene tunnel. The experimental results are presented in Section 5. Finally, the paper is concluded with suggestions for further research in Section 6.

2. Material

2.1. Problem Description

In modern tunneling operations, it is essential to acquire detailed information about the shotcrete thickness of lining and installed support structures to ensure safety, long-term stability, and quality control. The lining construction [24], including initial and secondary lining, is an important tunnel supporting structure. The initial lining refers to shotcrete after laying steel arch support [25], which is widely used as a support element in underground building construction.

Tunnel shotcrete spraying is a process of spraying concrete at a high velocity onto the surface of a rock or steel structure in order to prevent tunnel collapse during excavation [26]. Currently, shotcrete spraying is a manual process, which involves an operator manually operating the mechanical arm for spraying, as shown in Figure 1. This method is highly dependent on the operator's experience and skills, and the quality of the shotcrete largely depends on the operator. Furthermore, the manual operation exposes workers to a large amount of dust and heavy and dangerous work, which may cause health problems. Therefore, an automatic tunnel lining construction system, as shown in Figure 2, is necessary to automate the tunnel construction process.

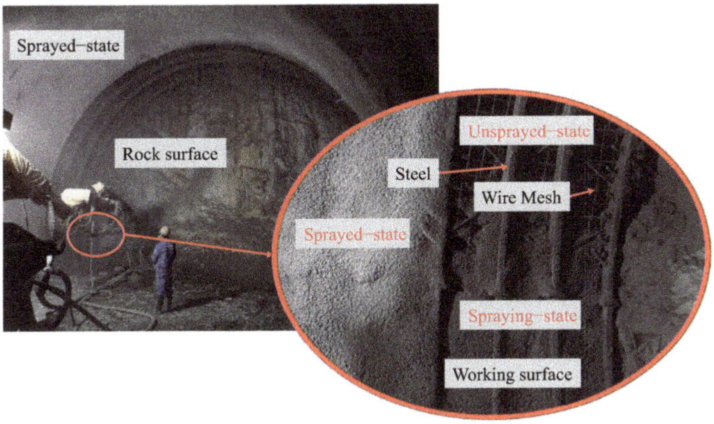

Figure 1. Current construction method and classification of wet shotcrete states. The red large circle indicates a magnification of the details in the small circle.

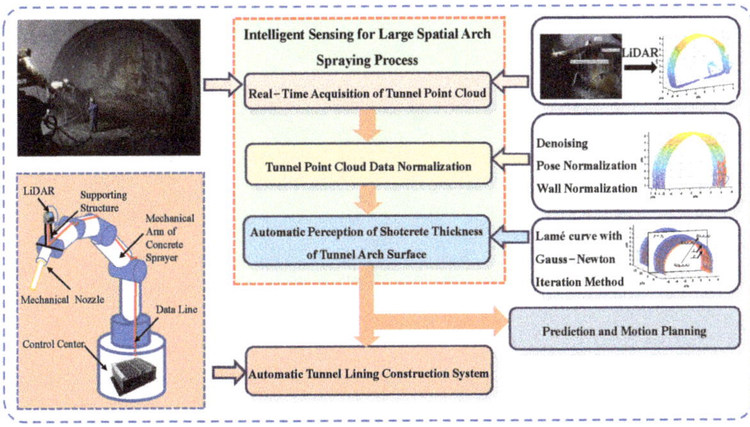

Figure 2. The solution of automatic tunnel lining construction system. The red box represents the components of "Intelligent Sensing for Large Spatial Arch Spraying Process", which include real-time point cloud acquisition, point cloud normalization, and intelligent perception for the steel arch. The arrows indicate the specific contents included in each component.

In the automatic tunnel lining construction system, the wet shotcrete process of the tunnel arch is divided into three states: the unsprayed-state, the spraying-state, and the sprayed-state [27,28]. The unsprayed-state is the state in which the shotcrete has not yet been sprayed, the spraying-state is the state in which the shotcrete is being sprayed, and the sprayed-state is the state in which the shotcrete has been applied, as shown in Figure 1. LiDAR is used to collect tunnel point cloud data in these three states, which is then processed using denoising, correction, clustering, and compensation techniques to extract the tunnel arch points.

Afterward, a wet shotcrete thickness description model is established to determine the depth of shotcrete needed for the tunnel arch surface to be sprayed continuously. With the prediction and motion planning of the mechanical arm, an automatic tunnel lining construction system is proposed. This paper focuses on the intelligent sensing for the large spatial arch spraying process in complex tunnel scenarios, which lays the foundation for prediction and motion planning of the system.

2.2. Process of Data Acquirement

The LiDAR used in this study is mounted on the mechanical arm of the concrete sprayer and integrates 16 laser/detector pairs in a compact housing for data collection [17]. In practice, the LiDAR is designed to be installed on the mechanical arm (as shown in Figure 3a). Since the shotcrete machine is generally oriented towards the working surface of the tunnel, during the wet spraying interval, when the mechanical arm stops moving, the LiDAR scans the tunnel to realize the measurement and perception of the environment. The original point cloud collected in reality is shown in Figure 3b, and there are mainly several issues with it:

(1) The point cloud is contaminated with noisy data due to severe dust pollution;
(2) The point cloud is tilted due to the movement of the mechanical arm and the installation of the LiDAR;
(3) The point cloud data contain redundant data;
(4) Some point cloud data are missing due to obstruction by obstacles.

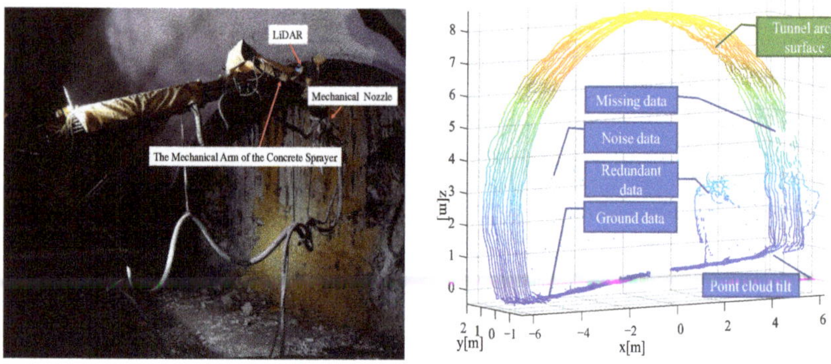

(**a**) Data collection site. (**b**) Data collected.

Figure 3. Data collection and analysis. (**a**) describes the data collection site. (**b**) displays a frame of tunnel point cloud data that includes various types of noise. The tunnel arch points are the target of what we want to extract.

To address the issues mentioned above with the collected tunnel point cloud data, an adaptive tunnel arch point extraction algorithm is proposed prior to sensing the shotcrete thickness of the lining.

3. Method

A point cloud is a vast collection of surface characteristic points of a target object that are directly obtained by LiDAR. However, due to the actual collection site conditions, noise interference is inevitable. Additionally, the presence of construction equipment, operators, rock waste, and other objects at the tunnel site may cause occlusion, which affects the quality of the point cloud to varying degrees. Moreover, the large volume of the tunnel point cloud and the existence of redundant information make it critical to extract the characteristics of the tunnel cross-section point cloud and reduce computational complexity. Therefore, standardizing the tunnel point cloud is crucial during the feature point extraction process to ensure accurate analysis and efficient computation.

3.1. Data Denoising

To filter out noise data caused by suspended concrete particles, we use a point cloud filtering algorithm based on the threshold neighborhood method proposed by Rusu and Cousins [29]. The algorithm selects neighboring points using a 3D Euclidean distance metric and terminates once a fixed number of neighbors n_{nbrs} or all points within a bounding sphere of radius r_{sph} have been found. The variables n_{nbrs} and r_{sph} control the size of the neighborhood selection. Subsequently, the mean μ and standard deviation σ of nearest neighbor distances are calculated, and points outside the $\mu \pm \vartheta \cdot \sigma$ range are removed, where the parameter ϑ adjusts the sensitivity of the threshold. In our implementation, we have found the optimal value of ϑ to be 1, and $n_{nbrs} = 30$.

Furthermore, based on experiments with multiple datasets, we have shown that the $\mu \pm \sigma$ thresholds are effective in removing noise, where around 1% of the points are considered as noise.

3.2. Adaptive Point Cloud Pose Normalization

During the process of collecting radar data, the tunnel point cloud may be inclined to different degrees due to various installation reasons, as illustrated in the left schematic diagram in Figure 4. To enhance the stability and accuracy of tunnel wet spray state perception, it is crucial to transform the coordinates of the tunnel point cloud before the three-dimensional reconstruction of the tunnel. The pose normalization of the tunnel point cloud needs to be performed to adjust all the tunnel point clouds to the position shown on the right side of Figure 4. This process enables the alignment of the tunnel point cloud with the reference frame, facilitating a more precise three-dimensional reconstruction of the tunnel.

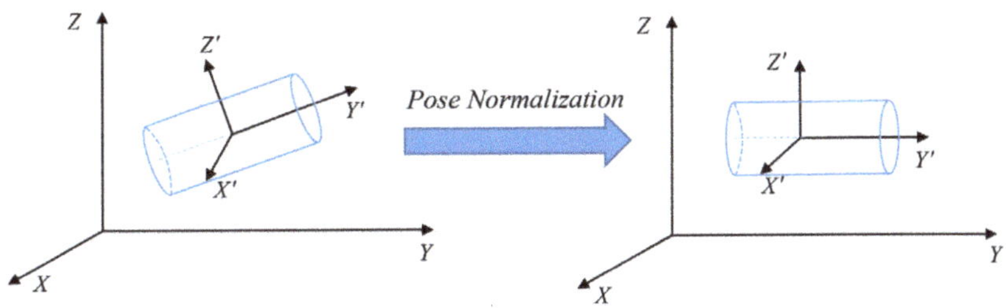

Figure 4. Schematic diagram of tunnel attitude standardization.

Considering the rigid body transformation characteristics of the tunnel, solving any part of the rotation transformation matrix can enable the complete rotation transformation of the tunnel point cloud. However, in the tunnel point cloud, the road point cloud exhibits significant planar characteristics, which can be utilized to correct the tunnel wall point cloud by solving the transformation matrix of the road point cloud. To extract the ground point cloud from the tunnel point cloud, we propose a planar extraction algorithm based on the M-estimator SAmple Consensus (MSAC) algorithm [30]. The MSAC algorithm optimizes the calculation method of the loss function of the RANSAC algorithm, addressing the RANSAC algorithm's sensitivity to the threshold T selection of interior points. The loss function value of the RANSAC algorithm is represented by C_1. For the points inside the model, the loss function is 0, whereas for the points outside the model, a constant penalty is incurred. Thus, setting the threshold too high can result in poor estimation, while setting it too low can affect the robustness of the model.

$$C_1 = \sum_i p_1(e_i^2) \quad (1)$$

where e is the error function, and p_1 is the robust scale parameter, defined as

$$p_1(e^2) = \begin{cases} 0 & e^2 < T^2 \\ \text{constant} & e^2 \geq T^2 \end{cases} \quad (2)$$

To address this issue, Torr and Zisserman [31] proposed a new loss function C_2 that can be minimized to obtain a more accurate model. The C_2 loss function is defined as

$$C_2 = \sum_i p_2(e_i^2) \quad (3)$$

where e is the error function, and p_2 is the robust standard error. The p_2 function assigns a weight to each data point based on its distance to the model, with larger weights assigned to points that are closer to the model. This effectively reduces the influence of outliers in the data and improves the accuracy of the model estimation.

$$p_2(e^2) = \begin{cases} e^2 & e^2 < T^2 \\ T^2 & e^2 \geq T^2 \end{cases} \quad (4)$$

To extract the plane of the point cloud data, the MSAC algorithm is utilized to obtain the plane equation, as shown in Equation (5). In this algorithm, a fixed penalty is given for out-of-model points, while for in-model points, the fitting effect of the model is considered to establish the most accurate model. Figure 5 shows the fitting result, which demonstrates that the extracted plane points correspond well with the road surface of the tunnel.

$$a_p x + b_p y + c_p z + d_p = 0 \quad (5)$$

where a_p, b_p, c_p, and d_p are the coefficients of the plane equation.

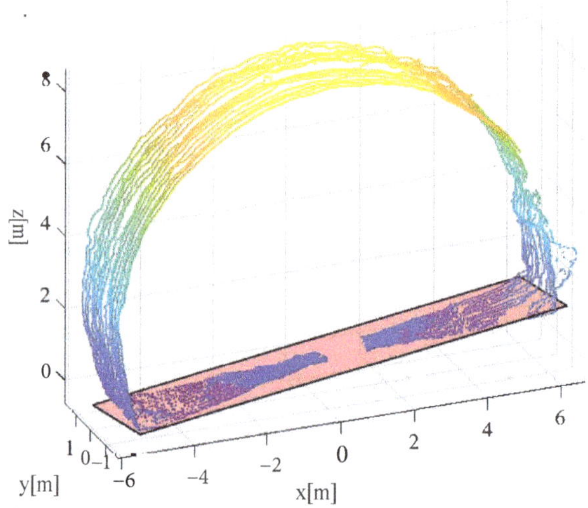

Figure 5. Fitting result of the MSAC algorithm for plane extraction. The pink plane represents the plane fitted by the algorithm.

In Figure 5, the point cloud is observed to be inclined at a certain angle with respect to the X, Y, and Z axes. However, if the normal vector of the road point cloud is found to be parallel to the Z-axis, it indicates that this part of the point cloud has already undergone the attitude standardization process.

In the cases where the normal vector of the road point cloud is not parallel to the Z-axis, a point cloud correction algorithm based on continuous projection is proposed in this section.

The proposed point cloud correction algorithm based on continuous projection involves a series of steps. Firstly, the point cloud is projected onto the YOZ, XOY, and XOZ planes sequentially. Next, the α-shape algorithm [32] is employed to determine the boundary points of the point cloud on each plane, which are then used to identify the center line of the point cloud in the plane. The rotation angle of the point cloud in the plane is determined by the declination angle from the coordinate axis, and the transformation matrix is obtained using the Rodrigue formula [33]. Finally, the point cloud is corrected according to the transformation matrix.

A common way to determine the boundary of a finite point set is through the α-shape algorithm [32]. For a finite point set, the algorithm forms a line segment between every two points and draws a circle with a diameter of the line segment. If one of the circles does not contain any other points except for the two points, then the two points are considered as two boundary points. The sum of these boundary points gives the boundary of the point cloud, as shown in Figure 6.

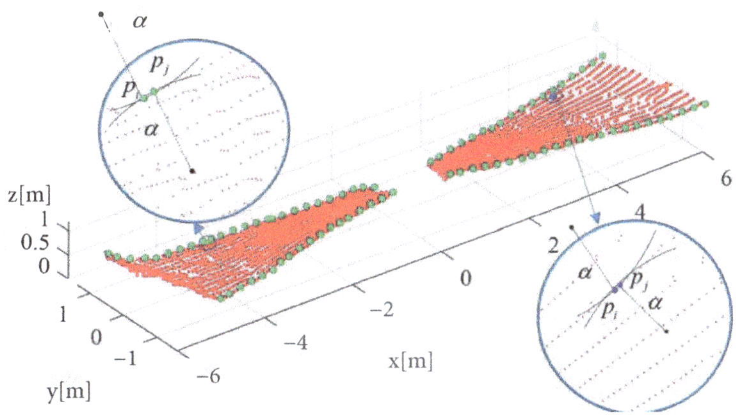

Figure 6. The alpha-shape algorithm criteria. The red points represent the point cloud data of the plane fitted by the MSAC algorithm in Figure 5, while the green points represent the boundary points detected by the α-shape algorithm. The points p_i and p_j on the left are the boundary points, whereas on the right, p_i and p_j are the internal points.

After obtaining the projected boundary points, a quadratic function is used to fit the boundary of the point cloud, as shown in Equation (6):

$$\begin{bmatrix} y_1 \\ y_2 \\ \vdots \\ y_n \end{bmatrix} = \begin{bmatrix} 1 & x_1 & x_1^2 \\ 1 & x_2 & x_2^2 \\ \vdots & \vdots & \vdots \\ 1 & x_n & x_n^2 \end{bmatrix} \begin{bmatrix} a_q \\ b_q \\ c_q \end{bmatrix} + \begin{bmatrix} \epsilon_1 \\ \epsilon_2 \\ \vdots \\ \epsilon_n \end{bmatrix} \tag{6}$$

where n is the number of boundary points; x_i and y_i are the coordinates of the i-th boundary point; a_q, b_q, and c_q are the coefficients of the quadratic function; and the ϵ_i represent the residual error between the fitted curve and the original point cloud data.

To obtain the midline of the plane, we first use the uniform sampling method to obtain 50 sampling point sets on each boundary. For each sampling point, we record the normal line perpendicular to the boundary and the intersection point set with the opposite boundary. Then, we recalculate the intersection of the normal line at the intersection point on the boundary with the original boundary and record the intersection point set. Next, we take the midpoint of each line segment between the two intersection points and then calculate the midpoint of the two midpoints. The resulting set of midpoints is used to fit the midline using Equation (7). The entire process is illustrated in Figure 7.

$$y = k_m x + b_m \tag{7}$$

where k_m and b_m are the parameters of a curve, and the RANSAC (RANdom SAmple Consensus) algorithm [34,35] is used to estimate the parameters of the model. Then, the inclination angle of the point cloud in the plane can be expressed by Equation (8).

$$\theta = \arctan(k_m) \tag{8}$$

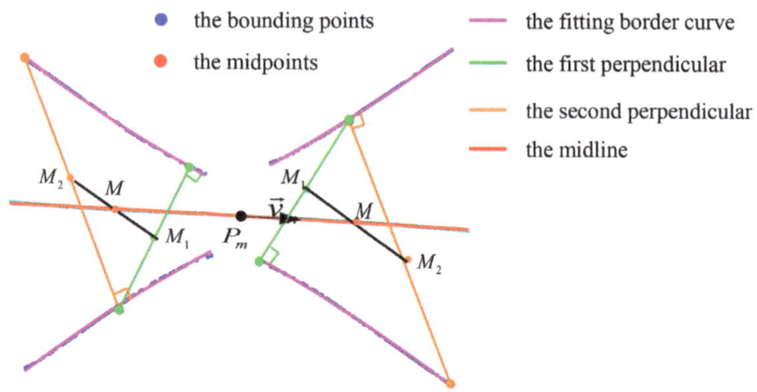

Figure 7. Determination of the midpoints on central axis. The green line represents one of the normal lines of the boundary. M_1 is the midpoint of the line segment between the intersection point of the normal line with the opposite boundary and the boundary itself. The orange line represents the normal line of the opposite boundary. M_2 is the midpoint of the line segment between the intersection point of the boundary's normal line with the original boundary and the boundary itself. M is the intersection point of the line segments between M_1 and M_2, which represents the point on the midline of the plane, and the direction vector is $\vec{v}$.

Based on the Rodriguez rotation formula, the rotated vector $\vec{v}_{rot}$ of any vector $\vec{v}_0$ in space, rotated by an angle θ around a given rotation axis $\vec{n}$, can be expressed by Equation (9). This equation ensures the accuracy of the rotation operation.

$$\vec{v}_{rot} = M_r(\vec{n}, \theta)\vec{v}_0 = \vec{v}_0 \cos\theta + (\vec{n} \times \vec{v}_0)\sin\theta + (1 - \cos\theta)(\vec{n} \cdot \vec{v}_0)\vec{n} \tag{9}$$

where M_r is the rotation matrix, which is defined as shown in Equation (10):

$$M_r(\vec{n}, \theta) = \begin{bmatrix} u^2(1-\cos\theta) + \cos\theta & uv(1-\cos\theta) - w\sin\theta & uw(1-\cos\theta) + v\sin\theta \\ uv(1-\cos\theta) + w\sin\theta & v^2(1-\cos\theta) + \cos\theta & uw(1-\cos\theta) - u\sin\theta \\ uw(1-\cos\theta) - v\sin\theta & vw(1-\cos\theta) + u\sin\theta & w^2(1-\cos\theta) + \cos\theta \end{bmatrix} \tag{10}$$

To investigate the impact of the continuous projection algorithm proposed in this section on point cloud correction, this paper conducts an analysis of the inclination angles and coordinate axes of 1000 frames of tunnel point clouds with varying inclination degrees, before and after correction. The results are presented in Figure 8: "·" represents the tilt angle before correction, and "*" represents the tilt angle after correction. From the comparison of 1000 point clouds, it can be seen that there were varying degrees of tilt before correction, with the maximum deviation angle exceeding 15° in all directions. After point cloud correction, the point cloud was corrected well in all directions, with a tilt angle not exceeding 2°. The local enlargement images of 400 frames to 1000 frames further demonstrate the effectiveness of the algorithm correction.

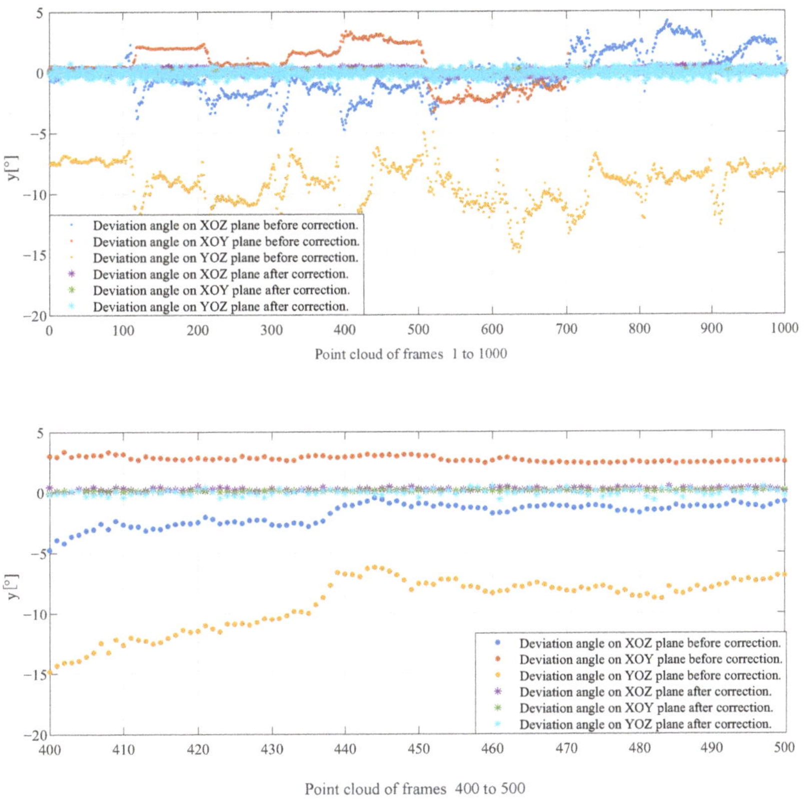

Figure 8. Comparison of deviation angle before and after correction. The x-axis represents different point cloud frames, while the y-axis represents the inclination angle.

3.3. Adaptive Point Cloud Wall Normalization

At the data collection site, construction equipment and workers are in continuous motion. Consequently, the point cloud collection process may result in some degree of obstruction of the tunnel wall, leading to data loss to varying extents. This can lead to subsequent issues in data processing and analysis. To address incomplete information, it is necessary to detect the area of the tunnel wall point cloud and interpolate missing data segments. This process is referred to as the wall standardization process of the tunnel point cloud in this paper.

In order to compensate for missing data in the tunnel wall, this paper presents an adaptive point cloud compensation algorithm based on an interpolation model. The algorithm consists of two main parts: automatic detection and automatic interpolation. During data collection, construction equipment and workers may obstruct the tunnel wall, leading to incomplete point cloud data. In order to address this issue, the proposed algorithm aims to automatically detect the missing areas of the tunnel wall point cloud and interpolate the missing data segments.

The adaptive point cloud compensation algorithm utilizes the first-order difference algorithm for the automatic detection of missing parts of the point cloud. Once the location of the missing point cloud is identified, the algorithm compensates for the missing segment through interpolation using the piecewise cubic Hermite interpolating polynomial (PCHIP) method, which is a type of piecewise polynomial interpolation that uses cubic Hermite polynomials to ensure the smoothness of the interpolated curve.

Assuming that $P_k(x_k, y_k, z_k)^T$ and $P_{k+1}(x_{k+1}, y_{k+1}, z_{k+1})^T$ are two points of tunnel arch points after clustering, with a missing area between P_k and P_{k+1}, the algorithm checks if $|z_k - z_{k+1}| \geq \delta$, where

$$\delta = \frac{\sum_{i=1}^{nums-1} |z_{i+1} - z_i|}{nums - 1} \tag{11}$$

Here, $nums$ is the number of clusters of tunnel arch points, and δ is the threshold value for determining the missing area.

The cubic Hermite interpolation polynomial H_3 is required to satisfy Equation (12):

$$\begin{cases} H_3(x_k) = \varphi(x_k), H_3(x_{k+1}) = \varphi(x_{k+1}) \\ \dot{H}_3(x_k) = \dot{\varphi}(x_k), \dot{H}_3(x_{k+1}) = \dot{\varphi}(x_{k+1}) \end{cases} \tag{12}$$

where $\varphi(x)$ is the interpolation function, and $H_3(x)$ is the basis function of the piecewise cubic Hermite interpolation polynomial, which can be expressed as

$$H_3(x) = \psi_k(x)\varphi(x_k) + \psi_{k+1}(x)\varphi(x_{k+1}) + \phi_k(x)\dot{\varphi}(x_k) + \phi_{k+1}(x)\dot{\varphi}(x_{k+1}) \tag{13}$$

where $\psi_k(x), \psi_{k+1}(x), \phi_k(x), \phi_{k+1}(x)$ is the cubic Hermitian interpolation basis function for nodes x_k and x_{k+1}, and they and their derivative must satisfy Equation (14).

$$\begin{bmatrix} \psi_k(x_k), & \psi_k(x_{k+1}), & \dot{\psi}_k(x_k), & \dot{\psi}_k(x_{k+1}) \\ \psi_{k+1}(x_k), & \psi_{k+1}(x_{k+1}), & \dot{\psi}_{k+1}(x_k), & \dot{\psi}_{k+1}(x_{k+1}) \\ \phi_k(x_k), & \phi_k(x_{k+1}), & \dot{\phi}_k(x_k), & \dot{\phi}_k(x_{k+1}) \\ \phi_{k+1}(x_k), & \phi_{k+1}(x_{k+1}), & \dot{\phi}_{k+1}(x_k), & \dot{\phi}_{k+1}(x_{k+1}) \end{bmatrix} = \mathbf{I}_4 \tag{14}$$

where $\mathbf{I}_4$ is a 4×4 identity matrix.

Under the constraints of Equation (14), $\psi_k(x)$ and $\phi_k(x)$ can be constructed as follows:

$$\begin{cases} \psi_k(x) = (a_\psi x + b_\psi)\left(\dfrac{x - x_{k+1}}{x_k - x_{k+1}}\right)^2 \\ \phi_k(x) = a_\phi(x - x_k)\left(\dfrac{x - x_{k+1}}{x_k - x_{k+1}}\right)^2 \end{cases} \tag{15}$$

The parameters a_ψ, b_ψ, and a_ϕ can be obtained by using Equations (14) and (15) under the constraint of satisfying Equation (14). The expressions of $\psi_{k+1}(x)$ and $\phi_{k+1}(x)$ can be constructed in a similar way. This results in an interpolation polynomial between the endpoints P_k and P_{k+1}, the interpolation results of which are shown in Figure 9. In Figure 9c,d, the red points represent the interpolated data points in the missing segments. It can be observed that the proposed algorithm not only preserves the general trend of the original data, but also achieves better data compensation accuracy with fewer wrongly interpolated data points, resulting in a superior data compensation effect.

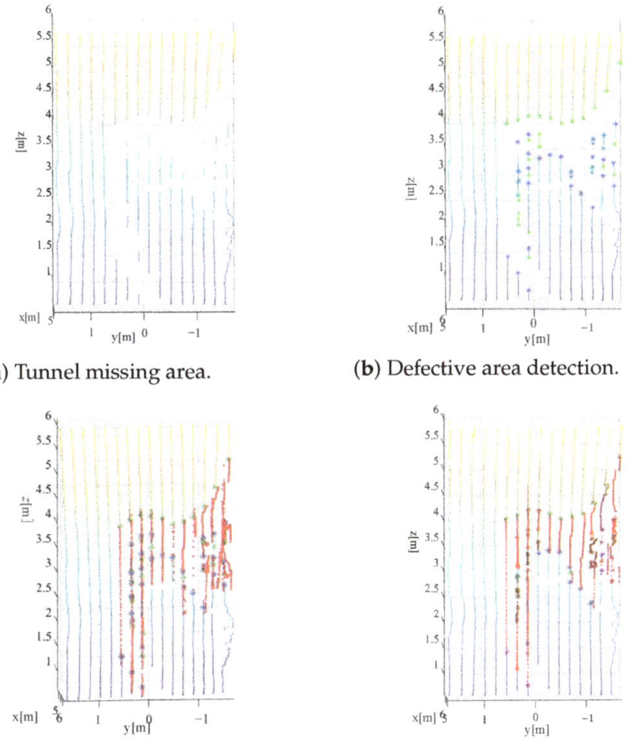

(**a**) Tunnel missing area. (**b**) Defective area detection.

(**c**) Piecewise linear interpolation. (**d**) Our algorithm.

Figure 9. Compensation of missing parts, the red dots represent the interpolation points of the algorithm. (**a**) shows an area of data loss, where the tunnel wall point cloud is missing. (**b**) shows the markers for the missing parts of the tunnel in the point cloud; specifically, the green asterisk (*) point represents the starting point of the missing segment, while the blue asterisk (*) point represents the end point of the missing segment. (**c**) shows the result of piecewise linear interpolation. (**d**) displays the interpolation result of our proposed algorithm.

4. Model

4.1. Theoretical Basis

4.1.1. The Gauss–Newton Iteration Method

The cross-section of a tunnel is typically designed as an ellipse [36]. There are two types of algorithms to obtain an ellipse equation: non-iterative and iterative algorithms. Examples of non-iterative algorithms include the Lagrange multiplier-based method proposed in [37]. Iterative algorithms include the Gauss–Newton algorithm-based method introduced in [38,39]. However, due to the complex tunnel environment during the construction phase, the collected point cloud data are not suitable for non-iterative algorithms. Therefore, in this study, Taylor series expansion was used to approximate the nonlinear regression model and improve the regression coefficients by multiple iterations and corrections until the minimum residual sum of squares was achieved. This method is referred to as the Gauss–Newton iteration method.

It is assumed that Equation (16) represents a nonlinear regression model of an elliptical cross-section of a tunnel:

$$\hat{y}_i = f(x_i, r) + \varepsilon_i, (i = 1, 2, \cdots, n) \tag{16}$$

where $r = (r_0, r_1, \cdots, r_{n_r})^T$ is an $n_r \times 1$ matrix of coefficients to be determined, and ε_i represents the error term, which follows a normal distribution. The total number of points to be fitted is denoted by n, and x_i is the x-coordinate of the i-th point, while $\hat{y}_i$ is the predicted value of x_i.

To obtain an initial value of the regression coefficient r, let $g^{(0)} = \left(g0^{(0)}, g_1^{(0)}, \cdots, g_{n_r}^{(0)}\right)^T$. Taylor expansion is used at $g^{(0)}$ in Equation (16), and the second order and above partial derivative terms are omitted to obtain Equation (17). This approach replaces the nonlinear regression model with a series expansion, and the regression coefficients of the nonlinear regression model are then iteratively updated and corrected until the minimum residual sum of squares is obtained using the Gauss–Newton iteration method.

$$f(x_i, r) = f(x_i, g^{(0)}) + \sum_{j=0}^{n_r-1} \left[\frac{\partial f(x_i, r)}{\partial r_j}\right]_{r=g^{(0)}} \left(r_j - g_j^{(0)}\right) \tag{17}$$

Equation (18) is obtained by combining Equations (16) and (17):

$$y_i^{(0)} \approx \sum_{j=0}^{n_r-1} G_{ij}^{(0)} b_j^{(0)} + \varepsilon_i, \quad (i = 1, 2, \cdots, n) \tag{18}$$

where

$$\begin{cases} y_i^{(0)} = y_i - f(x_i, g^{(0)}) \\ G_{ij}^{(0)} = \left[\frac{\partial f(x_i, r)}{\partial r_j}\right]_{r=g^{(0)}} \\ b_j^{(0)} = \left(r_j - g_j^{(0)}\right) \end{cases} \tag{19}$$

Equation (18) can be written in a more simplified matrix form as Equation (20):

$$Y^{(0)} \approx G^{(0)} b^{(0)} + \varepsilon \tag{20}$$

where

$$\begin{cases} Y_{n\times 1}^{(0)} = \begin{bmatrix} y_1 - f(x_1, g^{(0)}) \\ \cdots \\ y_n - f(x_n, g^{(0)}) \end{bmatrix} \\ G_{n\times n_r}^{(0)} = \begin{bmatrix} G_{10}^{(0)} & \cdots & G_{1n_r-1}^{(0)} \\ \vdots & & \vdots \\ G_{n0}^{(0)} & \cdots & G_{nn_r-1}^{(0)} \end{bmatrix} \\ b_{n_r\times 1}^{(0)} = \begin{bmatrix} b_0^{(0)} \\ \vdots \\ b_{n_r-1}^{(0)} \end{bmatrix} \end{cases} \tag{21}$$

Refine the correction coefficient $b^{(0)}$ using the least-squares method:

$$b^{(0)} = \left(G^{(0)T} G^{(0)}\right)^{-1} G^{(0)T} Y^{(0)} \tag{22}$$

The revised values for the regression coefficients $g^{(1)}$ can be obtained by using Equation (23):

$$g^{(1)} = g^{(0)} + b^{(0)} \tag{23}$$

To update the correction coefficient $b^{(s)}$ at the s-th iteration, we can use the least-squares method, as shown in Equation (24). Then, the updated regression coefficients $g^{(s+1)}$ at the $(s+1)$-th iteration can be obtained by adding $b^{(s)}$ to $g^{(s)}$. This iterative process is repeated until the SSR (sum of squares of residual) is below a certain tolerable error K, which is given by Equation (25). More specifically, the iterative process continues until $\left|\frac{SSR^{(s)}-SSR^{(s-1)}}{SSR^{(s)}}\right| \leq K$, where $SSR^{(s)}$ and $SSR^{(s-1)}$ are the SSR values at the s-th and $(s-1)$-th iterations, respectively:

$$b^{(s)} = \left(G^{(s)T}G^{(s)}\right)^{-1}G^{(s)T}Y^{(s)} \tag{24}$$

$$SSR^{(s)} = \sum_{i=0}^{n}\left[y_i - f\left(x_i, g^{(s)}\right)\right]^2 \tag{25}$$

where s represents the number of iterations.

4.1.2. Analysis of Common Fitting Models

In China, the majority of large-scale tunnels are constructed as arched structures. Consequently, when addressing the problem of fitting the inner contour of a tunnel, arched models, such as the circle, ellipse, and Lamé curve, are commonly employed.

(1) Circle: A circle is the most fundamental geometric shape. For any circular figure with center $O_c(x_c, y_c)$ and radius R_c, the standard equation of the circle is given by

$$(x - x_c)^2 + (y - y_c)^2 = R_c^2 \tag{26}$$

Based on the findings reported in [40], circular structures are known to exhibit excellent pressure-bearing capacity. Therefore, tunnels excavated using shield machines commonly adopt circular structures.

(2) Ellipse: The actual tunnel environment is complex, and various factors such as geotechnical characteristics and surrounding rock mechanics need to be considered. Additionally, the deformation of the tunnel during use must be addressed. The elliptical structure can adjust its load capacity by changing the eccentricity and is commonly used in practical engineering. The equation for an ellipse is as follows:

$$\frac{(x - x_c)^2}{a^2} + \frac{(y - y_c)^2}{b^2} = 1 \tag{27}$$

where $O_c(x_c, y_c)$ is the center of the ellipse, a is the semimajor axis, and b is the semiminor axis.

The load capacity of elliptical structural tunnels is closely related to the flatness of the ellipse, which can be described mathematically by the eccentricity e. In practical engineering, the eccentricity of the ellipse can be adjusted to change its load capacity. When e is closer to 0, the load capacity of the ellipse is stronger. Conversely, when e is closer to 1, the flatter the ellipse is, and the weaker its load capacity. This relationship between eccentricity and load capacity is important to consider when designing tunnels with elliptical cross-sections.

(3) The Lamé curve: This is also known as the hyperellipse [41], which is an extension of the ellipse. It has been widely used in tunnel engineering due to its adjustable shape parameters and excellent structural performance. The equation of the Lamé curve is given by

$$\left|\frac{x}{a}\right|^{\eta} + \left|\frac{y}{b}\right|^{\eta} = 1 \tag{28}$$

where a and b represent the major and minor axes of the Lamé curve, respectively, and η is the shape parameter that determines the shape of the curve.

By adjusting the values of a and b, symmetric closed curves such as rectangles, circles, and hyperellipses can be obtained. When $a = 6.0$ and $b = 4.0$, hyperellipse curves of different orders can be obtained, as shown in Figure 10.

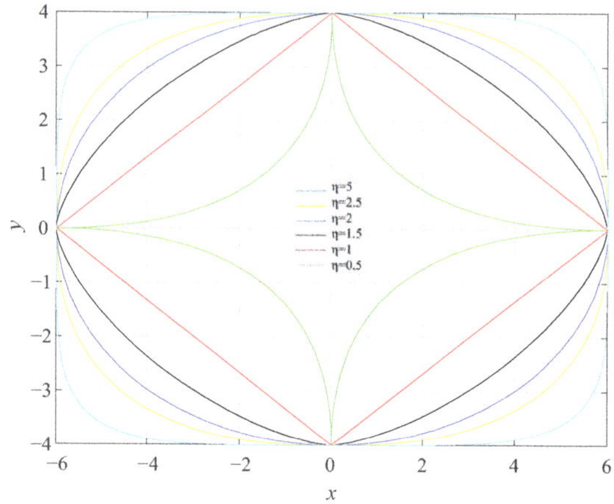

Figure 10. Lamé curve.

From Figure 10, it can be observed that when $0 < \eta < 1$, the curve is concave inward and takes the shape of a four-pointed star. When $\eta = 1$, the curve becomes a rhombus. For $1 < \eta < 2$, the curve is convex, and the curvature increases as it approaches the vertices. When $\eta = 2$, the curve becomes an ellipse, which is a circle if $a = b$. For $\eta > 2$, the curve becomes a rectangle with rounded corners, and as η increases, it approaches a rectangle, which is also referred to as an ellipse in this case.

Based on the analysis of the circle, ellipse, and Lamé curve, it can be concluded that the Lamé curve is more suitable for fitting the inner contour of the tunnel section, depending on the specific shape of the tunnel. Therefore, this paper adopts the Lamé curve to fit the tunnel section.

4.2. Extraction of Cross-Section Point Cloud

The calculation of the central axis is a crucial step in obtaining the tunnel section, as it describes the direction of the tunnel, and each section is perpendicular to it. The central axis of the tunnel is typically calculated by projecting the point cloud of the inner wall of the tunnel. There are four commonly used methods to obtain the tunnel center line:

(1) Manual acquisition: low efficiency and large errors.
(2) Extracting the rails: not suitable for tunnels without steel rails.
(3) Calculating the tunnel boundary through data model fitting: limited by the tunnel shape.
(4) Fitting boundary lines on both sides of the tunnel: adopted in this paper due to the easy determination of boundary lines. The width is obtained by determining boundary lines, which are then shifted to center and averaged to obtain the center line of the tunnel.

In this study, the fourth method was chosen due to its simplicity and effectiveness, as the boundary lines on both sides of the tunnel can be easily determined. The method involves determining the width of the tunnel by finding the boundary lines on both sides of the point cloud and then shifting each boundary line towards the center. The center line of the tunnel is finally obtained by taking the average value of the shifted boundary lines.

4.3. Fitting of Cross-Section Point Cloud

Based on the above analysis, it is apparent that the Lamé curve model is more suitable for fitting the inner contour of the tunnel in wet spraying due to its robustness. However, fitting the entire tunnel outline without distinction and estimating the error can increase the amount of calculation and produce a large fitting error due to the different wet spray conditions of each area. To account for these conditions, a method for fitting the inner contour of the tunnel using segmented Lamé curves is proposed in this paper.

Initially, the sequential sampling method is used to take n_p sampling points, and the Lamé curve is fitted to these points using the Gauss–Newton iteration method. The root mean square error (RMSE) of the fitting is then calculated, and if the threshold condition is met, the data are segmented, and sampling is performed from the left and right sides. The RMSE is recalculated until the threshold condition is exceeded, and the data segment fitted by the Lamé curve equation is obtained. Next, m sampling points are selected from the position where the data are interrupted, and this step is repeated until all points have been fitted.

$$RMSE = \sqrt{\frac{1}{n_p}\sum_{i=1}^{i=n_p}(observed_i - predicted_i)^2} \tag{29}$$

For the inner contour curve of a specific section during the wet spraying process of the tunnel, the results of the fitting are presented in Figure 11.

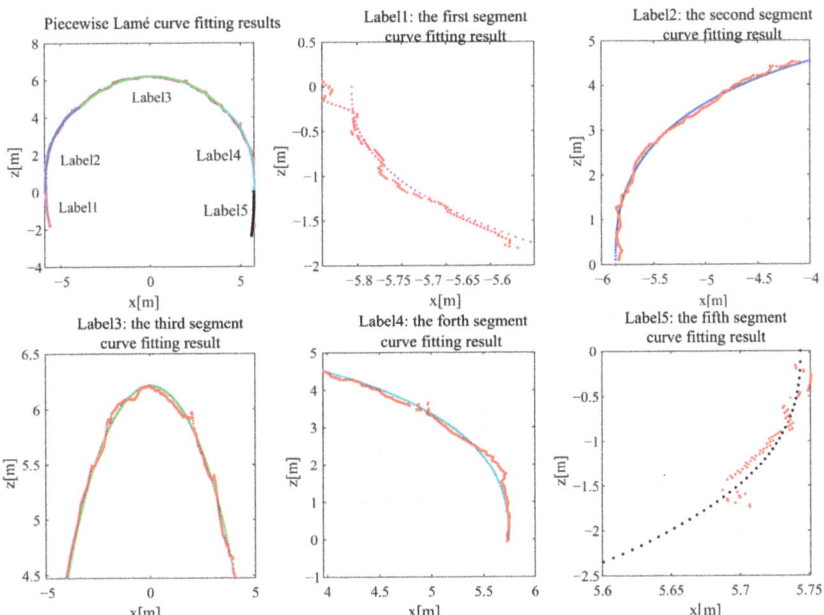

Figure 11. Piecewise Lamé curve fitting results. The labels 1 to 5 represent the fitting results of five curves. The red points indicate the original tunnel data point cloud, while the other color points represent the fitting curves of different regions.

In Figure 11, the red point cloud represents the actual inner contour curve of a section at a certain position in the tunnel, while the other lines of different colors represent different fitted Lamé curve segments. These curve segments are numbered from left to right, and Table 1 shows the fitting parameters and RMSE of these five curves.

Table 1. Piecewise Lamé curve fitting parameters.

ID	1	2	3	4	5
a	5.81	5.87	5.75	5.74	5.74
b	4.67	5.37	6.21	5.54	7.05
η	2.50	2.65	2.00	2.48	2.53
$RMSE$	0.0264	0.0476	0.0564	0.0570	0.0425

After analyzing the fitting parameters in Table 1, it can be observed that the Lamé curve coefficients vary for different sections. However, the root mean square error of each section is within the ideal range, indicating that the model fitting effect is satisfactory.

4.4. Thickness Perception Model

Figure 12 provides a front view of Figure 13, where the blue curve denotes the cross-section of the tunnel to be wet sprayed, while the red curve denotes the tunnel lining design line.

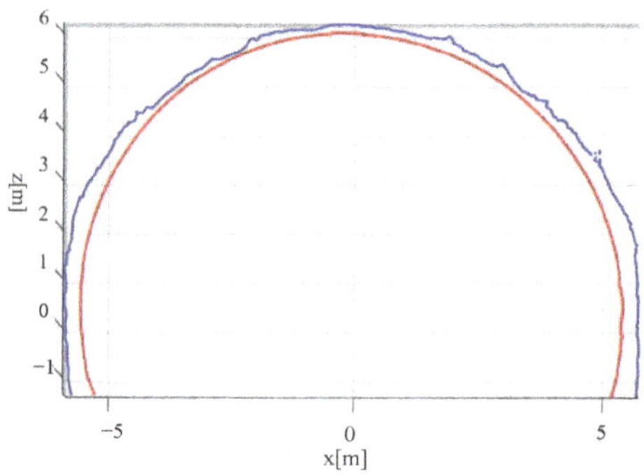

Figure 12. Front view of 3D wet shotcrete thickness description model. The blue curve denotes the cross-section of the tunnel to be wet sprayed, while the red curve denotes the tunnel lining design line.

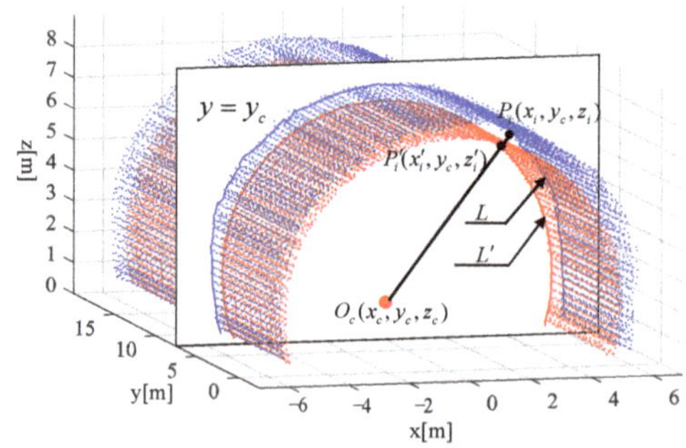

Figure 13. The 3D wet shotcrete thickness description model. The blue curve L represents the intersection curve of the unsprayed tunnel surface point cloud and the section, while the red curve L' represents the intersection curve of the designed surface and the section. Point P_i is a point in L, while point P_i' is the corresponding mapping point of P_i in L'. The center of the tunnel at cross-section $y = y_i$ is denoted by O_c.

In Figure 13, the depth d_i to be wet sprayed at a point P_i on the tunnel section can be observed. The blue curve L represents the outline of the tunnel section, while the red curve L' represents the inner outline of the tunnel lining design.

O_c denotes the central point of the section structure, while P_i' is the intersection point between $\overrightarrow{O_cP_i}$ and the inner contour line of the tunnel lining design. The thickness to be wet sprayed at point P_i' is then calculated as $d_i = \|P_iP_i{}'\|$.

During the fitting of the tunnel contour using the segmental Lamé curve, each point P_i on the contour was mapped to a point P_i' on the designed inner contour line of the tunnel. Therefore, P_i' is a point on the Lamé curve and satisfies the equation of the Lamé curve for its corresponding segment. From the 3D wet shotcrete thickness description model, we have that the thickness to be wet sprayed at point P_i' is given by $d_i = \|P_iP_i'\|$. Hence, P_i' satisfies both the equation of the segmental Lamé curve and the equation of the distance between P_i and P_i'. Therefore, we can write the two equations in a system of equations as follows:

$$\begin{cases} \left|\dfrac{x-x_c}{a}\right|^{\eta} + \left|\dfrac{z-z_c}{b}\right|^{\eta} = 1 \\ \overrightarrow{O_cP_i} = \lambda\overrightarrow{O_cP'_i} \end{cases} \tag{30}$$

where λ represents an arbitrary constant.

5. Experiment

The proposed algorithm has been implemented in a highway tunnel construction project in China, and the experimental results are presented in Figure 14.

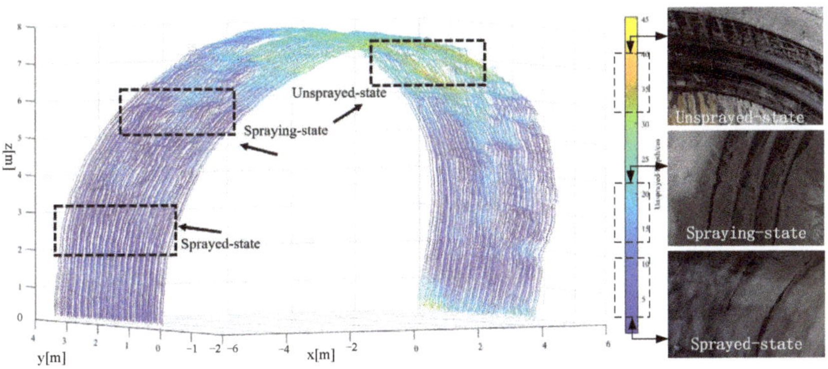

Figure 14. Experimental results.

As previously described, there are three stages of shotcreting, each requiring a different depth of concrete to be sprayed onto the tunnel surface. These depths are shown in Figure 14 for each respective stage. For the tunnel area in its three states of undried spraying, wet spraying, and dried spraying, we sampled a typical area of 1 m × 3 m × 0.2 m from the experimental results for verification and analysis.

(1) For the sampled areas in the unsprayed-state, which included 15,973 points, the average depth to be sprayed was 39.85 cm, which is close to the maximum design thickness of 40 cm for the concrete. Due to the varying depth of rock excavation in the unsprayed area, the depth to be sprayed for each point differed, and there was no specific pattern to follow. This reflects the actual construction conditions in industrial settings.

(2) In the sampled areas of the sprayed-state, which included 17,345 points, the average depth to be sprayed was 15.48 cm, with a maximum depth of 23.95 cm. This increase in depth from the bottom up is consistent with the wet spraying process, where spraying is done in a bottom-up sequence, and reflects the actual construction rules.

(3) For the sampled areas in the sprayed-state, which included 17,189 points, the average depth to be sprayed was 3.51 cm, with a maximum depth of 4.83 cm. In total, 90.43% of the sampling points were concentrated within the range of 3.5 ± 0.5 cm, and only

1.37% of the sampling points exceeded 4.5 cm, which is consistent with the on-site working conditions.

The consistency between the unsprayed depth of different states and the actual construction site indicates the reliability of the proposed algorithm.

To evaluate the accuracy of our model, the tunnel arch points extracted in the previous step were manually labeled, as shown in Figure 15.

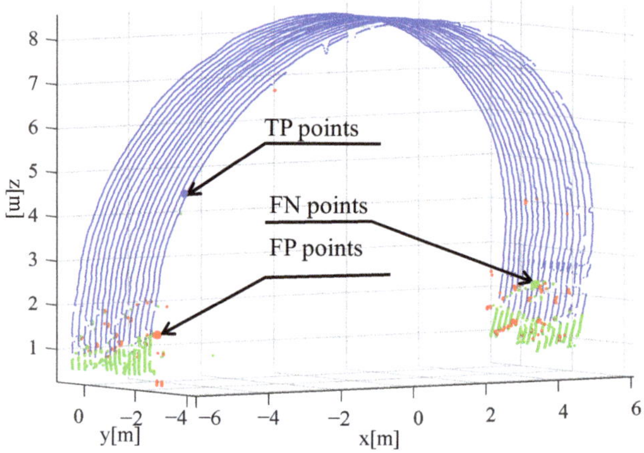

Figure 15. Different types of points in a certain section of the tunnel. The TP points and FP points represent tunnel surface points that were labeled correctly and incorrectly, respectively. The FN points were the points that were falsely labeled as non-tunnel surface points.

In Figure 15, the variables TP (True-Positive) marked in blue and FP (False-Positive) marked in red indicate the number of points that were labeled correctly and incorrectly as tunnel surface points, respectively. The variable FN (False-Negative) marked in green represents the number of points that were falsely labeled as non-tunnel surface points. To evaluate the performance of the model, the precision, recall, and F-score criteria used by Yang et al. [42,43] are adopted.

$$\begin{cases} precision = \dfrac{TP}{TP+FP}, \\ recall = \dfrac{TP}{TP+FN}, \\ F-score = 2 \cdot \dfrac{precision \cdot recall}{precision + recall}. \end{cases} \tag{31}$$

Table 2 presents the performance evaluation of the proposed algorithm for tunnel surface extraction and compensation using the criteria mentioned above, including precision, recall, and F-score.

Table 2. Assessment of the depth of concrete required to be sprayed on the tunnel surface in six sample areas.

ID	1	2	3	4	5	6	Average
Precision	0.918	0.924	0.987	0.933	0.941	0.912	0.936
Recall	0.906	0.915	0.934	0.920	0.903	0.917	0.916
F-score	0.912	0.920	0.960	0.927	0.922	0.915	0.926

Furthermore, we compared the average precision, recall, and F-score rates of the results obtained by different methods and obtained intuitive comparison results, as shown in Table 3. Based on the experimental results, the proposed method demonstrated higher precision and recall rates than the control group.

Table 3. Comparison of precision and recall rate of each method.

Method	Precision	Recall	F-Score
Region-growing	0.807	0.795	0.801
Elliptic cylindrical model	0.837	0.829	0.833
2D projection + BaySAC	0.792	0.801	0.796
Our method	0.936	0.916	0.926

The region-growing method [17] is applied to segment the rock surface from the tunnel point cloud. A curvature threshold is used with the region-growing algorithm to extract the rock surface of the tunnel. Additionally, the height threshold is used after DBSCAN to remove the miscellaneous points on the left and right walls of the tunnel. This method achieves an average precision, recall, and F-score of 80.7%, 79.5%, and 80.1%, respectively.

The elliptical cylinder model algorithm [36] differs from region growth algorithms in that it uses the central axis of the fit to divide the region into two parts. Subsequently, the elliptical fitting surface of the tunnel region is obtained through iteration to achieve the filtering of inner wall non-points. This method achieves an average precision, recall, and F-score of 83.7%, 82.9%, and 83.3%, respectively.

The continuous central axis is extracted by 2D projection [44], and then an interpolation algorithm based on quadric parametric surface fitting, using the BaySAC (Bayesian SAmpling Consensus) algorithm, is proposed to compute the cross-sectional point when it cannot be acquired directly from the tunnel points along the extraction direction of interest. This method achieves an average precision, recall, and F-score of 79.2%, 80.1%, and 79.6%, respectively.

By analyzing six sampled areas under different wet spraying conditions, our experimental results indicate that the algorithm achieved an average precision, recall, and F-score of 93.6%, 91.6%, and 92.6%, respectively. This demonstrates that our approach is better suited for analyzing complex wet spraying tunnel walls.

To verify the accuracy of the proposed method in determining the depth of concrete needed to be sprayed on the tunnel surface, a series of experiments was conducted in a real tunnel. The theoretical unsprayed depth $D_i{}'$ obtained from engineering design was compared to the algorithmic results D_i as shown in Figure 16.

The absolute average error MAD of the 1000 sampled areas is 0.989 cm, which demonstrates that the proposed algorithm meets the accuracy requirements of engineering design and confirms its reliability.

$$MAD = \frac{1}{n}\sum_{i=1}^{n}|D_i{}' - D_i| \tag{32}$$

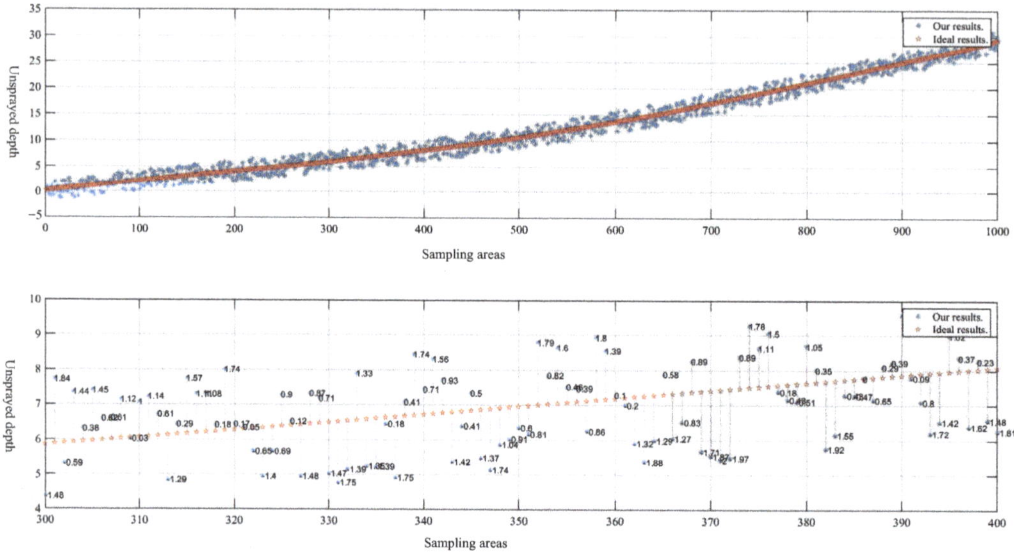

Figure 16. Comparison of the ideal unsprayed depth and the results obtained by our algorithm. The x-axis represents the sampling area, the y-axis represents the unsprayed depth, and the point marked with a star (☆) represents the ideal unsprayed depth. The point marked with an asterisk (*) represents the result obtained by our algorithm, and the number next to the point represents the absolute error between the unsprayed depth obtained by the proposed algorithm and the ideal unsprayed depth.

6. Conclusions

In this paper, we proposed an algorithm for analyzing the area of interest in a tunnel point cloud. We used a continuous projection point cloud correction algorithm to process the attitude of the inclined point cloud during acquisition and developed an adaptive point cloud compensation algorithm to overcome data loss caused by occlusion. Due to the irregular cross-sections of the large tunnel scene and large space, we proposed fitting the segment Lamé curve to mathematically describe the inner contour line of the tunnel, instead of using a simple elliptical cylinder or cylinder model. We then compared the tunnel design line to analyze the thickness of the tunnel to be wet sprayed, allowing for accurate assessment of the tunnel construction. The proposed algorithm has been shown to effectively evaluate the depth of concrete required to be sprayed on the tunnel surface, with an absolute average error of 0.989cm, meeting the precise requirements of engineering design and demonstrating its reliability.

We plan to continue to conduct research on tunnel wet spraying processes, specifically analyzing different scenarios and studying the more complex construction process of long bend tunnels. Our aim is to establish a comprehensive system for monitoring the depth of wet spraying in tunnels, as well as to investigate the positioning issue of mobile LiDAR data to achieve higher real-time detection algorithms.

Author Contributions: D.X.: Conceptualization, Funding acquisition, Supervision, Writing—review & editing; Q.S.: Formal analysis, Methodology, Software, Validation, Writing—original draft, Writing—review & editing; S.F.: Data curation, Investigation, Project administration; Y.G.: Resources. All authors have read and agreed to the published version of the manuscript.

Funding: This research was funded by National Natural Science Foundation of China under grant number 61973320, the Key Research and Development Program of Hunan Province under grant number 2022GK2059, and Xinjiang Province under grant number 2022294793.

Institutional Review Board Statement: Not applicable.

Informed Consent Statement: Not applicable.

Data Availability Statement: Data sharing not applicable.

Conflicts of Interest: We declare that we do not have any commercial or associative interest that represents a conflict of interest in connection with the work submitted.

References

1. Wu, H.; Yao, L.; Xu, Z.; Li, Y.; Ao, X.; Chen, Q.; Li, Z.; Meng, B. Road pothole extraction and safety evaluation by integration of point cloud and images derived from mobile mapping sensors. *Adv. Eng. Inform.* **2019**, *42*, 100936. [CrossRef]
2. Rato, D.; Santos, V. LIDAR based detection of road boundaries using the density of accumulated point clouds and their gradients. *Robot. Auton. Syst.* **2021**, *138*, 103714. [CrossRef]
3. Zhou, Z.; Gong, J. Automated residential building detection from airborne LiDAR data with deep neural networks. *Adv. Eng. Inform.* **2018**, *36*, 229–241. [CrossRef]
4. Shirowzhan, S.; Lim, S.; Trinder, J.; Li, H.; Sepasgozar, S.M. Data mining for recognition of spatial distribution patterns of building heights using airborne lidar data. *Adv. Eng. Inform.* **2020**, *43*, 101033. [CrossRef]
5. Wang, S.; Cai, G.; Cheng, M.; Marcato, J., Jr.; Huang, S.; Wang, Z.; Su, S.; Li, J. Robust 3D reconstruction of building surfaces from point clouds based on structural and closed constraints. *ISPRS J. Photogramm. Remote Sens.* **2020**, *170*, 29–44. [CrossRef]
6. Wang, Y.; Wang, J.; Chen, X.; Chu, T.; Liu, M.; Yang, T. Feature surface extraction and reconstruction from industrial components using multistep segmentation and optimization. *Remote Sens.* **2018**, *10*, 1073. [CrossRef]
7. Fekete, S.; Diederichs, M.; Lato, M. Geotechnical and operational applications for 3-dimensional laser scanning in drill and blast tunnels. *Tunn. Undergr. Space Technol.* **2010**, *25*, 614–628. [CrossRef]
8. Zhou, Y.; Wang, S.; Mei, X.; Yin, W.; Lin, C.; Hu, Q.; Mao, Q. Railway tunnel clearance inspection method based on 3D point cloud from mobile laser scanning. *Sensors* **2017**, *17*, 2055.
9. Ye, F.; Qin, N.; Liang, X.; Ouyang, A.; Qin, Z.; Su, E. Analyses of the defects in highway tunnels in China. *Tunn. Undergr. Space Technol.* **2021**, *107*, 103658. [CrossRef]
10. Zhao, Y.; Li, P. A statistical analysis of China's traffic tunnel development data. *Engineering* **2018**, *4*, 3–5. [CrossRef]
11. Walton, G.; Delaloye, D.; Diederichs, M.S. Development of an elliptical fitting algorithm to improve change detection capabilities with applications for deformation monitoring in circular tunnels and shafts. *Tunn. Undergr. Space Technol.* **2014**, *43*, 336–349. [CrossRef]
12. Li, D.; Xie, Q.; Gong, X.; Yu, Z.; Xu, J.; Sun, Y.; Wang, J. Automatic defect detection of metro tunnel surfaces using a vision-based inspection system. *Adv. Eng. Inform.* **2021**, *47*, 101206. [CrossRef]
13. Lemy, F.; Yong, S.; Schulz, T. A case study of monitoring tunnel wall displacement using laser scanning technology. In Proceedings of the Proceedings of 10th IAEG Congress' Engineering Geology for Tomorrow's Cities', Nottingham, UK, 6–10 September 2006; The Geological Society of London: London, UK, 2009; p. 482.
14. Fekete, S.; Diederichs, M.; Lato, M. Geotechnical applications of laser scanning in tunnels. In Proceedings of the ROCKENG09: Proceedings of the 3rd CANUS Rock Mechanics Symposium, Toronto, ON, Canada, 9–15 May 2009.
15. Gikas, V. Three-dimensional laser scanning for geometry documentation and construction management of highway tunnels during excavation. *Sensors* **2012**, *12*, 11249–11270. [CrossRef]
16. Amvrazis, S.; Voit, K.; Cordes, T.; Bergmeister, K. Drill and blast excavation forecasting using 3D laser scanning: Ausbruchprognose beim Sprengvortrieb mittels 3D-Laserscanning. *Geomech. Tunn.* **2017**, *10*, 298–316. [CrossRef]
17. Zhang, W.; Qiu, W.; Song, D.; Xie, B. Automatic tunnel steel arches extraction algorithm based on 3D LiDAR point cloud. *Sensors* **2019**, *19*, 3972. [CrossRef]
18. Luo, L.; Li, X.; Tao, M.; Dong, L. Mechanical behavior of rock-shotcrete interface under static and dynamic tensile loads. *Tunn. Undergr. Space Technol.* **2017**, *65*, 215–224. [CrossRef]
19. Song, K.I.; Cho, G.C. Bonding state evaluation of tunnel shotcrete applied onto hard rocks using the impact-echo method. *Ndt E Int.* **2009**, *42*, 487–500. [CrossRef]
20. Ginouse, N.; Jolin, M. Investigation of spray pattern in shotcrete applications. *Constr. Build. Mater.* **2015**, *93*, 966–972. [CrossRef]
21. Chun-Lei, L.; Hao, S.; Chun-Lai, L.; Jin-Yang, L. Intelligent Detection for Tunnel Shotcrete Spray Using Deep Learning and LiDAR. *IEEE Access* **2019**, *8*, 1755–1766. [CrossRef]
22. Ranjbarnia, M.; Zaheri, M.; Dias, D. Three-dimensional finite difference analysis of shallow sprayed concrete tunnels crossing a reverse fault or a normal fault: A parametric study. *Front. Struct. Civ. Eng.* **2020**, *14*, 998–1011. [CrossRef]
23. Oreste, P.; Spagnoli, G.; Luna Ramos, C.A. The Elastic Modulus Variation During the Shotcrete Curing Jointly Investigated by the Convergence-Confinement and the Hyperstatic Reaction Methods. *Geotech. Geol. Eng.* **2019**, *37*, 1435–1452. [CrossRef]

24. Yang, F.; Cao, S.R.; Qin, G. Mechanical behavior of two kinds of prestressed composite linings: A case study of the Yellow River Crossing Tunnel in China. *Tunn. Undergr. Space Technol.* **2018**, *79*, 96–109. [CrossRef]
25. Bloodworth, A.; Su, J. Numerical analysis and capacity evaluation of composite sprayed concrete lined tunnels. *Undergr. Space* **2018**, *3*, 87–108. [CrossRef]
26. Zhai, W.; Chapman, D.; Zhang, D.; Huang, H. Experimental study on the effectiveness of strengthening over-deformed segmental tunnel lining by steel plates. *Tunn. Undergr. Space Technol.* **2020**, *104*, 103530. [CrossRef]
27. Li, C.; Li, M.J.; Zhao, Y.G.; Liu, H.; Wan, Z.; Xu, J.C.; Xu, X.P.; Chen, Y.; Wang, B. Layer recognition and thickness evaluation of tunnel lining based on ground penetrating radar measurements. *J. Appl. Geophys.* **2011**, *73*, 45–48. [CrossRef]
28. Xisheng, D.; Tian, D.; Quan, Y.; Xin, Z. Tunnel lining thickness and voids detection by GPR. *Electron. J. Geotech. Eng.* **2015**, *20*, 2019–2030.
29. Rusu, R.B.; Marton, Z.C.; Blodow, N.; Dolha, M.; Beetz, M. Towards 3D point cloud based object maps for household environments. *Robot. Auton. Syst.* **2008**, *56*, 927–941. [CrossRef]
30. Pleansamai, K. M-estimator sample consensus planar extraction from image-based 3D point cloud for building information modelling. *Int. J. Geomate* **2019**, *17*, 69–76. [CrossRef]
31. Torr, P.; Zisserman, A. MLESAC: A New Robust Estimator with Application to Estimating Image Geometry. *Comput. Vis. Image Underst.* **2000**, *78*, 138–156. [CrossRef]
32. Edelsbrunner, H.; Kirkpatrick, D.; Seidel, R. On the shape of a set of points in the plane. *IEEE Trans. Inf. Theory* **1983**, *29*, 551–559. [CrossRef]
33. Hinrichsen, D.; Pritchard, A. An improved error estimate for reduced-order models of discrete-time systems. *IEEE Trans. Autom. Control.* **1990**, *35*, 317–320. [CrossRef]
34. Fischler, M.A.; Bolles, R.C. Random sample consensus:a paradigm for model fitting with applications to image analysis and automated cartography. *Commun. ACM* **1981**, *24*, 381–395. [CrossRef]
35. Derpanis, K.G. Overview of the RANSAC algorithm. *Image Rochester NY* **2010**, *4*, 2–3.
36. Zhu, N.; Jia, Y.; Luo, L. Tunnel point cloud filtering method based on elliptic cylindrical model. *Int. Arch. Photogramm. Remote Sens. Spat. Inf. Sci.* **2016**, *41*, 735. [CrossRef]
37. Ramos, P.M.; Janeiro, F.M.; Radil, T. Comparison of impedance measurements in a DSP using ellipse-fit and seven-parameter sine-fit algorithms. *Measurement* **2009**, *42*, 1370–1379. [CrossRef]
38. Stoer, J.; Bulirsch, R. *Introduction to Numerical Analysis*; Springer Science & Business Media: Berlin/Heidelberg, Germany, 2013; Volume 12.
39. Gratton, S.; Lawless, A.S.; Nichols, N.K. Approximate Gauss–Newton methods for nonlinear least squares problems. *SIAM J. Optim.* **2007**, *18*, 106–132. [CrossRef]
40. Sun, F.; Cai, X.; Zhu, Y. Analytical solution of internal force and displacement in multi-center circular arc tunnel lining based on initial parameter method. *Rock Soil Mech.* **2009**, *30*, 1127–1130.
41. Sun, S.; Hu, J.; Zhang, W. Shape optimization of openings on rotation shells based on super-elliptic function and sequential response surface method. *Acta Aeron. Astron. Sin.* **2015**, *36*, 3595–3607.
42. Yang, B.; Fang, L.; Li, J. Semi-automated extraction and delineation of 3D roads of street scene from mobile laser scanning point clouds. *ISPRS J. Photogramm. Remote Sens.* **2013**, *79*, 80–93. [CrossRef]
43. Liu, A.; Wang, Z.; Nie, W.; Su, Y. Graph-based characteristic view set extraction and matching for 3D model retrieval. *Inf. Sci.* **2015**, *320*, 429–442. [CrossRef]
44. Kang, Z.; Zhang, L.; Tuo, L.; Wang, B.; Chen, J. Continuous extraction of subway tunnel cross sections based on terrestrial point clouds. *Remote Sens.* **2014**, *6*, 857–879. [CrossRef]

sensors

MDPI

Article

Dynamic Photoresponse of a DNTT Organic Phototransistor

Marcello Campajola [1,*], Paolo Di Meo [1], Francesco Di Capua [1,2], Paolo Branchini [3] and Alberto Aloisio [1,2,4,5]

1 Istituto Nazionale di Fisica Nucleare (INFN), Sezione di Napoli, Via Cintia, 80126 Napoli, Italy
2 Department of Physics "E. Pancini", University of Naples "Federico II", Via Cintia, 80126 Napoli, Italy
3 Istituto Nazionale di Fisica Nucleare (INFN), Sezione di RomaTre, Via della Vasca Navale 84, 00146 Roma, Italy
4 CNR-SPIN, Via Campi Flegrei 34, 80078 Pozzuoli, Italy
5 Task Force di Bioelettronica, University of Naples "Federico II", Via Cintia, 80126 Napoli, Italy
* Correspondence: macampajola@na.infn.it

Abstract: The photosensitivity, responsivity, and signal-to-noise ratio of organic phototransistors depend on the timing characteristics of light pulses. However, in the literature, such figures of merit (FoM) are typically extracted in stationary conditions, very often from IV curves taken under constant light exposure. In this work, we studied the most relevant FoM of a DNTT-based organic phototransistor as a function of the timing parameters of light pulses, to assess the device suitability for real-time applications. The dynamic response to light pulse bursts at ~470 nm (close to the DNTT absorption peak) was characterized at different irradiances under various working conditions, such as pulse width and duty cycle. Several bias voltages were explored to allow for a trade-off to be made between operating points. Amplitude distortion in response to light pulse bursts was also addressed.

Keywords: organic field-effect transistors (OFET); organic phototransistor (OPT); dinaphtho-thieno-thiophene (DNTT); fast photoresponse; low-voltage operations OPT; random telegraph signals (RTS)

1. Introduction

Organic field-effect transistors (OFETs) are attracting the interest of the scientific community as a valid alternative to standard semiconductor-based devices. This is due to their several interesting properties, such as cost-effective fabrication over large areas, flexibility, and light weight. The recent progress in OFET fabrication techniques has led to the development of devices with high field-effect mobilities μ, low threshold voltages V_{th}, and high current on/off ratios $I_{on/off}$. OFETs can be successfully employed as building blocks in electronic circuits, as well as sensors in applications such as chemical and biological sensing, gas analysis, and pressure monitoring [1–4]. Several works also demonstrated how OFETs can be efficiently used as organic phototransistors (OPTs). The potential of such devices lies in the good photosensitive properties of organic semiconductors, combined with the transistor intrinsic amplification capability [5,6]. Fields of application of OPTs also include optical memory, light communication [7–9], and radiation detection and dosimetry in radiotherapy [10–12]. Several organic semiconductor materials have been investigated for OPT development, including TIPS-pentacene and dinaphtho[2,3-b:2′,3′-f]thieno[3,2-b′]thiophene (DNTT). For those materials, field-effect mobility comparable to that of amorphous silicon, typically μ~1 cm^2/Vs, low turn-on voltage, and on/off current ratio larger than 10^6 have been reported [13–15].

The origin of the photoresponse in OPTs is usually attributed to the increase in the minority carrier trap generation rate under light exposure [16,17]. For a p-type device (i.e., hole-transporting device), the accumulation of photogenerated electrons in trap states in the phototransistor channel causes a shift in the threshold voltage, resulting in multiple holes injected into the channel for a single photogenerated carrier pair. Such a mechanism is known as photoconductive gain. The transverse electric field induced by the gate plays the role of helping to segregate photogenerated electron–hole pairs, decreasing the probability of recombination. The photoconductive gain is recognized as a slow process:

Citation: Campajola, M.; Di Meo, P.; Di Capua, F.; Branchini, P.; Aloisio, A. Dynamic Photoresponse of a DNTT Organic Phototransistor. *Sensors* **2023**, *23*, 2386. https://doi.org/10.3390/s23052386

Academic Editors: Qibo Feng, Jiakun Li and Qixin He

Received: 17 January 2023
Revised: 12 February 2023
Accepted: 17 February 2023
Published: 21 February 2023

phototransistors that exhibit high photoconductive gain usually do not operate as fast as a photodiode with similar carrier transit time [16,18]. Hence, characterizing the dynamic photoresponse to transient light pulses is of paramount importance for devices of interest for real-time applications.

The organic thin-film charge photogeneration has been investigated in the range from picoseconds to seconds after excitation [19,20]. The optical response has been also investigated in depth in OPTs. Among others, we mention here studies on devices based on P3OT [21], PQT-12 [22], PBDFTDTBT [23], P3H3 [24], PDVT-8/PC_{61}BM [25], DNTT [26–28], and TIPS-pentacene [15]. Most of these works studied the performance of OPTs under quasi-static light conditions at different irradiance, whereas a characterization of the device dynamic response to short light pulses at low irradiance has been barely investigated. Quasi-static characterization is useful for comparing the performance of different materials and devices; however, it is not sufficient for assessing their suitability for detection of short light pulses.

In this paper, we present a systematic characterization of a DNTT-based OPT, specifically developed for real-time radiation detection and dosimetry applications [12,29,30], where fast photoresponse to weak, fast light pulses is expected to be a key feature. Specifically, we report on the response to ~470 nm light pulses, under various timing conditions, at different irradiances and operating points. Several figures of merit (FoM), such as photosensitivity, responsivity, and signal-to-noise ratio, have been investigated. In all the explored regions, these quantities strongly depend on the timing characteristics of the light pulses. In this work, we have characterized them as a function of the pulse width, frequency, and duty cycle, studying the correlation with the bias voltages. We also have studied stress and distortion effects in the photoresponse when light pulse bursts are applied, to evaluate the exploitation of such devices in applications where random light pulse sequences are expected.

2. Materials and Methods

2.1. Organic Phototransistor Layout

The device under characterization is a thin-film transistor based on DNTT. The device was designed, engineered, and realized at the CNR-IMM laboratory (Rome, Italy). Details on the fabrication process are given in Refs. [29,30]. Here, we only recall some useful information. The OPT was fabricated in a bottom-gate/top-contact configuration on a 100 µm thick substrate of polyethylene-naphthalate. The gate is made by a 70 nm thick Al layer. Source and drain are made by 30 nm thick interdigitated finger electrodes of Au. The semiconductor is 50 nm thick. A dielectric layer, made of a 600 nm thick fluoropolymer-based material (Cytop^{TM}), separates the semiconductor from the gate. The whole structure is encapsulated within a 240 nm thick layer of Cytop.

OPT are arranged in a 2 × 2 matrix (see Figure 1a). Each OPT has a channel length L = 100 µm and a width W = 40 mm. The active area A = L × W is 0.04 cm^2.

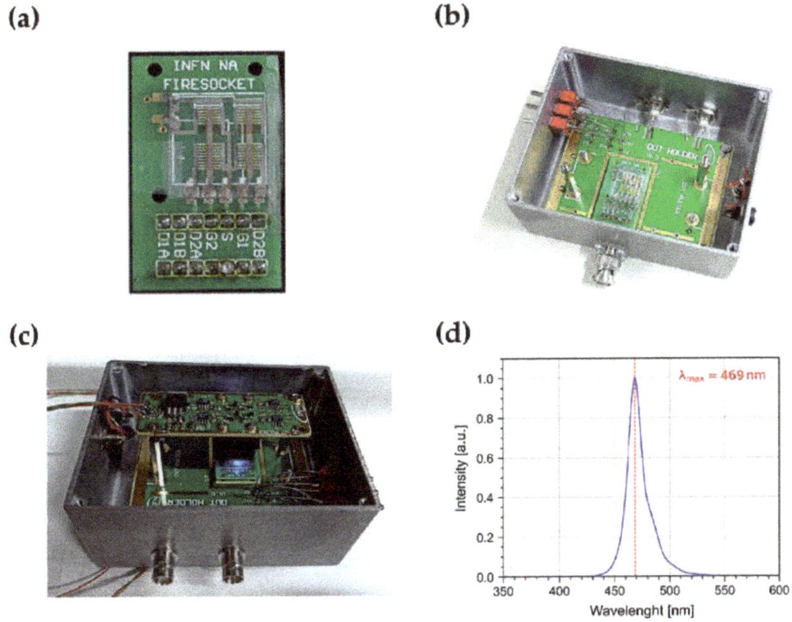

Figure 1. Pictures of (**a**) the 2 × 2 OPT matrix bonded on a custom-made socket; (**b**) the OPT matrix on the socket arranged on a motherboard within an aluminum box used as Faraday cage; (**c**) the aforementioned setup with a LED source and its driver circuit mounted on top of the OPT. (**d**) Emission spectrum of the LED source.

2.2. Electrical and Photoresponse Characterization Setup

The device was bonded on a custom-made socket (Figure 1a) arranged on a motherboard enclosed in a Faraday cage (Figure 1b), which also acts as a black box to shield the samples from environmental light.

The electrical characterization was performed by means of a B1500A Semiconductor Device Parameter Analyzer, equipped with three source meter units (SMUs) connected to the device source, gate, and drain. The SMUs were used to bias the OPT and to collect transfer and output curves.

For the dynamic photoresponse characterization, the device was illuminated with an LED source (Broadcom HLMP-KB45-A0000) installed on the top of the motherboard housing the device under test (DUT) within the Faraday cage (Figure 1c).

The LED emission spectrum (Figure 1d) was characterized by means of a CCD-based Ocean Insight Spectrometer with a sub-nm resolution. Such a LED was chosen due to its emission wavelength peak (469 nm) close to the absorption peak of the DNTT: ~450 nm (see Ref. [29]). The LED current was controlled by a driver circuit connected to a function generator used to supply control pulses with variable timing. An irradiance in the range from a few nW/cm^2 to about 10 μW/cm^2 was achieved with a collimated flat-top spot with a diameter of ~4 mm on the DUT surface. The irradiance was calibrated by means of a Thorlabs PM100USB power meter equipped with a S120C silicon photodiode.

The B1500A unit was used in the "I/V-t sampling measurement" mode [31] to sample at a rate of 20 Hz the OPT drain current trend under illumination.

All measurements were conducted under ambient atmospheric conditions at room temperature.

3. Results and Discussion

3.1. Electrical Characterization in Dark

We measured the output and transfer characteristics of the device in the dark. Output curves (Figure 2a) were acquired with a forward and reverse drain voltage V_{ds} scan in the range [+1 V, −12 V], for gate voltage V_{gs} values between −3 V and −12 V in steps of 3V. Transfer curves (Figure 2b,c) were acquired in linear (V_{ds} = −1 V) and saturation regimes (V_{ds} = −20 V) with a forward and reverse V_{gs} scan in the range [+4 V, −9 V].

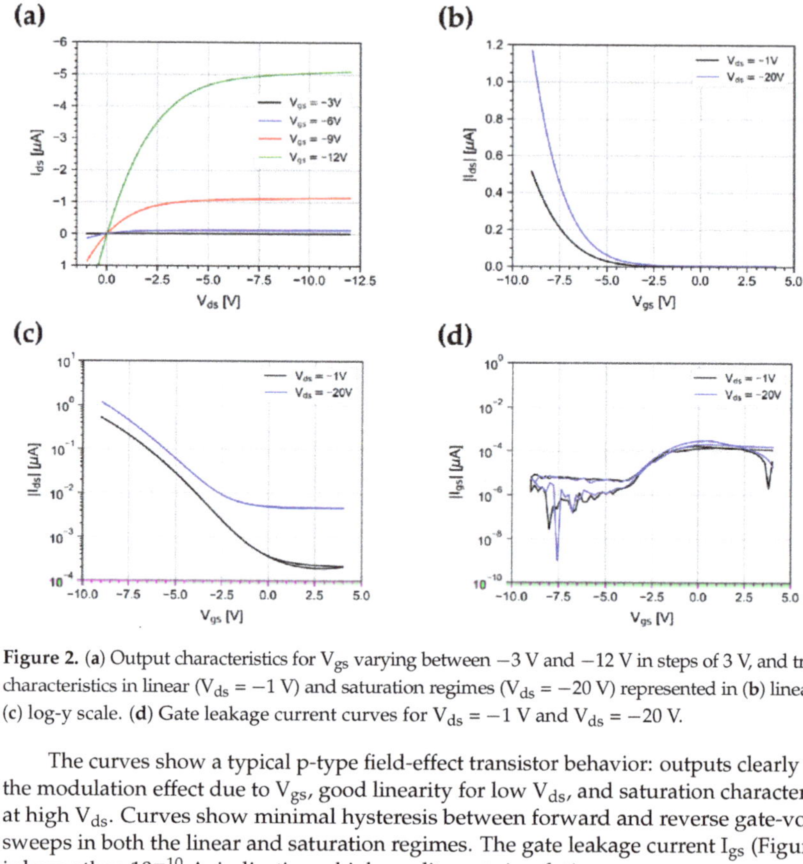

Figure 2. (**a**) Output characteristics for V_{gs} varying between −3 V and −12 V in steps of 3 V, and transfer characteristics in linear (V_{ds} = −1 V) and saturation regimes (V_{ds} = −20 V) represented in (**b**) linear and (**c**) log-y scale. (**d**) Gate leakage current curves for V_{ds} = −1 V and V_{ds} = −20 V.

The curves show a typical p-type field-effect transistor behavior: outputs clearly show the modulation effect due to V_{gs}, good linearity for low V_{ds}, and saturation characteristics at high V_{ds}. Curves show minimal hysteresis between forward and reverse gate-voltage sweeps in both the linear and saturation regimes. The gate leakage current I_{gs} (Figure 2d) is lower than 10^{-10} A, indicating a high-quality gate insulation.

From the curves, we extracted several characteristic parameters of the DUT [32]. The field-effect mobility, computed in the linear regime (V_{ds} = −1 V), is ~ 0.5 cm^2 V^{-1} s^{-1} at V_{gs} = −12 V. The threshold voltage is around −10 V. The onset voltage is in the range +2 V to −3 V. The subthreshold slope is ~2.4 V dec^{-1}. The log(I_{on}/I_{off}) ratio is around 5, where I_{on} is measured at V_{gs} = −12 V. All the extracted parameters in the explored ranges are in good agreement with values reported in the literature for DNTT-based OFETs [13,14,26–30].

3.2. Dynamic Photoresponse

We measured the device photoresponse to light bursts at ~470 nm. Upon illumination, the device shows a fast switching of the drain current. As an example, we show in Figure 3a the drain current measured when illuminating the device with a burst of 5 pulses at an irradiance of 500 nW/cm^2, with a repetition period T = 60 s (see Figure 3b) at V_{ds} = −5 V and V_{gs} = −5 V. The signal shows a temporal development that can be parameterized as the sum of two exponentials [33–36]. The measured characteristic times are $\tau_{fast} = (0.461 \pm 0.005)$ s and $\tau_{slow} = (4.02 \pm 0.02)$ s, as shown in Figure 3c. When the light is turned off, the device response drops very slowly, with characteristic times $\tau_{fast} = (2.71 \pm 0.03)$ s and $\tau_{slow} = (30.18 \pm 0.15)$ s, as shown in Figure 3d. Both rise and fall times are almost constant within the burst. The slow return of I_{ds} to the initial conditions when light is turned off is due to the time it takes for trapped photogenerated traps to recombine [33].

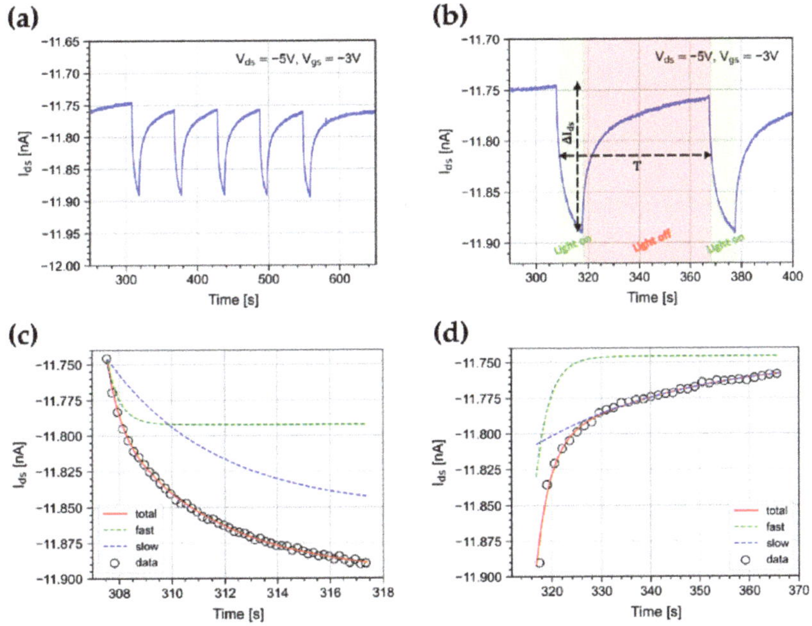

Figure 3. (**a**) I_{ds} current when illuminating the device with a 5-pulse burst with a period of T = 60 s. (**b**) Details of the first two pulses of the burst. Data and best-fit curve with details on the fast and slow exponential function components for the (**c**) signal growth and (**d**) decay.

We investigated several key FoMs, as well as studied the correlation of the pulse timing and burst structure with the irradiance and the bias voltages, in order to characterize the response in real-time applications.

The photosensitivity P and the photoresponsivity R were computed as follows [1,15]:

$$P = \frac{\Delta I_{ds}}{I_{ds}^{dark}}, \tag{1}$$

$$R = \frac{\Delta I_{ds}}{A \times IRR}, \tag{2}$$

where $\Delta I_{ds} = I_{ds}^{light} - I_{ds}^{dark}$, and I_{ds}^{light} and I_{ds}^{dark} are the drain current in dark conditions and under illumination, respectively; A is the active device area; and IRR is the irradiance of the light source. R and P depend on the transistor layout and on the polarization conditions [29].

Figure 4 shows the trend of R and P as a function of V_{gs} for $V_{ds} = -1$ V and for pulse widths up to 10 s at a constant irradiance of 500 nW/cm^2. The values shown in the plot are the average taken on bursts of 5 pulses. The responsivity has a minimum as V_{gs} approaches 0 V, while it gradually increases up to 0.03 AW^{-1} at $V_{gs} = -7$ V, as a result of the increasing exciton dissociation rate due to the transversal electrical field in the channel [27,37]. The maximum value of P is 0.024, reached at $V_{gs} = 0$ V. This is near the turn on voltage, as a result of the abundance of photogenerated carriers over scarce field-generated carriers. At more negative V_{gs} values, the photoresponse is subdued to the increasing number of field-induced charge carriers [27].

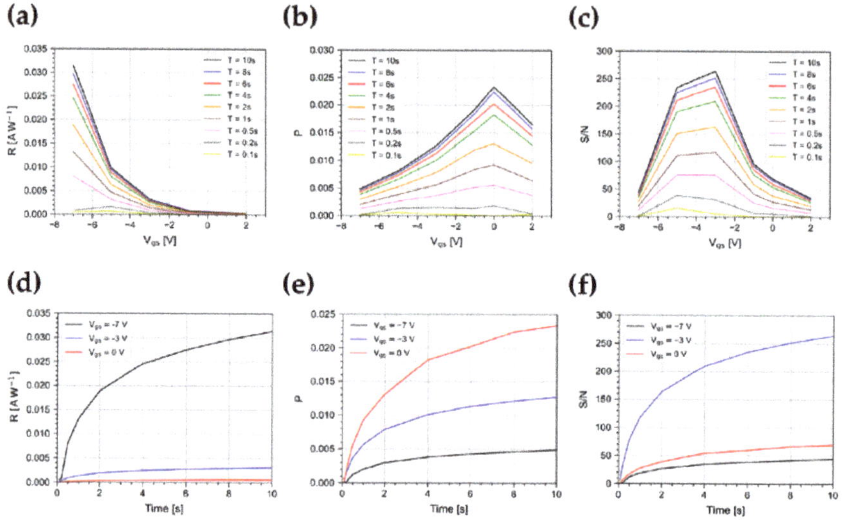

Figure 4. (**a**) Responsivity, (**b**) photosensitivity, and (**c**) signal-to-noise ratio as a function of V_{gs} for $V_{ds} = -1$ V for light pulse widths from 0.1 s to 10 s. (**d**) Responsivity, (**e**) photosensitivity, and (**f**) signal-to-noise ratio as a function of the light pulse width at $V_{gs} = -7, -3, 0$ V and $V_{ds} = -1$ V. All these FoMs are shown at an irradiance of 500 nW/cm^2.

Furthermore, we measured the signal-to-noise ratio S/N, where the noise *N* is evaluated as the RMS of the dark current measured immediately before the arrival of the light pulse. Figure 4c shows the trend of S/N as a function of V_{gs} and for different pulse widths. The maximum value is ~260, reached at $V_{gs} = -3$ V for a pulse width of 10 s. For a width shorter than 0.5 s, the photoresponse reduces, and the highest S/N ratio occurs at $V_{gs} = -5$ V.

As shown in Figure 4d–f, the magnitude of R, P, and S/N reduces for shorter pulse widths, because of the finite slew rate of I_{ds}. As an example, at $V_{gs} = -7$ V, by varying the light pulse width in the range from 100 ms to 1 s, R changes by a factor 30, while from 1 s to 10 s, R changes only by a factor ~2. A similar behavior is observed for P and S/N. As a consequence, the device characterizations achieved in quasi-stationary regimes are not representative of the device response to short light pulses, as expected in various photodetection applications.

We studied R, P and S/N for two more V_{ds} values (−5 V, −10 V) and compared the results, as shown in Figure 5a–c. The responsivity monotonically increases with the absolute values of gate and drain voltages. From the measured values of R, we derived the external quantum efficiency (EQE) of the device, i.e., the ratio between the number of incident photons and photogenerated carriers, as follows [1]:

$$\text{EQE} = \text{R}\frac{hc}{\lambda q}, \quad (3)$$

where q is the elementary charge, h is the Planck constant, and c is the speed of light in the vacuum. Values as high as 20% were measured at V_{gs} = −7 V and V_{ds} = −10 V.

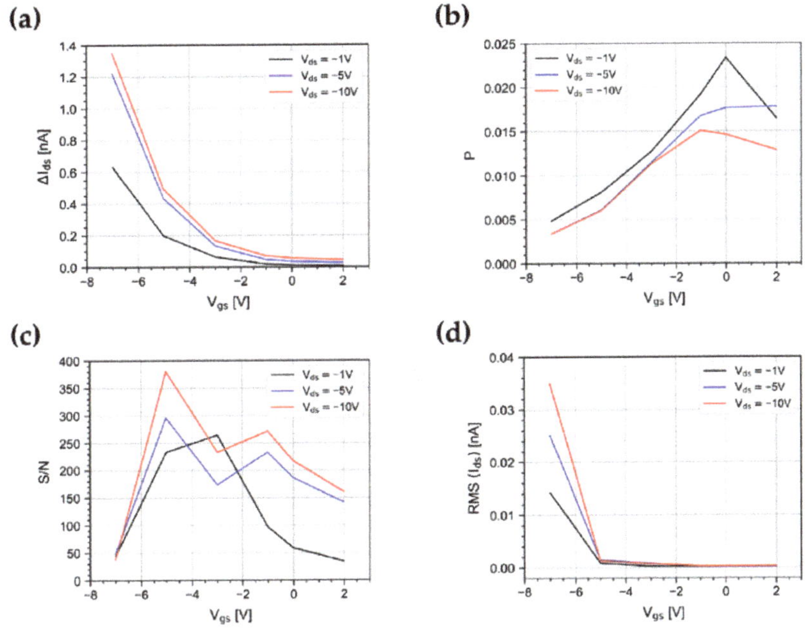

Figure 5. (**a**) Responsivity, (**b**) photosensitivity, (**c**) signal-to-noise ratio, and (**d**) root mean square of the I_{ds} dark current as a function of V_{gs} for V_{ds} = −1, −5, −10 V at an irradiance of 500 nW/cm^2.

It should be noted that as V_{gs} and V_{ds} increase, the dark current I_{ds} grows as well, as shown in transfer curves in Figure 2b. As a consequence, in applications where the OPT is DC coupled to the front-end circuit, and the read-out of short light pulses asks for high-gain trans-impedance amplifiers, the increment of the dark current could bring the amplifier in saturation.

Differently from R, both P and S/N correlate the photoresponse with the dark current and its root mean square (Figure 5d). The highest photosensitivity value was obtained with the lowest drain bias applied in our tests (V_{ds} = −1 V) and with a gate voltage V_{gs} = 0 V. On the contrary, the S/N plot suggests the best operating point to be at V_{gs} = −5 V and V_{ds} = −10 V. However, for high values of V_{ds}, it is well known that the high electrical field lowers the device stability [18]. Indeed, we have observed that random telegraph signal (RTS) phenomena arise. RTS consists of the discrete, fast fluctuation of the dark current I_{ds} between two or more values [38]. Figure 6 shows, as an example, a time window where the drain current was affected by RTS behaviors. The onset of RTS is clearly to avoid. The dark current fluctuations generate swift P variations and the step changes in the signal baseline could be interpreted as a photoresponse. A good compromise is then achieved by decreasing the gate and drain bias down to values as low as V_{ds} = −1 V and V_{gs} = −3 V. Taking all that into account, the optimal operating point clearly depends on the specific application and the experimental setup to be used.

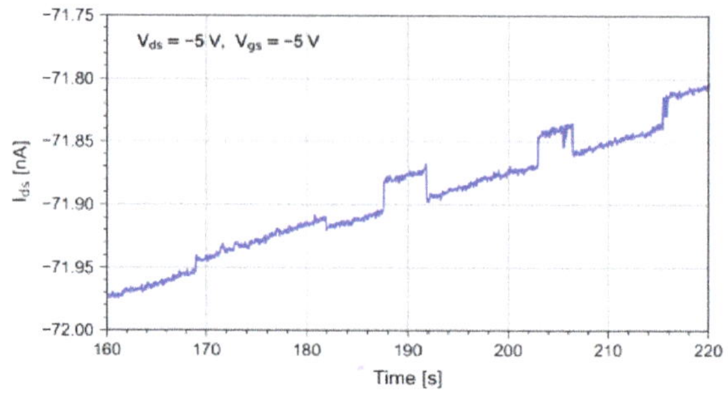

Figure 6. Random telegraph signal behavior in the drain dark current observed for the OPT polarization of $V_{ds} = -5$ V, $V_{gs} = -5$ V.

In order to study possible channel stress effects due to light exposure, we illuminated the DUT with bursts of 20 pulses for different repetition periods: T = 15, 20, 35, 60 s. Figure 7 shows the drain current trend when illuminating the device with bursts of light pulses with different periods. As noted in [29], the exposure to repeated light pulses causes a drift of the drain current. This can be explained by the pile-up of a persistent component of the photocurrent [33,39,40].

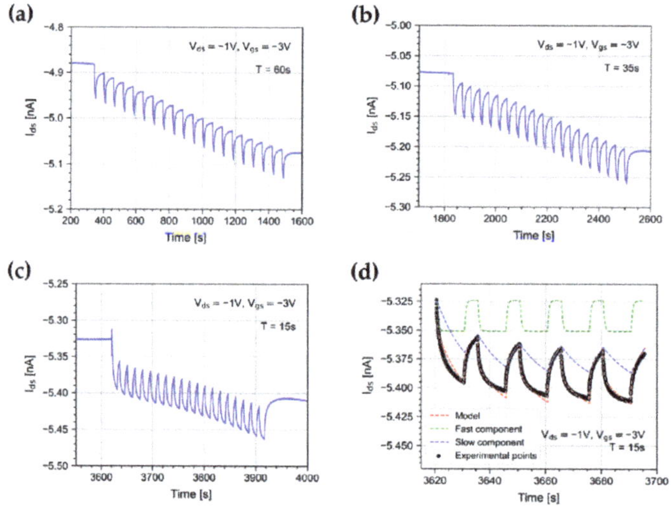

Figure 7. Drain current when illuminating the device with 10 s width light pulses with a period of (**a**) T = 60 s, (**b**) T = 35 s, and (**c**) T = 15 s. The OPT was polarized with $V_{ds} = -1$ V and $V_{gs} = -3$ V, and the irradiance was 500 nW/cm^2. (**d**) Drain current measured under 10 s light pulses with a period of 15 s (black points) and the eq. 6 model best-fit curve (red dotted line). The fitted slow and fast model components are shown in green and blue, respectively.

Moreover, a pulse height reduction between the first pulse of the burst and the following ones is observed. A similar effect was shown in [41]; however, it has neither been discussed nor interpreted. Figure 7a–c clearly shows that the shorter the period between pulses, the greater the pulse height reduction.

We found that such phenomena can be described by composing models taken from the literature [12,42,43]. As suggested in [42,43], we assumed that the photocurrent I'_o is linearly dependent on the total number of photogenerated minority carrier traps n:

$$I_{ph}(t) \propto n(t) \tag{4}$$

The time evolution of the defect density is governed by a rate equation:

$$\frac{dn}{dt} = a - bn(t), \tag{5}$$

where a is the defect generation rate, which is ultimately related to irradiance, and b is the defect recombination rate.

The phenomena we observed require the presence of two populations of defects with densities n_x and n_y, whose activation energies are different [12]. In such an approach, n_x is responsible for the fast photocurrent component, while n_y for the persistent photocurrent component. Differently from [12], where two defect populations with continuous activation energy values were considered, here we simply assumed two discrete values. The overall photocurrent is hence given by:

$$\mathrm{I_{ph}}(t) \propto n_x(t) + n_y(t) \tag{6}$$

where both n_x and n_y follow a rate equation in the form of Equation (5).

Figure 7d shows the measured drain current (in black) and superimposed the Equation (6) model best-fit curve (red dotted curve). The fast and persistent photocurrent components due to the n_x and n_y defect densities' evolution are shown in green and blue, respectively.

The model reproduces the decrease of the signal height observed in data between the first pulse and the following ones well. This phenomenon happens because of the build-up of the persistent photocurrent component, which after the first burst does not return to zero. The wander of the drain current baseline changes the response to the light pulses in the burst as a function of the burst length.

Figure 8a,c show the trend of R and P for the pulses of the burst for different periods. R and P show a marked reduction between the first and the second pulse, while they remain nearly constant afterward. Figure 8b,d show R and P values averaged over the burst pulses and normalized to the value achieved at 60 s, as a function of the light pulse period. They clearly show that the shorter the pulse period, the smaller are R and P. A similar trend is observed at different gate polarizations.

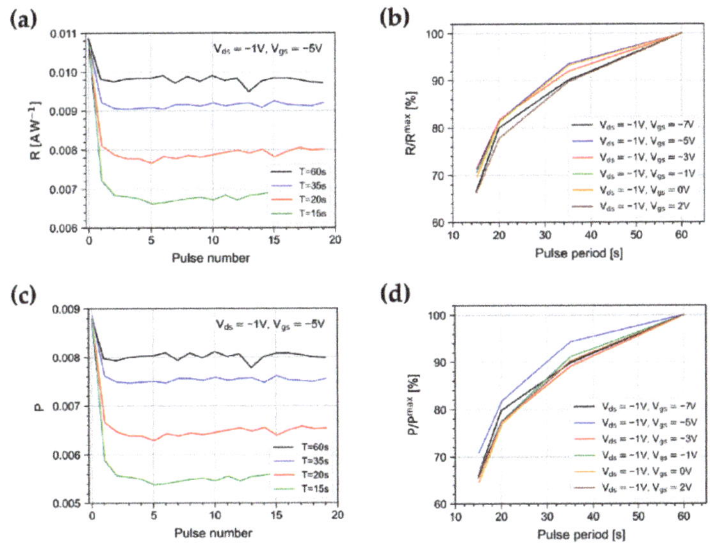

Figure 8. Trends of (**a**) R and (**c**) P for pulses in the burst, for different periods, at V_{ds} = −1 V and V_{gs} = −5 V. Averages of (**b**) R and (**d**) P in the burst, normalized to the values at 60 s, as a function of the light pulse period for different gate polarizations.

Finally, we have investigated the OPT photoresponse to 10 s light pulses as a function of the irradiance in the range from 100 nW/cm^2 to 1900 nW/cm^2. As shown in Figure 9, ΔI_{ds} increases with the irradiance and values as high as ~120 pA are reached at 1900 nW/cm^2. On the other hand, light pulses at an irradiance as low as 100 nW/cm^2 still produce a detectable response of ~20 pA. A nonlinear behavior is observed in Figure 9a, which is likely to be attributed to the limiting nature of charge transport mechanisms in the polymer [44]. As a consequence, at the largest irradiance applied in our test, R reduces, and values as low as 1.5 mAW^{-1} are reached (Figure 9b).

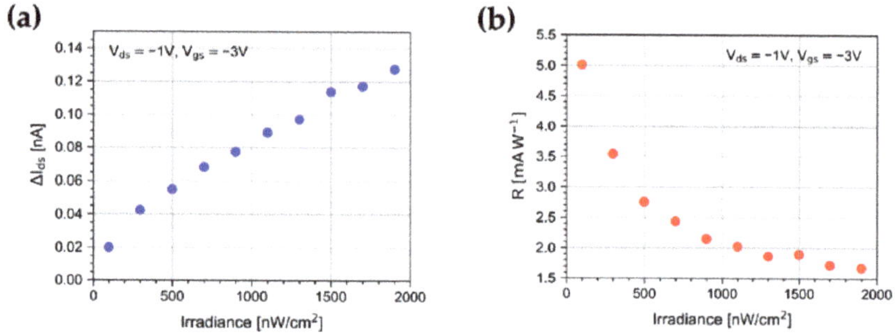

Figure 9. Trend of (**a**) ΔI_{ds} and (**b**) R for 10 s light pulses as a function of the irradiance at V_{gs} = −3 V and V_{ds} = −1 V.

4. Conclusions

Our results aim to evaluate the dynamic performance of a DNTT-based OPT in the view of deployment in applications that foresee the detections of fast, random light pulse sequences, such as radiation detection, dosimetry, and visible light communication. We have characterized the dynamic photoresponse to ~470 nm light pulses, close to the DNTT absorption peak.

We investigated several FoMs, such as photosensitivity, responsivity, and signal-to-noise ratio, under various timing conditions, at different irradiances and operating points. In all the explored regions, we observed that photosensitivity, responsivity, and signal-to-noise ratio strongly depend on the timing characteristics of the light pulses. R changes by more than an order of magnitude from 100 ms to 1 s, while R changes only by a factor ~2 from 1 s to 10 s. Stationary conditions are reached only after an exposition of a couple of minutes. P and S/N behave in a similar fashion.

The maximum of the responsivity is obtained at the higher gate and drain voltages explored. Differently, the highest photosensitivity value is obtained with the lowest drain bias applied in our tests and with the gate and source shorted to ground. The S/N plots suggest increasing both V_{gs} and V_{ds} to achieve the best operating point. It is noticeable, however, that for high values of V_{ds}, random telegraph signal effects in the drain current arise. Their occurrence makes it questionable to quantify the S/N ratio, and moreover, is clearly to be avoided, because such step-like fluctuations could be interpreted as a true signal. In applications where signal amplitude is paramount, both V_{gs} and V_{ds} should be increased to the limit allowed by the stability of the device operation and bias stress effects. On the other hand, in order to avoid saturation in DC-coupled high-gain trans-impedance amplifiers, a low dark current is mandatory, and hence low bias voltages are required.

We also studied stress and distortion effects in the photoresponse when pulse bursts are applied. We observed a reduction in the ΔI_{ds} photoresponse, between the first and the second pulse, while it remains approximately constant in the following pulses. We found that such an effect in the data is well reproduced by composing models taken from the literature, in which the photocurrent is assumed to be proportional to the defect density. Supposing the presence of just two kinds of defects with discrete activation energies allowed us to reproduce the experimental data accurately in a time window of minutes.

In future works, we aim to correlate the OPT detectivity and the limit of detection to the timing of the incoming light pulses.

Author Contributions: Conceptualization and methodology, A.A. and M.C.; measurements, A.A., M.C. and P.D.M.; data analysis, M.C.; writing—original draft preparation, A.A. and M.C.; writing—review and editing, A.A. and M.C.; visualization, A.A., M.C., P.B., F.D.C. and P.D.M. All authors have read and agreed to the published version of the manuscript.

Funding: This research has been funded by the CSN V of INFN, in the framework of the FIRE experiment and by Horizon 2020 Marie Sklodowska-Curie RISE project JENNIFER2 (European grant agreement No. 822070).

Institutional Review Board Statement: Not applicable.

Informed Consent Statement: Not applicable.

Data Availability Statement: The data presented in this study are available on request from the corresponding author.

Acknowledgments: The authors would like to thank Antonio Anastasio (INFN-Napoli), Alfonso Boiano (INFN-Napoli), and Lorenzo Roscilli (INFN-Napoli) for the design of equipment used in this paper.

Conflicts of Interest: The authors declare no conflict of interest.

References

1. Luo, L.; Liu, Z. Recent progress in organic field-effect transistor-based chem/bio-sensors. *VIEW* **2022**, *3*, 20200115. [CrossRef]
2. Liu, K.; Ouyang, B.; Guo, X.; Guo, Y.; Liu, Y. Advances in flexible organic field-effect transistors and their applications for flexible electronics. *npj Flex Electron* **2022**, *6*, 1. [CrossRef]
3. Yuvaraja, S.; Nawaz, A.; Liu, Q.; Dubal, D.; Surya, S.G.; Salama, K.N.; Sonar, P. Organic field-effect transistor-based flexible sensors. *Chem. Soc. Rev.* **2020**, *49*, 3423–3460. [CrossRef] [PubMed]
4. Tang, W.; Huang, Y.; Han, L.; Liu, R.; Su, Y.; Guo, X.; Yan, F. Recent progress in printable organic field effect transistors. *J. Mater. Chem. C* **2019**, *7*, 790–808. [CrossRef]
5. Ostroverkhova, O. Organic Optoelectronic Materials: Mechanisms and Applications. *Chem. Rev.* **2016**, *116*, 13279–13412. [CrossRef] [PubMed]
6. Wakayama, Y.; Hayakawa, R.; Seo, H.S. Recent progress in photoactive organic field-effect transistors. *Sci. Technol. Adv. Mater.* **2014**, *15*, 024202. [CrossRef] [PubMed]
7. Tavasli, A.; Gurunlu, B.; Gunturkun, D.; Isci, R.; Faraji, S. A Review on Solution-Processed Organic Phototransistors and Their Recent Developments. *Electronics* **2022**, *11*, 316. [CrossRef]
8. Vega-Colado, C.; Arredondo, B.; Torres, J.C.; López-Fraguas, E.; Vergaz, R.; Martín-Martín, D.; Del Pozo, G.; Romero, B.; Apilo, P.; Quintana, X.; et al. An All-Organic Flexible Visible Light Communication System. *Sensors* **2018**, *18*, 3045. [CrossRef]
9. Manousiadis, P.P.; Yoshida, K.; Turnbull, G.A.; Samuel, I.D.W. Organic semiconductors for visible light communications. *Phil. Trans. R. Soc.* **2020**, *A378*, 20190186. [CrossRef]
10. Griffith, M.J.; Cottam, S.; Stamenkovic, J.; Posar, J.A.; Petasecca, M. Printable Organic Semiconductors for Radiation Detection: From Fundamentals to Fabrication and Functionality. *Front. Phys.* **2020**, *8*, 22. [CrossRef]
11. Zeidell, A.M.; Ren, T.; Filston, D.S.; Iqbal, H.F.; Holland, E.; Bourland, J.D.; Anthony, J.E.; Jurchescu, O.D. Organic Field-Effect Transistors as Flexible, Tissue-Equivalent Radiation Dosimeters in Medical Applications. *Adv. Sci.* **2020**, *7*, 2001522. [CrossRef] [PubMed]
12. Calvi, S.; Basiricò, L.; Carturan, S.M.; Fratelli, I.; Valletta, A.; Aloisio, A.; De Rosa, S.; Pino, F.; Campajola, M.; Ciavatti, A.; et al. Flexible fully organic indirect detector for MeV proton beams. *npj Flex Electron* **2023**, *7*, 5. [CrossRef]
13. Za'aba, N.K.; Morrison, J.J.; Taylor, D.M. Effect of relative humidity and temperature on the stability of DNTT transistors: A density of states investigation. *Org. Electron.* **2017**, *45*, 174–181. [CrossRef]
14. Kraft, U.; Takimiya, K.; Kang, M.J.; Rödel, R.; Letzkus, F.; Burghartz, J.N.; Weber, E.; Klauk, H. Detailed analysis and contact properties of low-voltage organic thin- film transistors based on dinaphtho[2,3-b:20,30-f]thieno[3,2-b] thiophene (DNTT) and its didecyl and diphenyl derivatives. *Org. Electron.* **2016**, *18*, 33–40. [CrossRef]
15. Bharti, D.; Raghuwanshi, V.; Varun, I.; Mahato, A.K.; Tiwari, S.P. Photo-Response of Low Voltage Flexible TIPS-Pentacene Organic Field-Effect Transistors. *IEEE Sensors* **2017**, *17*, 12. [CrossRef]
16. Pierre, A.; Arias, A.C. Solution-processed image sensors on flexible substrates. *Flex. Print. Electron.* **2016**, *1*, 043001. [CrossRef]
17. Baeg, K.-J.; Binda, M.; Natali, D.; Caironi, M.; Noh, Y.-Y. Organic Light Detectors: Photodiodes and Phototransistors. *Adv. Mater.* **2013**, *25*, 4267–4295. [CrossRef] [PubMed]
18. Bai, S.; Li, R.; Huang, H.; Qi, Y.; Xu, Y.; Song, J.; Yao, F.; Sandberg, O.J.; Meredith, P.; Armin, A.; et al. Transient analysis of photomultiplication-type organic photodiodes. *Appl. Phys. Rev.* **2022**, *9*, 021405. [CrossRef]

19. Ostroverkhova, O.; Shcherbyna, S.; Cooke, D.G.; Egerton, R.; Tykwinski, R.R.; Anthony, J.E.; Hegmann, F.A. Fast photoresponse in organic semiconductors: Understanding the mechanisms and structure-property relationships. In *Linear and Nonlinear Optics of Organic Materials IV, Proceedings of the Optical Science and Technology, the SPIE 49th Annual Meeting, Denver, CO, USA, 2–6 August 2004*; SPIE: Bellingham, WA, USA, 2004.
20. Day, J.; Subramanian, S.; Anthony, J.E.; Lu, Z.; Twieg, R.J.; Ostroverkhova, O. Photoconductivity in organic thin films: From picoseconds to seconds after excitation. *J. Appl. Phys.* **2008**, *103*, 123715. [CrossRef]
21. Narayan, K.S.; Kumar, N. Light responsive polymer field-effect transistor. *Appl. Phys. Lett.* **2001**, *79*, 1891. [CrossRef]
22. Wasapinyokul, K.; Milne, W.I.; Chu, D.P. Photoresponse and saturation behavior of organic thin film transistors. *J. Appl. Phys.* **2009**, *105*, 024509.
23. Huang, W.; Yang, B.; Sun, J.; Liu, B.; Yang, J.; Zou, Y.; Xiong, J.; Zhou, C.; Gao, Y. Organic field-effect transistor and its photoresponse using a benzo[1,2-b:4,5-b0]difuran-based donor–acceptor conjugated polymer. *Org. Electron.* **2014**, *15*, 1050–1055. [CrossRef]
24. Kösemen, Z.A.; Kösemen, A.; Öztürk, S.; Canımkurbey, B.; Erkovan, M.; Yerli, Y. Performance improvement in photosensitive organic field effect transistor by using multi-layer structure. *Thin Solid Films* **2019**, *672*, 90–99. [CrossRef]
25. Zhong, J.; Wu, X.; Lan, S.; Fang, Y.; Chen, H.; Guo, T. High Performance Flexible Organic Phototransistors with Ultrashort Channel Length. *ACS Photonics* **2018**, *5*, 3712–3722. [CrossRef]
26. Za'aba, N.K.; Taylor, D.M. Photo-induced effects in organic thin film transistors based on dinaphtho [2,3-b:2′,3′-f] Thieno[3,2-b′] thiophene (DNTT). *Org. Electron.* **2019**, *65*, 39–48. [CrossRef]
27. Milvich, J.; Zaki, T.; Aghamohammadi, M.; Rödel, R.; Kraft, U.; Klauk, H.; Burghartz, J.N. Flexible low-voltage organic phototransistors based on air-stable dinaphtho[2,3-b:20,30-f]thieno[3,2-b]thiophene (DNTT). *Org. Electron.* **2015**, *20*, 63–68. [CrossRef]
28. Yu, F.; Wu, S.; Wang, X.; Zhang, G.; Lu, H.; Qiu, L. Flexible and low-voltage organic phototransistor. *RSC Adv.* **2017**, *7*, 11572. [CrossRef]
29. Calvi, S.; Rapisarda, M.; Valletta, A.; Scagliotti, M.; De Rosa, S.; Tortora, L.; Branchini, P.; Mariucci, L. Highly sensitive organic phototransistor for flexible optical detector arrays. *Organic Electronics* **2022**, *102*, 106452. [CrossRef]
30. Becharguia, H.; Mahdouani, M.; Bourguiga, R.; Branchini, P.; Fabbri, A.; De Rosa, S.; Calvi, S.; Mariucci, L.; Valletta, A.; Tortora, L. Effects of illumination on the electrical characteristics in organic thin-film transistors based on dinaphtho [2,3-b:2′,3′-f] thieno[3,2-b] thiophene (DNTT): Experiment and modeling. *Synthetic Metals* **2022**, *283*, 116985. [CrossRef]
31. Available online: https://www.manualslib.com/manual/1273314/Agilent-Technologies-B1500a.html?page=34#manual (accessed on 10 January 2023).
32. Klauk, H. Organic thin-film transistors. *Chem. Soc. Rev.* **2010**, *39*, 2643–2666. [CrossRef]
33. Jia, R.; Wu, X.; Deng, W.; Zhang, X.; Huang, L.; Niu, K.; Chi, L.; Jie, J. Unraveling the Mechanism of the Persistent Photoconductivity in Organic Phototransistors. *Adv. Funct. Mater.* **2019**, *29*, 1905657. [CrossRef]
34. Jin, Z.; Gao, L.; Zhou, Q.; Wang, J. High-performance flexible ultraviolet photoconductors based on solutionprocessed ultrathin ZnO/Au nanoparticle composite films. *Sci. Rep.* **2015**, *4*, 4268.
35. Sun, Z.; Liu, Z.; Li, J.; Tai, G.; Lau, S.; Yan, F. Infrared Photodetectors Based on CVD-Grown Graphene and PbS Quantum Dots with Ultrahigh Responsivity. *Adv. Mater.* **2012**, *24*, 5878–5883. [CrossRef]
36. Dey, A.; Singh, A.; Das, D.; Iyer, P.K. Photosensitive organic field effect transistors: The influence of ZnPc morphology and bilayer dielectrics for achieving a low operating voltage and low bias stress effect. *Phys. Chem. Chem. Phys.* **2016**, *18*, 32602–32609. [CrossRef] [PubMed]
37. Bhargava, K.; Singh, V. High-sensitivity organic phototransistors prepared by floating film transfer method. *APEX* **2016**, *9*, 091601. [CrossRef]
38. Simoen, E.; Claeys, C. *Random Telegraph Signals in Semiconductor Devices*; IOP Publishing: Bristol, UK, 2016.
39. Singh, S.; Mohapatra, Y.N. Persistent photocurrent (PPC) in solution-processed organic thin film transistors: Mechanisms of gate voltage control. *J. Appl. Phys.* **2016**, *120*, 045501. [CrossRef]
40. Lutsyk, P.; Janus, K.; Mikołajczyk, M.; Sworakowski, J.; Boratyński, B.; Tłaczała, M. Long-lived persistent currents in poly(3-octylthiophene) thin film transistors. *Org. Electron.* **2010**, *11*, 490–497. [CrossRef]
41. Kösemen, Z.A.; Kösemen, A.; Öztürk, S.; Canimkurbey, B.; San, S.E.; Yerli, Y.; Tunç, A.V. Effect of intrinsic polymer properties on the photo sensitive organic field-effect transistors (Photo-OFETs). *Microelectron. Eng.* **2016**, *161*, 36–42. [CrossRef]
42. Mullenbach, T.K.; Curtin, I.J.; Zhang, T.; Holmes, R.J. Probing dark exciton diffusion using photovoltage. *Nat. Commun.* **2017**, *8*, 14215. [CrossRef]
43. Street, R.A.; Yang, Y.; Thompson, B.C.; McCulloch, I. Capacitance Spectroscopy of Light Induced Trap States in Organic Solar Cells. *J. Phys. Chem. C* **2016**, *120*, 22169–22178. [CrossRef]
44. Hamilton, M.C.; Martin, S.; Kanicki, J. Thin-Film organic polymer phototransistors. *IEEE Trans. Electron. Dev.* **2004**, *51*, 877–885. [CrossRef]

sensors

MDPI

Communication

Speckle Measurement for Small In-Plane Vibration Using GaAs

Jiongye Gao, Bin Zhang *, Qibo Feng, Xu Shen, Yong Xue and Jiacheng Liu

Key Laboratory of Luminescence and Optical Information, Ministry of Education, Beijing Jiaotong University, Beijing 100044, China

* Correspondence: bzhang@bjtu.edu.cn

Abstract: In this study, the measurement characteristics of speckles based on the photoinduced electromotive force (photo-emf) effect for high-frequency, small-amplitude, and in-plane vibration were theoretically and experimentally studied. The relevant theoretical models were utilized. A GaAs crystal was used as the photo-emf detector for experimental research, as well as to study the influence of the amplitude and frequency of the vibration, the imaging magnification of the measuring system, and the average speckle size of the measuring light on the first harmonic of the induced photocurrent in the experiments. The correctness of the supplemented theoretical model was verified, and a theoretical and experimental basis was provided for the feasibility of using GaAs to measure in-plane vibrations with nanoscale amplitudes.

Keywords: in-plane vibration; GaAs; speckle measurement; photoinduced electromotive force

Citation: Gao, J.; Zhang, B.; Feng, Q.; Shen, X.; Xue, Y.; Liu, J. Speckle Measurement for Small In-Plane Vibration Using GaAs. *Sensors* **2023**, *23*, 2724. https://doi.org/10.3390/s23052724

Academic Editor: Ali Passian

Received: 29 January 2023
Revised: 28 February 2023
Accepted: 1 March 2023
Published: 2 March 2023

1. Introduction

The measurement of high-frequency micro-vibrations plays a critical role in multiple areas, such as micro-electro-mechanical systems, structural health monitoring, laser ultrasonic technology, materials science, and biomedicine [1–8]. Optical measurement methods with non-contact characteristics have been widely studied and applied [7–10]. Compared with the measurement of out-of-plane vibration, where the measuring light is parallel to the vibration direction, the measurement of in-plane vibration is relatively complex because the measuring light cannot be incident on the vibration surface in the direction parallel to the in-plane vibration direction. Speckle interferometry is commonly used for in-plane displacement measurements [11–13]. However, owing to the high-speed performance limitation of charge-coupled devices in the imaging process, they cannot measure higher frequencies. In addition, there is a low tolerance for the presence of speckles in laser beams from rough vibration surfaces for some traditional laser interferometers, leading to low measurement sensitivity.

Laser interferometry based on photorefractive crystals has become a research focus because of its advantages of wavefront matching and low-frequency cutoff [14,15] and offers some potential research directions and applications with many emerging materials [16–19]. The corresponding vibration measurement methods can restrain the influence of rough surface speckles and environmental disturbances on the measurements, thereby improving the measurement sensitivity. Among them, laser interferometry based on the photoinduced electromotive force (photo-emf) effect was first reported by Stepanov et al. [20]. In this case, the photo-emf effect is usually produced by projecting a non-steady two-beam interference light pattern on a photorefractive crystal. The interference pattern is created by the interference of a probe beam scattered by the measured sample with a coherent reference beam. The vibration of this pattern subsequently produces an ac electric current in the crystal. Stepanov et al. successively conducted theoretical research on the photo-emf effects of different photorefractive materials, as well as experimental research on interferometry under external electric field modulation for out-of-plane vibration measurement [21–28].

Because of the non-steady characteristics of the photo-emf, as long as the beam with harmonic oscillation shines on the photorefractive material, an alternating photocurrent will be generated. This characteristic makes it possible to measure in-plane vibration with a non-interfering speckle beam. Meanwhile, due to the non-reference beam path, it has the advantage of a simple optical arrangement. Mosquera and Frejlich conducted research on the speckle pattern measurement of in-plane vibration using a BTO crystal, and they proposed that the photocurrent signal can reach a maximum value under a certain vibration amplitude [29]. Santos conducted a theoretical study on the photocurrent signal generated by a transverse-vibration-modulated speckle light incident on a GdTe:V crystal and proposed a corresponding mathematical model and the application of this photo-emf signal to evaluate the response time of the CdTe:V crystal [30–33]. Heinz and Garmire used an array of semi-insulating GaAs photoconductive sensors to detect intensity variations caused by the transverse movement of speckles [34]. Salazar conducted experimental research based on BSO crystals and analyzed the influence of speckle size on the photocurrent [35]. Bryushinin studied the strains and stresses of their measured medium, caused by mechanical vibration, using a photo-emf sensor and measured the resonant frequency of the mechanical system and the distribution of the phase modulation amplitude on the measured surface [36].

In this study, a speckle measurement method for in-plane vibration based on a GaAs crystal photo-emf was investigated. A concrete calculation of the response time was added to the corresponding theoretical model. With the GaAs crystal as the photo detector, the effects of the vibration amplitude and frequency, as well as those of the imaging magnification of the measurement system and the average speckle size of the measuring light on the induced photocurrent, were explored. The numerical simulation and experimental results were in good agreement with those of the modified theoretical model. The measurement of in-plane vibrations of nanoscale amplitude in the frequency range of thousands of hertz was achieved.

2. Theoretical Analysis

The photo-emf effect is a weak non-linear optical phenomenon. A photorefractive crystal is used to receive scattered light from a remotely measured object with vibration information (as shown in Figure 1). A short-circuited crystal is illuminated by a speckle pattern with a speckle size of d. When the oscillatory movement of a speckle with a certain frequency occurs, a photocurrent is generated. The photocurrent is caused by the generation of charge carrier distribution in the conduction or valence band and the formation of a space charge field in the photoionization process; the charge carrier distribution will vibrate with the optical pattern vibration. If the time required for the spatial charge field buildup of the material is significantly lower than the vibration frequency, the vibration of the photogenerated charge carrier density can be considered to occur in the static space charge field. This behavior causes a vibrational phase shift between the static space charge field and the carrier density, resulting in the generation of a non-steady-state photocurrent.

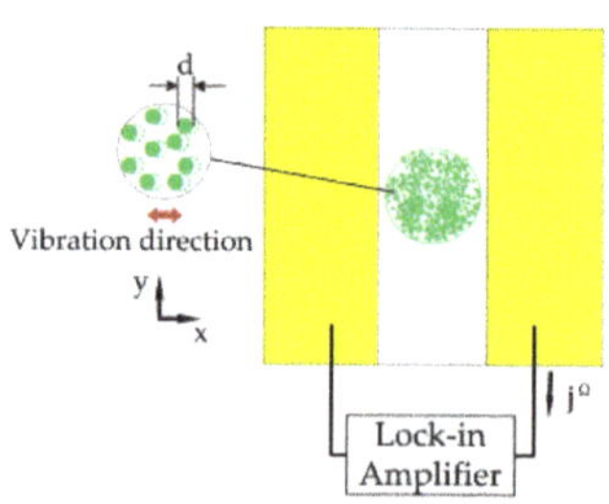

Figure 1. Principle of in-plane vibration measurements based on GaAs.

The intensity of a speckle pattern can be approximated by considering it as a set of Gaussian beams whose diameters are normally distributed around a diameter equal to the average speckle size. Each Gaussian beam represents a speckle particle, and the diameters of these Gaussian beams are approximately equal to the average speckle size [35]

$$d = 1.22\lambda \frac{d_i}{D_p}, \tag{1}$$

where d is the subjective speckle size, d_i is the image distance, λ is the wavelength, and D_p is the imaging system aperture. The light intensity of speckled particles can be expressed as

$$I = I_0 e^{-(x^2+y^2)}, \tag{2}$$

where $w = d/2$, $x = X/\sqrt{2}w$, and $y = Y/\sqrt{2}w$ are the coordinates normalized by the spatial coordinates X and Y, respectively; $w = d/2$ is the Gaussian beam radius. When the measured object vibrates along the transverse direction x with amplitude δ and angular frequency Ω, the speckle light pattern follows the measured object vibration with angular frequency Ω and normalized amplitude $A = M\delta/w$, where M is the magnification of the imaging system. Equation (2) changes to

$$I = I_0 e^{-[(x+A\sin\Omega t)^2+y^2]}, \tag{3}$$

Assuming that the material response time τ_{sc} is significantly longer than the period of the vibration pattern $2\pi/\Omega$ ($\tau_{sc}\Omega \gg 1$), which is significantly longer than the photoelectron lifetime τ ($\tau\Omega \ll 1$), and considering cyclic frontier conditions, the average current density along the electrode space L can be obtained. The time-dependent term is as follows [30].

$$\bar{j}_{xt}(t) = \frac{j_D}{l}\int_{-\frac{l}{2}}^{\frac{l}{2}} e^{-[(x+A\sin\Omega t)^2+y^2]} \frac{E_0/E_D + \frac{\Omega}{2\pi}\int_{-\frac{2\pi}{\Omega}}^{\frac{2\pi}{\Omega}} e^{-[(x+A\sin\Omega t)^2+y^2]}\,(2x+2A\sin\Omega t)dt}{\frac{\Omega}{2\pi}\int_{-\frac{2\pi}{\Omega}}^{\frac{2\pi}{\Omega}} e^{-[(x+A\sin\Omega t)^2+y^2]}dt + R_d} dx, \tag{4}$$

where $j_D = e\mu E_D N_0$ and $E_D = D/\mu w$. μ and $D = k_B T\mu/e$ are the mobility and diffusion coefficients of the charge carriers, respectively; k_B is the Boltzmann constant; T is the absolute temperature; $E_0 = j_0/e\mu N_0$ is the external electric field applied between the two electrodes; e is the electronic charge; j_0 is the average current density along the x-direction x; $R_d = N_d/N_0$ is the dark-to-bright conductivity parameter; and $l = L/w$.

The photocurrent was obtained by the average current density multiplied by the area of the transverse electrode. Figure 2 shows the relationship between the amplitude and photocurrent at different high frequencies, which satisfy $\tau_{sc}\Omega \gg 1$.

If τ_{sc} is not significantly longer than the vibration period $2\pi/\Omega$, Equation (4) should be modified to

$$\bar{j}_{xt}(t) = \frac{j_D}{l}\int_{-\frac{l}{2}}^{\frac{l}{2}} \frac{E_0/E_D + \frac{\Omega}{2\pi}\int_{-\frac{2\pi}{\Omega}}^{\frac{2\pi}{\Omega}} e^{-[(x+A\sin\Omega t)^2+y^2]}\,(2x+2A\sin\Omega t)dt}{\frac{\Omega}{2\pi}\int_{-\frac{2\pi}{\Omega}}^{\frac{2\pi}{\Omega}} e^{-[(x+A\sin\Omega t)^2+y^2]}dt + R_d}\, e^{-[(x+A\sin\Omega t)^2+y^2]} \frac{\tau_{sc}\Omega}{\sqrt{1+(\tau_{sc}\Omega)^2}} dx, \tag{5}$$

where $\tau_{sc} = \frac{\varepsilon\varepsilon_0}{\sigma_0}$, ε is the dielectric constant, ε_0 is the vacuum permittivity, and σ_0 is the photoconductivity of the crystal [32].

The response time τ_{sc} depends on the material itself, as well as on the incident light intensity and geometric parameters, such as the speckle size in this case. The formula of τ_{sc} should be further clarified for sufficiently accurate photocurrent simulation. By analogy with the response time expression in the existing interferometric measurement mode [37], the specific formula for τ_{sc} in the speckle measurement mode can be expressed as

$$\tau_{sc}^{-1} = \frac{\sigma_0}{\varepsilon\varepsilon_0} = \frac{e\mu\alpha P_l}{\varepsilon\varepsilon_0 h\nu\left(\frac{L}{L_D}\right)^2} = \frac{\alpha P_l}{\varepsilon\varepsilon_0\left(\frac{h\nu}{e}\right)\left(\frac{k_B T}{e}\right)\left(\frac{L}{L_D}\right)^2}, \quad (6)$$

where α is the optical absorption coefficient, h is Planck's constant, ν is the light frequency, P_l is the incident light power, and $L_D = d/2\pi$ is the average diffusion length of photoinduced carriers.

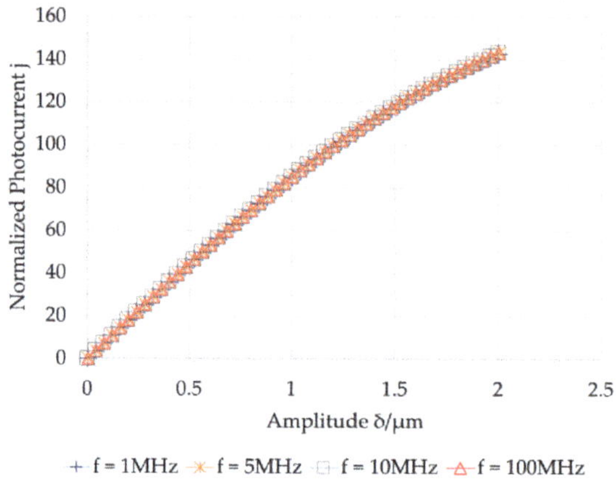

Figure 2. Relation between the photocurrent and vibration amplitude under $\tau_{sc}\Omega$ >> 1.

3. Numerical and Experimental Results

3.1. Experimental Arrangement

The experimental setup is illustrated in Figure 3. A GaAs crystal (10 mm × 10 mm × 0.8 mm, crystal orientation [100], Molecular Technology GmbH, Berlin, Germany) with two parallel striped gold electrodes with an inter-electrode space of 2 mm on the front surface served as the detector. Its main properties include a dielectric constant of 13.1, mobility of 5.2×10^3 $cm^2/(V{\cdot}s)$, lattice constant of 0.56534 nm, and band gap of 1.4 eV. The measured object was a small and thin scattering glass plate, which was firmly adhered to a shear piezoelectric chip (PL5FBP3, Thorlabs, Newton, New Jersey, United States) and could generate transverse vibrations.

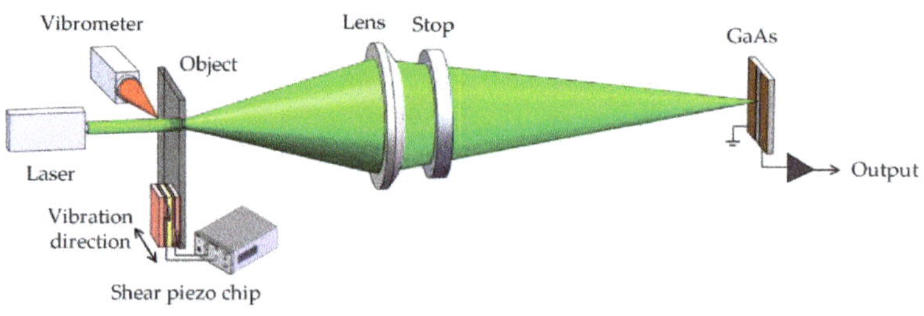

Figure 3. Schematic of experimental system.

The scattered glass plate was illuminated by a laser beam with a wavelength of 532 nm, 200 mW. The transmitted light was collected through a lens with a focal length of 50.8 mm, and the speckle pattern was obtained on the GaAs detector. A stop placed adjacent to the lens was used to control the transmitted light aperture. When the vibration frequency is sufficiently high, the space charge field will not match with the photoconductive phase of GaAs, resulting in a photocurrent of the corresponding frequency. The photocurrent generated by the GaAs was measured using a lock-in amplifier. A laser vibrometer (PDV100, Polytec, Baden-Württemberg, Germany) was placed perpendicular to the optical measuring path. The measuring light hits the side of the object to obtain the corresponding out-of-plane vibration, which is also in-plane vibration. This information was used to verify the in-plane vibration and calibrate the vibration information of the measured object.

The shear piezoelectric plate was driven by 200 V at a 20 kHz sinusoidal voltage. The time-domain comparison measurement results are shown in Figure 4. Signal 1 (yellow) was obtained using PDV100, and signal 2 was obtained using GaAs (blue). The measurement results are consistent, which verifies the feasibility of the system.

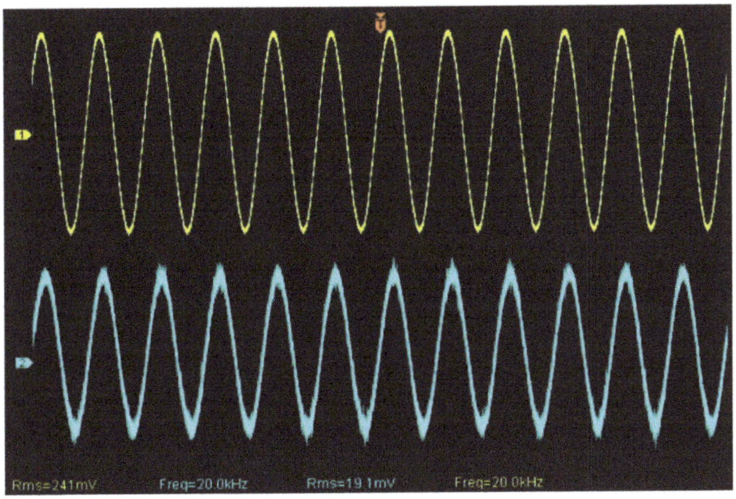

Figure 4. Time-domain comparison measurement results at 20 kHz.

The small in-plane vibrations at 50 kHz, 60 kHz, 70 kHz, and 75 kHz were measured separately using this system, and the frequency-domain signals were obtained using a spectrometer. The results are shown in Figure 5. It can be observed that the measured vibration frequency is consistent with the signal frequency loaded on the shear piezoelectric chip.

3.2. Results

The amplitude of the measured object was calibrated with the PDV100 laser vibrometer. The effects of the amplitude and frequency of the vibration, imaging magnification of the measuring system, and average speckle size of the measuring light on the induced photocurrent were explored.

3.2.1. Effect of Vibration Amplitude

The aperture diameter was set to D_p = 50.8 mm, the lens focal length was f = 50.8 mm, and the object distance (from the diffusing element to the lens) was d_s = 76.81 mm. We calculated the image distance (from the lens to the crystal) to be d_i = 150 mm and the average speckle size to be d = 1.961 μm. At the same frequency, the amplitude of the measured object was changed by varying the input voltage of the shear piezoelectric chip. The measurement results and theoretical simulation results are shown in Figure 6.

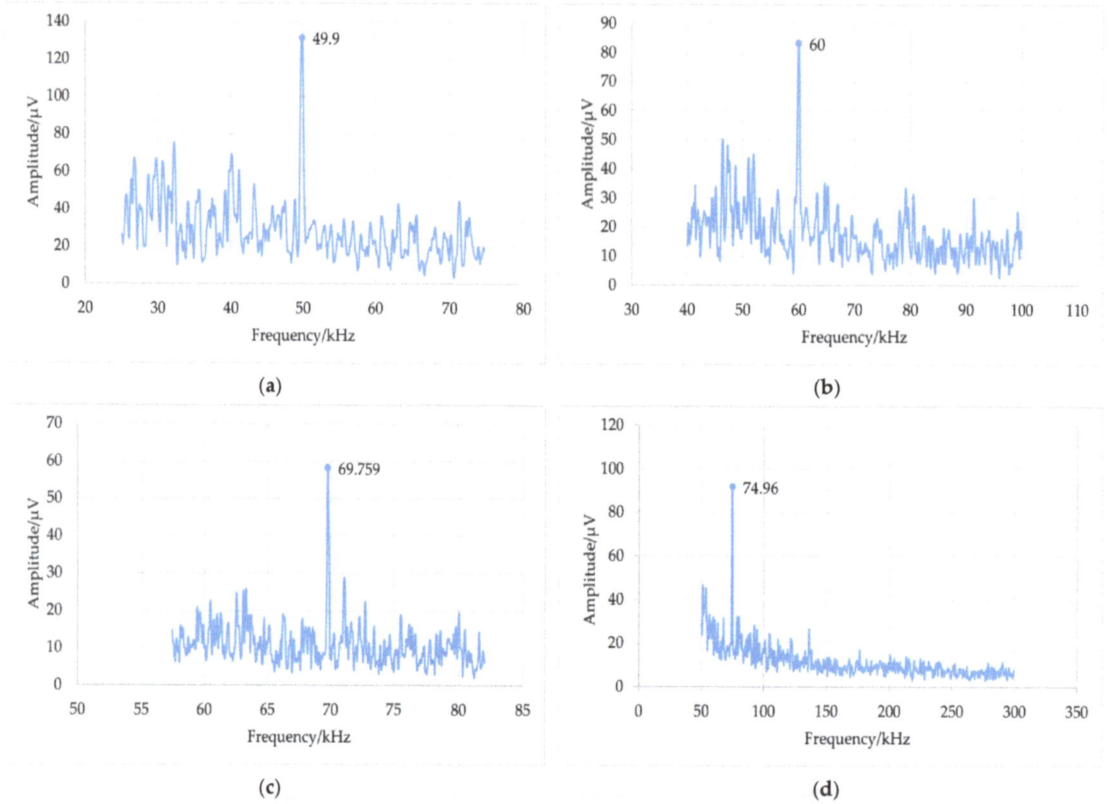

Figure 5. Measurement results of the system at different frequencies: (**a**) 50 kHz; (**b**) 60 kHz; (**c**) 70 kHz; (**d**) 75 kHz.

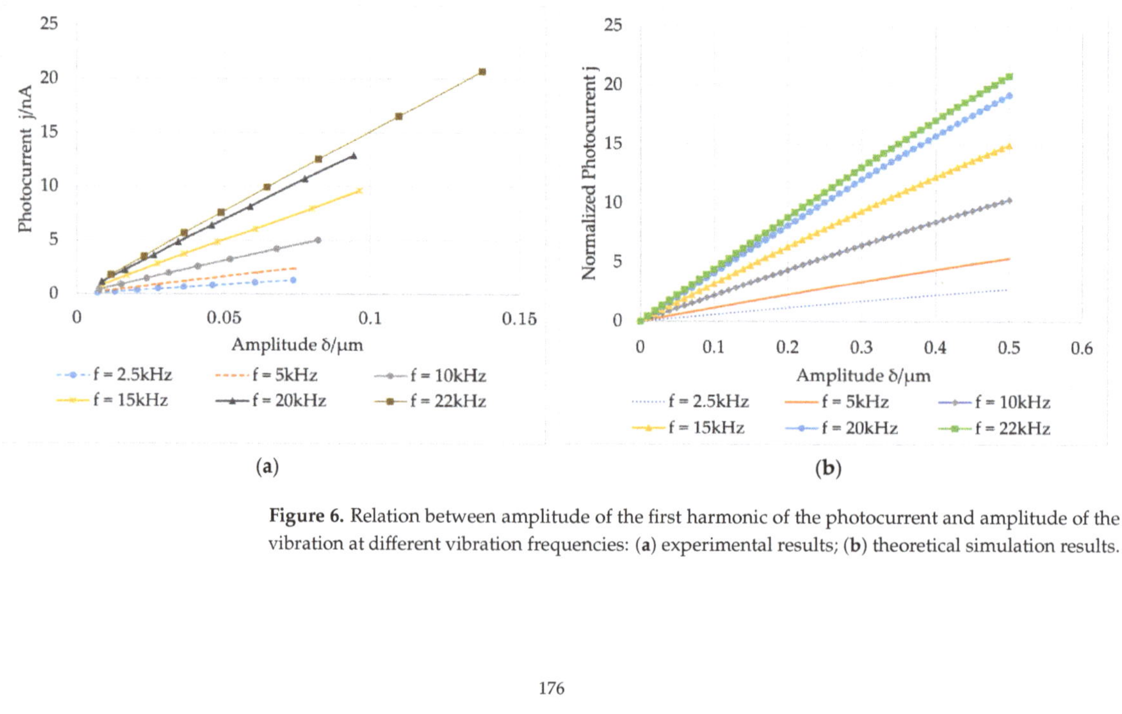

Figure 6. Relation between amplitude of the first harmonic of the photocurrent and amplitude of the vibration at different vibration frequencies: (**a**) experimental results; (**b**) theoretical simulation results.

As shown in Figure 6a, under the same vibration frequency, the induced photocurrent increases with increasing amplitude, which is the same as the simulation curve in Figure 6b. It can also be observed that the measurement system can measure in-plane vibrations of nanoscale amplitude. Although the response time is affected by the incident light intensity and the speckle size, GaAs is a fast response material. $\tau_{sc}\Omega \gg 1$ can be met when the vibration frequency reaches several megahertz or even tens of megahertz. According to the previous statement, the photocurrent is independent of the frequency and response time in this case. For a low vibration frequency of kHz, as shown in Figure 6, $\tau_{sc}\Omega \gg 1$ is not satisfied. When the vibration amplitude is small enough compared to the speckle size, the photocurrent is larger at higher frequencies and shows a monotonic increasing relationship.

3.2.2. Effect of Vibration Frequency

Under the same conditions that were applied in the previous experiment and the same voltage drive, different frequencies were loaded on the shear piezoelectric chip. Experiments and relevant theoretical simulations were performed to determine the relationship between the vibration frequency and photoinduced current, as shown in Figure 7.

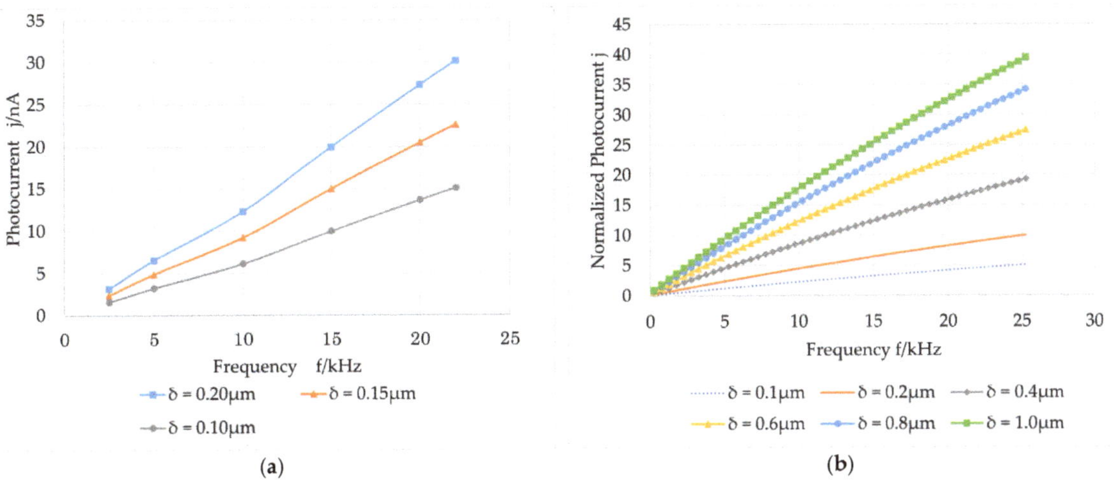

Figure 7. Relation between the amplitude of the first harmonic of the photocurrent and vibration frequency at different vibration amplitudes: (**a**) experimental results; (**b**) theoretical simulation results.

According to Figure 7a, the photocurrent increases with increasing vibration frequency, which is the same as the tendency of the simulation results in Figure 7b. When $\tau_{sc}\Omega \gg 1$ is not satisfied, the photocurrent will monotonously increase with the vibration frequency.

3.2.3. Effect of Imaging Magnification

The magnification of the imaging system was $M = d_i/d_s = \frac{d_s}{d_s - f} = \frac{d_s - f}{f}$, and the average speckle size was $d = 1.22\lambda\frac{d_i}{D_p}$. The image distance d_i must be controlled to maintain the average speckle size d = 1.961 μm while changing the imaging magnification.

The vibration frequency was maintained at 22 kHz, and the focal length of the lens was varied to obtain different imaging magnifications. The vibration amplitude of the speckle changed with the imaging magnification. The relationship between the photocurrent and in-plane vibration amplitude of the shear piezoelectric chip is shown in Figure 8, where the magnification M of the imaging system is 0.5, 1, and 1.952.

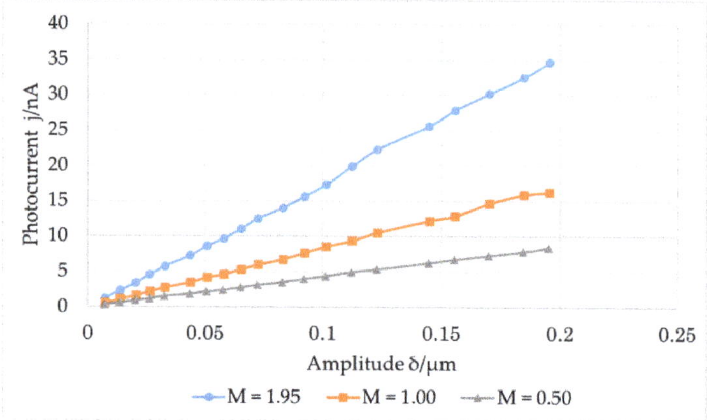

Figure 8. Relation between the amplitudes of the first harmonic of the photocurrent and vibration at different magnifications.

As shown in Figure 8, when the imaging magnification of the measurement system increases, the photocurrent also increases. This behavior occurs because the measured object will be magnified on the image plane after passing through the imaging system, and the amplitude of its vibration will also be magnified; thus, the photocurrent generated will also increase.

3.2.4. Effect of Average Speckle Size

The average speckle size was changed by changing the aperture of the imaging system. The focal length of the lens was 50.8 mm. A variable stop was placed behind the lens to change the size of the transmitted light. An optical attenuator was used to maintain the average intensity of the image surface unchanged. The experimental results obtained by adjusting the aperture to 10, 20, 30, and 40 mm are shown in Figure 9a, and the theoretical simulation results are shown in Figure 9b.

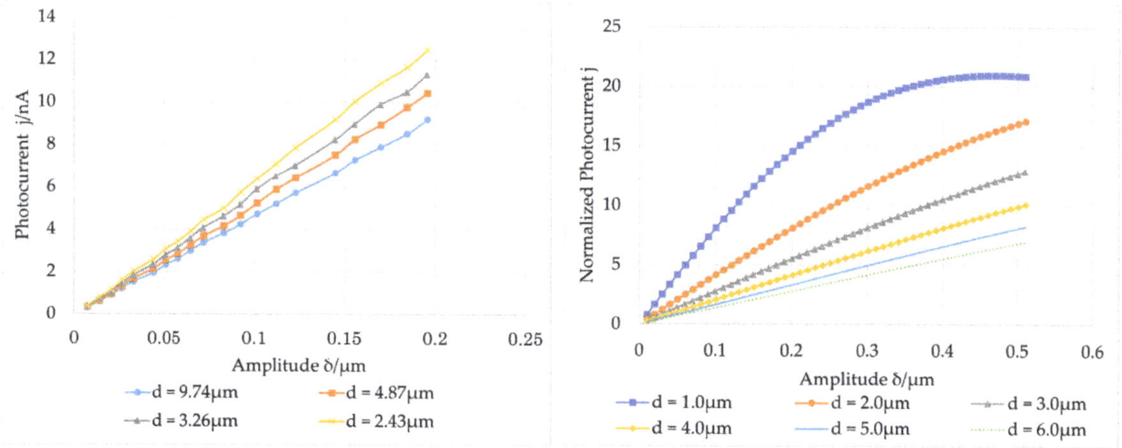

Figure 9. Relation curve between the amplitude of the first harmonic of the photocurrent and amplitude of the vibration at different average speckle sizes: (**a**) experimental results; (**b**) theoretical simulation results.

It can be observed that the smaller the average speckle size, the larger the output photocurrent. This finding is consistent with the simulation results shown in Figure 9b. As previously mentioned, the maximum value of j^{ω} will appear when the amplitude δ approaches the speckle radius w if $\tau_{sc}\Omega \gg 1$. However, the material response time should be considered for a case that does not meet $\tau_{sc}\Omega \gg 1$. The relationship between the photocurrent and amplitude in this case can be obtained by approximately estimating the response time. The photocurrent also monotonously increased with the in-plane vibration amplitude.

4. Conclusions

In this study, we developed a speckle measurement method for in-plane vibrations based on the photo-emf effect of crystal GaAs. The relevant theoretical models were utilized. Through experimental and theoretical simulation, the measurement sensitivity of nanoscale vibration was confirmed and the influences of the vibration amplitude, the vibration frequency, the imaging magnification, and the average speckle size on the first harmonic of the induced photocurrent when $\tau_{sc}\Omega \gg 1$ was not satisfied were determined. The accuracy of the supplementary response-time calculation formula was verified. This study provides a theoretical simulation and experimental basis for the application of GaAs in the measurement of in-plane, high-frequency submicron vibrations.

Author Contributions: Conceptualization, J.G. and B.Z.; methodology, J.G.; software, J.G.; validation, J.G., B.Z. and X.S.; formal analysis, J.G.; investigation, J.G.; resources, X.S.; data curation, J.G.; writing—original draft preparation, J.G.; writing—review and editing, B.Z. and Q.F.; visualization, J.G.; supervision, Y.X. and J.L.; project administration, B.Z.; funding acquisition, Q.F. All authors have read and agreed to the published version of the manuscript.

Funding: This research was funded by the National Natural Science Foundation of China (No. 51775034).

Institutional Review Board Statement: Not applicable.

Informed Consent Statement: Not applicable.

Data Availability Statement: Not applicable.

Acknowledgments: This work was supported by the National Natural Science Foundation of China (No. 51775034).

Conflicts of Interest: The authors declare no conflict of interest.

References

1. Gao, X.; Zhang, B.; Feng, Q.; Xie, X.; Yang, L. Nano-vibration measurements using the photoelectromotive force effect in the GaAs crystal. *Instrum. Exp. Tech.* **2016**, *59*, 470–475. [CrossRef]
2. Li, H.; Jiang, W.; Deng, J.; Yu, R.; Pan, Q. A Sensitive Frequency Range Method Based on Laser Ultrasounds for Micro-Crack Depth Determination. *Sensors* **2022**, *22*, 7221. [CrossRef]
3. Ghafoor, I.; Tse, P.; Rostami, J.; Ng, K.-M. Non-Contact Inspection of Railhead via Laser-Generated Rayleigh Waves and an Enhanced Matching Pursuit to Assist Detection of Surface and Subsurface Defects. *Sensors* **2021**, *21*, 2994. [CrossRef]
4. Kang, K.; Park, K. Noncontact Laser Ultrasound Detection of Cracks Using Hydrophone. *Sensors* **2021**, *21*, 3371. [CrossRef]
5. Passian, A.; Evans, P.G.; Varma, V.K.; Ferrell, T.L.; Thundat, T. Piezoresistive detection of acoustic waves. *Rev. Sci. Instrum.* **2003**, *74*, 1031–1035. [CrossRef]
6. Zaharov, V.; Farahi, R.H.; Snyder, P.; Davison, B.; Passian, A.; Blouin, B.C.C.N. Karhunen-Loeve treatment to remove noise and facilitate data analysis in sensing, spectroscopy and other applications. *Analyst* **2014**, *139*, 5927–5935. [CrossRef]
7. Yu, B.; Tola, K.D.; Lee, C.; Park, S. Improving the Ability of a Laser Ultrasonic Wave-Based Detection of Damage on the Curved Surface of a Pipe Using a Deep Learning Technique. *Sensors* **2021**, *21*, 7105. [CrossRef]
8. Choi, S.; Ota, T.; Nin, F.; Shioda, T.; Suzuki, T.; Hibino, H. Rapid optical tomographic vibrometry using a swept multi-gigahertz comb. *Opt. Express* **2021**, *29*, 16749. [CrossRef]
9. Cao, Y.; Xiong, J.; Liu, X.; Xia, Z.; Wang, W.; Yadav, N.P.; Liu, W. Sensing of ultrasonic fields based on polarization parametric indirect microscopic imaging. *Chin. Opt. Lett.* **2019**, *17*, 41702. [CrossRef]
10. Volikova, A.M.; Lobach, I.A.; Kablukov, S.I. Laser vibrometer-rangefinder based on self-sweeping fiber laser. *Opt. Express* **2022**, *30*, 22025. [CrossRef]
11. Souto Janeiro, A.; Fernández López, A.; Chimeno Manguan, M.; Pérez-Merino, P. Three-dimensional digital image correla-tion based on speckle pattern projection for non-invasive vibrational analysis. *Sensors* **2022**, *22*, 9766. [CrossRef]

12. Zhao, Y.; Zhang, H. Displacement measurement method based on laser self-mixing interference in the presence of speckle. *Chin. Opt. Lett.* **2020**, *18*, 51201. [CrossRef]
13. Gao, X.; Yang, L.; Wang, Y.; Zhang, B.; Dan, X.; Li, J.; Wu, S. Spatial phase-shift dual-beam speckle interferometry. *Appl. Opt.* **2018**, *57*, 414. [CrossRef] [PubMed]
14. Bryushinin, M.; Kulikov, V.; Sokolov, I.; Delaye, P.; Pauliat, G. Non-steady-state photo-EMF in semi-insulating GaAs under frequency-modulated illumination. *Europhys. Lett.* **2014**, *105*, 64003. [CrossRef]
15. Mansurova, S.; Zarate, P.M.; Rodriguez, P.; Stepanov, S.; Köber, S.; Meerholz, K. Non-steady-state photoelectromotive force effect under linear and periodical phase modulation: Application to detection of Doppler frequency shift. *Opt. Lett.* **2012**, *37*, 383–385. [CrossRef]
16. Kour, R.; Arya, S.; Verma, S.; Singh, A.; Mahajan, P.; Khosla, A. Review-recent advances and challenges in indium gallium nitride (InxGa1-xN) materials for solid state lighting. *ECS J. Solid State Sci. Technol.* **2019**, *9*, 15011. [CrossRef]
17. Young, S.-J.; Liu, Y.-H.; Shiblee, M.D.N.I.; Ahmed, K.; Lai, L.-T.; Nagahara, L.; Thundat, T.; Yoshida, T.; Arya, S.; Furukawa, H.; et al. Flexible Ultraviolet Photodetectors Based on One-Dimensional Gallium-Doped Zinc Oxide Nanostructures. *ACS Appl. Electron. Mater.* **2020**, *2*, 3522–3529. [CrossRef]
18. He, Z.; Asare-Yeboah, K.; Zhang, Z.; Bi, S. Manipulate organic crystal morphology and charge transport. *Org. Electron.* **2022**, *103*, 106448. [CrossRef]
19. He, Z.; Zhang, Z.; Bi, S. Tailoring the molecular weight of polymer additives for organic semiconductors. *Mater. Adv.* **2022**, *3*, 1953–1973. [CrossRef]
20. Stepanov, S.I.; Sokolov, I.A.; Trofimov, G.S.; Vlad, V.I.; Popa, D.; Apostol, I. Measuring vibration amplitudes in the picometer range using moving light gratings in photoconductive GaAs:Cr. *Opt. Lett.* **1990**, *15*, 1239–1241. [CrossRef]
21. Stepanov, S. Chapter 6—Photo-electromotive-force effect in semiconductors. In *Handbook of Advanced Electronic and Photonic Materials and Devices*; Singh Nalwa, H., Ed.; Academic Press: Burlington, MA, USA, 2001; pp. 205–272.
22. Stepanov, S.; Montero, P.R.; Flores, M.A.C.; Mixcoatl, J.C.; Lopez, A.A.; Carrasco, L.A.; Sanchez, M.S. Interferometric applications of GaAs adaptive photo-EMF detectors. *J. Opt. Technol.* **2002**, *69*, 428. [CrossRef]
23. Korneev, N.; Mansurova, S.; Rodriguez, P.; Stepanov, S. Fast and slow processes in the dynamics of near-surface space-charge grating formation in GaAs. *J. Opt. Soc. Am. B* **1997**, *14*, 396. [CrossRef]
24. Korneev, N.A.; Mansurova, S.S.; Stepanov, S.I.; Hall, T.J.; Powell, A.K. Non-steady-state photoelectromotive force in semiconductor photorefractive crystals biased by dc field. *J. Opt. Soc. Am. B* **1996**, *13*, 2278. [CrossRef]
25. Sokolov, I.A.; Stepanov, S.I. Detection of optical signals with high-amplitude phase modulation by adaptive photodetectors. *Appl. Opt.* **1993**, *32*, 1958–1964. [CrossRef]
26. Sokolov, I.A.; Stepanov, S.I. Non-steady-state photoelectromotive force in crystals with long photocarrier lifetimes. *J. Opt. Soc. Am. B* **1993**, *10*, 1483. [CrossRef]
27. Korneev, N.A.; Stepanov, S.I. Non-steady-state photoelectromotive force in semiconductor crystals with high light absorption. *J. Appl. Phys.* **1993**, *74*, 2736–2741. [CrossRef]
28. Korneev, N.; Stepanov, S. Measurement of Small Lateral Vibrations of Speckle Patterns Using a Non-steady-state Photo-EMF in GaAs: Cr. *J. Mod. Opt.* **1991**, *38*, 2153–2158. [CrossRef]
29. Mosquera, L.; Frejlich, J. Self-calibrating speckle photo-electromotive force for large vibration amplitude measurement. *J. Opt. A Pure Appl. Opt.* **2004**, *6*, 1001–1004. [CrossRef]
30. Dos Santos, T.O.; Launay, J.C.; Frejlich, J. Photo-electromotive-force from volume speckle pattern vibration with large amplitude. *J. Appl. Phys.* **2008**, *103*, 113104. [CrossRef]
31. Dos Santos, T.O.; Frejlich, J.; Shcherbin, K. Photo electromotive force in CdTe:Ge: Manifestation of two photorefractive centers. *Appl. Phys. B* **2010**, *99*, 701–707. [CrossRef]
32. Santos, T.; Launay, J.C.; Odoulov, S.G.; Frejlich, J. The speckle photo-electromotive force on a vanadium-doped CdTe crystal. *J. Opt. A Pure Appl. Opt.* **2008**, *10*, 104007. [CrossRef]
33. Dos Santos, T.O.; Frejlich, J.; Launay, J.C.; Shcherbin, K. Speckle photo electromotive force in CdTe:V and CdTe:Ge for measurement of vibration with large amplitude. *Appl. Phys. B* **2009**, *95*, 627–632. [CrossRef]
34. Heinz, P.; Garmire, E. Optical vibration detection with a photoconductance monitoring array. *Appl. Phys. Lett.* **2004**, *84*, 3196–3198. [CrossRef]
35. Salazar, A. Speckle photo-electromotive force in Bi12SiO20: Effect of the speckle size. *Opt. Commun.* **2013**, *298–299*, 207–212. [CrossRef]
36. Bryushinin, M.; Kulikov, V.; Petrov, A.; Sokolov, I.A.; Romashko, R.V.; Kulchin, Y.N. Non-steady-state photo-EMF interferometer for detection of mechanical oscillations in transparent scattering objects. *Appl. Opt.* **2020**, *59*, 2370. [CrossRef] [PubMed]
37. Gao, X. Study of Ultrasonic Interference Measurement Method Based on the Photoemf Effect. Doctoral Dissertation, Beijing Jiaotong University, Beijing, China, 2017.

sensors

MDPI

Article

Research on Small-Scale Detection Instrument for Drinking Water Combined Laser Spectroscopy and Conductivity Technology

Zhaoshuo Tian, Hao Chen, Qiping Ding *, Xiaohua Che, Zongjie Bi and Ling Wang

Institute of Ship and Ocean Opto-Elec Equipment, Harbin Institute of Technology, Weihai 264209, China
* Correspondence: fsytzs@126.com

Abstract: In order to realize rapid and accurate evaluation of drinking water quality, a small-scale water quality detection instrument is designed in this paper that can detect two representative water quality parameters: the permanganate index and total dissolved solids (TDS). The permanganate index measured by the laser spectroscopy method can achieve the approximate value of the organic matter in water, and the TDS measured by the conductivity method can obtain the approximate value of the inorganic matter in water. In addition, to facilitate the popularization of civilian applications, the evaluation method of water quality based on the percent-scores proposed by us is presented in this paper. The water quality results can be displayed on the instrument screen. In the experiment, we measured the water quality parameters of the tap water as well as those after the primary and secondary filtration in Weihai City, Shandong Province, China. The testing results show that the instrument can quickly detect dissolved inorganic and organic matter, and intuitively display the water quality evaluation score on the screen. The instrument designed in this paper has the advantages of high sensitivity, high integration, and small volume, which lays the foundation for the popularity of the detection instrument.

Keywords: water quality; drinking water; laser spectroscopy; conductivity; small-scale detection instrument

Citation: Tian, Z.; Chen, H.; Ding, Q.; Che, X.; Bi, Z.; Wang, L. Research on Small-Scale Detection Instrument for Drinking Water Combined Laser Spectroscopy and Conductivity Technology. *Sensors* **2023**, *23*, 2985. https://doi.org/10.3390/s23062985

Academic Editors: Qibo Feng, Jiakun Li and Qixin He

Received: 26 January 2023
Revised: 23 February 2023
Accepted: 7 March 2023
Published: 9 March 2023

1. Introduction

Water is a necessary material condition for human survival and development, and drinking water safety is an important prerequisite to ensuring human health. Unqualified drinking water quality will lead to a variety of diseases [1–4]. Therefore, it is of great significance to achieve rapid monitoring of drinking water quality for people's health. Soluble and insoluble matter in water can be divided into organic matter and inorganic matter according to their chemical characteristics. There are many kinds of organic matter in water, mainly humus, protein, fat, amino acids, carbohydrates, and synthetic organic compounds (SOC). In addition, algae, bacteria, and viruses in water also belong to the category of organic matter. Inorganic matter dissolved in water mainly exists in the form of ions, among which inorganic cations include Ca, Mg, Na, K., etc., and inorganic anions include F^-, Cl^-, NO^{2-}, Br^-, $NO3^-$, $PO_4{}^{3-}$, $SO_3{}^{2-}$, $SO_4{}^{2-}$, etc. It is impossible for rapid monitoring of drinking water quality to contain every substance in water, so comprehensive parameters are generally used for water quality monitoring. The organic matter parameters include permanganate index, chemical oxygen demand (COD), total organic carbon (TOC), dissolved organic matter (DOM), etc. The inorganic matter parameters include conductivity, salinity, etc.

With the development of the world economy, people require that the quality of drinking water be increasingly improved [5]. More and more research has been conducted on the rapid monitoring system of domestic drinking water. Punit et al. [6] developed a sustainable water quality monitoring system that measures five water quality parameters (pH, oxidation reduction potential (ORP), dissolved oxygen (DO), electrical conductivity (EC), and temperature). Partial least squares regression (PLSR) model was used to establish the comprehensive evaluation index of water quality. The R^2 between the estimated value and the predicted value of the water quality parameter was 0.93. The results show that the system can replace traditional water quality monitoring technology. Arif Ul Alam et al. [7] developed a multi-parameter water quality monitoring system (MWQMS) that includes an array of low-cost, easy-to-use, high-sensitivity electrochemical sensors. The proposed MWQMS system can simultaneously monitor pH, free chlorine, and temperature with sensitivities of 57.5 mV/pH, 186 nA/ppm, and 16.9 mV/°C, respectively, as well as BPA with a <10 nM limit of detection. Anand et al. [8] established a practical water quality monitoring system that integrated wireless sensor networks and different information and communication technologies by using an online water quality monitoring network for data acquisition, data processing, and data visualization. It measures the physical and chemical parameters of drinking water, such as pH, turbidity, conductivity, and temperature, within a preset time interval. The above-mentioned drinking water quality monitoring systems can realize the real-time measurement of water quality parameters, but they are still in the laboratory stage and the equipment is complex and expensive, so they are not suitable for civilian application at present. Currently, TDS sensors are the most common rapid detection equipment for drinking water quality in the market. TDS refers to the total amount of all solutes in water, including the content of both inorganic and organic matter [9]. Its measurement unit is milligrams per liter (mg/L). A higher TDS value indicates that there is more dissolved matter in the water. The TDS sensor can give the TDS value by measuring the conductivity of water, which has the advantages of simplicity, convenience, small volume, and low price. Its measured value reflects the total ion concentration in water, which can be approximately converted into the salt content in water. However, since most organic matter such as proteins, microorganisms, colloids, bacteria, and viruses dissolved in water, are non-conductive, the TDS value obtained by the conductivity method cannot reflect their content. Therefore, the TDS detection commonly used by the conductance method is one-sided and inaccurate in evaluating water quality.

In order to reflect the content of organic matter in drinking water, instruments for detecting COD and TOC parameters in water by ultraviolet absorption spectrometry have appeared on the market at present [10], but there exist some problems such as low sensitivity, a high detection limit, and short service life. Fluorescence spectrum detection technology has a good application prospect in water quality detection due to its advantages of fast analysis speed and remarkable sensitivity and selectivity [11–14]. The sensitivity of this method for detecting organic matter is 10–1000 times higher than that of the light absorption method [15]. Therefore, in this paper, we use fluorescence technology to realize organic matter measurement. In order to obtain higher fluorescence intensity in a small size, a violet semiconductor laser is used as the excitation light source. In addition, the laser spectra are detected by our self-made, highly sensitive micro spectrometer, which can realize the highly sensitive and rapid detection of dissolved organic matter in water. The permanganate index of water quality can be obtained by the analysis and calculation of the fluorescence spectrum, and the relatively accurate measurement of organic matter in water is realized. On this basis, we have developed a small drinking water quality detection instrument that integrates organic matter detection by laser spectroscopy and inorganic matter detection by conductivity. The instrument can be conveniently installed in the tap water pipeline, water dispenser, or water purifier and directly give the water quality evaluation scores as well as water quality parameters. In addition, it can give an alarm on unqualified water as an indicator to replace the filter element of the water purifier in time. The instrument has the advantages of high sensitivity, small volume, and low price, which open up a wide range of practical application prospects.

2. Instrument Structure and Detection Principle

2.1. Detection Standard

In March 2022, China released the Standards for Drinking Water Quality (GB5749-2022), which stipulates that the Maximum Contaminant Levels (MCL) of total dissolved solids in drinking water are 1000 mg/L. The MCL of TDS issued by the U.S. Environmental Protection Agency (EPA) is 500 mg/L [16]. The TDS sensor commonly used in the market is used to measure the content of total dissolved solids. The sensor is based on the conductivity method and mainly measures the content of inorganic salts in water, including calcium, magnesium, sodium, potassium, and some minerals. Water containing certain concentrations of inorganic salts, such as calcium and magnesium ions, is beneficial to human health, so it is not that the lower the total soluble solids in water, the better the water quality. However, the TDS sensor cannot detect the content of organic matter. Considering that some dissolved organic matter, microorganisms, bacteria, and viruses in water can emit fluorescence excited by ultraviolet lasers, the combination of laser spectroscopy and the conductivity method can realize a comprehensive detection of drinking water (the detection parameters are shown in Table 1). The content of organic matter in water (including microorganisms, bacteria, viruses, and so on) can be expressed by the permanganate index in Table 1. However, the chemical analysis method used in the Chinese standard to detect organic matter has the disadvantages of having a complex operation, causing secondary pollution, and being time-consuming, which makes it unsuitable for the rapid detection of civil drinking water. The laser spectroscopy method proposed by us has the advantages of high detection sensitivity, fast detection speed, and simple and convenient operation, which makes it suitable for rapid detection of drinking water quality at low organic content.

Table 1. Comparison between the detection method in this paper and the standard detection method in China.

Detection Parameter	Drinking Water Standard	Standard Detection Method	Detection Method in This Paper	Remarks
TDS	≤1000 mg/L [1] (≤500 mg/L [2])	Gravimetric method	Conductance method	Reflect inorganic content
Permanganate index	≤3 mg/L	Chemical analysis	Laser spectroscopy	Reflect organic content

[1] MCL of TDS in drinking water in China. [2] MCL of TDS in drinking water regulated by the U.S. EPA.

2.2. Instrument Structure

The structure of the small-scale drinking water quality detecting instrument designed by us is shown in Figure 1, which includes five parts as follows: a laser emission unit, a spectral detection unit, a conductance detection unit, a signal processing unit, and a water sample detection unit. The laser emission unit is composed of a laser driving power supply, a 405 nm semiconductor laser (NDV4312, NICHIA, China), and a laser collimating lens. The output laser can be modulated by the power supply to achieve laser output with different pulse widths so as to adjust the average laser output power (10–100 mW). The spectrum detection unit includes a long-wave pass filter (Yulai Optics, China) and a self-made miniature spectrometer. The filter with a cut-off wavelength of 420 nm is used to eliminate the laser scattering interference. The wavelength measurement range of the spectrometer is 400–760 nm, and the spectral resolution is 2 nm. The conductance detection unit consists of a conductance probe (BA01, AtomBit, China), an amplifier circuit (BA111, AtomBit, China), a single-chip microcomputer (STM32, STMicroelectronics, Italy), and an output serial port. The signal processing unit adopts a microcomputer, which receives the conductance digital signal through a serial port to control the laser power supply and emit laser pulses. It also carries out spectral calculation and analysis after the spectral signal is received through the USB interface. The calculated water quality parameter results can be displayed on the LCD screen. The water sample detection unit is made up of a transparent glass tube. The laser is incident into the tube at a certain inclination angle.

The entrance of the spectrometer is close to the side wall of the water tube to receive the fluorescence and Raman signals excited by the laser in the water. A conductance probe is inserted into the tee, which is near the inlet end of the water tube so that the water can flow through the conductance electrode. As shown in Figure 1, the five parts of the units are integrated into the aluminum metal shell. The instrument size is $18 \times 12 \times 10$ cm^3. The LCD screen embedded in the surface of the aluminum shell can be used to set the instrument parameters and display the water quality measurement parameters. The inlet and outlet of the water tube are connected to the water pipe, and whenever the water in the tube is flowing or stationary, the water quality can be detected. The laser is incident on the thick water tube at a certain angle. The excited fluorescence and Raman signals are received by the spectrometer. The spectral signal is transmitted to the microcomputer through the USB interface. The spectral curve after being denoised is analyzed by the computer, and the water quality parameter value can be obtained. Some important parameters can be displayed on the LCD screen, such as the permanganate index, TDS value, water quality score value, etc. The instrument can work continuously for 24 h, and the period of each data display is 2 s.

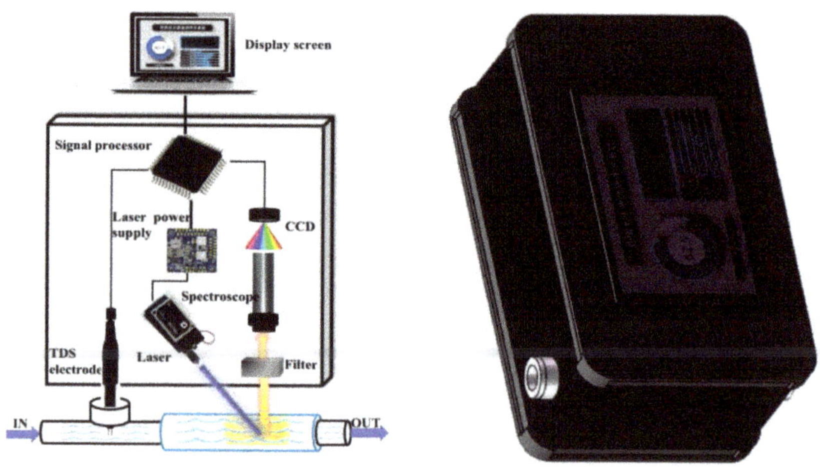

(a) (b)

Figure 1. Composition and outline of the small drinking water quality instrument. (**a**) Structure schematic of the detecting instrument; (**b**) Outline picture of the detecting instrument.

2.3. Detection Principle

2.3.1. Laser Spectroscopy Detection

When a laser is incident on the water, a scattered laser, Raman signal, and fluorescence signal are generated. The fluorescence comes mainly from DOM in water, including humic-like matter, esters, polycyclic aromatic hydrocarbons, protein-like matter, etc. The higher the concentration of organic matter in water, the stronger the fluorescence signal that is generated; thus, it can be considered that the permanganate index is proportional to fluorescence intensity [17]:

$$C = A \times I_F + B \quad (1)$$

where C is the value of the permanganate index, I_F is the intensity of the DOM fluorescence signal. The coefficient A and the limit of detection B are constants, which can be determined experimentally.

We measure the permanganate index parameter of water quality using the Laser Fluorescence–Raman ratio (LFRR) method, which is defined as laser-induced DOM fluorescence at water divided by the intensity of the water Raman peak [17]. The formula is as follows:

$$I_T = \frac{I_F}{I_r} \tag{2}$$

where I_T is the ratio of Laser Fluorescence–Raman, I_r is the intensity of the water Raman signal after deducting the fluorescence background. In practical application, I_F in Equation (1) can be replaced by I_T to achieve the accurate permanganate index measurement.

2.3.2. TDS Detection

The TDS detection in this paper adopts the conductance method. A special chip in the conductance detection system is integrated with a high precision oscillating circuit, an A/D conversion circuit, and a floating-point arithmetic unit. The chip is equipped with the company's patented conductivity-to-TDS conversion algorithm and temperature correction algorithm to quickly realize the detection of TDS. Automatic temperature correction can be realized in a wide temperature range, reducing the measurement error caused by the change in TDS value with temperature. The measuring range is 0~3000 mg/L and the detection limit is 1 mg/L.

2.3.3. Water Quality Evaluation Score

For ordinary consumers, the TDS and permanganate index are not intuitive and difficult to understand. It is necessary to directly express the water quality evaluation parameters in a way that is easy for ordinary people to understand. In our published paper [18], we gave a percentage method for evaluating water quality parameters by laser spectroscopy, and the water quality evaluation score can be directly displayed on the screen of the detecting instrument. The formula for the water quality evaluation parameters is as follows:

$$S_w = \frac{I_R - I_F}{I_R - I_0} \times 100 \tag{3}$$

where S_w is the score of water quality evaluation, I_R is the water Raman signal generated by inelastic scattering of the excitation light, I_0 is background noise signal. With the purpose of achieving a user-friendly result, a constant of 100 is induced in the formula to convert the S_w into the range from 0 to 100.

3. Experimental Measurement Results

3.1. Preliminary Experiment

The tap water of Weihai City, Shandong Province, China, is chosen as the testing object. In the preliminary experiment, we prepared seven diluted water samples by diluting tap water with different volumes of deionized water. The mixing ratios of tap water and deionized water were as follows: 1:0, 3:1, 1:1, 1:3, 1:7, 1:15, and 0:1, respectively, which were numbered from no.1 to no.7 in sequence. Then we measured the concentration for the 7 samples according to the water quality—determination of permanganate index (GB/T 11892-1989 issued by China) [19]. The permanganate indices for the seven samples were 4.15, 3.12, 2.04, 0.99, 0.51, 0.26, and 0 mg/L, respectively. In addition, we also tested the 7 samples using the instrument we developed under the same test conditions. The laser spectrum for each sample was collected 50 times. The laser-excited spectra of different water samples are shown in Figure 2a. It can be seen that there are three emission fluorescence peaks in this spectroscopy curve, including the laser source peak at 405 nm, the Raman peak of water at 471 nm, and the fluorescence peak of DOM at 525 nm (mainly comes from esters and aromatics). Under identical laser power, the intensity of the DOM fluorescence increases as the permanganate index increases. The correlation between the ratio of Laser Fluorescence–Raman and the concentration of permanganate index is shown in Figure 2b. The ratio of Laser Fluorescence–Raman (I_T) increases with permanganate index at 0~5 mg/L, and a linear regression equation with a correlation coefficient R^2 = 0.9881 is obtained:

$$C = 3.0253 \times I_T + 0.1072 \quad (4)$$

where C is the value of the permanganate index, and the constant 0.1072 is the detection limit. In the practical application, we can calculate the permanganate index from Equation (4).

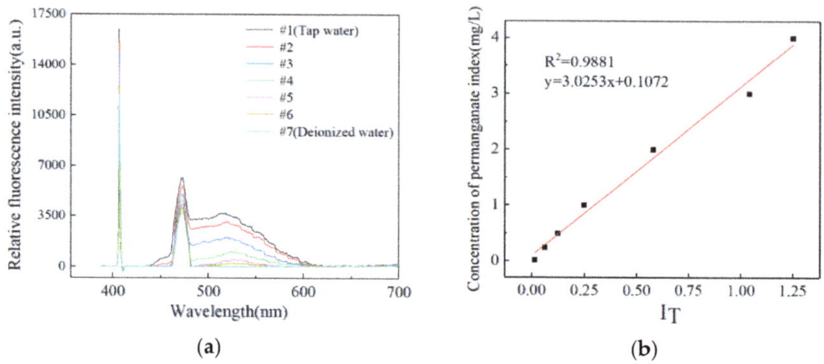

Figure 2. (**a**) The laser-excited spectra of different water samples; (**b**) The correlation between the ratio of Laser Fluorescence–Raman and the concentration of permanganate index.

In addition, we also measured the TDS values by using our designed instrument for the 7 water samples and obtained the water quality evaluation scores by Equation (3). The results of the water quality measurement are shown in Table 2. We can see that, with the increase in the dilution ratio of tap water, the permanganate index and TDS value decrease, and the water quality evaluation parameters gradually increase. Table 2 shows that the permanganate index value determined by laser spectroscopy is very close to the chemical analysis value, and their relative errors remain below 9%. Overall, the result indicates that the laser spectroscopy proposed in this paper can meet the demand for permanganate index measurements of tap water with relatively high accuracy.

Table 2. The results of water quality measurement.

Sample Number	Permanganate Index (mg/L)			TDS (mg/L)	Water Quality Rating (Points)
	Laser Spectroscopy (mg/L)	Chemical Analysis (mg/L)	Relative Error (%)		
#1	4	4.15	3.75	125	59.9
#2	2.97	3.12	5.05	92	68.8
#3	2.01	2.04	1.5	66	78.2
#4	0.97	0.99	2	30	87.9
#5	0.48	0.51	6.25	17	93.1
#6	0.24	0.26	8.33	8	95.4
#7	0.04	0	/	0	100

3.2. Practical Application Experiment

Using the detection instrument shown in Figure 3, we measured the quality of tap water before and after purification by water purifiers. The water purifier (NFX-OP model made by Meishui Company of Japan), includes a two-stage filtration system. Its structure is shown in Figure 2. Firstly, the detection instrument is placed in front of the water purifier and used to measure the tap water. The flow velocity of tap water is 2.15 L/min. The spectral signals of the water excited by the laser and measured by the spectrometer are shown in the upper part of Figure 3. As shown in Figure 3a, it can be seen that the spectral curve includes the laser scattering peak of 405 nm (mainly from the glass tube wall and the water scatter), the Raman peak of water at 471 nm, and the fluorescence peak of organic matter in water at 525 nm. Then the instrument was inserted between the primary filter and the secondary filter, and the spectral curve from the primary filtered water was measured as shown in Figure 3b. The Raman peak intensity was almost unchanged. However, the fluorescence peak of 525 nm was significantly reduced due to the filtration effect of the water purifier, which indicated that most organic matter in the water had been filtered and removed. Finally, the instrument was placed at the back end of the secondary filtration, and the spectral curve from the secondary filtration water was measured as shown in Figure 3c. The fluorescence peak almost disappeared, which indicated that the organic matter in the water was almost completely removed after the secondary filtration.

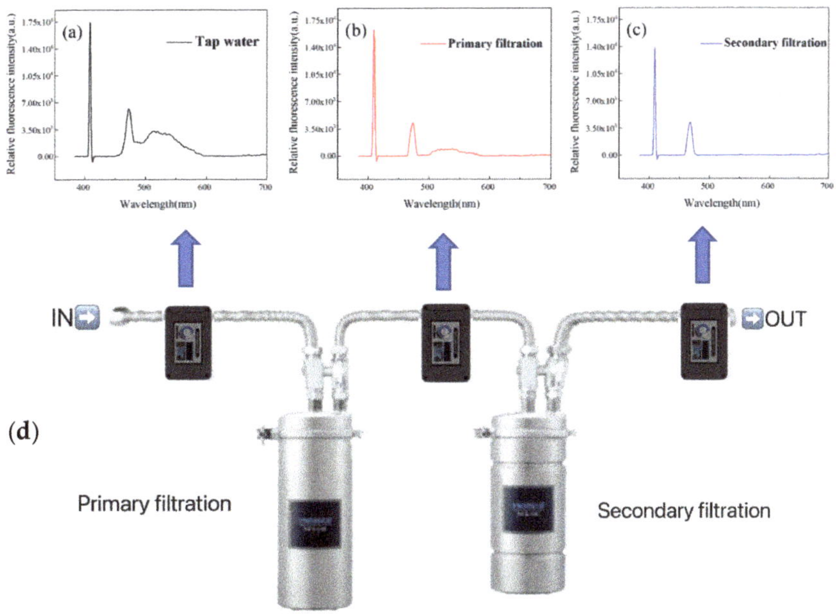

Figure 3. Laser spectral comparison of tap water, primary filtered water, and secondary filtered water. (**a**) Laser spectra of tap water; (**b**) Laser spectra of primary filtered water; (**c**) Laser spectra of secondary filtered water; (**d**) The two-stage water filtration system and three water quality monitoring spots.

The results of the permanganate index, TDS, and water quality scores can be directly displayed on the screen of the instrument, and the measurement results in the three spots are shown in Table 3. It can be seen that the TDS parameters of tap water are 128 mg/L; after primary filtration and secondary filtration, their values are 138 mg/L and 156 mg/L, respectively, which indicate that the content of inorganic salts in tap water before and after filtration meets the drinking water standard. However, the permanganate index in tap water is more than 3 mg/L, which is over the drinking water standard. The permanganate index decreased to 1.41 mg/L after primary filtration, which indicated that although there was some organic matter in the filtered water body, it was below the standard for drinking water. After secondary filtration, the permanganate index is close to 0 mg/L, which indicates that the organic matter in the water body is almost filtered out and that it reached a better drinking water quality. The water quality scores of the three places were measured to be 59.9, 81.2, and 94.8 respectively, according to our proposed percentage method for evaluating water quality parameters by laser spectroscopy [19]. The scores are higher after filtration, which indicates that the quality of tap water has improved. The scores exceeded 90 points after secondary filtration, so the purified water was suitable for people to drink [19].

Table 3. Water quality measurement results for tap water, primary filtered water, and secondary filtered water.

Water Sample	Permanganate Index (mg/L)	TDS (mg/L)	Water Quality Scores
Tap water	4	125	59.9
Primary filtered water	1.41	138	81.2
Secondary filtered water	0.23	156	94.8

The permanganate index is a key parameter for drinking water quality evaluation. In order to verify its accuracy, we also tested the permanganate index of three spots using the chemical analysis method; their comparison results are shown in Table 4. It can be seen that the relative error of the two methods are 3.75%, 8.51%, and 8.70%, respectively. The average relative error is 6.99%. The result indicates that the water quality detection instrument designed in this paper has relatively high accuracy and sensitivity. In addition, to validate the long-term stability of the instrument, we set up a water quality monitoring demonstration system in our laboratory, as shown in Figure 4. It includes an in-and-out water monitoring screen. The water quality parameters before and after purification can be intuitively displayed on the two screens. The values of two water quality parameters only drift within ±5% in more than 3 months of testing time, and their measurement results are independent of flow velocity. The experimental results prove that the water quality monitoring instrument in the demonstration system has high stability.

Table 4. The comparison of the permanganate index measurement.

Water Sample	Laser Spectroscopy (mg/L)	Chemical Analysis (mg/L)	Relative Error (%)
Tap water	4	4.15	3.75
Primary filtered water	1.41	1.53	8.51
Secondary filtered water	0.23	0.25	8.70

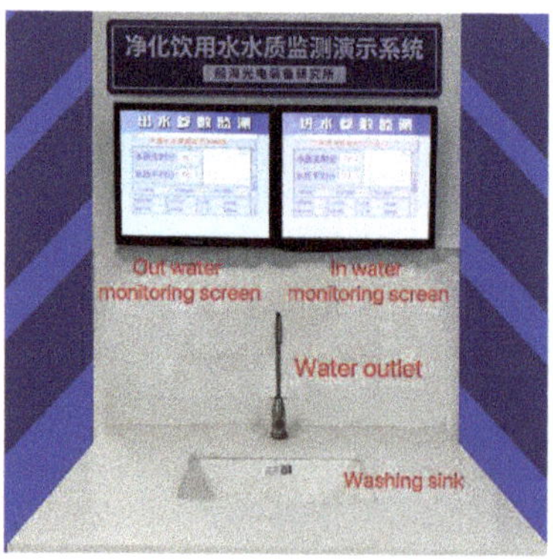

Figure 4. The drinking water quality monitoring demonstration system.

4. Conclusions

In this paper, a small-scale drinking water quality monitoring instrument is designed that can detect permanganate parameters by laser spectroscopy method and obtain TDS parameters by conductivity method. To a large extent, the two parameters represent the content of organic matter and inorganic matter in water. For the convenience of civil applications, we proposed a percentage method to evaluate drinking water quality quickly and intuitively. In the experiment, using the instrument, we measured the water quality parameters in three different spots of a water purifier made by the Japan Meishui Company.

The corresponding TDS values as well as the permanganate index are obtained, and the water quality parameters and corresponding evaluation scores can be displayed on the screen. The testing results showed that after the tap water was filtered, the permanganate value decreased, the water quality score increased, and the TDS value was almost unchanged. In conclusion, the water quality detection instrument developed by us has a high sensitivity, a fast detection speed, and a small volume, which can reflect the drinking water quality in a more comprehensive way. It has the following characteristics:

(1) The small-scale laser spectrometer can be used to detect the organic matter in water with high sensitivity, which makes up for the shortcoming that the TDS sensor is unfit for measuring organic matter in water. It makes the rapid detection of drinking water quality possible.
(2) The TDS value exceeds the standard, which indicates that the water quality is poor and unfit for drinking. In fact, the TDS value of tap water in China rarely exceeds 500 mg/L, which indicates that TDS meets the water quality standard even without water purification. In addition, the TDS value is not the smaller the better, because the water contains a certain concentration of ions, such as Ca and Mg ions, that are beneficial to the human body. If the TDS parameter measured by the conductivity method is lower than a certain value, it can be considered that the inorganic matter in the water meets the standard.
(3) Since TDS mainly reflects inorganic salt content, its value measured by the conductance method cannot reflect water quality accurately. We adopt high-sensitivity laser spectroscopy technology to detect organic matter and give a percentage value to evaluate the water quality. If the water quality score is 100 points, the concentration of organic matter in the water is 0 mg/L, such as in deionized water. A score above 90 indicates that the water has less organic matter, making it suitable for human consumption.

Author Contributions: Conceptualization and methodology, Z.T. and Q.D.; software, L.W. and Z.B.; experiments, X.C. and H.C.; writing, Z.T. and Q.D. All authors have read and agreed to the published version of the manuscript.

Funding: This research was funded by National Natural Science Foundation of China (U22A2008); the China National Key R&D Program (Department of Science and Technology in China), (2020YFE0201500).

Institutional Review Board Statement: Not applicable.

Informed Consent Statement: Not applicable.

Data Availability Statement: Not applicable.

Acknowledgments: The authors thank Weihai Zhixin Photoelectric Technology Co., Ltd. for its support and help in the experimental part. The authors thank the reviewers for their valuable comments.

Conflicts of Interest: The authors declare no conflict of interest.

Abbreviations

The following abbreviations are used in this manuscript:

TDS	Total Dissolved Solids
SOC	Synthetic Organic Compounds
COD	Chemical Oxygen Demand
TOC	Total Organic Carbon,
DOM	Dissolved Organic Matter
ORP	Oxidation Reduction Potential
EC	Electrical Conductivity
PLSR	Partial Least Squares Regression
MCL	Maximum Contaminant Levels
EPA	Environmental Protection Agency
LFRR	Laser Fluorescence–Raman ratio

References

1. Koppanen, M.; Kesti, T.; Kokko, M.; Rintala, J.; Palmroth, M. An online flow-imaging particle counter and conventional water quality sensors detect drinking water contamination in the presence of normal water quality fluctuations. *Water Res.* **2022**, *213*, 118149. [PubMed]
2. Sinitsa, S.; Sochen, N.; Mendlovic, D.; Borisover, M.; Lew, B.; Sela-Saldinger, S.; Vladimir, Y.; Nadia, B.; Yulia, K.; Lavi, R.; et al. Optical sensor system for early warning of inflow organic matter breach in large-scale irrigation systems and water treatment systems. *IEEE Sens. J.* **2022**, *22*, 1680–1691. [CrossRef]
3. Banerjee, A.; Singh, S.; Ghosh, R.; Hasan, M.N.; Bera, A.; Roy, L.; Bhattacharya, N.; Halder, A.; Chattopadhyay, A.; Mukhopadhyay, S.; et al. A portable spectroscopic instrument for multiplexed monitoring of acute water toxicity: Design, testing, and evaluation. *Rev. Sci. Instrum.* **2022**, *93*, 115105. [CrossRef]
4. Chang, W.L.; Sun, I.M.; Tsai, J.A.; Meng, H.F.; Zan, H.W.; Chen, L.Y.; Lu, C.J. Rapid quality test for drinking water by vertical-channel organic semiconductor gas sensor. *Anal. Chim. Acta* **2022**, *1206*, 339729. [CrossRef]
5. Selvi, A.G.; Vibhithra, S.; John, C.S.A. IoT Based Water Quality Monitoring System for Smart Cities. *J. Trend Sci. Res. Dev.* **2021**, *5*, 986–991.
6. Punit, K.; Karunesh, K.G.; Raj, K.G.; Panchariya, P.C. Towards the Green Analytics: Design and Development of Sustainable Drinking Water Quality Monitoring System for Shekhawati Region in Rajasthan. *MAPAN* **2021**, *36*, 1–15.
7. Alam, A.U.; Clyne, D.; Deen, M.J. A Low-Cost Multi-Parameter Water Quality Monitoring System. *Sensors* **2021**, *21*, 3775. [CrossRef] [PubMed]
8. Anand, B.R.; Nagalakshmi, K.; Suresh, V.M.; Vaishnavee, V. Potnet: Online Potable Water Quality Monitoring Network for Overhead Water Tanks in Rural Water Supply Schemes in India. *Int. J. Sens. Wirel. Commun. Control* **2021**, *11*, 872–887.
9. Sandip, K.R.; Preeta, S. Application of machine learning for real-time evaluation of salinity (or TDS) in drinking water using photonic sensors. *Drink. Water Eng. Sci.* **2016**, *9*, 37–45.
10. Shi, Z.; Chow, C.W.K.; Fabris, R.; Liu, J.; Jin, B. Applications of Online UV-Vis Spectrophotometer for Drinking Water Quality Monitoring and Process Control: A Review. *Sensors* **2022**, *22*, 2987. [CrossRef]
11. Junchen, L.; Ye, Y.; Zheng, D.; Guangyu, Z.; Sune, S. Short-range remote sensing of water quality by a handheld fluorosensor system. *Appl. Opt.* **2020**, *59*, C1–C7.
12. Carstea, E.M.; Popa, C.L.; Baker, A.; Bridgeman, J. In situ fluorescence measurements of dissolved organic matter: A review. *Sci. Total Environ.* **2020**, *699*, 134361. [CrossRef]
13. Sorensen, J.P.R.; Vivanco, A.; Ascott, M.J.; Gooddy, D.C.; Lapworth, D.J.; Read, D.S.; Rushworth, C.M.; Bucknall, J.; Herbert, K.; Karapanos, I.; et al. Online fluorescence spectroscopy for the real-time evaluation of the microbial quality of drinking water. *Water Res.* **2018**, *137*, 301–309. [CrossRef] [PubMed]
14. Li, L.; Wang, Y.; Zhang, W.; Yu, S.; Wang, X.; Gao, N. New advances in fluorescence excitation-emission matrix spectroscopy for the characterization of dissolved organic matter in drinking water treatment: A review. *Chem. Eng. J.* **2020**, *381*, 122676. [CrossRef]
15. Bridgeman, J.; Baker, A.; Brown, D.; Boxall, J.B. Portable LED fluorescence instrumentation for the rapid assessment of potable water quality. *Sci. Total Environ.* **2015**, *524*, 338–346. [CrossRef]
16. Haskell, D.; Heo, J.; Park, J.; Dong, C. Hydrogeochemical Evaluation of Groundwater Quality Parameters for Ogallala Aquifer in the Southern High Plains Region, USA. *Int. J. Environ. Res. Public Health* **2022**, *19*, 8453. [CrossRef] [PubMed]
17. Xiaohua, C.; Zhaoshuo, T.; Fenghao, S.; Qingcao, L.; Zongjie, B.; Hao, C.; Zihao, C. Research on chemical oxygen demand based on laser Fluorescence-Raman spectroscopy. *Front. Phys.* **2022**, *10*, 1116. [CrossRef]
18. Zhaoshuo, T.; Zongjie, B.; Yiwei, W.; Hongyan, Z. Rapid evaluation method based on DOM for water quality by microlaser fluorescence spectrometer. *Appl. Phys. B Lasers Opt.* **2020**, *126*, 205–219.
19. Sun, B.; Meng, X.; Niu, Y.; Li, J.; Zou, N. Application of instrumental method for determination of water environmental permanganate index. *Environ. Prot. Oil Gas Fields* **2022**, *32*, 45–49.

sensors

MDPI

Article

Validation of a Lumped Parameter Model of the Battery Thermal Management System of a Hybrid Train by Means of Ultrasonic Clamp-On Flow Sensor Measurements and Hydronic Optimization

Raffaele De Rosa [1], Luca Romagnuolo [1,*], Emma Frosina [2], Luigi Belli [3] and Adolfo Senatore [1]

[1] Department of Industrial Engineering, University of Naples "Federico II", Via Claudio, 21, 80125 Napoli, Italy
[2] Department of Engineering, University of Sannio, Via Roma 21, 82100 Benevento, Italy
[3] Hitachi Rail STS S.p.A., Via Argine 425, 80147 Napoli, Italy
* Correspondence: luca.romagnuolo@unina.it

Abstract: Electrification of the field of transport is one of the key elements needed to reach the targets of greenhouse gas emissions reduction and carbon neutrality planned by the European Green Deal. In the railway sector, the hybrid powertrain solution (diesel–electric) is emerging, especially for non-electrified lines. Electric components, especially battery power systems, need an efficient thermal management system that guarantees the batteries will work within specific temperature ranges and a thermal uniformity between the modules. Therefore, a hydronic balancing needs to be realized between the parallel branches that supply the battery modules, which is often realized by introducing pressure losses in the system. In this paper, a thermal management system for battery modules (BTMS) of a hybrid train has been studied experimentally, to analyze the flow rates in each branch and the pressure losses. Since many branches of this system are built inside the battery box of the hybrid train, flow rate measurements have been conducted by means of an ultrasonic clamp-on flow sensor because of its minimal invasiveness and its ability to be quickly installed without modifying the system layout. Experimental data of flow rate and pressure drop have then been used to validate a lumped parameter model of the system, realized in the Simcenter AMESim® environment. This tool has then been used to find the hydronic balancing condition among all the battery modules; two solutions have been proposed, and a comparison in terms of overall power saved due to the reduction in pressure losses has been performed.

Keywords: battery thermal management system (BTMS); hybrid train propulsion; ultrasonic flow sensor; clamp-on sensor; hydronic balancing methods

Citation: De Rosa, R.; Romagnuolo, L.; Frosina, E.; Belli, L.; Senatore, A. Validation of a Lumped Parameter Model of the Battery Thermal Management System of a Hybrid Train by Means of Ultrasonic Clamp-On Flow Sensor Measurements and Hydronic Optimization. *Sensors* **2023**, *23*, 390. https://doi.org/10.3390/s23010390

Academic Editors: Qibo Feng, Jiakun Li and Qixin He

Received: 14 November 2022
Revised: 15 December 2022
Accepted: 26 December 2022
Published: 30 December 2022

1. Introduction

The transport sector is one of the main causes of CO_2 emissions [1], and trains are responsible for 4.6% of the greenhouse gas emissions from transportation because, in several countries, there are still non-electrified lines with trains powered by diesel engines [2]. Electrification costs time and money; battery power, on the other hand, is the cleanest zero–emission solution to replace diesel trains and start achieving climate change targets, which will instantly improve air quality in cities and non-electrified stations. This technology will allow for travel beyond electrified routes, ensuring seamless journeys [3]. Moreover, battery-powered trains can also use regenerative braking, making them much more environmentally friendly than diesel railcars [4].

However, a battery power system is a temperature-sensitive technology, where performance is influenced by the temperature in terms of efficiency, lifetime, and safety [5,6]. For lithium-ion batteries, the optimum operating temperature is between 20 and 40 °C, and the temperature difference between battery cells of the same pack should be less than 5 °C. Furthermore, the improper distribution of temperature can cause significant performance degradation and may also lead to overheating and thermal runaway [7–9]. For these reasons, the battery thermal management system (BTMS) is one of the most crucial elements of an electric train. Air cooling, liquid cooling, and heat pipe-based cooling are the most common principles on which the BTMS is built [10–15], but there are also new technologies such as phase change material-based cooling [16] and thermoelectric element-based cooling [17–20].

Unlike other sectors, such as automotive and aerospace, where it is by now consolidated, battery-based electric powertrain in the railway field is still in the embryonic stage. Only in recent years, due to the Sustainable and Smart Mobility strategy [21] that is encouraged by the EU and other countries, researchers and big players are developing a program that involves removing diesel engines and replacing them with a battery-based electric powertrain to operate on non-electrified lines [3,22,23]. Those systems must be correctly controlled in terms of temperature, especially the battery pack. Studies on BTMS have also been widely developed in fields such as the automotive and aerospace. Xiong et al. [24] developed an AMESim model of a liquid cooling system for a power battery of a plug-in hybrid electric vehicle in order to analyze the thermal behavior. Sun et al. [16] conducted a numerical analysis into the inhibiting effects of a novel hybrid BTMS, combining active and passive cooling on thermal runaway propagation caused by single cell. Kellerman et al. [25] developed a numeric model of BTMS for a hybrid electric aircraft under the assumption that the ambient temperature may be higher than the allowed battery operating temperature.

In the railway field, a different approach is needed, especially with regard to liquid cooling systems. Batteries in the railway applications must satisfy an energy accumulation in the order of MWh [26] while, in the automotive sector, it is in the order of tens of kWh [27,28]. For this reason, in railway applications, the overall dimension of the battery packs is much larger and the refrigerant flow rate requirement is also higher. Therefore, given the large flow rates involved, hydraulic optimization can lead to considerable energy savings.

Previous research on BTMS for battery powered train mainly focused on thermal aspects. Iwase et al. [4] conducted a thermal simulation to confirm the feasibility of natural air cooling for train battery storage systems. Kang et al. [29] proposed a thermal prediction model of a 1S18P battery pack classified into joules heating with equivalent resistance, reversible heat, and heat dissipation. Teng and Yeow [30] analyzed the thermal performance of two battery module cooling methods with the indirect liquid cooling system with three types of tubular cooling plates between cells with interior fluid; they concluded that the structural layout of a multiple parallel-channels cold plate resulted in a lower coolant pressure and temperature gradient.

The aim of this research is the study of a liquid cooling system for train batteries, which goes from the inlet manifold to the outlet manifold of the hydraulic system and to optimize it from a hydronic point of view in order to find the condition that entails the balancing of the flow rate in all the branches, to minimize the pressure drops and to guarantee to all modules have the same operating conditions. To achieve this goal, a BTMS is studied experimentally and numerically. From the results of the experimental campaign, the Lumped Parameter Model (LPM) is validated. LPMs simplify spatially distributed systems into discrete entities, in which radial or axial gradients are not considered and interactions with the surroundings occur through ports on the boundary. The key advantages of this approach are computational efficiency and simplification of the mathematical formulation, even for more complex systems. However, in some cases the 0D-1D approach alone is not sufficient, and, for this reason, 3D modeling is required for some system components. The validation takes place in two phases. The first involves the validation of the hydraulic system without batteries. The second provides the hydraulic characterization of the battery module on the bench. Thanks to the validated model, numerical tests are then carried out that are aimed at finding the aforementioned conditions. These are described in detail in Section 4.2.

This paper follows the steps of the LPM first introduced in [31], in which the priority was the cooling of the power converters. In this case, our study focuses on the hydronic balancing of the liquid cooling system for battery modules.

2. Materials and Methods

2.1. System Architecture and Description

The cooling system represented in Figure 1 shows a battery box made of sixteen battery modules arranged in three parallel branches, in order to obtain the best performance. Each module must be cooled by the same coolant flow rate and supplied at a constant temperature; therefore, the fluid is first cooled in a chiller and then supplied to the modules at 25 °C.

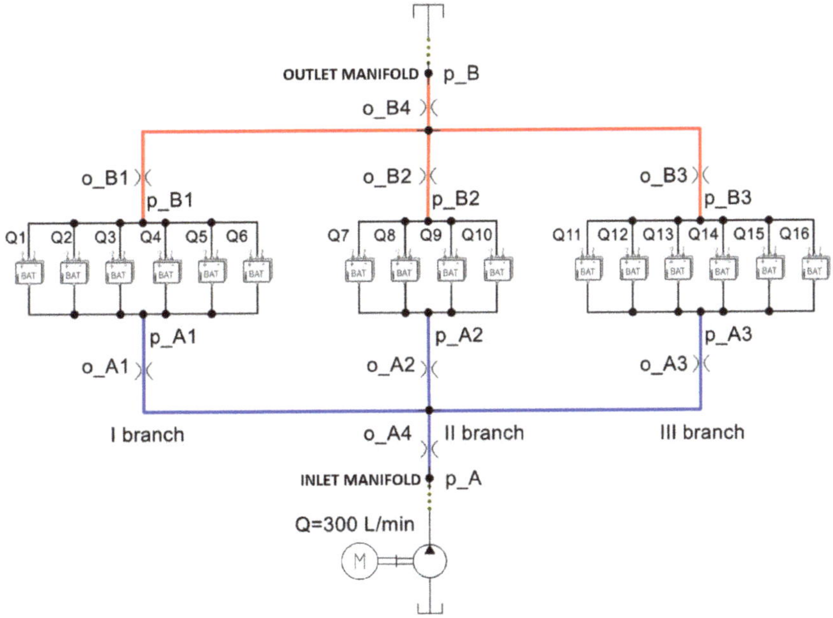

Figure 1. Battery liquid cooling system.

Figure 2 describes the functioning of the battery module, which is composed of the battery cells, a heat exchanger, and a fan placed underneath. First, the coolant passes through a chiller, in which it reaches 25 °C, and then, thanks to an air to liquid heat exchanger, it cools the air that is conveyed to the battery cells with a fan.

In the first stage of this research the cooling system section was modeled without the batteries because all sixteen battery modules were not available at first; therefore, the cooling system without batteries was the only prototype available for experimental validation. The proposed hydraulic model was validated via comparing simulations and experimental results that were carried out via ultrasonic clamp-on flow measurements. In the second stage, the single battery module was hydraulically characterised, and the previous validated model was completed with battery modules. Thanks to the completed model, it was possible to evaluate the unbalance of flow rates in the individual battery modules and then choose the best way to achieve the balance.

In this work, two solutions were proposed: The first involves the use of calibrated orifices in the branches of the cooling system with greater flow rate. The second involves a layout modification without introducing secondary losses. Finally, a comparison between the two solutions was carried out.

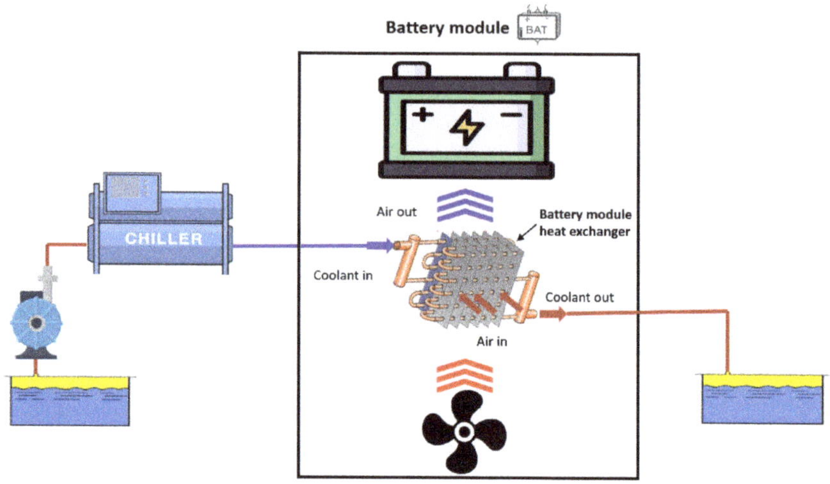

Figure 2. Framework of the battery cooling system.

2.2. Ultrasonic Clamp-On Flow Measurements

Experimental flow measurements were carried out with the ultrasonic clamp-on sensor Keyence FD-Q32C for DN32 pipes, powered by a 24 V power supply. This sensor measures the time it takes to transmit a signal ultrasound from the emitter to the receiver. If the flow rate increases, the signal is accelerated, that is, less time is required for transmission from the emitter to the receiver. By employing the correlation between duration and flow velocity, the sensor measures the instantaneous flow rate [32].

A clamp-on type sensor was chosen because it allows us to easily evaluate the flow rate of the various branches of the system without modifying the system structure, since it was installed outside the pipe for a completely non-wetted measurement. This prevented any risk of adverse effects on the liquid and eliminated the need for piping work. On the other hand, this sensor must be calibrated by adjusting the flow rate signal with a calibration coefficient, whose value was obtained from the comparison with the measurement carried out by a turbine flowmeter (Signet Flow Controller GF George Fischer 3-9010.111) that was positioned downstream of the pump and has an accuracy of ±0.5% [33]. The optimal value of the adjusting flow rate, equal to 1.07, was calculated as the slope of the last-squares linear regression of the data obtained with the two sensors, as shown in Figure 3. In this case, from the tests carried out, the optimal value of the adjusting flow rate span was found to be 1.07.

2.3. Numerical Model of the Prototype Available

The first analyses were carried out on the available prototype that included the hydraulic system without the battery modules. This is because once the batteries were mounted in the battery box enclosure it was not possible to measure the flow rate in the various branches, since there is not enough space left to mount the clamp-on flow sensor. The hydraulic system object of this research is a closed-cycle circuit with three branches in parallel of different lengths which mainly consists of a cooling unit (pump and tank), inlet and outlet manifolds, delivery and return pipes, flexible hoses, quick couplings, and an orifice with a diameter of 10 mm and a thickness of 6 mm in the first and second branches resulting from a preliminary sizing of the system. The circuit was implemented in Simcenter AMESim, a multi-domain, lumped parameter simulation software, suitable for performing simulations of the system as a whole, in which the arrangement of components refers to that in the physical prototype (Figure 4).

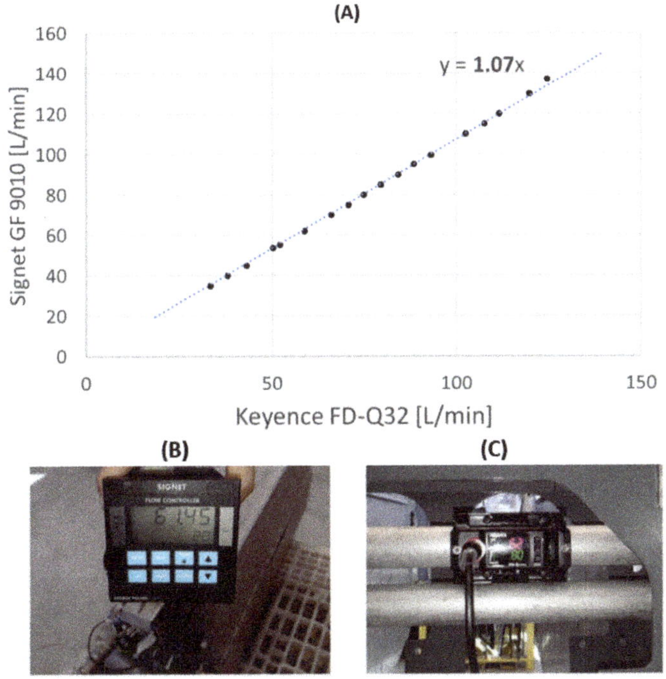

Figure 3. (**A**) Least-squares linear regression of the data obtained with the two sensors; (**B**) George Fischer measurement; (**C**) Keyence measurement with adjusting flow rate span.

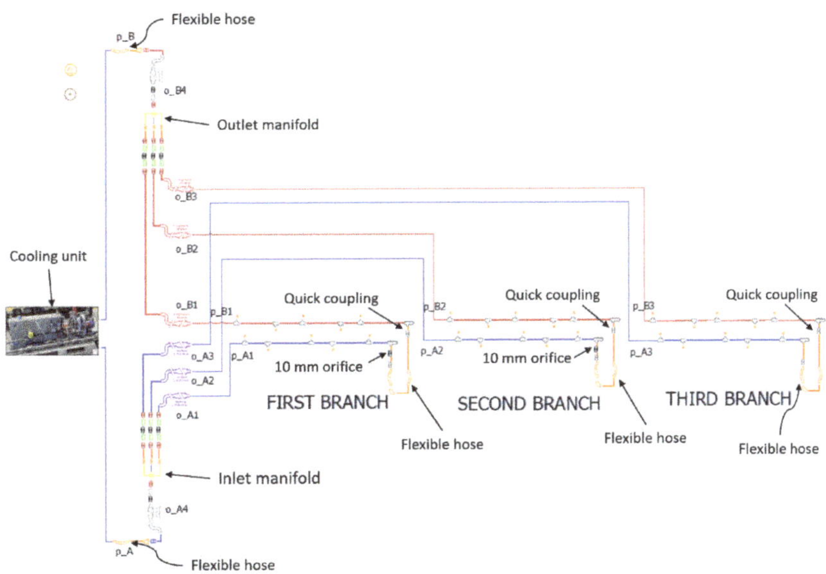

Figure 4. Simcenter AMESim model of the prototype available for testing.

The mainly used libraries were Thermal Hydraulic and Thermal Hydraulic Resistance. The branches of different lengths ($l_1 < l_2 < l_3$) were modeled using the thermal–hydraulic modular piping that includes straight pipes, direction changes, and diameter changes. This submodel calculated the pressure drops while taking into account the compressibility of the fluid and expansion of the pipe wall with pressure. The heat exchange and the influence of temperature on the fluid viscosity were also considered. The methodology described by the flowchart in Figure 5 was followed in order to select the AMESim submodels of the pipelines of the cooling system. This line selection method considers the parameters of the line and the fluid properties, and then it provides an analysis involving the following values:

- The aspect ratio A_r is the ratio between the length of the section l and the hydraulic diameter d_h:

$$A_r = l/d_h \tag{1}$$

- The dissipation number D_n is defined as:

$$D_n = \frac{4lv}{cd_h^2} \tag{2}$$

where v is the kinematic viscosity and c is the speed of sound;
- The wave travel time T_{wave} is the time that pressure disturbance takes to cross the pipe:

$$T_{wave} = l/c. \tag{3}$$

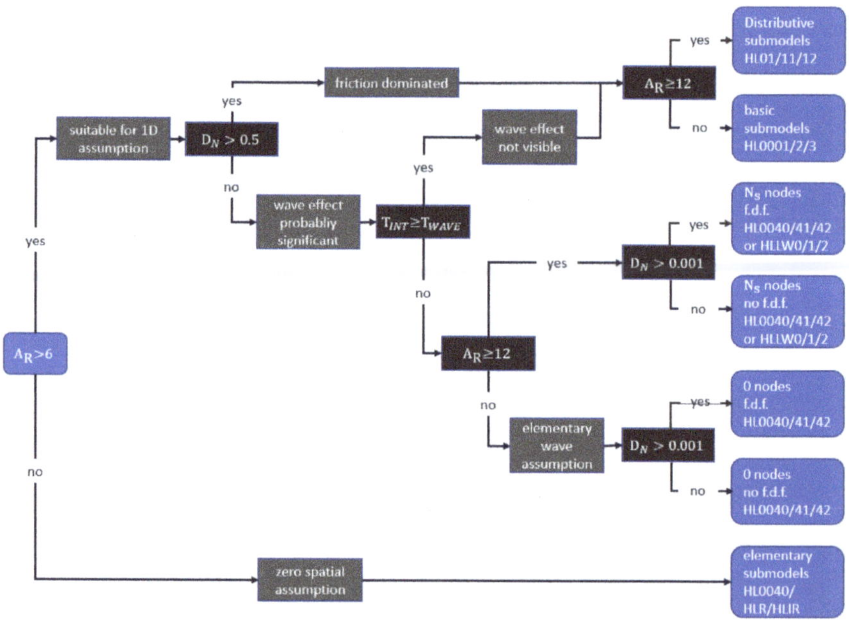

Figure 5. Flow chart for the choice of submodels of the hydraulic lines [34].

The coolant used was the Antifrogen N-39, produced by Clariant, which consists of a mixture of 60% water and 40% ethylene glycol added with corrosion inhibitors. It was modeled by using the Media Property Assistant tool, in which parameters such as density, viscosity, and heat capacity, obtained from the datasheet (available on the supplier's website [35]) were inserted. The cooling unit was modeled as a super component which mainly contains a 60 L tank and a centrifugal pump with an impeller with a diameter of 139 mm.

The flow was then split by a three-way manifold. This component was characterized using another tool called Simerics MP+, a commercial 3D CFD simulation software, in order to evaluate how the inlet flow rate (138 L/min) was distributed to the manifold outlets.

In this study, a steady-state flow analysis was performed that includes a standard $k-\epsilon$ turbulence model with a Converge Criterion of 10^{-4}. The boundary conditions were an inlet volumetric flux of 138 L/min and an outlet pressure of 101,325 Pa. Figure 6 provides a synthesis of the results derived from this 3D CFD simulation.

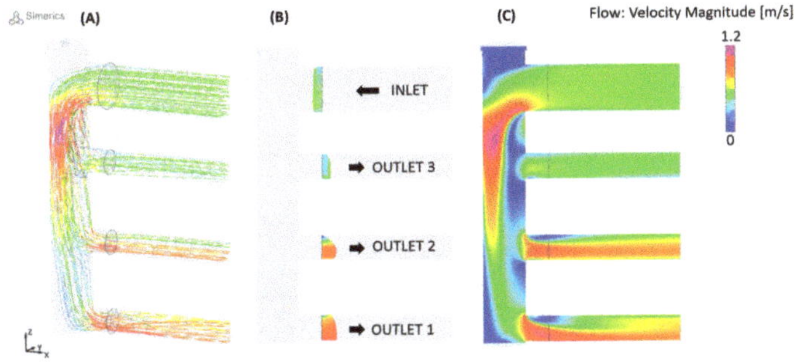

Figure 6. The 3D CFD simulation of flow rate distribution in the manifold. (**A**) Streamlines view; (**B**) vectors view; (**C**) contour view.

Mesh sensitivity analysis was performed for the outlet flow rates $Q_{outlet1}$, $Q_{outlet2}$, $Q_{outlet3}$ where medium mesh and fine mesh had, respectively, 0.43×10^6 and 3.4×10^6 cells. The results of the mesh sensitivity analysis are shown in Figure 7, from which it is clear that a good level of approximation can be obtained with a medium mesh with a lower computational effort. A workstation, equipped with a 64 GB RAM and Intel® Xeon® CPU E5-2699 v3 2.30 GHz processor, was used to perform the analyses. The calculation times in the case of medium mesh and fine mesh using 8 cores are, respectively, 468 s and 1093 s.

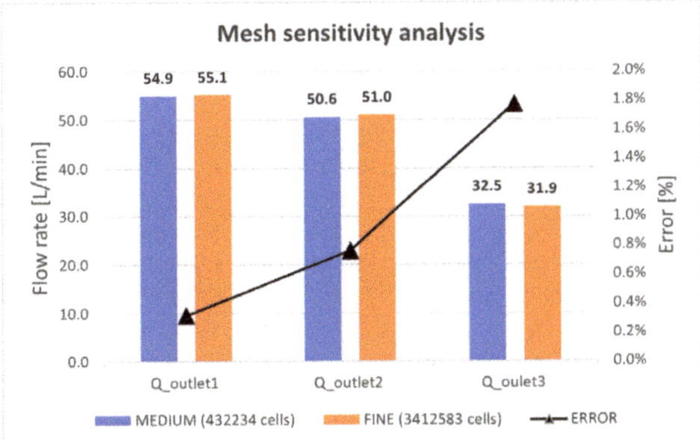

Figure 7. Mesh sensitivity analysis.

The manifold was then modeled in Simcenter AMESim (Figure 8) by means of a thermo-hydraulic volume with one inlet and three outlets, and by inserting an equivalent orifice at each outlet. The cross-sectional area of these orifices (respectively, Ω_1, Ω_2, and Ω_3) was evaluated in order to obtain the same flow distribution found in the 3D CFD characterization, as described below.

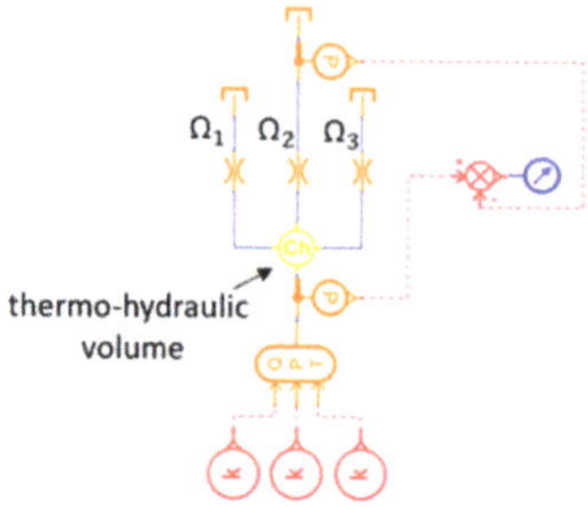

Figure 8. Manifold lumped parameter model.

The three orifices are in parallel [36]; therefore, Equation (4) holds:

$$\Omega_{eq} = \Omega_1 + \Omega_2 + \Omega_3. \tag{4}$$

From the known orifice Equation [36], knowing Q_{inlet} and Δp, it is possible to obtain Ω_{eq}:

$$\Omega_{eq}[\text{mm}^2] = \frac{Q_{inlet}[\text{L/min}]}{18.97 \cdot C_f} \cdot \sqrt{\frac{\rho}{2\Delta p[bar]}} \tag{5}$$

where ρ is the coolant density at 25 °C, equal to 1050 kg/m^3, and C_f, equal to 0.611, is the von Mises' theoretical value for a circular sharp edge orifice [36]. Finally, from the flow rate ratios $Q_{outlet1}/Q_{outlet2}$ and $Q_{outlet1}/Q_{outlet3}$ and from Equation (4) it is possible to derive the sections Ω_1, Ω_2, and Ω_3, respectively, equal to 781 mm^2, 1221 mm^2, and 1343 mm^2.

3. Results for the Model Validation

In this section, the numeric hydraulic results are shown.

Although the system works in a single operating point, to better test the model, it was validated by comparing the numerical results with the experimental ones in three different operating conditions, as shown in Table 1. The first three operating conditions provided the passage of fluid only in one branch at a time and excluded the other two. In the fourth operating condition the third branch was excluded, whereas in the fifth condition, which was the only condition that simulated the real operating conditions of the system, the coolant flowed in all branches at the same time. The experimental measurements were repeated ten times and are reported in the box-plots shown in Figure 9. From the last condition, it is possible to notice, as expected, that when the first two branches are shorter, they had a lower volumetric flow rate than the third due to the presence of the 10 mm orifices.

Table 1 also summarizes the comparison between the numerical results and the mean values of the experimental results obtained by the Keyence FD-Q32C sensor in all operating conditions.

Table 1. Comparison between the numerical and the experimental results.

Operating Condition	First Branch			Second Branch			Third Branch		
	Numeric (L/min)	Exp. (L/min)	Error %	Numeric (L/min)	Exp. (L/min)	Error %	Numeric (L/min)	Exp. (L/min)	Error %
1	53	52	2	0	0	0	0	0	0
2	0	0	0	53	52	2	0	0	0
3	0	0	0	0	0	0	67	62	8
4	50	49	2	50	49	2	0	0	0
5	42	43	2	42	42	0	54	52	4

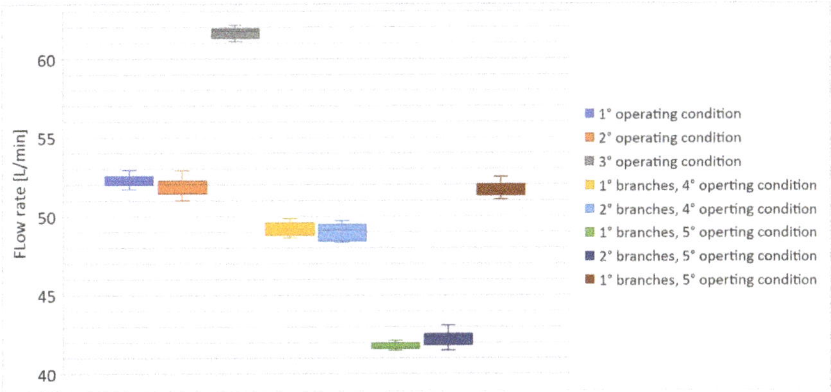

Figure 9. Box plots of the experimental measurements. Number of tests repeated for each case: 10.

The results show that in most cases the error falls within the uncertainty range of the flow sensor, equal to 1 L/min [37]. For the third branch only, a slightly higher error was encountered. Nevertheless, the model is considered sufficiently reliable and can be used for further investigations and optimization of the system.

4. Complete Hydraulic Model with Battery Modules

4.1. Battery Cooling System Description and Characterization

The previously validated model was extended by implementing a submodel that introduced the pressure drop due to the air to liquid heat exchanger, integrated in the battery module. The heat exchanger was experimentally characterized as a pressure loss on the test bench. Figure 10 shows the $(Q - \Delta p)$ curve that was derived from these experimental data. This curve was assigned to an orifice that simulates the hydraulic behavior of the battery module and then was implemented using the previous validated numerical model (Figure 11).

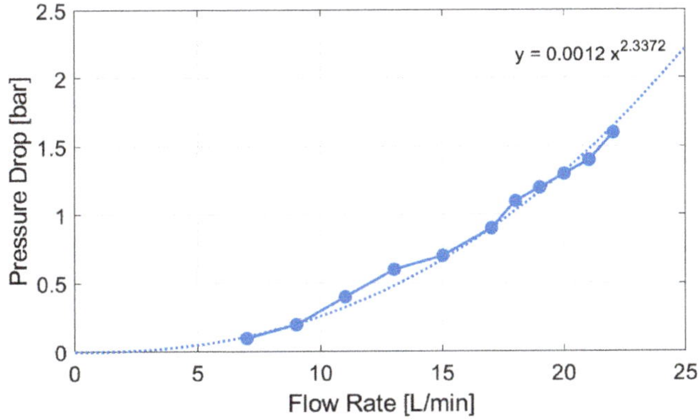

Figure 10. $(Q \div \Delta p)$ curve of the battery module heat exchanger.

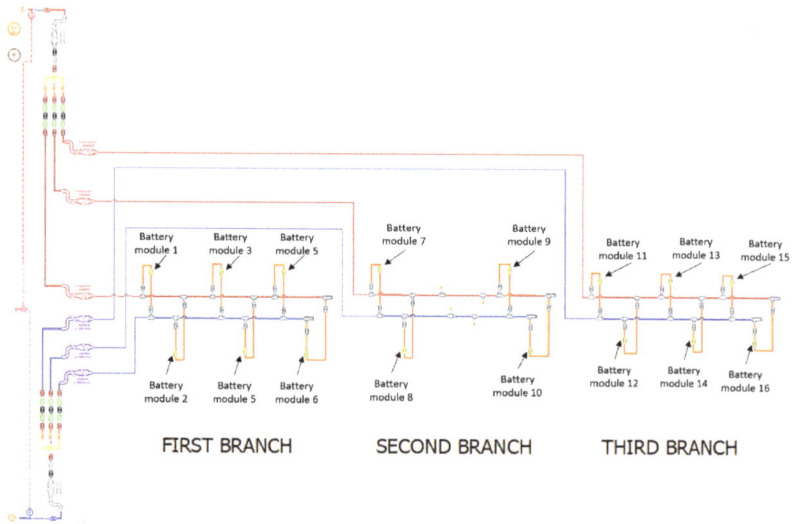

Figure 11. Numerical model with battery modules.

4.2. Hydronic Balancing

As previously said, each of the 16 battery modules that compose the electrical energy storage system must be refrigerated with a specific flow rate at a specific temperature. The system architecture (Figure 1) is made by three branches with the modules arranged in parallel, in order to favor an equal distribution of the inlet flow Q of the coolant and, therefore, to ensure the same thermal conditions for all users. Even if all the users are in parallel, from a physical point of view these are arranged in sequence in order to reach each battery by a line section of different length. The first and third branches serve six users while the second serves only four users. As can be seen from Figure 12, these conditions favor the unbalancing of the flow rates; therefore, actions are needed to obtain the optimum condition of hydronic balancing:

$$Q_1 \cong Q_2 \cong \ldots \cong Q_{16}. \tag{6}$$

In this case two solutions are proposed, the first provides calibrated orifices in the branches with greater flow rate; the second involves a layout modification.

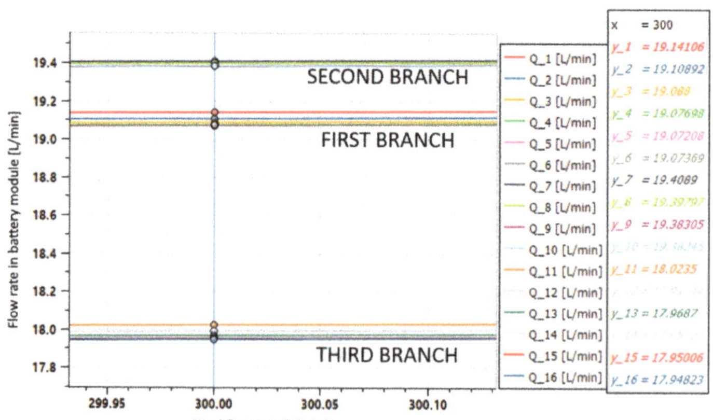

Figure 12. Flow rates in the individual battery modules for an inlet flow of 300 L/min.

4.2.1. Solution 1: Calibrated Orifices

From the diagram of the individual flow rates (Figures 12 and 13A), it can be seen that the flow rate in the first two branches was higher than the one in third branch. In order to balance the flow rates, the calibrated orifices can be placed upstream of the battery modules of the first two branches. Figure 13B shows the results obtained when using 10 mm orifices in the first two branches of the cooling system, while Figure 13C shows the results obtained when using 9 mm orifices in the first branch and 8.5 mm orifice in the second branch. With these measures, a fairly good hydronic balance was achieved; the difference between the higher and the lower flow rate was reduced by 90%.

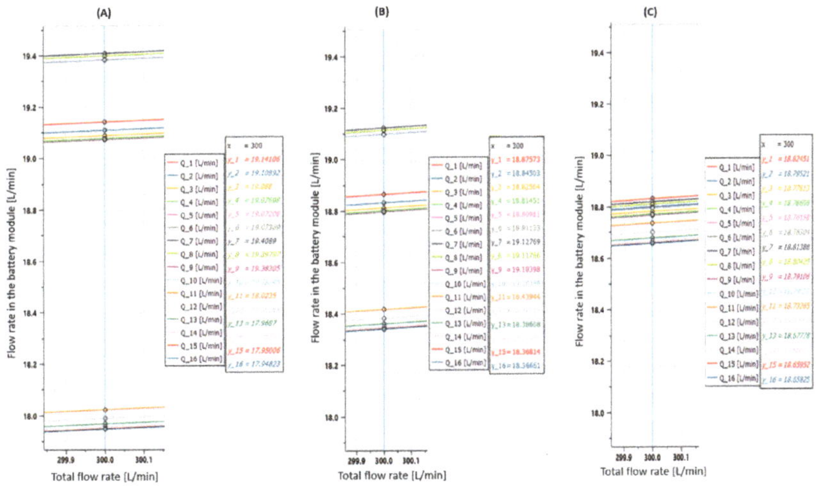

Figure 13. (**A**) Flow rates without calibrated orifices. (**B**) Flow rates with 10 mm orifice in the first and second branches. (**C**) Flow rates with 9 mm orifice in the first branch and 8.5 mm orifice in the second branch.

4.2.2. Solution 2: Layout Modification

Assuming the pressure drop across a single battery module can be described as the pressure drop occurring in an equivalent orifice with a section Ω, the flow rates that circulate in the three branches of the hydraulic system were evaluated according to Equation (7):

$$Q = \sqrt{\frac{\Delta p}{R}} \tag{7}$$

where R indicates the hydraulic resistance introduced by the equivalent orifice:

$$R = \frac{\rho}{2C_f^2\Omega^2}. \tag{8}$$

Since the three branches were in parallel, the pressure drop across the hydraulic circuit was always equal to $\Delta p_{A,B}$, and, since all the battery modules are the same, the sections of the equivalent orifices with which they were replaced were equal. According to the definition of the orifices in parallel, the first and the third branches, which deliver coolant to six battery modules, had an equivalent orifice section of 6Ω, while the second branch, with four battery modules, had an equivalent orifice section of 4Ω. From Equations (7) and (8), it is possible to evaluate the flow rates in the three branches Q_I, Q_{II}, and Q_{III} as follows:

$$Q_I = Q_{III} = C_f \cdot 6\Omega\sqrt{\frac{2\Delta p_{A,B}}{\rho}} \tag{9}$$

$$Q_{II} = C_f \cdot 4\Omega\sqrt{\frac{2\Delta p_{A,B}}{\rho}}. \tag{10}$$

Therefore:

$$\frac{Q_I}{Q_{II}} = \frac{Q_{III}}{Q_{II}} = 1.5 \tag{11}$$

which is consistent with the results found before.

Regarding the pressure drop of the three branches it can be observed that (Figure 1):

$$\Delta p_{A,B} = \Delta p_{A1,B1} + \Delta p_{A,A1} + \Delta p_{B1,B} \tag{12}$$

$$\Delta p_{A,B} = \Delta p_{A2,B2} + \Delta p_{A,A2} + \Delta p_{B2,B} \tag{13}$$

$$\Delta p_{A,B} = \Delta p_{A3,B3} + \Delta p_{A,A3} + \Delta p_{B3,B}. \tag{14}$$

Assuming:

$$\Delta p_{A,A1} + \Delta p_{B1,B} = \Delta p_1 \tag{15}$$

$$\Delta p_{A,A2} + \Delta p_{B2,B} = \Delta p_2 \tag{16}$$

$$\Delta p_{A,A3} + \Delta p_{B3,B} = \Delta p_3 \tag{17}$$

it is possible to rewrite Equations (12)–(14) as:

$$\Delta p_{A,B} = \Delta p_{A1,B1} + \Delta p_1 \tag{18}$$

$$\Delta p_{A,B} = \Delta p_{A2,B2} + \Delta p_2 \tag{19}$$

$$\Delta p_{A,B} = \Delta p_{A3,B3} + \Delta p_3 \tag{20}$$

where, from the Darcy–Weisbach equation:

$$\Delta p_i = \frac{\lambda_i l_i}{d_h}\frac{\rho Q_i^2}{2A^2} \ \ with\ i = 1,2,3\ \ (for\ Q,\ i = I, II, III) \tag{21}$$

in which

- λ is the friction factor for a relative unitary length stretch of pipeline l/d_h; it depends on the Reynolds number and the relative roughness of the inner surface of the pipe;
- d_h is the hydraulic diameter;
- l is the length of the pipeline section for each branch;
- A is the section of the pipe.

Analyzing Equation (21), from Equation (11) results:

$$Q_I \cong Q_{III} > Q_{II}. \tag{22}$$

In addition, given the arrangement of the branches:

$$l_1 < l_2 < l_3 \tag{23}$$

while, for high Reynolds numbers (as in the present case, where Re is of the order of 10^6), λ is almost constant.

Given the relations (22) and (23), and since Δp_i depends on the square of the flow rate, while only linearly on the length of the stroke, it is observed that:

$$\Delta p_3 > \Delta p_1 > \Delta p_2 \tag{24}$$

from which, given (18)–(20), it follows that:

$$\Delta p_{A2,B2} > \Delta p_{A1,B1} > \Delta p_{A3,B3} \tag{25}$$

and finally, given (7), results:

$$Q_{7,8,9,10} > Q_{1,2,3,4,5,6} > Q_{11,12,13,14,15,16} \quad (26)$$

which is consistent with the results shown in Figure 12.

From the previous considerations it appears that if the Δp_i are the same, the flow rates in the individual modules are also equal; therefore, in order to improve the balancing of the flow rates without introducing further pressure drops, Equation (21) must be considered, in which all terms are constant, except for l and Q.

Q depends on the number of modules placed in a branch, as can be seen from (9) and (10), while l depends on the position in which the modules are arranged. To equal the Δp_i it is necessary to increase l when Q is small; in other words, the branch with fewer users must be placed as far away as possible as well as in the model in Figure 14. In this condition it is possible to see the results in terms of flow rate in the individual modules in Figure 15.

Thanks to this solution, a maximum flow rate imbalance of 0.4 L/min was obtained, which was much lower than the maximum flow rate imbalance of 1.5 L/min of the initial layout. In this case the difference between the higher and the lower flow rate was reduced by 73%.

4.2.3. Comparison between the Two Solutions

In terms of flow rates, the solution with the calibrated orifices involves a better balance than the layout modification. Indeed, while the first reduces the imbalance by 90%, the second is limited to 73%. In contrast, in terms of overall pressure drop between the inlet manifold and the outlet manifold $\Delta p_{A,B}$ (Figure 1), the first solution provided a $\Delta p_{A,B1}$ of 1.47 bar, while the second, since it introduced secondary losses due to orifices, provided a higher $\Delta p_{A,B2}$ of 1.64 bar. This difference in pressure drops, multiplied by the nominal flow rate Q of 300 L/min, caused a global power saving P_s for the whole system of 585 W.

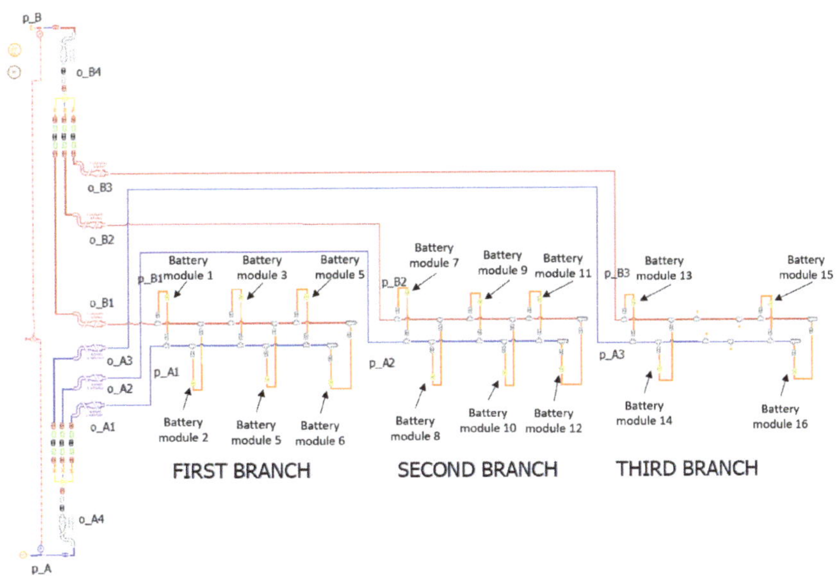

Figure 14. Layout modification: the branch with fewer users is placed as last.

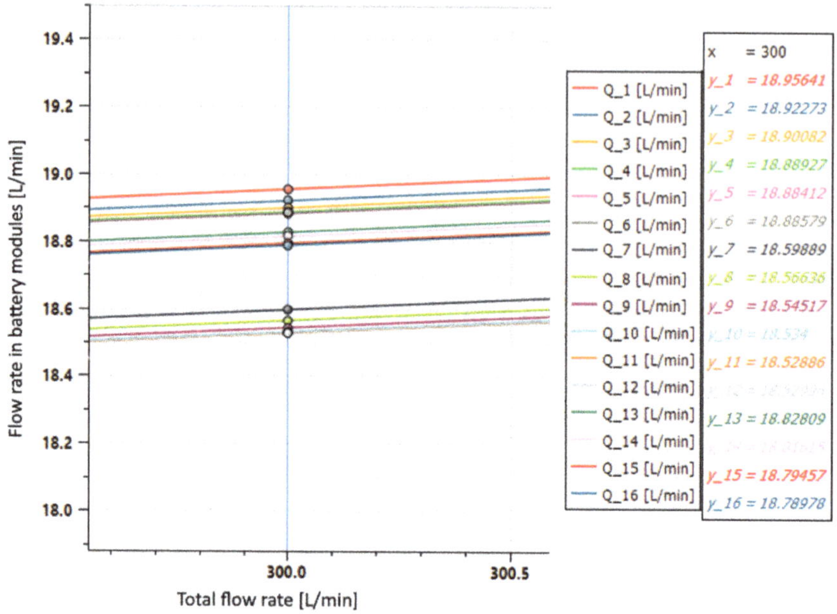

Figure 15. Flow rates in the individual battery modules in case of layout modification.

5. Conclusions

In this paper, a lumped parameter model of the three branches liquid cooling system of a battery pack for railway traction was initially validated. The validation was carried out in two phases. First, a model of a cooling system without batteries was used, in which the numerical results were compared with the experimental results obtained using an ultrasonic clam-on type flow sensor. The results obtained showed a maximum error of 2% in the first two branches, while, in the third branch, there was a slightly higher error which will be investigated by inserting additional pressure transducers in the next battery box prototype. Then, the battery modules, previously characterized on a test bench, were also included in the model. The validated model was used to find the hydronic balancing condition among all the battery modules. To distribute the flow equally, two solutions were proposed. The first involved calibrated orifices upstream of the modules with greater volumetric flow rates, while the second modified the layout by placing the branch with fewer users in the last position. Finally, the two solutions were compared to evaluate the energy savings obtained by using the second solution instead of the first. From the comparison, using the second solution, a power saving for the pump of almost 600 W was estimated.

This power saving method is based on the equalization of the flow rates without introducing secondary losses. This is possible if the branch in which there is the lowest flow rate, the one with the least utilities, is also the longest one. Thus, the product of the two terms l and Q in the Darcy–Weisbach equation tends to be the same for every branch.

This model represents a significant advantage and provides great support in the optimization phase of the cooling system in terms of energy saving and rapid and low-cost experimentation, thus increasing the efficiency of the R&D phase in product manufacturing. Indeed, thanks to this approach it has been possible to compare several solutions without building expensive prototypes, both when considering the calibration of the orifices and the assembly of the system.

In this case, the main challenge facing liquid BTMS was to obtain a criterion for energy savings derived only from the hydraulic system. In order to achieve this, different flow measurements were necessary at different points in the system. The use of a clamp-on type sensor made the experimental campaign easier and more flexible, since with just one sensor it was possible to detect the flow rate in different branches without making any changes to the system.

Future developments will provide the implementation of a model for the thermal dissipation of the batteries in order to create a complete digital twin of the cooling system of the battery thermal management that will allow us to perform numerical tests aimed at finding the best cooling method to reduce energy consumption for safe, sustainable, and comfortable collective mobility.

Author Contributions: Conceptualization, E.F., L.B. and A.S.; Methodology, R.D.R. and E.F.; Software, R.D.R. and L.R.; Validation, R.D.R.; Formal analysis, R.D.R.; Investigation, R.D.R. and L.B.; Resources, L.B. and A.S.; Data curation, R.D.R.; Writing—original draft, R.D.R.; Writing—review & editing, L.R., E.F. and A.S.; Visualization, L.R.; Supervision, L.R., E.F. and A.S.; Project administration, L.R., E.F. and A.S.; Funding acquisition, A.S. All authors have read and agreed to the published version of the manuscript.

Funding: This research received no external funding.

Institutional Review Board Statement: Not applicable.

Informed Consent Statement: Not applicable.

Data Availability Statement: Data available on request due to restrictions, e.g., privacy or ethical.

Conflicts of Interest: The authors declare no conflict of interest.

Nomenclature

A	Pipe cross-sectional area (m^2)	Q	Volumetric flow rate (m^3/s)
A_r	Aspect ratio (-)	R	Hydraulic resistance (Pa s^2/m^6)
c	Speed of sound (m/s)	Re	Reynolds number (-)
C_f	Orifice coefficient (-)	rr	Relative roughness (-)
d_h	Hydraulic diameter (m)	T_{wave}	Wave travel time (s)
D_n	Dissipation number (-)	λ	Friction factor (-)
l	Length of a pipeline (m)	ν	Kinematic viscosity (m^2/s)
p	Pressure (Pa)	ρ	Density (kg/m^3)
P	Power (-)	Ω	Orifice cross-sectional area (m^2)

References

1. Alataş, S. Do environmental technologies help to reduce transport sector CO_2 emissions? Evidence from the EU15 countries. *Res. Transp. Econ.* **2022**, *91*, 101047. [CrossRef]
2. Jiang, C.; Wan, Y.; Yang, H.; Zhang, A. Impacts of high-speed rail projects on CO_2 emissions due to modal interactions: A review. *Transp. Res. Part Transp. Environ.* **2021**, *100*, 103081. [CrossRef]
3. Hitachi. Battery, A Zero–Emission Solution Ensuring Seamless Journeys and an Integrated Passenger Experience. 2022. Available online: https://www.hitachirail.com/products-and-solutions/rolling-stock/battery/ (accessed on 29 November 2022).
4. Iwase, T.; Kawamura, J.; Tokai, K.; Kageyama, M. Development of battery system for railway vehicle. In Proceedings of the 2015 International Conference on Electrical Systems for Aircraft, Railway, Ship Propulsion and Road Vehicles (ESARS), Aachen, Germany, 3–5 March 2015. [CrossRef]
5. Lu, M.; Zhang, X.; Ji, J.; Xu, X.; Zhang, Y. Research progress on power battery cooling technology for electric vehicles. *J. Energy Storage* **2020**, *27*, 101155. [CrossRef]
6. Thakur, A.K.; Prabakaran, R.; Elkadeem, M.R.; Sharshir, S.W.; Arıcı, M.; Wang, C.; Zhao, W.; Hwang, J.Y.; Saidur, R. A state of art review and future viewpoint on advance cooling techniques for Lithium–ion battery system of electric vehicles. *J. Energy Storage* **2020**, *32*, 101771. [CrossRef]
7. Panchal, S.; Khasow, R.; Dincer, I.; Agelin-Chaab, M.; Fraser, R.; Fowler, M. Thermal design and simulation of mini-channel cold plate for water cooled large sized prismatic lithium-ion battery. *Appl. Therm. Eng.* **2017**, *122*, 80–90. [CrossRef]
8. Malik, M.; Dincer, I.; Rosen, M.A.; Mathew, M.; Fowler, M. Thermal and electrical performance evaluations of series connected Li-ion batteries in a pack with liquid cooling. *Appl. Therm. Eng.* **2018**, *129*, 472–481. [CrossRef]

9. Song, L.; Zheng, Y.; Xiao, Z.; Wang, C.; Long, T. Review on Thermal Runaway of Lithium-Ion Batteries for Electric Vehicles. *J. Electron. Mater.* **2022**, *51*, 30–46. [CrossRef]
10. Akinlabi, A.A.; Solyali, D. Configuration, design, and optimization of air-cooled battery thermal management system for electric vehicles: A review. *Renew. Sustain. Energy Rev.* **2020**, *125*, 109815. [CrossRef]
11. Zhao, G.; Wang, X.; Negnevitsky, M.; Zhang, H. A review of air-cooling battery thermal management systems for electric and hybrid electric vehicles. *J. Power Sources* **2021**, *501*, 230001. [CrossRef]
12. Tang, X.; Guo, Q.; Li, M.; Wei, C.; Pan, Z.; Wang, Y. Performance analysis on liquid-cooled battery thermal management for electric vehicles based on machine learning. *J. Power Sources* **2021**, *494*, 229727. [CrossRef]
13. Tang, Z.; Liu, Z.; Li, J.; Cheng, J. A lightweight liquid cooling thermal management structure for prismatic batteries. *J. Energy Storage* **2021**, *42*, 103078. [CrossRef]
14. Liang, J.; Gan, Y.; Li, Y. Investigation on the thermal performance of a battery thermal management system using heat pipe under different ambient temperatures. *Energy Convers. Manag.* **2018**, *155*, 1–9. [CrossRef]
15. Zhou, H.; Dai, C.; Liu, Y.; Fu, X.; Du, Y. Experimental investigation of battery thermal management and safety with heat pipe and immersion phase change liquid. *J. Power Sources* **2020**, *473*, 228545. [CrossRef]
16. Sun, Z.; Guo, Y.; Zhang, C.; Xu, H.; Zhou, Q.; Wang, C. A Novel Hybrid Battery Thermal Management System for Prevention of Thermal Runaway Propagation. *IEEE Trans. Transp. Electrif.* **2022**. [CrossRef]
17. Zhao, Y.; Zou, B.; Zhang, T.; Jiang, Z.; Ding, J.; Ding, Y. A comprehensive review of composite phase change material based thermal management system for lithium-ion batteries. *Renew. Sustain. Energy Rev.* **2022**, *167*, 112667. [CrossRef]
18. Sanker, S.B.; Baby, R. Phase change material based thermal management of lithium ion batteries: A review on thermal performance of various thermal conductivity enhancers. *J. Energy Storage* **2022**, *50*, 104606. [CrossRef]
19. Siddique, A.R.M.; Mahmud, S.; Heyst, B.V. A comprehensive review on a passive (phase change materials) and an active (thermoelectric cooler) battery thermal management system and their limitations. *J. Power Sources* **2018**, *401*, 224–237. [CrossRef]
20. Liu, X.; Zhang, C.F.; Zhou, J.G.; Xiong, X.; Wang, Y.P. Thermal performance of battery thermal management system using fins to enhance the combination of thermoelectric Cooler and phase change Material. *Appl. Energy* **2022**, *322*, 119503. [CrossRef]
21. European Commission. Communication from the Commission to the European Parliament, the Council, the European Economic and Social Committee and the Committee of the Regions. Sustainable and Smart Mobility Strategy. 2020. Available online: https://eur-lex.europa.eu/legal-content/EN/TXT/?uri=COM:2020:789:FIN (accessed on 29 November 2022).
22. Popovich, N.; Phadke, A. Big batteries on wheels: Converting diesel trains to battery electric can provide signiicant economic, environmental, and grid resilience beneets. *Preprint* **2021**. [CrossRef]
23. Bombaridier. Bombardier Talent 3 Battery-Powered Train, Germany. 2022. Available online: https://www.railway-technology.com/projects/bombardier-talent-3-battery-train/ (accessed on 29 November 2022).
24. Xiong, F.; Zeng, J.; Ouyang, C.; Zhong, J.; Nishida, R.; Shinshi, T.; Dan, D.; Yao, C.; Zhang, H. *Design and Simulation of Liquid Cooled System for Power Battery of PHEV*; IOP Publishing: Bristol, UK, 2017. [CrossRef]
25. Kellermann, H.; Fuhrmann, S.; Shamiyeh, M.; Hornung, M. Design of a Battery Cooling System for Hybrid Electric Aircraft. *J. Propuls. Power* **2022**, 38, 1–16. [CrossRef]
26. Royston, S.J.; Gladwin, D.T.; Stone, D.A.; Ollerenshaw, R.; Clark, P.; Rail, S.; Siemens, E.; York, M. Development and Validation of a Battery Model for Battery Electric Multiple Unit Trains. In Proceedings of the Development and Validation of a Battery Model for Battery Electric Multiple Unit Trains, Lisbon, Portugal, 14–17 October 2019.
27. Wang, Y.; Gao, Q.; Wang, G.; Lu, P.; Zhao, M.; Bao, W. A review on research status and key technologies of battery thermal management and its enhanced safety. *Int. J. Energy Res.* **2018**, *42*, 4008–4033. [CrossRef]
28. Temporelli, A.; Carvalho, M.L.; Girardi, P. Life Cycle Assessment of Electric Vehicle Batteries: An Overview of Recent Literature. *Energies* **2020**, *13*, 2864. [CrossRef]
29. Kang, T.; Park, S.; Lee, P.Y.; Cho, I.H.; Yoo, K.; Kim, J. Thermal Analysis of a Parallel-Configured Battery Pack (1S18P) Using 21700 Cells for a Battery-Powered Train. *Electronics* **2020**, *9*, 447. [CrossRef]
30. Teng, H.; Yeow, K. Design of direct and indirect liquid cooling systems for high-capacity, high-power lithium-ion battery packs. *SAE Int. J. Altern. Powertrains* **2012**, *1*, 525–536. [CrossRef]
31. De Rosa, R.; Belli, L.; Frosina, E.; Venanzio, P.; Romagnuolo, L.; Senatore, A. Predictive model of cooling system for railway electric propulsion: Validation of design choices and last mile analysis. *J. Phys. Conf. Ser.* **2022**, *2385*, 012062. [CrossRef]
32. Conrad, K.; Lynnworth, L. Fundamentals of Ultrasonic Flow Meters. In Proceedings of the American School of Gas Measurement Technology, Houston, TX, USA, 19–22 September 2002; pp. 52–61.
33. Fischer, G. GF SIGNET 9010 Intelek-Pro Flow Controller Instruction Manual. Available online: https://www.gfps.com/it-it.html (accessed on 15 March 2022).
34. *Simcenter Amesim Thermal-Hydraulic Library User's Guide*; Technical Report; Volupe: Askim, Sweden, 2021.
35. Antifrogen® N. Heat Transfer Fluid. Available online: https://www.clariant.com/en/Solutions/Products/2013/12/09/18/25/Antifrogen-N (accessed on 15 March 2022).

36. Vacca, A.; Franzoni, G. *Hydraulic Fluid Power: Fundamentals, Applications, and Circuit Design*; Wiley: Hoboken, NJ, USA, 2021; 682p.
37. Keyence User Manual. Basic Clamp-on Flow Sensor FD-Q Series Setting Guide. Available online: https://www.keyence.it/products/process/flow/fd-q/specs/ (accessed on 15 March 2022).

sensors

MDPI

Article

Optimized Replication of ADC-Based Particle Counting Algorithm with Reconfigurable Multi-Variables in Pseudo-Supervised Digital Twining of Reference Dust Sensor Systems

Seungmin Lee, Jisu Kwon and Daejin Park *

School of Electronic and Electrical Engineering, Kyungpook National University, Daegu 41566, Republic of Korea; lsm1106@knu.ac.kr (S.L.); kjisu96@knu.ac.kr (J.K.)
* Correspondence: boltanut@knu.ac.kr; Tel.: +82-10-7529-1231

Abstract: As the application fields for digital twins have expanded, various studies have been conducted with the objective of optimizing the costs. Among these studies, research on low-power and low-performance embedded devices has been implemented at a low cost by replicating the performance of existing devices. In this study, we attempt to obtain similar particle count results in a single-sensing device replicated from the particle count results in a multi-sensing device without knowledge of the particle count acquisition algorithm of the multi-sensing device. Through filtering, we suppressed the noise and baseline movements of the raw data of the device. In addition, in the process of determining the multi-threshold for obtaining the particle counts, the existing complex particle count determination algorithm was simplified to make it possible to utilize the look-up table. The proposed simplified particle count calculation algorithm reduced the optimal multi-threshold search time by 87% on average and the root mean square error by 58.5% compared to existing method. In addition, it was confirmed that the distribution of particle count from optimal multi-thresholds has a similar shape to that from multi-sensing devices.

Keywords: digital twin; dust sensing; particle count; ADC filter; embedded device

Citation: Lee, S.; Kwon, J.; Park, D. Optimized Replication of ADC-Based Particle Counting Algorithm with Reconfigurable Multi-Variables in Pseudo-Supervised Digital Twining of Reference Dust Sensor Systems. *Sensors* **2023**, *23*, 5557. https://doi.org/10.3390/s23125557

Academic Editors: Qibo Feng, Jiakun Li and Qixin He

Received: 8 May 2023
Revised: 11 June 2023
Accepted: 12 June 2023
Published: 14 June 2023

1. Introduction

A digital twin replicates real-world environments and simulates prediction results using a computer. The applications of digital twins are expanding throughout the industry owing to their advantages such as safety, repeatability, and the low cost of predicting results [1]. Digital twins are widely used in manufacturing to predict the results of a product, and recently, the scope of autonomous driving has expanded through the digitization of cities [2–8].

In particular, Industry 4.0 further emphasizes the importance of digital twins in smart factories that have intelligent production systems [9–12]. Additionally, with the increase in the average life expectancy of people, interest in health is increasing. Recently, due to respiratory diseases caused by the coronavirus, interest in air quality among living environments is on rise [13–15].

Dust-sensing devices digitize and provide dust concentration information in the air. The use of a digital twin of dust concentration is important in minimizing the occurrence of defects in ultra-fine processes in the industrial field [16], and in the case of buildings, it is used for periodic internal air circulation [17,18]. The importance of digital twins in the continuous management and conservation of energy consumption, particularly in relation to the heating, ventilation, and air conditioning systems of buildings, is consistently increasing [19]. In daily life, the increase in indoor fine dust concentration due to air pollution is measured and used to establish an IoT environment linked to the automatic operation of air purifiers and fans.

Digital-twinned dust sensing systems are used in both industry and daily life, and their field of application is expanding [20,21]. Accordingly, to cover long-term operation and wide measurement areas, existing devices are being replaced by low-power and low-cost embedded devices, and existing algorithms are being improved and optimized to be made suitable for low-memory embedded devices [22,23].

In this study, we minimized the cost by replicating the particle count (PC) of an existing dust sensing device. The existing device uses multiple sensors to measure the number of particles according to the size of dust. In this study, we attempt to replicate the performance of a multi-sensor device (reference device) through a single-sensor device (test device) to minimize the power consumption and reduce the cost of the dust sensing system.

To this end, this study was conducted in three parts: (1) analog-to-digital converter (ADC) filter design, (2) multi-threshold search, and (3) PC similarity analysis.

First, the light scattered by the dust was measured in terms of voltage using a photodiode. The larger the dust particle, the more light is scattered, resulting in a higher voltage. The change in the magnitude of the voltage measured by the photodiode is small. Thus, if the baseline of the ADC fluctuates owing to the power supply noise, an error occurs in the PC calculation. To improve this, in this study, we minimized the effect of power noise by detecting baseline fluctuations using an average filter and then removing them. In addition, because the test device is vulnerable to noise, an average filter was additionally applied to suppress noise.

After filtering the ADC, determining a multi-threshold for measuring the PC according to the size of the dust particle for the filtered ADC is necessary. In general, it is to generate a look-up table (LUT) for a PC measurement algorithm because the threshold used in the PC measurement process is not independent of other thresholds. Therefore, the searching time for optimal multi-thresholds increases exponentially with the number of dust size categories. Furthermore, the general PC measuring algorithm considers the width of the voltage pulse in the ADC. This makes the algorithm more complex and makes it difficult to generate LUTs, which further increases the searching time for optimal multi-thresholds. The simplification of optimal multi-threshold searching algorithm is required because run-time execution is important in digital twinning using light weight embedded devices. In this study, we improved the searching time and root mean square error (RMSE) using LUTs by simplifying various variables used for the PC measurement. We reduced search time by 87% and RMSE by 58.5% compared to the existing method.

In searching for the optimal multi-threshold, the suitability of the given multi-threshold is determined by comparing the PCs from the test device (TPCs) with that from the reference device (RPCs). However, the TPCs are generally lower than the RPCs because the test device is a single-sensor device. Thus, comparing the two sets of PC data was difficult. To solve this, we normalized RPCs and TPCs in two steps. Firstly, we normalized RPCs and TPCs according to the maxima of RPC0.3 and TPC0.3, which are generally the highest values between PCs. This process normalizes the scale between the RPCs and TPCs. After the first normalization, a secondary normalization is performed for each RPC and TPC with the corresponding maxima of the RPCs. This process normalizes the internal scales of RPCs and TPCs. After normalization, the similarity was compared using the RMSE.

By repeating the update of the multi-threshold and similarity measurement, the optimal multi-threshold with the highest similarity are determined. Then, we can obtain the TPCs using the optimal multi-threshold that replicates the RPCs.

Figure 1 is a schematic of the proposed method.

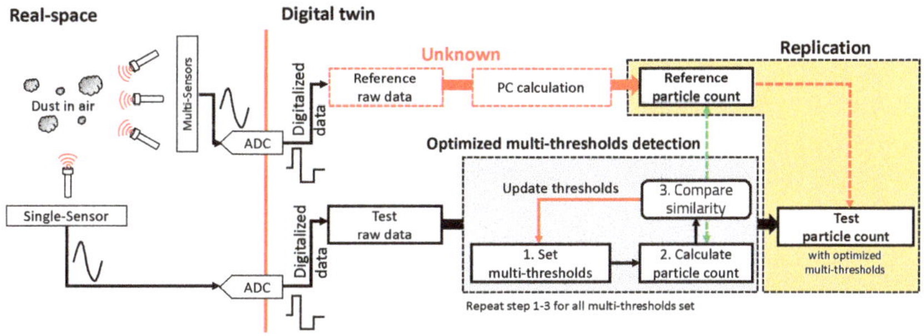

Figure 1. Digital twin in dust sensing.

In this manner, without a known ADC and PC measuring algorithm for the reference device, the optimal multi-thresholds are determined such that the TPCs replicate the RPCs.

The remainder of this paper is organized as follows. Section 2 introduces the structure of the dust-sensing system and explains the ADC filtering. Section 3 explains the TPCs acquisition processes using optimal multi-thresholds. After confirming that the RPCs are replicated well through experiments on the acquisition of TPCs in Section 4, concluding statements are provided in Section 5.

2. Dust Sensing System

2.1. Light Scattering Method

PM encompasses particles of sulfate, black carbon, dust from erosion, pollen, and so on [24]. Various methods are available for measuring dust concentration. Existing representative dust measurement methods for PM2.5 include the gravimetric method and the beta-ray absorption method [25]. The gravimetric method manually measures the weight of the collected dust, while the beta-ray absorption method automatically measures the concentration using the amount of beta-rays absorbed by the dust.

Although the gravimetric method is accurate, it requires considerable time to collect dust and has the disadvantage of needing to maintain a consistent temperature and humidity during the measurement process. The beta-ray absorption method measures the dust concentration at intervals of one hour, but its accuracy is relatively lower compared to the gravimetric method. This measurement method is suitable for confirming the concentration of dust in everyday life on a daily or hourly basis; however, it is not suitable for real-time measurements.

Recently, various sensor-based measurement devices have been developed to overcome the limitations of traditional dust measurement devices, even though their reliability has not been fully achieved. Among these devices, the light scattering method-based measurement device is lightweight, compact, and capable of providing measurements within a short time frame, ranging from one second to one minute [26,27].

Table 1 presents the comparison of characteristics of different dust sensing methods.

Table 1. Comparison of characteristics of dust sensing methods.

Sensing Method	Accuracy	Time	Measurement	Portability	Cost
Gravimetric	Very high	24 h	Manual	Low	High
Bete-ray	high	1 h	Automatic	Low	High
Light scattering	Low	1 s ∼1 min	Automatic	High	Low

The light scattering method has a limitation in that it cannot be used for administrative statistics due to its low accuracy compared to other methods. However, it is the most suitable for digital twinning of dust sensing due to its advantages such as its portability, low power consumption, low cost, and real-time processing.

The light-scattering method measures the amount of light scattered by dust. Figure 2 illustrates the concept of the light scattering method.

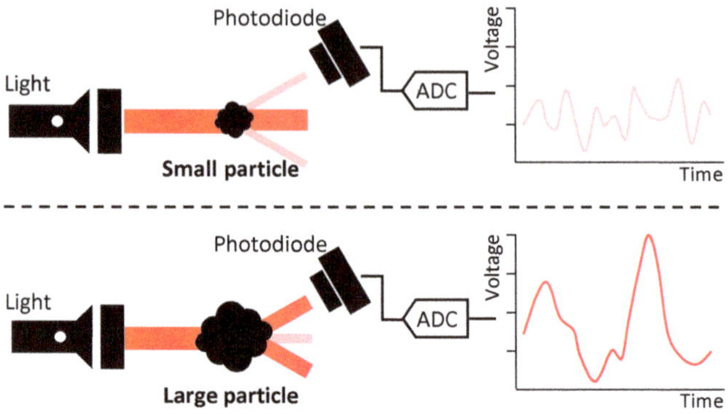

Figure 2. Light scattering method.

The larger the dust, the more light is scattered, resulting in a higher intensity of the detected light and wider voltage pulses. Based on this characteristic, the PC categories were classified according to the magnitude of the voltage, and the number of samples falling within each category range was counted. The sampling frequencies of the ADC and TPCs used in this study were 50 kHz and 1/6 Hz, respectively. One TPC sample for each PC category was calculated by counting 300,000 ADC samples.

Typically, the number of dust particles based on size is primarily counted for each category of differential PC (dPC), and the actual PC value is subsequently measured by accumulating the dPC values. For instance, dPC0.3 consists of particle counts smaller than 0.5 μm, and the threshold of dPC is determined by the next category of dPC.

Moreover, dPC0.5, dPC1.0, $\cdots$, and dPC10.0 are calculated in a similar way. PC0.3 is the accumulation of dPC0.3 to dPC10.0, and PC0.5 is the accumulation of dPC0.5 to dPC10.0, as shown in (1):

$$\begin{aligned} PC0.3 &= dPC0.3 + dPC0.5 + \cdots + dPC10.0 \\ PC0.5 &= dPC0.5 + dPC1.0 + \cdots + dPC10.0 \\ &\vdots \\ PC10.0 &= dPC10.0 \end{aligned} \tag{1}$$

Thus, PC0.3 $\geq$ PC0.5 $\geq \cdots \geq$ PC10.0.

2.2. ADC Filtering

The light scattered by dust was measured using a photodiode. The voltage measured increased as the size of the dust particles increased. The PC is categorized into various categories based on the size of the dust, such as PC0.3, PC0.5, and PC1.0. To obtain the PC for each category, thresholds are required. In general, a global threshold was used assuming no changes in environmental conditions. However, when the global threshold is used, variations in the baseline caused by power supply noise can lead to erroneous detection. Additionally, it becomes challenging to use the existing global threshold when there are overall voltage fluctuations due to external environmental factors during the measurement. To address these issues, baseline variations and bases were eliminated, creating an environment suitable for the application of global thresholds. Figure 3 illustrates the raw ADC data, showing the fluctuations in the baseline.

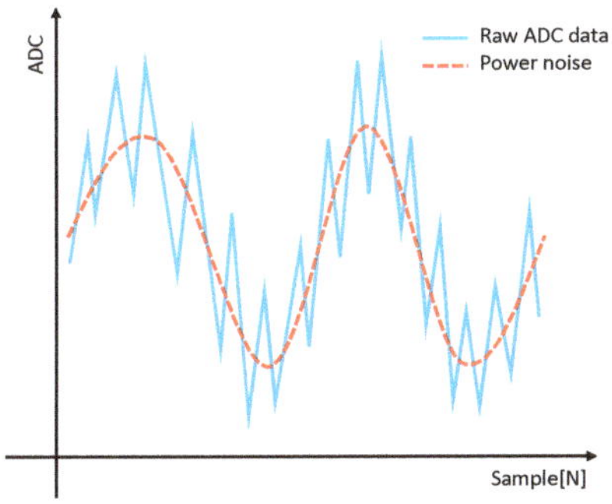

Figure 3. ADC raw data and baseline fluctuation noise.

As shown in Figure 3, baseline fluctuations caused by power supply noise were observed. These fluctuations posed challenges when detecting particles based on the global threshold. Excessive particle detection occurred when the baseline was low, while all samples exceeded the threshold when the baseline was high, resulting in no particle detection. Therefore, a preprocessing step is necessary to suppress baseline fluctuations.

Baseline fluctuations are commonly mitigated using a high-pass filter. Another approach involves acquiring baseline information through a low-pass filter and then removing the low-pass-filtered signal to suppress baseline fluctuations. In this study, we designed a suitable filter, considering both finite impulse response (FIR) filters and infinite impulse response (IIR) filters, to effectively suppress baseline fluctuations.

2.2.1. FIR Average Filter

The test device used in this study can apply to an eight-tap FIR filter. Based on this, we apply a seven-tap moving average filter as in (2):

$$y[n] = \sum_{i=0}^{6} \frac{1}{7} x[n-i] \tag{2}$$

Figure 4 illustrates the input and filtered ADCs.

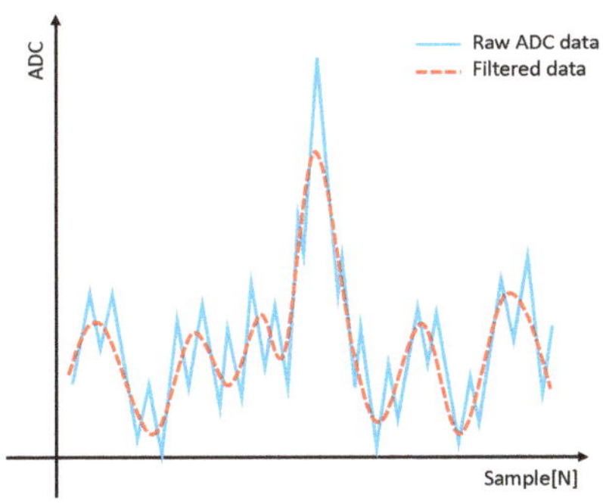

Figure 4. Input and filtered ADCs using 7-tap FIR moving average filter.

As a result, the FIR moving average filter, which uses a small number of taps, is highly responsive to baseline changes and can be influenced by high ADC samples. Consequently, it has a disadvantage in preserving high ADC samples generated by a large amount of dust.

On the other hand, if the window size of the moving average is increased, the filter becomes less sensitive to baseline changes, leading to improved preservation of high ADC samples. However, increasing the number of taps in the FIR filter significantly increases the cost, presenting a challenge.

2.2.2. IIR Average Filter

The IIR filter offers the advantage of achieving a similar effect to that of a large number of taps in an FIR filter, even with a small number of taps. The transfer function of the IIR filter is defined using the Z-transform, as shown in (3):

$$H(z) = \frac{Y(z)}{X(z)} = \frac{\sum_{i=0}^{P} b_i z^{-i}}{\sum_{j=0}^{Q} a_j z^{-j}}, \tag{3}$$

where P is the feedforward filter order, Q is the feedback filter order, b_i represents the feedforward filter coefficients, and a_j represents the feedback filter coefficients.

In this paper, we designed a second-order IIR filter to produce a similar result to that of an FIR filter with a window size of 50, as shown in (4):

$$\begin{aligned} b_i &= [0.02, 0.000417, 0.000417] \\ a_j &= [1, -0.97917, 0] \end{aligned} \tag{4}$$

Figure 5 illustrates the experimental results using the IIR filter.

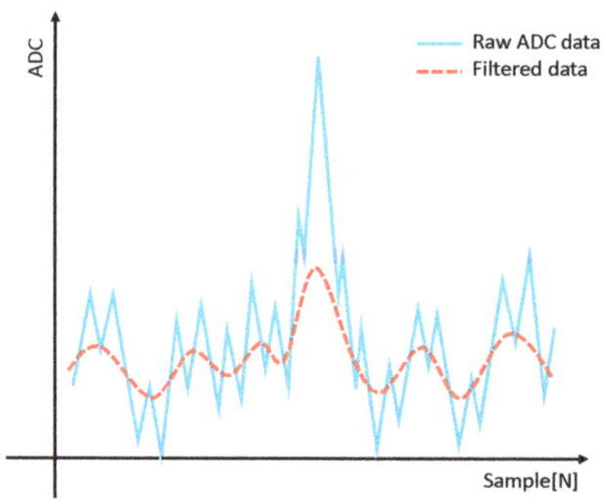

Figure 5. Input and filtered ADCs using second-order IIR filter.

The results of the 49-tap FIR filter using MATLAB were compared, and it was confirmed that the results of the second-order IIR filter were sufficiently similar.

2.2.3. Composite FIR and IIR Filters

From the previous experiment, it was confirmed that the IIR filter can effectively detect baseline fluctuations, even with a small number of taps. In order to further improve the filtering performance, a 4-tap lowpass filter was applied to the FIR filter to suppress extreme high-frequency noise signals. This means that the FIR filter suppresses the high-frequency component, while the IIR filter captures the baseline and suppresses it, resulting in an overall effect similar to a bandpass filter that suppresses the low-frequency component.

2.3. General PC Calculation Algorithm

From the filtered ADC data, the number of voltage pulses was counted using each threshold to measure each dPC. Subsequently, each PC was obtained by accumulating the corresponding dPC values. Therefore, multiple thresholds are required, one for each category of PC.

In general, the amount of light scattering increases with the size of the dust, resulting in higher voltage measurements on the photodiode. This indicates a larger amount of dust being observed. As a result, the ADC threshold gradually increases depending on the size of the dust: Thr0.3 $\leq$ Thr0.5 $\leq \cdots \leq$ Thr10.0. Therefore, when searching for a threshold, the search can be performed within a range larger than the threshold of the previous category. Using a LUT can also be effective in minimizing overlapping operations since previously calculated thresholds often overlap during the multi-threshold search process.

However, the general PC calculation algorithm is not simply based on applying the threshold to the ADC values. It takes into account factors such as the change in states (increase or decrease) between previous and current samples, category changes based on the threshold, and the width of the voltage pulse. Figure 6 illustrates the conceptual diagram of the general PC calculation algorithm.

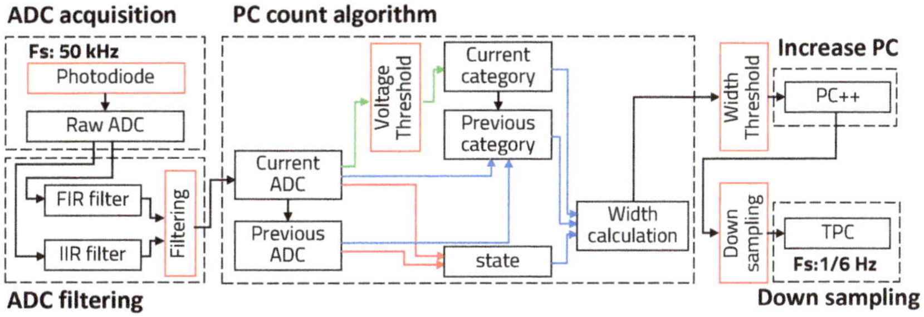

Figure 6. Algorithm scheme of general PC calculation.

It is necessary to consider not only the optimization of multiple thresholds for the ADC voltage but also multiple thresholds for the pulse width. The measurement results of PC according to these thresholds are not independent of each other. Therefore, applying an LUT to the general PC calculation algorithm becomes difficult.

3. Proposed Algorithm

3.1. Simplified PC Calculation Algorithm

In this study, we simplified the algorithm by incrementing each PC when the voltage of a sample exceeded a given threshold. In this simplified case, the same PC is obtained regardless of the category for which the threshold is used to calculate the PC. As a result, the PC values obtained according to the thresholds can be generated as an LUT. During the optimal multi-threshold searching process, when calculating the similarity of TPCs for a given set of multi-thresholds, there is no need to calculate the TPCs each time based on the given thresholds. Instead, we can refer to the TPCs in the LUT and compare the similarity with the RPCs.

Figure 7 illustrates the distributions of the LUT based on different thresholds and an example of the TPC at thresholds 10, 20, and 40.

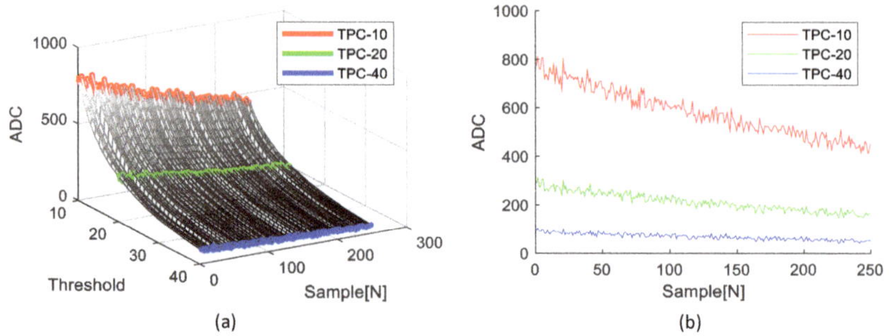

Figure 7. The LUT distribution and the example of TPCs according to thresholds: (**a**) LUT distribution; (**b**) TPCs with thresholds 10, 20, and 40.

Figure 8 shows the simplified TPCs calculating algorithm using LUT.

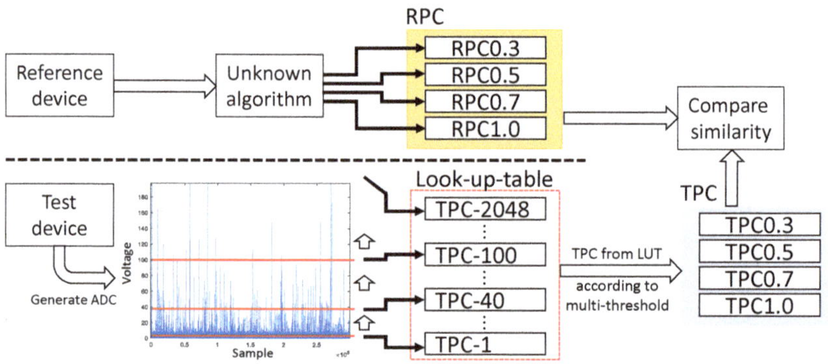

Figure 8. The simplified TPC calculating algorithm using LUTs.

Compared to Figure 6, the redundancy of the calculation of the TPCs according to the threshold is removed, resulting in a simplified process.

3.2. PC Similarity Measurement

In general, the RMSE is widely used to measure the similarity of two signals.

For two given signals X and Y, the RMSE is calculated as in (5):

$$RMSE(X,Y) = \sqrt{\frac{\sum_{i=1}^{N}(x_i - y_i)^2}{N}}, \tag{5}$$

where N is the length of the two signals.

RMSE is commonly used to measure the similarity between two signals when their scales are similar.

However, comparing the similarity between RPCs and TPCs becomes challenging due to the significant difference in their scales. In such cases, the Pearson similarity method [28] is often employed as an alternative. The Pearson similarity is calculated using (6), which is suitable for comparing signals with different scales:

$$\rho(X,Y) = \frac{1}{N-1}\sum_{i=1}^{N}\left(\frac{X-\mu_X}{\sigma_X}\frac{Y-\mu_Y}{\sigma_Y}\right), \tag{6}$$

where μ and σ represent the mean and standard deviation of the signal, respectively.

The Pearson similarity is a robust measure that takes into account baseline movements and scale changes between signals. It achieves this by normalizing the signals through the subtraction of their mean values and division by their standard deviations. This normalization process helps to mitigate the impact of baseline shifts and variations in scale, allowing for a more accurate comparison of the signals' similarity.

In this study, instead of comparing two single-channel signals, it is necessary to compare multiple multi-channel signals corresponding to the number of PC categories. As a result, normalizing RPCs and TPCs using the Pearson similarity becomes challenging.

In this study, we calculate the RMSE after normalizing the scales of RPCs and TPCs through two-step normalization using (7) and (8):

$$\begin{aligned} RPC0.3' &= RPC0.3/max(RPC0.3) \\ TPC0.3' &= TPC0.3/max(TPC0.3) \\ RPC0.5' &= RPC0.5/max(RPC0.3) \\ TPC0.5' &= RPC0.5/max(TPC0.3) \\ &\vdots \\ RPC10.0' &= RPC10.0/max(RPC0.3) \\ TPC10.0' &= RPC10.0/max(TPC0.3) \end{aligned} \tag{7}$$

$$\begin{aligned} RPC0.3'' &= RPC0.3'/max(RPC0.3') \\ TPC0.3'' &= TPC0.3'/max(RPC0.3') \\ RPC0.5'' &= RPC0.5'/max(RPC0.5') \\ TPC0.5'' &= RPC0.5'/max(RPC0.5') \\ &\vdots \\ RPC10.0'' &= RPC10.0'/max(RPC10.0') \\ TPC10.0'' &= RPC10.0'/max(RPC10.0') \end{aligned} \tag{8}$$

Figure 9 shows the process of primary and secondary normalized RPCs and TPCs for measuring the RMSE.

Equation (7) and Figure 9b depict the first normalization of RPCs and TPCs, denoted as RPCs′ and TPCs′. This first normalization is performed by dividing RPCs and TPCs by their respective maxima, specifically RPC0.3 and TPC0.3. By doing so, the scale difference between the RPCs and TPCs is addressed. Equation (8) and Figure 9c illustrate the secondary normalization of RPCs and TPCs, denoted as RPCs″ and TPCs″. This secondary normalization is carried out to address the scale difference within each multi-channel signal, namely RPCs′ and TPCs′. During the secondary normalization process, RPCs′ and TPCs′ are divided by their corresponding maxima in RPCs, allowing the RMSE to be calculated based on the magnitude of RPCs.

After normalizing the RPCs and TPCs, the similarity between them can be accurately measured by utilizing the RMSE metric.

Figure 10 shows the process of measuring the RMSE through normalization based on PC0.3.

For the RPCs, fixed data were obtained from the reference device. However, in the case of TPCs, even if the same PCs are obtained from the LUT using the same threshold, the normalized PC values can differ due to variations in the maximum value of PC0.3. As a result, the LUT stores TPCs before normalization and measures their similarity after normalizing the TPCs obtained from the LUT. This approach allows for accurate comparison and assessment of the TPCs′ similarity.

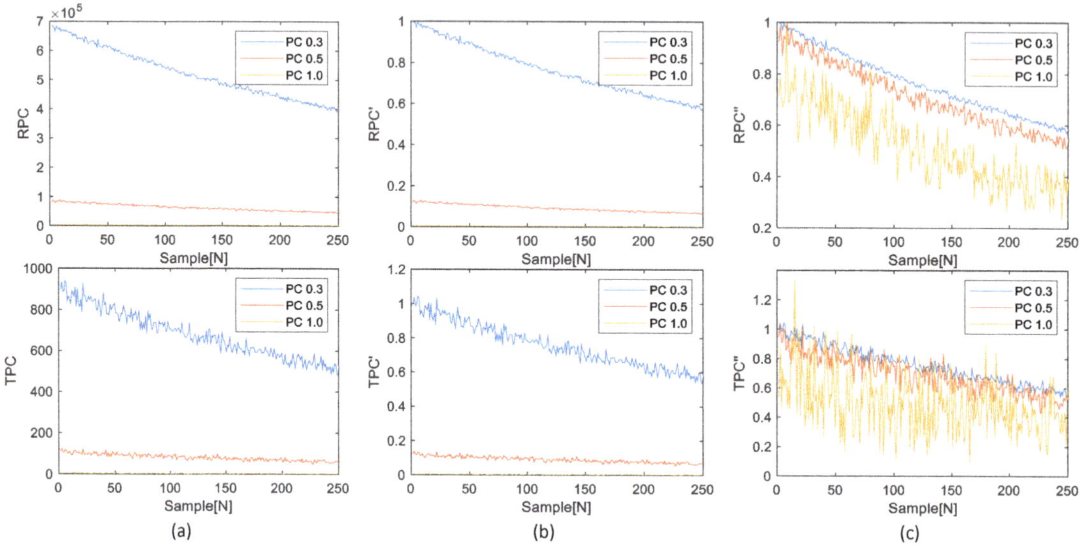

Figure 9. Example of normalization process for calculating RMSE: (**a**) RPCs and TPCs with given multi-thresholds, (**b**) first normalization using maxima of RPC0.3 and TPC0.3, and (**c**) secondary normalization using maxima of RPCs′.

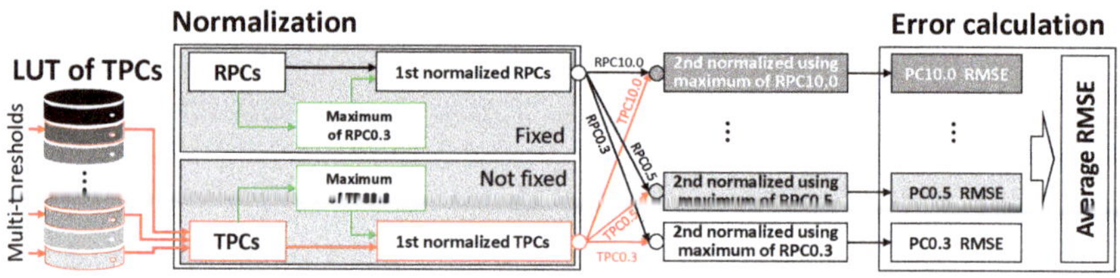

Figure 10. RMSE between RPCs and TPCs using two-step normalization.

4. Experiments

4.1. Experimental Environment

In this study, two particle sensors were used as test and reference devices. Figure 11 illustrates the two devices studied in the test chamber.

The two sensors were positioned in close proximity to each other within the test chamber to enable simultaneous sensing and measurement of a stable change in particle concentration. In this study, incense smoke was utilized as a surrogate for particulate matter (PM). Following the smoking of incense for a specified duration, the process of reducing the dust concentration through ventilation was observed and measured. To assess the feasibility of digitally twinning the reference device, the temperature and humidity levels inside the chamber were maintained at a constant level to minimize the influence of external environmental variables. The test device employed in the experiment captured the ADC readings at a sampling frequency of 50 kHz, followed by the acquisition of PCs at a sampling frequency of 1/6 Hz, matching the sampling frequency of the RPCs.

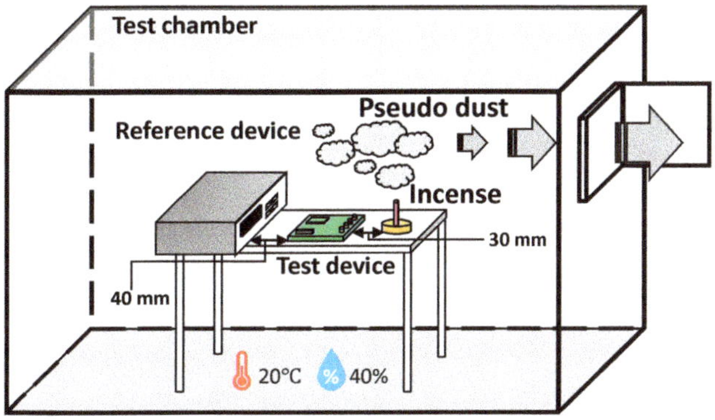

Figure 11. Arrangement of the two sensor devices and test environment.

4.2. Test Dataset Generation

For the experiments, three datasets were acquired from the chamber. The first dataset assumed a high-concentration condition; the second dataset assumed a complex dataset, including changes in the ventilation rate; and the third dataset assumed a low-concentration condition. Figure 12 shows the RPCs of the three datasets obtained from the reference device.

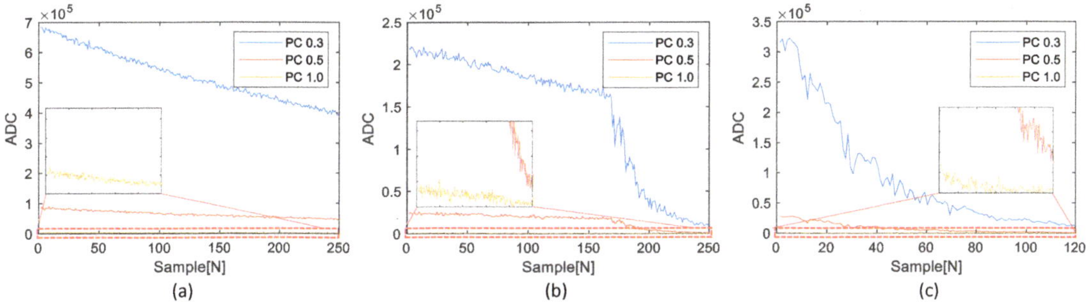

Figure 12. Three RPC dataset from the reference device: (**a**) first dataset with high concentration, (**b**) second dataset with ventilation rate change, (**c**) third dataset with low concentration.

The small plot boxes inside each figure show the enlarged results of each PC1.0. The reference device provides seven RPCs: RPC0.3, RPC0.5, RPC1.0, RPC2.5, RPC4.0, RPC7.0, and RPC10.0. However, Figure 12 displays only the three RPCs, RPC0.3, RPC0.5, and RPC1.0, because the test device can reliably replicate these RPCs. The possibility of the reliable replication of RPC2.5 or higher RPCs was analyzed through the following experiment.

Figure 13 shows the example of the change in RMSE between the RPC and TPC obtained according to the threshold for the first dataset in Figure 12a.

In the case of TPC0.3, TPC0.5, and TPC1.0, it is possible to detect the optimal threshold where the RMSE has a minimum value. However, in the case of TPC2.5, the RMSE continuously decreases as the threshold increases, and the optimal threshold of TPC2.5 continuously increases until all TPC2.5 values become 0. This phenomenon occurs because the expected result value is less than 1 when normalizing TPC2.5 according to the ratio of RPC0.3 to RPC2.5. Each value of TPC2.5 represents the number of samples larger than the given threshold Thr2.5. Therefore, having a value less than 1 means that there are no samples larger than Thr2.5, and the RMSE is minimized when all TPC2.5 values become 0.

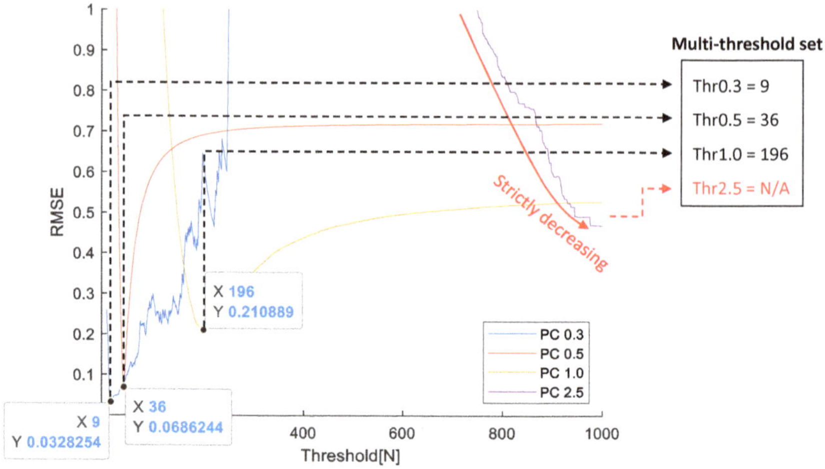

Figure 13. RMSE distribution of each PC according to the thresholds.

This limitation arises from the fact that the test device used in this study is a low-cost, single-sensing device. As a result, we observed that reliable replication of the RPC calculation algorithm is feasible for PC0.3, PC0.5, and PC1.0 using the test device. Since the focus of this study is to generate TPCs that replicate RPCs, PCs with values exceeding PC2.5 were excluded from the analysis.

4.3. Experiments

The experimental results involved a comparison between the RPCs and the corresponding TPCs. The TPCs obtained from the test device were generated using both the existing general multi-threshold approach and the proposed simplified multi-threshold approach. Figures 14–16 illustrate the comparisons of the results for the three datasets shown in Figure 12.

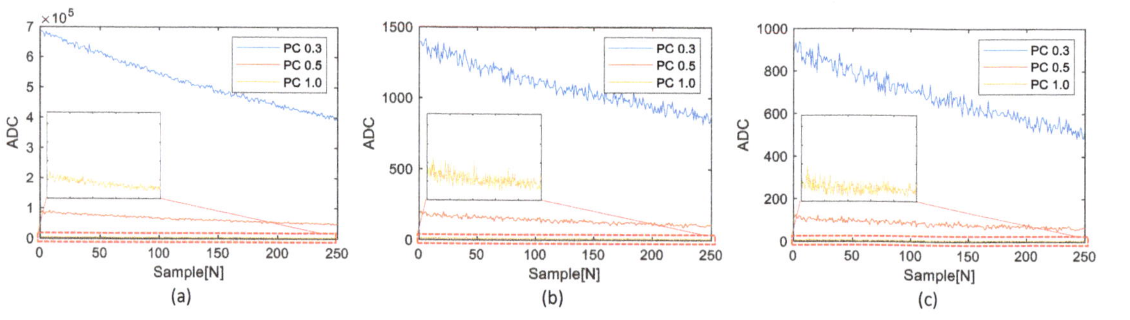

Figure 14. Comparison of results of the first dataset: (**a**) RPCs, (**b**) TPCs determined by the existing general multi-threshold detection, (**c**) TPCs determined by the proposed simplified multi-threshold detection.

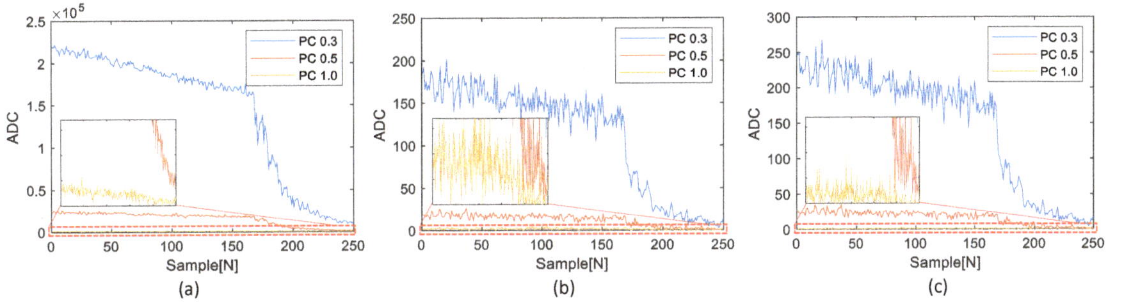

Figure 15. Comparison of results of the second dataset: (**a**) RPCs, (**b**) TPCs determined by the existing general multi-threshold detection, (**c**) TPCs determined by the proposed simplified multi-threshold detection.

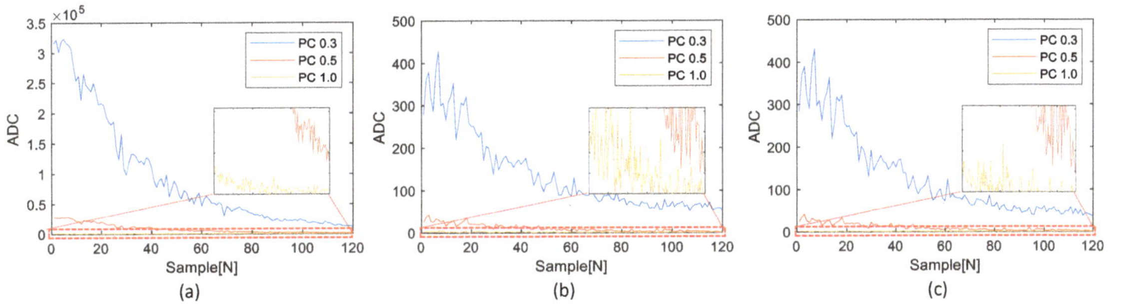

Figure 16. Comparison of results of the third dataset: (**a**) RPCs, (**b**) TPCs determined by the existing general multi-threshold detection, (**c**) TPCs determined by the proposed simplified multi-threshold detection.

The small plot boxes inside each figure show the enlarged results of each PC1.0. As shown in Figures 14–16, without knowledge of the ADC and PCs generation algorithm of the reference device, we replicated the RPCs and the TPCs.

Table 2 shows a detailed comparison of the RMSE results in Figures 14–16.

Table 2. RMSE comparison between existing method and proposed method.

Method	Data	RMSE			
		PC0.3	**PC0.5**	**PC1.0**	**Average**
Existing method	First dataset	0.046512	0.072553	0.312398	0.143821
	Second dataset	0.048505	0.095210	0.960891	0.368202
	Third dataset	0.068228	0.092008	0.982927	0.381054
	Average	0.054415	0.086590	0.752072	0.297692
Proposed method	First dataset	0.032825	0.068624	0.210889	0.104113
	Second dataset	0.046248	0.088694	0.241081	0.125341
	Third dataset	0.062146	0.080355	0.281287	0.141263
	Average	0.047073	0.079224	0.244419	0.123572

Through our evaluation, we observed a notable improvement in the digital twinning performance, as the overall RMSE in the proposed method was significantly reduced. Specifically, the overall RMSE decreased from 0.297692 to 0.123572, indicating a substantial reduction of approximately 58.5%. Of particular significance is the improvement observed in PC1.0, where the average RMSE decreased from 0.752072 to 0.244419, representing a significant reduction in RMSE of about 67.5%. This reduction in RMSE highlights the effectiveness of the proposed method in accurately replicating PC1.0 values. These findings demonstrate the enhanced performance and accuracy achieved by our proposed method in digital twinning, providing more reliable replication of PC values and reducing the discrepancy between the reference device and the embedded device.

Figure 17 shows a comparison of the processing times of the optimal multi-threshold searching algorithms for the three test datasets.

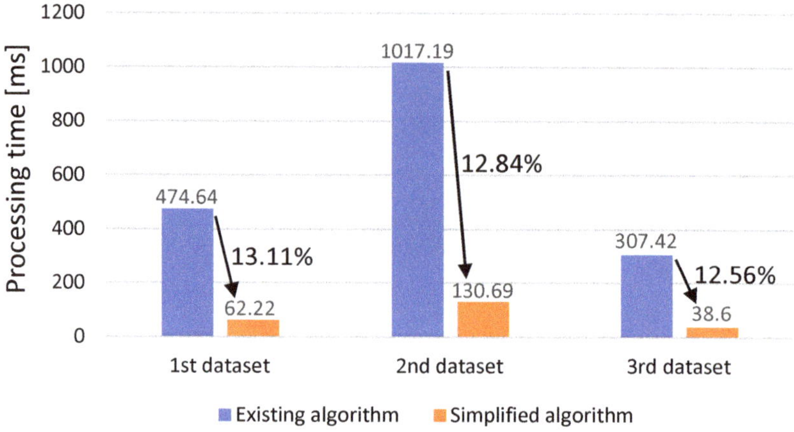

Figure 17. Comparison of processing time of optimal multi-threshold searching algorithms.

The execution time of the simplified algorithm was reduced by an average of approximately 87% from 474.64 to 62.22 for the first dataset, as shown in Figure 14; from 1017.19 to 130.69 for the second dataset, as shown in Figure 15; and from 307.42 to 38.60 for the third dataset, as shown in Figure 16.

Compared to the existing algorithm that has to repeat the calculation of the PC for the 50 kHz sampling frequency data, the proposed simplified algorithm that requires only one calculation through the LUT can considerably reduce the execution time. Therefore, it can be confirmed that the proposed simplified algorithm using LUT not only significantly reduces the execution time, but also maintains the distribution of PCs, as shown in Figures 14–16.

5. Conclusions

In this study, we aimed to replicate the PCs of a high-cost reference device using a low-power and low-cost embedded device with a single sensor. To optimize the execution time, we proposed a simplified multi-threshold search process for calculating TPCs using an LUT. The experimental results demonstrated that the TPCs calculated using the LUT effectively preserved the distribution of RPCs and existing TPCs while significantly reducing the execution time and RMSE by approximately 87% and 58.5% on average, respectively.

The RMSE analysis based on the LUT thresholds revealed that the proposed method reliably replicated PC0.3, PC0.5, and PC1.0 with optimal thresholds. However, in the case of PC2.5 or higher PCs, digital twinning was not feasible due to the limited number of samples in the single-sensor device compared to the reference device. To expand the applicability of digital twinning, future research should focus on increasing the number of channels or signal amplification to enable the detection of appropriate thresholds for higher PCs.

Furthermore, the proposed method has limitations in that it was tested under constant temperature and humidity conditions to minimize the influence of external factors. To achieve more generalized digital twinning, it is necessary to explore the preprocessing of RPCs and raw ADC data or post-processing of TPCs to account for changes in temperature and humidity. Developing a robust PC calculation algorithm that can handle environmental variations is crucial, along with analyzing sensor response time, PC calculation execution time, and minimum processor performance requirements for real-time processing.

The proposed algorithm successfully obtained TPCs by replicating RPCs. As a next step, future research should focus on obtaining test particulate matter (TPM) by replicating the unknown transfer function of reference particulate matter (RPM) from RPCs. Considering the scale difference between RPCs and TPCs in the TPM calculation process will ensure that RPMs and TPMs have similar scales and distributions. Since PMs are widely used as indicators of dust concentration, this future research will represent the final stage in achieving digital twinning of dust sensing systems.

Author Contributions: Conceptualization, D.P.; methodology, S.L.; software, S.L.; validation, S.L.; formal analysis, S.L.; investigation, J.K.; resources, J.K.; data curation, J.K.; writing—original draft preparation, S.L.; writing—review and editing, J.K.; visualization, S.L.; supervision, D.P.; project administration, D.P.; funding acquisition, D.P. All authors have read and agreed to the published version of the manuscript.

Funding: This study was supported by the BK21 FOUR project funded by the Ministry of Education, Korea (4199990113966), and the Basic Science Research Program through the National Research Foundation of Korea (NRF) funded by the Ministry of Education (NRF-2018R1A6A1A03025109, 30%, NRF-2020R1I1A1A01072343, 10%, NRF-2022R1I1A3069260, 10%) and by the MSIT (Ministry of Science and ICT) (2020M3H2A1078119), Korea, under the Innovative Human Resource Development for Local Intellectualization support program (IITP-2022-RS-2022-00156389, 10%), supervised by the IITP (Institute for Information and Communications Technology Planning and Evaluation). This work was partly supported by an Institute of Information and Communications Technology Planning and Evaluation (IITP) grant funded by the Korean government (MSIT) (No. 2021-0-00944, Metamorphic approach of unstructured validation/verification for analyzing binary code, 30%) and (No. 2022-0-01170, PIM Semiconductor Design Research Center, 10%). The EDA tool was supported by the IC Design Education Center (IDEC), Korea.

Institutional Review Board Statement: Not applicable.

Informed Consent Statement: Not applicable.

Data Availability Statement: Not applicable.

Conflicts of Interest: The authors declare no conflict of interest. The funders had no role in the design of the study; in the collection, analyses, or interpretation of data; in the writing of the manuscript; or in the decision to publish the results

Abbreviations

The following abbreviations are used in this manuscript:

ADC	Analog-to-digital converter
RMSE	Root mean square error
LUT	Look-up table
FIR filter	Finite impulse response filter
IIR filter	Infinite impulse response filter
PC	Particle count
RPC	Reference particle count
TPC	Test particle count
PM	Particulate matter
RPM	Reference particulate matter
TPM	Test particulate matter

References

1. Seok, M.G.; Tan, W.J.; Cai, W.; Park, D. Digital-Twin Consistency Checking Based on Observed Timed Events With Unobservable Transitions in Smart Manufacturing. *IEEE Trans. Ind. Inform.* **2023**, *19*, 6208–6219. [CrossRef]
2. Seok, M.G.; Cai, W.; Park, D. Hierarchical Aggregation/Disaggregation for Adaptive Abstraction-Level Conversion in Digital Twin-Based Smart Semiconductor Manufacturing. *IEEE Access* **2021**, *9*, 71145–71158. [CrossRef]

3. Zou, R.; Liang, X.; Chen, Q.; Wang, M.; Zaghloul, M.A.S.; Lan, H.; Buric, M.P.; Ohodnicki, P.R.; Chorpening, B.; To, A.C.; et al. A Digital Twin Approach to Study Additive Manufacturing Processing Using Embedded Optical Fiber Sensors and Numerical Modeling. *J. Light. Technol.* **2020**, *38*, 6402–6411. [CrossRef]
4. Zhang, S.; Dong, H.; Maschek, U.; Song, H. A digital-twin-assisted fault diagnosis of railway point machine. In Proceedings of the 2021 IEEE 1st International Conference on Digital Twins and Parallel Intelligence (DTPI), Beijing, China, 15 July–15 August 2021; pp. 430–433. [CrossRef]
5. Wang, Y.; Cao, Y.; Wang, F.Y. Anomaly Detection in Digital Twin Model. In Proceedings of the 2021 IEEE 1st International Conference on Digital Twins and Parallel Intelligence (DTPI), Beijing, China, 15 July–15 August 2021 ; pp. 208–211. [CrossRef]
6. Wang, Z.; Gupta, R.; Han, K.; Wang, H.; Ganlath, A.; Ammar, N.; Tiwari, P. Mobility Digital Twin: Concept, Architecture, Case Study, and Future Challenges. *IEEE Internet Things J.* **2022**, *9*, 17452–17467. [CrossRef]
7. O'Connell, E.; O'Brien, W.; Bhattacharya, M.; Moore, D.; Penica, M. Digital Twins: Enabling Interoperability in Smart Manufacturing Networks. *Telecom* **2023**, *4*, 265–278. [CrossRef]
8. Xie, J.; Wan, J. Digital Twin Four-Dimension Fusion Modeling Method Design and Application to the Discrete Manufacturing Line. *Big Data Cogn. Comput.* **2023**, *7*, 89. [CrossRef]
9. Aheleroff, S.; Xu, X.; Zhong, R.Y.; Lu, Y. Digital Twin as a Service (DTaaS) in Industry 4.0: An Architecture Reference Model. *Adv. Eng. Inform.* **2021**, *47*, 101225. [CrossRef]
10. Qin, H.; Wang, H.; Zhang, Y.; Lin, L. Constructing Digital Twin for Smart Manufacturing. In Proceedings of the 2021 IEEE 24th International Conference on Computer Supported Cooperative Work in Design (CSCWD), Dalian, China, 5–7 May 2021; pp. 638–642. [CrossRef]
11. Hoang, T.M.; Dinh-Van, S.; Barn, B.; Trestian, R.; Nguyen, H.X. RIS-Aided Smart Manufacturing: Information Transmission and Machine Health Monitoring. *IEEE Internet Things J.* **2022**, *9*, 22930–22943. [CrossRef]
12. Khalaj, O.; Jamshidi, M.B.; Hassas, P.; Mašek, B.; Štadler, C.; Svoboda, J. Digital Twinning of a Magnetic Forging Holder to Enhance Productivity for Industry 4.0 and Metaverse. *Processes* **2023**, *11*, 1703. [CrossRef]
13. Specht, J.P.; Esfahani, S.; Xing, Y.; Köck, A.; Cole, M.; Gardner, J.W. Thermally Modulated CMOS-Compatible Particle Sensor for Air Quality Monitoring. *IEEE Trans. Instrum. Meas.* **2022**, *71*, 1–13. [CrossRef]
14. Tancev, G.; Toro, F.G. Towards a Digital Twin for Air Quality Monitoring Networks in Smart Cities. In Proceedings of the 2022 IEEE International Smart Cities Conference (ISC2), Pafos, Cyprus, 26–29 September 2022; pp. 1–4. [CrossRef]
15. Solimini, A.; Filipponi, F.; Fegatelli, D.A.; Caputo, B.; De Marco, C.M.; Spagnoli, A.; Vestri, A.R. A global association between Covid-19 cases and airborne particulate matter at regional level. *Sci. Rep.* **2021**, *11*, 6256. [CrossRef] [PubMed]
16. Wang, Y.; Ma, L.; Jiu, M.; Jiang, H. Detection of Conductive Particles in TFT-LCD Circuit Using Generative Adversarial Networks. *IEEE Access* **2020**, *8*, 101338–101350. [CrossRef]
17. Yang, B.; Lv, Z.; Wang, F. Digital Twins for Intelligent Green Buildings. *Buildings* **2022**, *12*, 856. [CrossRef]
18. Englezos, D.; Hadjidemetriou, L.; Papadopoulos, P.; Timotheou, S.; Polycarpou, M.; Panayiotou, C. A Digital Twin Architecture for Smart Buildings. In Proceedings of the 2022 IEEE International Smart Cities Conference (ISC2), Pafos, Cyprus, 26–29 September 2022; pp. 1–7. [CrossRef]
19. Hodavand, F.; Ramaji, I.J.; Sadeghi, N. Digital Twin for Fault Detection and Diagnosis of Building Operations: A Systematic Review. *Buildings* **2023**, *13*, 1426. [CrossRef]
20. Kristiani, E.; Kuo, T.Y.; Yang, C.T.; Pai, K.C.; Huang, C.Y.; Nguyen, K.L.P. PM2.5 Forecasting Model Using a Combination of Deep Learning and Statistical Feature Selection. *IEEE Access* **2021**, *9*, 68573–68582. [CrossRef]
21. Song, S.; Lam, J.C.K.; Han, Y.; Li, V.O.K. ResNet-LSTM for Real-Time PM2.5 and PM10 Estimation Using Sequential Smartphone Images. *IEEE Access* **2020**, *8*, 220069–220082. [CrossRef]
22. Lee, S.; Jeong, Y.; Kwak, J.; Park, D.; Park, K.H. Advanced Real-Time Dynamic Programming in the Polygonal Approximation of ECG Signals for a Lightweight Embedded Device. *IEEE Access* **2019**, *7*, 162850–162861. [CrossRef]
23. Lee, S.; Park, D. Improved Dynamic Programming in Local Linear Approximation Based on a Template in a Lightweight ECG Signal-Processing Edge Device. *J. Inf. Process. Syst.* **2022**, *18*, 97–114. [CrossRef]
24. Lombardo, L.; Parvis, M.; Angelini, E.; Grassini, S. An Optical Sampling System for Distributed Atmospheric Particulate Matter. *IEEE Trans. Instrum. Meas.* **2019**, *68*, 2396–2403. [CrossRef]
25. Zhao, Z.; Li, D.; Liu, G.; Wu, F.; Sui, J. An Improved Dust-Concentration Measurement Algorithm Based on Multifeature Fusion of β-Ray Intensity Fluctuations. *IEEE Trans. Instrum. Meas.* **2020**, *69*, 6420–6428. [CrossRef]
26. Pedersini, F. Improving a Commodity Dust Sensor to Enable Particle Size Analysis. *IEEE Trans. Instrum. Meas.* **2019**, *68*, 177–188. [CrossRef]
27. Chellasivalingam, M.; Imran, H.; Pandit, M.; Boies, A.M.; Seshia, A.A. Weakly Coupled Piezoelectric MEMS Resonators for Aerosol Sensing. *Sensors* **2020**, *20*, 3162. [CrossRef] [PubMed]
28. Lee, S.; Park, D. Efficient Template Cluster Generation for Real-Time Abnormal Beat Detection in Lightweight Embedded ECG Acquisition Devices. *IEEE Access* **2021**, *9*, 70596–70605. [CrossRef]

sensors

MDPI

Article

The Image Definition Assessment of Optoelectronic Tracking Equipment Based on the BRISQUE Algorithm with Gaussian Weights

Ning Zhang [1] and Cui Lin [1,2,*]

[1] Changchun Institute of Optics, Fine Mechanics and Physics, Chinese Academy of Sciences, Changchun 130033, China
[2] University of Chinese Academy of Sciences, Beijing 100049, China
* Correspondence: clin20008@163.com

Abstract: Defocus is an important factor that causes image quality degradation of optoelectronic tracking equipment in the shooting range. In this paper, an improved blind/referenceless image spatial quality evaluator (BRISQUE) algorithm is formulated by using the image characteristic extraction technology to obtain a characteristic vector (CV). The CV consists of 36 characteristic values that can effectively reflect the defocusing condition of the corresponding image. The image is evaluated and scored subjectively by the human eyes. The subjective evaluation scores and CVs constitute a set of training data samples for the defocusing evaluation model. An image database that contains sufficiently many training samples is constructed. The training model is trained to obtain the support vector machine (SVM) model by using the regression function of the SVM. In the experiments, the BRISQUE algorithm is used to obtain the image feature vector. The method of establishing the image definition evaluation model via SVM is feasible and yields higher subjective and objective consistency.

Keywords: optoelectronic tracking equipment; image definition; defocus; BRISQUE algorithm; support vector machine

Citation: Zhang, N.; Lin, C. The Image Definition Assessment of Optoelectronic Tracking Equipment Based on the BRISQUE Algorithm with Gaussian Weights. *Sensors* **2023**, *23*, 1621. https://doi.org/10.3390/s23031621

Academic Editors: Qibo Feng, Jiakun Li and Qixin He

Received: 30 November 2022
Revised: 12 January 2023
Accepted: 28 January 2023
Published: 2 February 2023
Corrected: 12 July 2023

1. Introduction

The image, which is an important carrier of information, has been widely used in health, medical community, consumer electronics, etc. However, distortions are inevitably induced during image acquisition, transmission, processing, and display. The distortions cause the image quality degradation [1]. Evaluating, comparing, and optimizing the image quality effectively has gradually become a research hotspot in many fields, such as visual psychology, image processing, pattern recognition, and artificial intelligence [2–4].

Image distortion occurs, to a certain extent, in the process of acquisition, processing, compression, transmission, and display. Therefore, it is necessary to establish objective and effective quality assessment methods to evaluate the image quality [5–7]. At present, the image quality assessment includes subjective assessment and objective assessment. Image quality is evaluated by the subjective perception of the human eyes in a subjective evaluation method. As an objective evaluation method of the image quality, the mathematical models of image quality assessment are established [8,9].

The objective methods of image quality assessment include full reference image quality assessment (FR-IQA), reduced reference image quality assessment (RR-IQA), and no reference image quality assessment (NR-IQA), according to whether the reference image is needed. In the paper, NR-IQA is used to evaluate the image quality [10,11].

The main factors which affect the quality of optical measurement images include atmospheric disturbance, atmospheric extinction, optical diffraction of optical lens, defocusing, image motion, camera jitters, noise of image sensor, and so on. The image quality assessment method of defocused image is mainly studied in this paper.

If the external noises can be ignored, defocus is an important factor in image blur in the image tracking process of optoelectronic tracking equipment. To estimate the defocus severity of optoelectronic tracking equipment, image quality is evaluated objectively via an image quality evaluation algorithm [12,13]. At the same time, image characteristic values, which reflect the image quality, are obtained. The values can provide a condition for establishing the model for evaluating the focus performance based on the correlations between image characteristic values and defocus state parameters.

The optical system of optoelectronic tracking equipment can be regarded as a low-pass filter and an increase in the defocus is equivalent to a reduction in the filter cut-off frequency [14–18].

This paper mainly studies image evaluation indices in the defocusing state of optoelectronic tracking equipment and a method for obtaining the image characteristic values based on the indices. The characteristic values that are obtained via an image evaluation algorithm can be used to repair the image quality degradation that is caused by defocused equipment. The result of the image evaluation algorithm should be consistent with the subjective perception of the human eyes [19–21].

The causes of image blur also include interference factors, such as image motion of equipment and data compression, in addition to the defocus of the imaging system. A general referenceless image evaluation algorithm should be selected instead of a referenceless image evaluation algorithm with known distortion [22–24].

Comparisons are performed from two aspects: the theory and the performance of the evaluation algorithm. The main referenceless image quality evaluation algorithms that perform well are as follows: (1) Moorthy's blind image quality index (BIQI) algorithm, which is implemented in the wavelet domain [25]; (2) Moorthy's distortion-identification-based image verity and integrity evaluation (DIIVINE) algorithm, which is based on the BIQI algorithm [7]; (3) Saad's distortion-identification-based image verity and integrity evaluation (DIIVINE) algorithm [26] and the BLIINDS-II improved algorithm [27]; (4) Mittal's BRISQE algorithm [28] and the natural image quality evaluator (NIQE) algorithm, which is referenceless [29]; (5) Li's general regression neural network (GRNN) algorithm [30]; and (6) Lintao Han's combining convolution and self-attention for image quality assessment network [31].

Spatial distortion directly affects the visual quality of an image. By considering effective spatial characteristics, image quality evaluation can achieve increased consistency with subjective evaluation. At the same time, the characteristic values that are obtained via spatial characteristic extraction lay the foundation for the study of building an evaluation model for the defocused state.

Ruderman et al. found that the luminance of natural image normalization tends to follow a normal (Gaussian) distribution [32]. They posit that the distortion of an image changes the statistical characteristics of the normalization coefficient. By measuring the changes in the statistical characteristics, the distortion type can be predicted and the image visual quality can be evaluated [33]. Based on this theory, Mr. Mittal put forward the BRISQUE algorithm [28], which is based on the image spatial statistical characteristics. Ronin Institute et al. apply a broad spectrum of statistics of local and global features to characterize the variety of possible video distortions [34].

Based on the image defocus characteristics of optoelectronic tracking equipment in this paper, an improved BRISQUE algorithm is used with image characteristic extraction technology to obtain a characteristic value (CV). The CV includes 36 characteristic values that effectively reflect the defocus condition of the image [35]. The image is evaluated by the human eyes and scored subjectively. Subjective evaluation scores and feature vectors constitute a set of training data samples of the defocus evaluation model. A sufficient amount of training samples is obtained by calculating the CVs of the image database. Then, the evaluation model is obtained by using a machine learning method that is based on SVM to train the samples [36].

Many studies have employed machine learning models for prediction or classification in many fields. A convolutional neural network (CNN) is used for robust classification of PV panel faults [37]. A support vector machine (SVM) has become a common method of discrimination. In the field of machine learning, it is usually used for pattern recognition, classification, and regression analysis. For example, CNN- and SVM-based models can provide doctors with the detection of heart failure using electrocardiogram signals [38]. The SVM and general regression neural networks (GRNN) were used for the diagnosis of malfunction [39]. The adaptive support vector machine (A-SVM) was introduced for classification together with the ORICA-CSP method [40].

The defocused image sequences of the optoelectronic equipment are computed via the BRISQUE algorithm to obtain the CVs. The CVs are inputted into the evaluation model to calculate the prediction scores. The image sequences are evaluated by the human eyes subjectively. By considering the subjective and objective consistency of the results of the evaluation algorithm, the effectiveness of the evaluation algorithm is assessed.

2. Acquiring the CV via the Improved BRISQUE Algorithm

The image database is built and the CVs of image samples from the image database are obtained via the improved BRISQUE algorithm, which is weighted by a Gaussian function. The image samples are evaluated subjectively by the human eyes and used as SVM model training samples.

2.1. Training Image Sample Selection and Database Establishment

Many preliminary studies and experiments have demonstrated that if an image sequence of the optoelectronic tracking equipment is used for training directly, the training model will be inaccurate, which will lead to the failure of forecast evaluation. The main reason is that it is impossible to cover various details because the target and background tracking are too monotonous. Using public database images for training is proposed. We have used three public databases, namely, Laboratory for Image & Video Engineering (IVE), Categorical Subjective Image Quality (CSIQ), and Tampere Image Database (TID2013). Table 1 lists the databases that are used in this article and their data types.

Table 1. Image databases for training the model.

Name	Num. of Distorted Images	Num. of Reference Images	Image Type
IVE	235	10	Grey and color images
TID2013	1700	25	Color images
CISQ	866	30	Color images

According to the defocus characteristics of the device tracking image, an image database that includes images in a sequence that ranges from defocused to focused and back to defocused is established and each image is subjectively evaluated and scored. The scoring principle is that a severely defocused image is assigned a low score and a better focused image has a higher score. The results of model training demonstrate that the size of the database should exceed 1000 pictures and the quality of the database directly affects the application stability.

2.2. BRISQUE Algorithm

Two important advantages of using the BRISQUE algorithm are that the image definition evaluation score that is obtained by the algorithm can effectively reflect the defocus state, and the obtained image characteristic vector facilitates the subsequent training and evaluation of the machine learning model.

From an image, the BRISQUE algorithm is used to extract 36 characteristic values, which include the variances of the image brightness and the mean value. These features are called local normalized brightness statistical characteristics.

Given an intensity image *I(i,j)*, an operation that subtracts the image mean can be applied to the image to obtain the mean subtracted contrast normalized (MSCN) image $\hat{I}(i,j)$:

$$\hat{I}(i,j) = \frac{I(i,j) - \mu(i,j)}{\sigma(i,j) + C} \tag{1}$$

where $i = 1, \ldots, M$ and $j = 1, \ldots, N$ are spatial indices; *M* and *N* are the image height and width, respectively; *C* is a constant that prevents instabilities from occurring when the denominator tends to zero; and $\mu(i,j)$ and $\sigma(i,j)$ are the local mean and standard deviation, respectively, of *I(i,j)*.

We model the statistical relationship between neighboring pixels using the empirical distributions of the pairwise products of neighboring MSCN coefficients along four orientations: horizontal (*H*), vertical (*V*), main diagonal (*D*1), and secondary diagonal (*D*2).

$$H(i,j) = \hat{I}(i,j)\hat{I}(i+1,j) \tag{2}$$

$$V(i,j) = \hat{I}(i,j)\hat{I}(i,j+1) \tag{3}$$

$$D1(i,j) = \hat{I}(i,j)\hat{I}(i+1,j+1) \tag{4}$$

$$D2(i,j) = \hat{I}(i,j)\hat{I}(i+1,j-1) \tag{5}$$

The statistical properties of the MSCN coefficients are affected by the presence of distortion. Quantifying these changes will make it possible to predict the type of distortion that affects an image and its perceptual quality. According to [24], a generalized Gaussian distribution (GGD) can be used to effectively capture a broader spectrum of distorted image statistics. The GGD with zero means is expressed as follows:

$$f\left(x;\alpha,\sigma^2\right) = \frac{\alpha}{2\beta\Gamma(1/\alpha)}\exp\left(-\left(\frac{|x|}{\beta}\right)^{\alpha}\right) \tag{6}$$

where

$$\beta = \sigma\sqrt{\frac{\Gamma(1/\alpha)}{\Gamma(3/\alpha)}} \tag{7}$$

and $\Gamma(\cdot)$ is the gamma function:

$$\Gamma(a) = \int_0^{\infty} t^{a-1}e^{-t}dt \quad a > 0 \tag{8}$$

The shape parameter, which is denoted as α, controls the 'shape' of the distribution, while σ^2 control the variance. The parameters of the GGD (α,σ^2) are estimated via the moment-matching-based approach that was proposed in [41].

The appropriate values of α and σ are calculated via the moment-matching-based method and are two of the 36 characteristic values to be obtained. The parameters (ν,σ_l,σ_r) and η are calculated based on Equations (9) and (12) for the other four images: *H*, *V*, *D*1, and *D*2.

$$f\left(x;\alpha,\sigma^2\right) = \begin{cases} \frac{\nu}{(\beta_l+\beta_r)\Gamma(1/\nu)}\exp\left(-\left(\frac{-x}{\beta_l}\right)^{\nu}\right), & x < 0 \\ \frac{\nu}{(\beta_l+\beta_r)\Gamma(1/\nu)}\exp\left(-\left(\frac{-x}{\beta_r}\right)^{\nu}\right), & x \geq 0 \end{cases} \tag{9}$$

where

$$\beta_l - \sigma_l\sqrt{\frac{\Gamma(1/\nu)}{\Gamma(3/\nu)}} \tag{10}$$

$$\beta_r = \sigma_r \sqrt{\frac{\Gamma(1/\nu)}{\Gamma(3/\nu)}} \tag{11}$$

$$\eta = (\beta_r - \beta_l) \sqrt{\frac{\Gamma(2/\nu)}{\Gamma(1/\nu)}} \tag{12}$$

The details of the calculation process are presented in [24]. Via Equations (2)–(12), we obtain 16 + 2 = 18 characteristic values. The other 18 characteristic values must be calculated in other ways. The original image is down-sampled with a sampling

The characteristic values of the down-sampled image are calculated by following the given steps again and we obtain another 18 characteristic values. Now, the calculation of the 36 characteristic values is complete.

2.3. Improved BRISQUE Algorithm That Is Weighted by a Gaussian Function

Preliminary model training and prediction studies demonstrate that the characteristic values that were directly obtained via the BRISQUE algorithm cannot stably evaluate the defocused image sequence. For this particular situation, an improved BRISQUE algorithm that is weighted by a Gaussian function is selected in this paper.

The pixels of the training image are scanned by using a Gaussian function template and the center pixel value of the template is replaced with the weighted average gray value of the pixels in the neighborhood that is determined by the template. The template parameters of the Gaussian function are shown in Table 2. The image that is obtained by weighting the training image by the Gaussian function is denoted as VarI. The characteristic values of the new image are calculated by following the specified steps and we obtain 36 characteristic values, which are the input of machine learning training.

Table 2. Template of the weighted Gaussian function.

Weightiness	1	2	3	4	5	6	7
1	0.000157	0.00099	0.003	0.0043	0.003	0.00099	0.000157
2	0.00099	0.0062	0.0187	0.027	0.0187	0.0062	0.00099
3	0.0043	0.027	0.0813	0.1174	0.0813	0.027	0.003
4	0.003	0.0187	0.0563	0.0813	0.0563	0.0187	0.003
5	0.00099	0.0062	0.0187	0.027	0.0187	0.0062	0.00099
6	0.000157	0.00099	0.003	0.0043	0.003	0.00099	0.000157
7	0.000157	0.00099	0.003	0.0043	0.003	0.00099	0.000157

3. Support Vector Machine Model and Training

SVM is one of the basic methods of machine learning and the most important branch of machine learning theory [42–44]. It plays an important role in the practical applications of machine learning. SVM, which is a supervised learning model, is commonly used for pattern recognition, classification, and regression analysis.

This paper uses the regression function of SVM. The improved BRISQUE algorithm is used to calculate the CVs and subjective evaluation scores of images in the image database as the model training sample for obtaining the SVM model. The image definition CVs of the image database, which are calculated via the improved BRISQUE algorithm, are the independent variables. The scores of the subjective evaluation are the dependent variables. The independent and dependent variables are used as model training samples to obtain the SVM model. The image CVs of optoelectronic tracking equipment are input into the SVM model and predicted to obtain image evaluation scores. By comparing with the subjective evaluation of the human eyes, the accuracy and reliability of the evaluation are assessed. If the evaluation result does not meet the requirements, the above process can be iterated until a subjective and objective SVM evaluation model is obtained. Another image database can be used to calculate the characteristic vectors as needed and the image quality is scored for the inputs of the new training model via SVM.

This paper calls the LIBSVM library function, which was developed by Professor Chi-Jen Lin [45] to train and test the SVM model. The LIBSVM library function version is libsvm-3.23. In this paper, the support vector regression model "ε-SVR" is used in SVM.

The specified training sample can be represented as $\{(x_1,z_1), \ldots \ldots, (x_l,z_l)\}$, where $x_i \in R^n$ is the characteristic vector, which is obtained via the improved BRISQUE algorithm and composed of 36 characteristic values, and $z_i \in R^1$ denotes the subjective evaluation score of the image, which is the target output of the training model. When the penalty parameter $C > 0$ and the parameter $\varepsilon > 0$, the standard form of the SVR is as expressed in Equation (13):

$$\min_{w,b,\xi,\xi^*} \left(\frac{1}{2} w^T w + C \sum_{i=1}^{l} \xi_i + C \sum_{i=1}^{l} \xi_i^* \right) \tag{13}$$

s.t.

$$w^T \phi(x_i) + b - z_i \le \varepsilon + \xi_i \tag{14}$$

$$z_i - w^T \phi(x_i) - b \le \varepsilon + \xi_i \tag{15}$$

$$\xi_i, \xi_i^* \ge 0, i = 1, \cdots\cdots, l \tag{16}$$

According to the principle of SVM, Equation (13) is converted to a dual problem to calculate α. The radial basis function (RBF) is selected as the kernel function, which is denoted as $K(x,z) = \phi^T(x)\phi(z)$; the form of the RBF is as follows:

$$K(\|x - z\|) = e^{-\frac{\|x-z\|^2}{(2\times\sigma)^2}} \tag{17}$$

where σ is set to 0.5.

The training parameters of the LIBSVM library function are set as follows: penalty parameter C is set to 1024, the probability estimate is set to 1, and other parameters use the default parameter values of the LIBSVM function.

The samples from the image database of Table 1 are input into the SVM model and model training is completed. The number of support vectors, which is denoted as *total_sv*, is 772, and the bias b is −118.247.

4. Defocused Image Acquisition and Image Evaluation Test

4.1. Defocused Image Sequence Acquisition

In the process of tracking the real target using the optoelectronic tracking equipment, to ensure that the target can be tracked effectively, the focus state cannot be adjusted. The acquired image samples typically do not contain all image definition features, which makes it impossible to fully evaluate the performance of the SVM model.

To identify the test images that meet the requirements, in the process of evaluating the imaging quality of the optoelectronic tracking device, an imaging system is built for obtaining image samples of various defocus states. A photo of the system is shown in Figure 1. A Nikon 800 mm/F5.6 fixed-focus lens from the Nikon Corporation of Japan is used in the imaging system. The piA2400-17 visible light camera is from BASLER Corporation of Germany. The main properties of the camera are as follows: pixel size: 3.45 μm × 3.45 μm; and the number of pixels: 2448 × 2050.

4.2. Predictive Test of Definition Evaluation of Defocused Images

In this paper, a series of defocused and focused images with continuous change were obtained by manually controlling the defocused position of the optical lens in the imaging system. The images are used to test the effectiveness of the definition evaluation algorithm of defocused images. At the same time, they are also used for algorithm comparison.

Figure 1. Photo of the imaging system.

To acquire stable evaluation scores, static scenes are photographed using the imaging system. Therefore, the image sequences in this paper are very similar to human visual perception. The major differences between the images are definition and edge sharpness. Serial numbers of the clear images are given in advance.

The image sequences are inputted into the trained SVM model, and the image definition evaluation scores of defocused image sequences are the outputs of SVM model. Because the image-focusing process and the serial numbers of the clear images are known, the image definition evaluation scores can be compared with the defocused states of the image sequences.

For the image sequences, the larger the score, the clearer the image is. Due to the evaluation scores related to the CVs obtained by the BRISQUE algorithm, they are not fixed values. However, the scores can reflect the definition of the image sequence with the same scene. The image definition scores vary greatly among the image sequences with different scenes.

4.2.1. Single-Peak Defocused Image Test

The indoor image sequence that was obtained by the experimental imaging system is shown in Figure 2. The shooting process is from defocus to focus and back to defocus. The 9th image of the 12 images in Figure 2 has the best visual effect. In the predictive evaluation test of the 12 pictures via the SVM model, we obtained the curve that is shown in Figure 3. The *X*-axis of the curve represents the serial numbers of the pictures and the *Y*-axis represents the corresponding image definition evaluation values. The first image has the largest defocused position, and its evaluation score is only −3.34. The ninth image with the highest definition has the highest score of 20. The curve is consistent with the clarity of the real image.

4.2.2. The Test of Algorithm Comparison

The structural similarity (SSIM) is compared with the SVM model in this paper. As shown in Figure 4, the first image in the image sequence has the largest defocus, and it is the most blurred image to human visual perception. As the serial number increases, the image has a higher definition with defocused decreasing. The 14th image is the clearest to human perception. The evaluation curves with SSIM and the SVM trained model are shown in Figures 5 and 6, respectively. Due to the different calculation principles of the two algorithms, the evaluation scores cannot be directly compared.

As shown in Figure 5, the evaluation scores with SSIM increase monotonously in the range of the first image to the eleventh image, which is consistent with the subjective evaluation by human eyes. However, the evaluation scores start to fall from the 12th image, and it is inconsistent with subjective evaluation. As shown in Figure 6, the evaluation scores with the SVM model increase with the serial numbers of the images in the sequence. Image

1 has the lowest score of 10.4, and image 14 has the highest score of 65.5. The evaluation with SVM is completely consistent with human subjective evaluation.

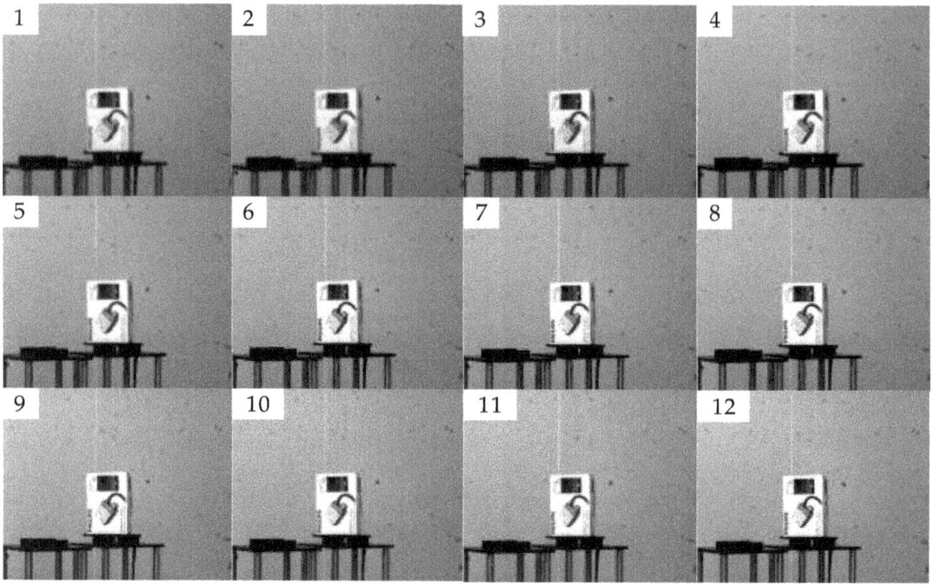

Figure 2. Indoor defocused image sequence. Sub-figures (1–12) represent the imaging results of the laboratory imaging system for the same target. The shooting process is from defocus to focus and back to defocus. The 9th picture shows the focus state.

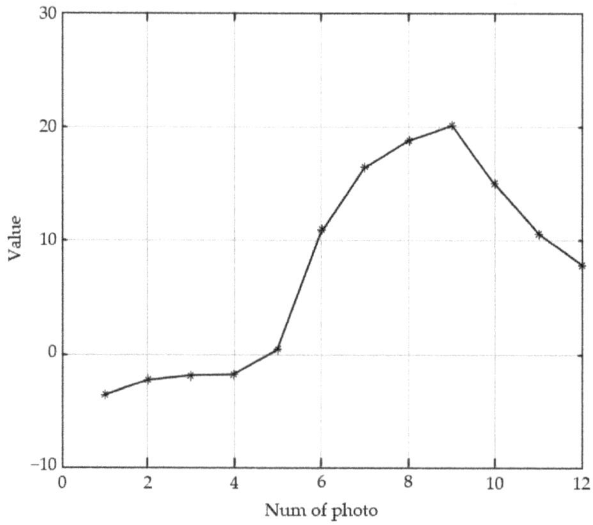

Figure 3. Predictive scores of the defocused image sequence.

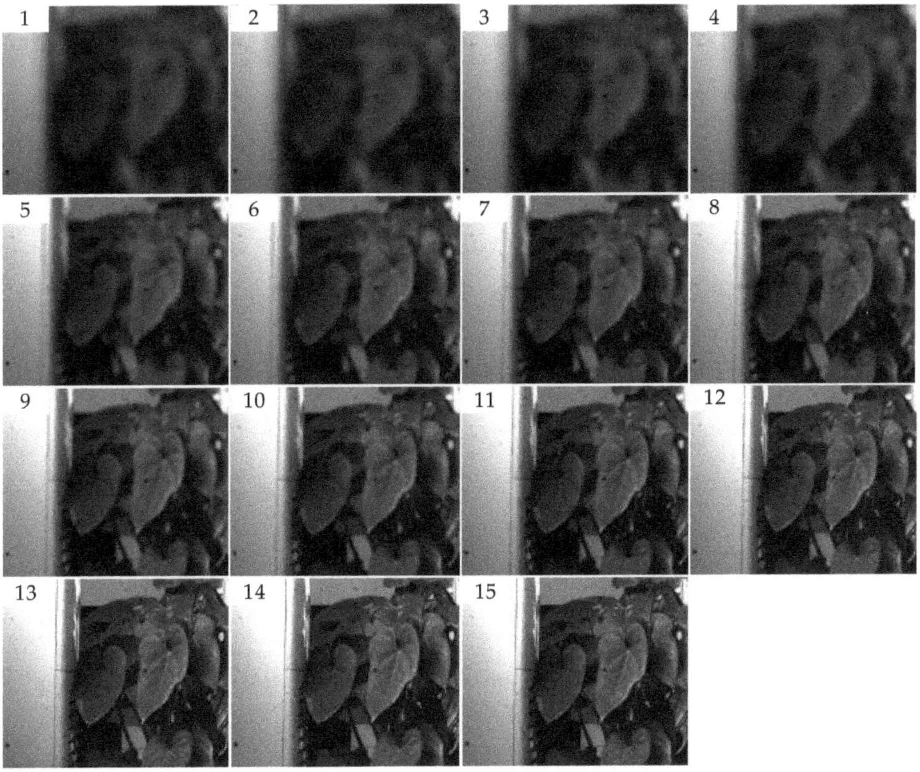

Figure 4. Defocused image sequence for comparison. Sub-figures (1–15) shows the imaging effect of the same target at different degrees of defocus. The first image is the most defocused, and the 15th image is the clearest.

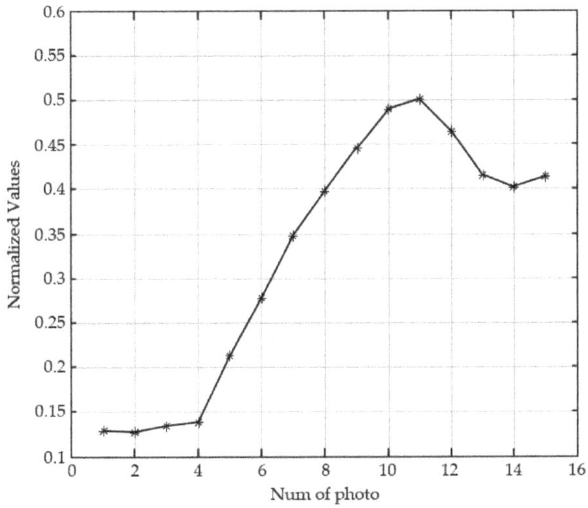

Figure 5. Evaluation scores of the image sequence in Figure 4 with SSIM.

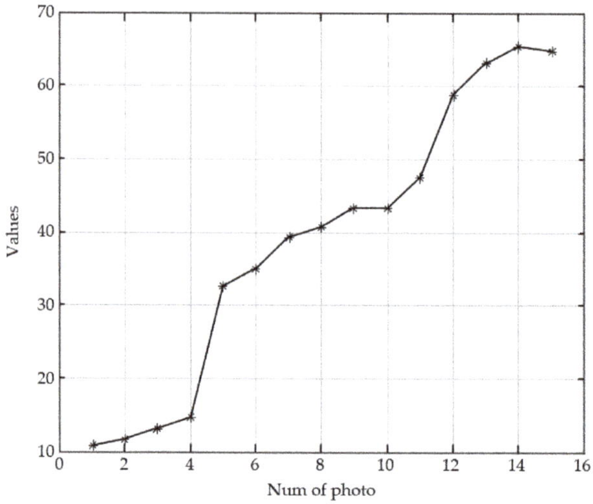

Figure 6. Evaluation scores of the image sequence in Figure 4 with the SVM model.

4.2.3. Dual-Peak Defocused Image Test

The dual-peak defocused image sequence is shown in Figure 7. The shooting process is focus, defocus, focus, and defocus in image 8 and image 21, respectively. According to the predictive evaluation test of the 28 pictures using the SVM model, the curve in Figure 8 is obtained. The X-axis of the curve represents the serial numbers of the pictures and the Y-axis represents the corresponding image definition evaluation values. We marked the two focused peak images with red hexagonal stars. The score of the 8th image is 58.3, and the score of the 22nd image is 55. The curve is consistent with the subjective evaluation by the human eyes of the test images. The curve also exhibits dual peaks, which demonstrates the convergence of the prediction model.

4.2.4. Repeatability Testing of Dual-Peak Defocused Image

Repeated tests were carried out to check the generalization performance of the SVM model. Another 29 images were acquired by changing the imaging scene and imaging process. The images were captured in order of focus, defocus, focus, defocus, and focus. Two randomly selected images in this image sequence are shown in Figure 9, and the definition evaluation scores of the sequence with the SVM model are shown in Figure 10. The result shows that the evaluation scores change by the focusing and defocused order, and the definition evaluation with the SVM model shows stability consistent with perspective evaluation. The SVM model has good generalization performance.

Through many test experiments, the image feature characteristic vectors are calculated via the improved BRISQUE algorithm and the evaluation model that is established via the SVM algorithm is used to evaluate the definition evaluation prediction. The evaluation results are highly consistent with the subjective evaluation results of the human eyes.

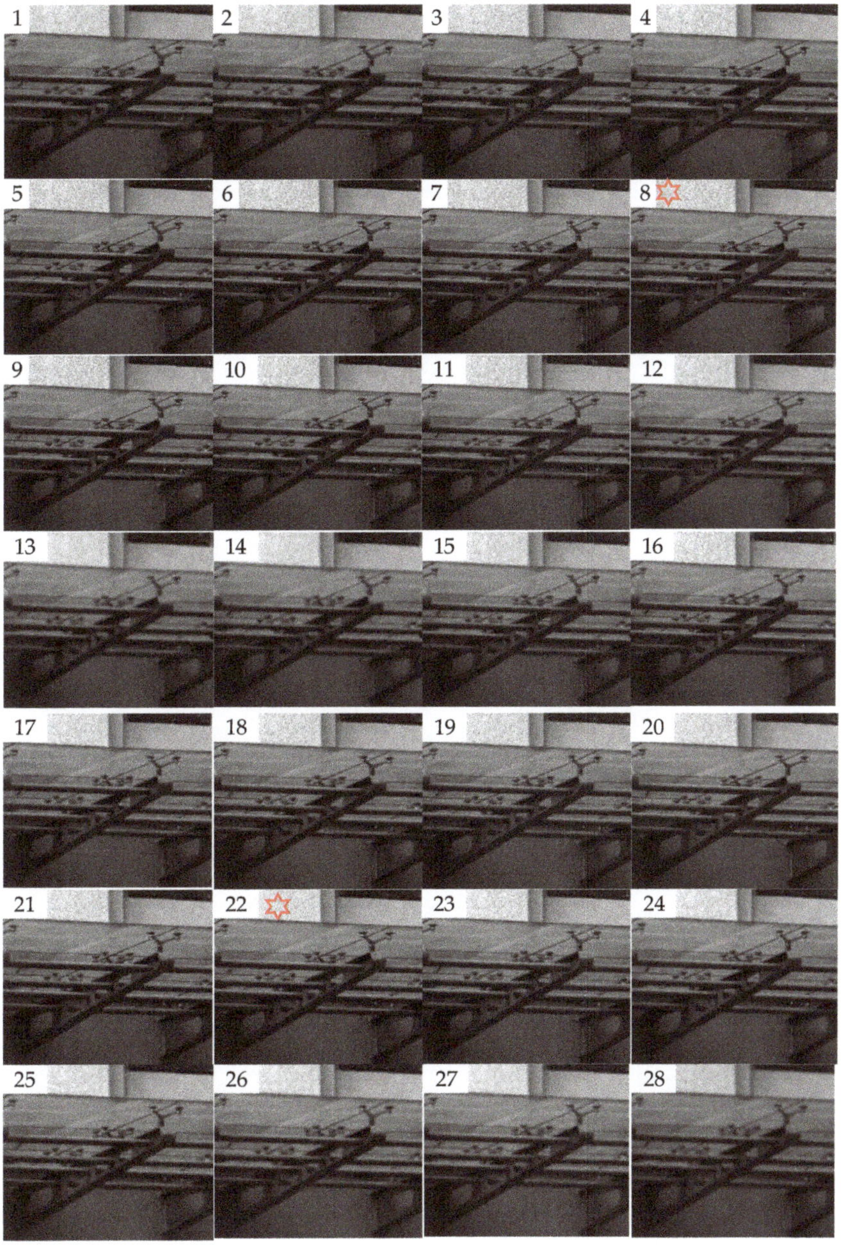

Figure 7. Dual-peak defocused image sequence (1–28). The shooting process is defocus, focus, defocus, focus, and defocus respectively. Image 8 and image 22 are in focus.

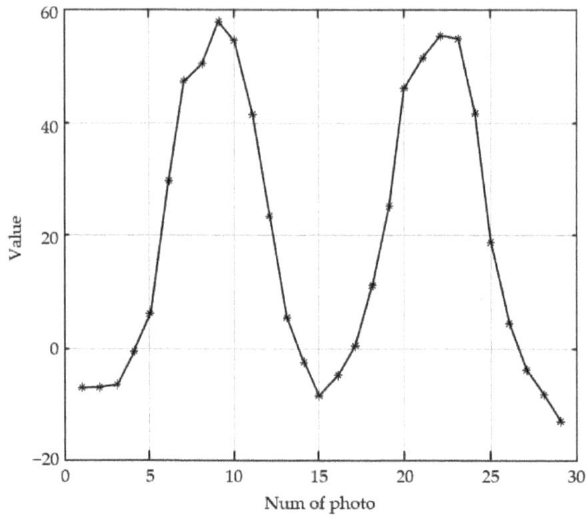

Figure 8. Predictive scores of the dual-peak defocused image sequence.

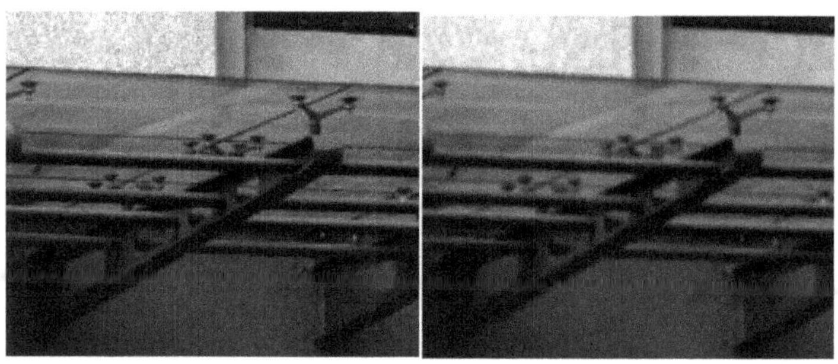

Figure 9. Defocused image sequence for repeatability testing (two randomly selected images).

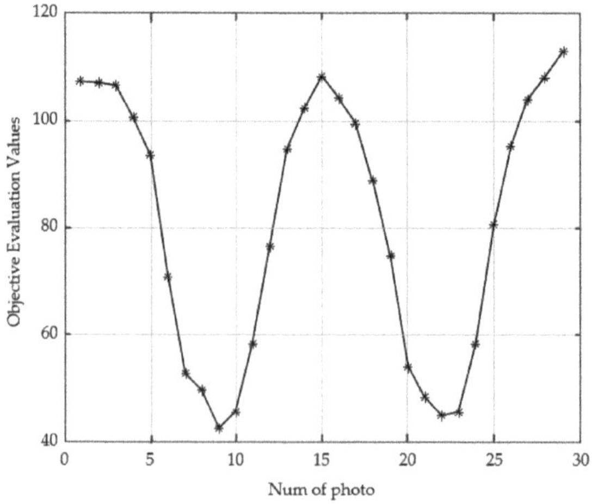

Figure 10. Evaluation scores with the SVM model for repeatability testing.

5. Discussion

For the purpose of increasing the effect of model training, we improved the BRISQUE algorithm that is weighted by the Gaussian function, and other weighted functions and parameters can also be researched in the future. The kernel function is selected as the radial basis function (RBF) in this paper, and other kernel functions can also be tried. In the future, the research of image objective evaluation model training based on machine learning will focus on two aspects. First, we should research and improve the new methods, which is to characterize the image spatial statistical characteristics. Second, we can introduce new machine learning algorithms, such as deep learning algorithms, which lead to a model with stronger self-learning ability.

6. Conclusions

Aiming at the problem of defocusing on large-scale optoelectronic tracking equipment in the shooting range, the use of image definition indicators for evaluation is proposed in this paper. An improved BRISQUE algorithm is used to objectively evaluate a defocused image and a CV that consists of 36 characteristic values are obtained. The CV is input into a previously trained SVM model to obtain an image definition evaluation score. Many image samples were obtained using the established imaging experimental system and experimental tests were carried out. The experimental results demonstrate that the image definition evaluation method that is used in this paper can effectively evaluate the defocusing condition of an optoelectronic tracking device, and the obtained image CV can effectively reflect the image defocus state.

Author Contributions: Conceptualization, N.Z. and C.L.; methodology, N.Z.; software, N.Z. and C.L.; validation, N.Z. and C.L.; formal analysis, N.Z. and C.L.; investigation, N.Z.; resources, N.Z.; data curation, C.L.; writing—original draft preparation, N.Z.; writing—review and editing, C.L.; visualization, N.Z. and C.L.; supervision, N.Z. and C.L.; project administration, N.Z.; funding acquisition, N.Z. All authors have read and agreed to the published version of the manuscript.

Funding: This research was supported by the National Natural Science Foundation of China (NSFC) (Grant No. 61905243) and Jilin Province Science & Technology Development Program Project in China (Grant No. 20190103157JH).

Institutional Review Board Statement: Not applicable.

Informed Consent Statement: Not applicable.

Data Availability Statement: Data are contained within the article.

Acknowledgments: The authors would like to thank the anonymous reviewers for their constructive comments.

Conflicts of Interest: The authors declare no conflict of interest.

References

1. Xu, N.; Ma, D.; Ren, G.; Huang, Y. BM-IQE: An Image Quality Evaluator with Block-Matching for Both Real-Life Scenes and Remote Sensing Scenes. *Sensors* **2020**, *20*, 3472. [CrossRef]
2. Takam Tchendjou, G.; Simeu, E. Visual Perceptual Quality Assessment Based on Blind Machine Learning Techniques. *Sensors* **2021**, *22*, 175. [CrossRef] [PubMed]
3. Ponomarenko, N.; Lukin, V.; Zelensky, A.; Egiazarian, K.; Carli, M.; Battisti, F. TID2008—A Database for Evaluation of Full-Reference Visual Quality Assessment Metrics. *Adv. Mod. Radioelectron.* **2009**, *10*, 30–45.
4. Wei, M.-S.; Xing, F.; You, Z. A Real-Time Detection and Positioning Method for Small and Weak Targets Using a 1D Morphology-Based Approach in 2D Images. *Light Sci. Appl.* **2018**, *7*, 18006. [CrossRef]
5. Moorthy, A.K.; Bovik, A.C. Blind Image Quality Assessment: From Natural Scene Statistics to Perceptual Quality. *IEEE Trans. Image Process.* **2011**, *20*, 3350–3364. [CrossRef]
6. Tran, V.L.; Lin, H.-Y. Extending and Matching a High Dynamic Range Image from a Single Image. *Sensors* **2020**, *20*, 3950. [CrossRef]
7. Rahmani, B.; Loterie, D.; Konstantinou, G.; Psaltis, D.; Moser, C. Multimode Optical Fiber Transmission with a Deep Learning Network. *Light Sci. Appl.* **2018**, *7*, 69. [CrossRef]

8. Stępień, I.; Obuchowicz, R.; Piórkowski, A.; Oszust, M. Fusion of Deep Convolutional Neural Networks for No-Reference Magnetic Resonance Image Quality Assessment. *Sensors* **2021**, *21*, 1043. [CrossRef] [PubMed]
9. Xiao, Q.; Bai, X.; Gao, P.; He, Y. Application of Convolutional Neural Network-Based Feature Extraction and Data Fusion for Geographical Origin Identification of Radix Astragali by Visible/Short-Wave Near-Infrared and Near Infrared Hyperspectral Imaging. *Sensors* **2020**, *20*, 4940. [CrossRef]
10. Zhang, L.; Zhang, L.; Mou, X.; Zhang, D. FSIM: A Feature Similarity Index for Image Quality Assessment. *IEEE Trans. Image Process.* **2011**, *20*, 2378–2386. [CrossRef] [PubMed]
11. Zhang, X.-X.; Chen, B.; He, F.; Song, K.-F.; Zhang, P.; Wang, J.-S. Wide-Field Auroral Imager Onboard the Fengyun Satellite. *Light Sci. Appl.* **2019**, *8*, 47. [CrossRef]
12. Capodiferro, L.; Jacovitti, G.; Di Claudio, E.D. Two-Dimensional Approach to Full-Reference Image Quality Assessment Based on Positional Structural Information. *IEEE Trans. Image Process.* **2012**, *21*, 505–516. [CrossRef] [PubMed]
13. Golestaneh, S.A.; Chandler, D.M. No-Reference Quality Assessment of JPEG Images via a Quality Relevance Map. *IEEE Signal Process. Lett.* **2014**, *21*, 155–158. [CrossRef]
14. Olson, J.T.; Espinola, R.L.; Jacobs, E.L. Comparison of Tilted Slit and Tilted Edge Superresolution Modulation Transfer Function Techniques. *Opt. Eng.* **2007**, *46*, 01640. [CrossRef]
15. Bentzen, S.M. Evaluation of the Spatial Resolution of a CT Scanner by Direct Analysis of the Edge Response Function. *Med. Phys.* **1983**, *10*, 579–581. [CrossRef]
16. Bao, Y.; Yu, Y.; Xu, H.; Guo, C.; Li, J.; Sun, S.; Zhou, Z.-K.; Qiu, C.-W.; Wang, X.-H. Full-Colour Nanoprint-Hologram Synchronous Metasurface with Arbitrary Hue-Saturation-Brightness Control. *Light Sci. Appl.* **2019**, *8*, 95. [CrossRef] [PubMed]
17. Li, L.; Shuang, Y.; Ma, Q.; Li, H.; Zhao, H.; Wei, M.; Liu, C.; Hao, C.; Qiu, C.-W.; Cui, T.J. Intelligent Metasurface Imager and Recognizer. *Light Sci. Appl.* **2019**, *8*, 97. [CrossRef] [PubMed]
18. Nijhawan, O.P.; Gupta, S.K.; Hradaynath, R. Polychromatic MTF of Electrostatic Point Symmetric Electron Lenses. *Appl. Opt.* **1983**, *22*, 2453–2455. [CrossRef]
19. Seghir, Z.A.; Hachouf, F.; Morain-Nicolier, F. Blind Image Quality Metric for Blurry and Noisy Image. In Proceedings of the 2013 IEEE Second International Conference on Image Information Processing (ICIIP-2013), Shimla, India, 9–11 December 2013; pp. 193–197.
20. Gu, K.; Zhai, G.; Liu, M.; Yang, X.; Zhang, W.; Sun, X.; Chen, W.; Zuo, Y. FISBLIM: A FIve-Step BLInd Metric for Quality Assessment of Multiply Distorted Images. In Proceedings of the SiPS 2013 Proceedings, Taipei, Taiwan, 16–18 October 2013; pp. 241–246.
21. Ponomarenko, N.; Lukin, V.; Egiazarian, K. HVS-Metric-Based Performance Analysis of Image Denoising Algorithms. In Proceedings of the 3rd European Workshop on Visual Information Processing, Paris, France, 4–6 July 2011; pp. 156–161.
22. Ye, P.; Doermann, D. No-Reference Image Quality Assessment Based on Visual Codebook. In Proceedings of the 2011 18th IEEE International Conference on Image Processing, Brussels, Belgium, 11–14 September 2011; pp. 3089–3092.
23. Wang, Z.; Sheikh, H.R.; Bovik, A.C. No-Reference Perceptual Quality Assessment of JPEG Compressed Images. In Proceedings of the International Conference on Image Processing, Rochester, NY, USA, 22–25 September 2002; Volume 1, pp. 477–480.
24. Campisi, P.; Carli, M.; Giunta, G.; Neri, A. Blind Quality Assessment System for Multimedia Communications Using Tracing Watermarking. *IEEE Trans. Signal Process.* **2003**, *51*, 996–1002. [CrossRef]
25. Moorthy, A.K.; Bovik, A.C. A Two-Step Framework for Constructing Blind Image Quality Indices. *IEEE Signal Process. Lett.* **2010**, *17*, 513–516. [CrossRef]
26. Saad, M.A.; Bovik, A.C.; Charrier, C. A DCT Statistics-Based Blind Image Quality Index. *IEEE Signal Process. Lett.* **2010**, *17*, 583–586. [CrossRef]
27. Saad, M.A.; Bovik, A.C.; Charrier, C. DCT Statistics Model-Based Blind Image Quality Assessment. In Proceedings of the 2011 18th IEEE International Conference on Image Processing, Brussels, Belgium, 11–14 September 2011; pp. 3093–3096.
28. Mittal, A.; Moorthy, A.K.; Bovik, A.C. No-Reference Image Quality Assessment in the Spatial Domain. *IEEE Trans. Image Process.* **2012**, *21*, 4695–4708. [CrossRef]
29. Mittal, A.; Soundararajan, R.; Bovik, A.C. Making a "Completely Blind" Image Quality Analyzer. *IEEE Signal Process. Lett.* **2013**, *20*, 209–212. [CrossRef]
30. Li, C.; Bovik, A.C.; Wu, X. Blind Image Quality Assessment Using a General Regression Neural Network. *IEEE Trans. Neural Netw.* **2011**, *22*, 793–799. [PubMed]
31. Han, L.; Lv, H.; Zhao, Y.; Liu, H.; Bi, G.; Yin, Z.; Fang, Y. Conv-Former: A Novel Network Combining Convolution and Self-Attetion for Image Quality Assessment. *Sensors* **2023**, *23*, 427. [CrossRef] [PubMed]
32. Ruderman, D.L. The Statistics of Natural Images. *Netw. Comput. Neural Syst.* **1994**, *5*, 517–548. [CrossRef]
33. Simoncelli, E.P.; Freeman, W.T.; Adelson, E.H.; Heeger, D.J. Shiftable Multiscale Transforms. *IEEE Trans. Inf. Theory* **1992**, *38*, 587–607. [CrossRef]
34. Varga, D. No-Reference Video Quality Assessment Using the Temporal Statistics of Global and Local Image Features. *Sensors* **2022**, *22*, 9696.
35. Lasmar, N.-E.; Stitou, Y.; Berthoumieu, Y. Multiscale Skewed Heavy Tailed Model for Texture Analysis. In Proceedings of the 2009 16th IEEE International Conference on Image Processing (ICIP), Cairo, Egypt, 7–10 November 2009; pp. 2281–2284.

36. Adankon, M.M.; Cheriet, M.; Biem, A. Semisupervised Learning Using Bayesian Interpretation: Application to LS-SVM. *IEEE Trans. Neural Netw.* **2011**, *22*, 513–524. [CrossRef] [PubMed]
37. Memon, S.A.; Javed, Q.; Kim, W.-G.; Mahmood, Z.; Khan, U.; Shahzad, M. A Machine-Learning-Based Robust Classification Method for PV Panel Faults. *Sensors* **2022**, *22*, 8515. [CrossRef]
38. Botros, J.; Mourad-Chehade, F.; Laplanche, D. CNN and SVM-Based Models for the Detection of Heart Failure Using Electrocardiogram Signals. *Sensors* **2022**, *22*, 9190. [CrossRef]
39. Chu, W.-L.; Lin, C.-J.; Kao, K.-C. Fault Diagnosis of a Rotor and Ball-Bearing System Using DWT Integrated with SVM, GRNN, and Visual Dot Patterns. *Sensors* **2019**, *19*, 4806. [CrossRef]
40. Antony, M.J.; Sankaralingam, B.P.; Mahendran, R.K.; Gardezi, A.A.; Shafiq, M.; Choi, J.-G.; Hamam, H. Classification of EEG Using Adaptive SVM Classifier with CSP and Online Recursive Independent Component Analysis. *Sensors* **2022**, *22*, 7596. [CrossRef] [PubMed]
41. Sharifi, K.; Leon-Garcia, A. Estimation of Shape Parameter for Generalized Gaussian Distributions in Subband Decompositions of Video. *IEEE Trans. Circuits Syst. Video Technol.* **1995**, *5*, 52–56. [CrossRef]
42. Wu, J.; Yang, H. Linear Regression-Based Efficient SVM Learning for Large-Scale Classification. *IEEE Trans. Neural Netw. Learn. Syst.* **2015**, *26*, 2357–2369. [CrossRef] [PubMed]
43. Baldeck, C.A.; Asner, G.P. Single-Species Detection with Airborne Imaging Spectroscopy Data: A Comparison of Support Vector Techniques. *IEEE J. Sel. Top. Appl. Earth Obs. Remote Sens.* **2015**, *8*, 2501–2512. [CrossRef]
44. Sun, J.; Li, Q.; Lu, W.; Wang, Q. Image Recognition of Laser Radar Using Linear SVM Correlation Filter. *Chin. Opt. Lett.* **2007**, *5*, 549–551.
45. Chang, C.-C.; Lin, C.-J. LIBSVM: A Library for Support Vector Machines. *ACM Trans. Intell. Syst. Technol.* **2011**, *2*, 1–27. [CrossRef]

sensors

MDPI

Article

Detection and Classification of Cotton Foreign Fibers Based on Polarization Imaging and Improved YOLOv5

Rui Wang [1], Zhi-Feng Zhang [1,*], Ben Yang [1], Hai-Qi Xi [1], Yu-Sheng Zhai [1], Rui-Liang Zhang [1], Li-Jie Geng [1], Zhi-Yong Chen [2] and Kun Yang [1]

[1] School of Physics and Electronic Engineering, Zhengzhou University of Light Industry, Zhengzhou 450002, China
[2] Fiber Inspection Bureau in Henan Province, Zhengzhou 450002, China
* Correspondence: fengzz9276@sina.com

Abstract: It is important to detect and classify foreign fibers in cotton, especially white and transparent foreign fibers, to produce subsequent yarn and textile quality. There are some problems in the actual cotton foreign fiber removing process, such as some foreign fibers missing inspection, low recognition accuracy of small foreign fibers, and low detection speed. A polarization imaging device of cotton foreign fiber was constructed based on the difference in optical properties and polarization characteristics between cotton fibers. An object detection and classification algorithm based on an improved YOLOv5 was proposed to achieve small foreign fiber recognition and classification. The methods were as follows: (1) The lightweight network Shufflenetv2 with the Hard-Swish activation function was used as the backbone feature extraction network to improve the detection speed and reduce the model volume. (2) The PANet network connection of YOLOv5 was modified to obtain a fine-grained feature map to improve the detection accuracy for small targets. (3) A CA attention module was added to the YOLOv5 network to increase the weight of the useful features while suppressing the weight of invalid features to improve the detection accuracy of foreign fiber targets. Moreover, we conducted ablation experiments on the improved strategy. The model volume, mAP@0.5, mAP@0.5:0.95, and FPS of the improved YOLOv5 were up to 0.75 MB, 96.9%, 59.9%, and 385 f/s, respectively, compared to YOLOv5, and the improved YOLOv5 increased by 1.03%, 7.13%, and 126.47%, respectively, which proves that the method can be applied to the vision system of an actual production line for cotton foreign fiber detection.

Keywords: deep learning; foreign fiber detection; YOLOv5; polarization imaging; line laser

Citation: Wang, R.; Zhang, Z.-F.; Yang, B.; Xi, H.-Q.; Zhai, Y.-S.; Zhang, R.-L.; Geng, L.-J.; Chen, Z.-Y.; Yang, K. Detection and Classification of Cotton Foreign Fibers Based on Polarization Imaging and Improved YOLOv5. *Sensors* **2023**, *23*, 4415. https://doi.org/10.3390/s23094415

Academic Editor: Vittorio Passaro

Received: 29 March 2023
Revised: 27 April 2023
Accepted: 28 April 2023
Published: 30 April 2023

1. Introduction

Cotton is the largest natural fiber in the textile industry. During the processes of cotton cultivation, harvesting, transportation, and processing, a large number of foreign fibers is inevitably mixed in due to various factors, such as cotton hulls, leaves, mulch films, chemical fibers, and paper pieces. These foreign fibers have adverse effects on the textile products, not only reducing the spinning efficiency, but also causing fabric defects and reducing product grade [1]. Therefore, the detection of foreign cotton fibers is an important and necessary step before spinning. It is time-consuming and inefficient to rely on workers to manually detect foreign fibers from cotton, and the detection accuracy of foreign fibers is low [2,3]. In recent years, numerous detection methods for foreign fibers have been developed, including photoelectric, ultrasonic, and optical detection, according to the detection principle [4,5]. However, photoelectric detection technology can only detect colored foreign fibers but not white transparent foreign fibers [6]. Ultrasonic detection technology can only detect foreign fibers in a large area, and its speed is slower [7]. Presently, foreign fiber detection mainly uses machine vision technology with high recognition rate, high detection speed, and low cost. Machine vision technology can be divided into

X-ray imaging technology, ultraviolet fluorescence imaging technology, infrared imaging technology, line laser imaging technology, hyperspectral imaging technology, and polarized light imaging technology [8]. Pai et al. [9] identified and classified three types of foreign fibers in cotton using an X-ray microtomographic imaging system. However, the imaging speed of the X-ray imaging method is slow, and the equipment cost is high. Luo et al. [10] proposed a machine vision method combined with UV fluorescence to sort foreign fibers in cotton. However, UV fluorescence imaging is less effective in detecting foreign fibers that are similar to cotton in color and without a fluorescence effect. Cai et al. [11] imaged cotton using 12 types of foreign fibers in the near-infrared band. The infrared imaging method has a better detection effect for foreign fibers with a significant difference in the absorption between the near-infrared band and cotton; however, the infrared spectrum camera is slow and expensive, and the relevant technology is still in the laboratory research stage. Hua et al. [12] proposed a method to identify foreign fibers based on line laser imaging; solid-state lasers have been widely used in machine vision detection owing to their low cost, small volume, and ease of operation. Mustafic et al. [13] employed hyperspectral fluorescence imaging to identify foreign fiber in cotton; however, hyperspectral imaging technology requires a high external environment, and the devices are expensive. Zhang et al. [14] utilized polarization imaging technology to increase the detection rate of transparent films.

The foreign fiber detection algorithm is the core part of foreign fiber recognition and classification and can be divided into traditional image algorithms and deep learning image algorithms. Traditional image algorithms rely on the artificial design of foreign fiber characteristics by the algorithm designer and utilize image preprocessing, feature extraction, feature selection [15–17], image segmentation [18,19], and image classification [20–22] to achieve foreign fiber detection. However, traditional image algorithms have limited ability to recognize and classify multiple types of foreign fiber targets and cope with complex scenes, whereas deep learning image algorithms have the ability to learn excellent complex features. He et al. [23–25] achieved the recognition of foreign fibers in seed cotton images based on a Faster-RCNN. Du et al. [26] and Dong et al. [27] used ResNet-50 and Inception-ResNet-V2 instead of the original VGG16 of Faster-RCNN to extract the features of foreign fibers, and the K-means++ algorithm was used to improve the size and number of candidate boxes to achieve the classification and localization of multiscale foreign fibers. Wu et al. [28] introduced the MobileNets network and constructed the MobileNets YOLOv3 model to detect foreign fibers in cotton. Wei [29] implemented a real-time intelligent classifier for foreign fiber images. On a dataset of 20,000 foreign fiber images, the classification accuracy reached 95%. Wu et al. [30] combined traditional convolution with depth-wise separable convolution and introduced a convolutional layer attention mechanism to establish a deep learning model for recognizing foreign fibers in cotton. The recognition accuracy for five types of foreign fibers, such as plastic ropes and human hairs, was 91.93% on the test set. Zhang et al. [31] introduced the residual network as a feature extraction network and combined it with the feature pyramid to propose an improved Faster R-CNN network for the detection of six types of foreign fibers, such as feathers and waste paper. The accuracy and recall rate of this network were 97.6% and 82.4%, respectively, which were higher than those of the VGG16 and ResNet50 networks. Zhang et al. [32] utilized the YOLOv5 neural network to perform classification and recognition of weeds, blackjack, and other foreign fibers that were segmented from images. The content of various foreign fibers was also measured, and the recognition accuracy reached 98%.

The actual production line of cotton requires an extremely high detection speed and a lightweight network with a smaller volume and faster detection speed. Moreover, cotton on the actual production line is carded through a carding machine, and foreign fibers mixed in the cotton are broken into smaller foreign fibers. These methods fail to consider the effective detection of foreign fibers in small targets. Therefore, this study is based on the YOLOv5 algorithm, and improved methods of Shufflenetv2 and PANet are introduced into YOLOv5. An improved YOLOv5 algorithm combined with an attention mechanism module (CA) network is proposed in this paper, which can realize the real-time detection of foreign fibers of multiple types of small targets.

The following contributions are made by our work:

- A polarization imaging device of cotton foreign fiber was constructed using line laser polarization imaging technology.
- In order to reduce the model volume and improve the detection speed, the lightweight network Shufflenetv2 with Hard-Swish function was added as the backbone feature extraction network.
- In order to increase the detection accuracy of foreign fibers in small targets, an improved PANet was added to YOLOV5.
- The CA module was added before the Head of YOLOv5 to allocate the weight of the channel features and spatial features to improve the accuracy of foreign fiber recognition and classification.

In summary, the line laser polarization imaging approach proposed in this study has an important guiding value for the online identification and classification of cotton foreign fibers and the control of foreign fiber generation in cotton planting and picking. Compared with other typical object detection algorithms, our proposed algorithm has a higher detection speed, smaller model size, and higher detection accuracy and is more suitable for foreign fiber detection tasks.

2. Materials and Methods

2.1. Experiment Materials

The cotton and foreign fiber samples used in the experiment were provided by the Henan Fiber Inspection Bureau and originated from the Xinjiang Uygur Autonomous Region, China. The experiment was conducted using 20 common types of foreign fibers in cotton, as shown in Figure 1, and the sizes of foreign fibers were categorized as 0.5 mm^2, 1 mm^2, 1.5 mm^2, 3 mm^2, and 5 mm^2. Group 1 comprised colored foreign fibers, and it was easier to distinguish them in cotton, whereas Group 2 comprised white transparent foreign fibers that were more difficult to detect because they are extremely similar to cotton fiber in color and appearance.

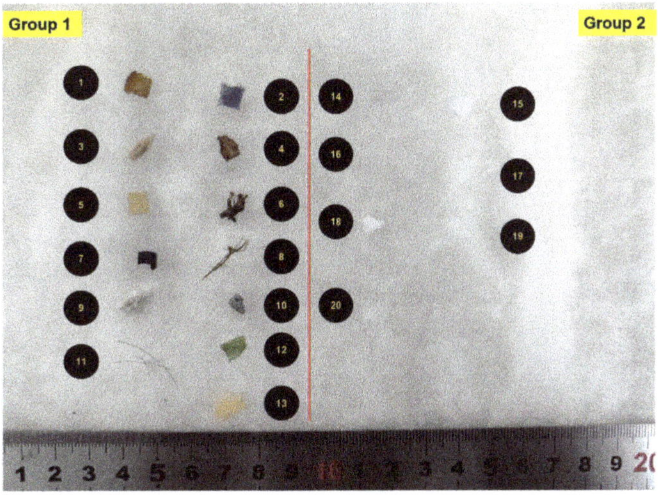

Figure 1. Foreign fiber and cotton samples. (1) Dead leaves; (2) cloth; (3) hemp rope; (4) bark; (5) kraft paper; (6) stalk; (7) PVC; (8) yarn; (9) cottonseed; (10) stone; (11) hair; (12) leaf; (13) sponge; (14) polypropylene; (15) white paper; (16) polyethylene; (17) feather; (18) foam; (19) chemical fiber; (20) mulch film.

2.2. Experiment Equipment

In actual detection, cotton containing foreign fibers was first made into a thin layer with a width of approximately 10 cm and thickness of approximately 2 mm. The cotton thin layer sample was irradiated by a uniform line laser, and the scattered light of cotton was mist-like. Mulch film, plastic and paper pieces, and other white foreign fibers are mostly dense materials, and the reflected light is approximately a mirror reflection [12].

The experiment found that the characteristic information of cotton foreign fiber image was the most prominent when the incident angle of the line laser was about 45°. For example, when the laser incident angle was 45°, the average gray value (M(X)) of the foreign fiber image was larger, and the contrast value (Var(x, y)) was the largest, as shown in Table 1.

Table 1. Average gray value and contrast value of foreign fiber.

Parameter	Angle of Incidence			
	45°	60°	75°	90°
M(X)	85.0056	87.1309	79.0224	96.0166
Var(x, y)	5866.0319	5287.1670	4263.1522	2042.0076

Because of the different polarization characteristics of different foreign fibers, the reflected light waves have polarization information of the foreign fibers, and different types of foreign fibers can be distinguished through polarization imaging [14].

A physical image of the cotton foreign fiber polarization imaging detection device is shown in Figure 2. The sensor of the camera (MV-CH050-10UP, HIKROBOT) was equipped with four-way (0, 45, 90, 135) pixel-level polarization filters with a resolution of 2448 × 2048 and a target surface size of 2/3″ using USB power output. The light source was a 405 nm line laser (SL-405-35-S-B-90-24V, OSELA) with a power of 35 mW.

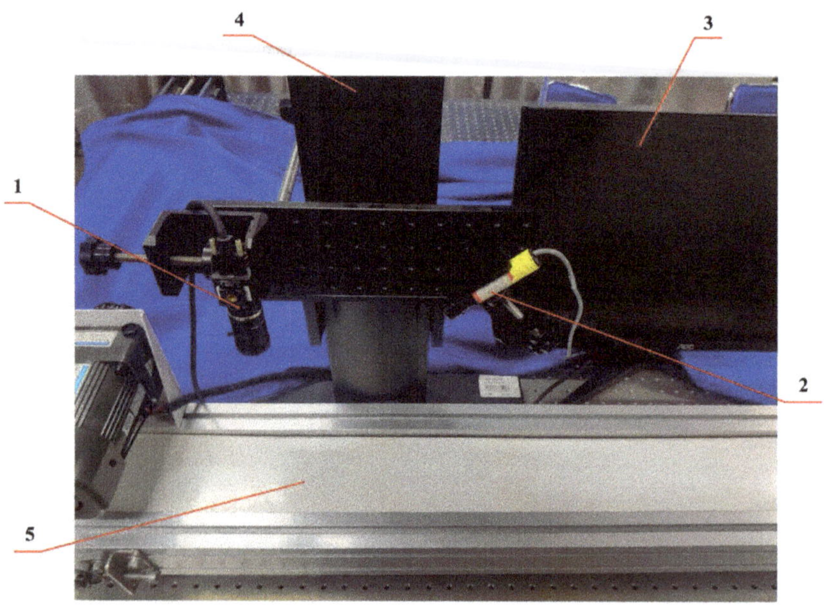

Figure 2. Physical image of cotton foreign fiber polarization imaging detection device. (1) Polarized camera; (2) 405 nm line laser; (3) computer; (4) machine vision frame; (5) electric conveyor belt.

2.3. Dataset, Environment, and Parameters

The target detection dataset in this study was acquired using the image acquisition system shown in Figure 2, containing a total of 3944 foreign fiber target images of 20 categories, which were divided into training, validation, and test sets. The data were enhanced by gaussian blur, affine transformation, brightness transformation, dropping pixel transformation, and flip transformation [33,34]. The enhanced dataset consisted of 21,381 images, and the data format was JPG. Table 2 lists the statistical information of the dataset.

Table 2. Target statistics of the cotton foreign fiber.

Categories	Number of Targets		
	Training Set	**Validation Set**	**Testing Set**
Polypropylene	967	112	121
Cloth	966	132	126
Mulch film	551	56	59
Sponge	954	109	137
Chemical fiber	976	111	113
Stem	994	108	98
Dead leaves	1002	102	114
PVC	1005	88	113
Hair	499	59	72
Hemp rope	1025	91	120
Cotton seed	989	108	121
Kraft paper	429	54	51
Foam	972	113	139
Yarn	108	11	16
Stone	974	109	117
Bark	991	101	108
Leaf	968	105	127
Polyethylene	985	115	136
Feather	987	120	117
Paper	975	121	134
Total	17,317	1925	2139

The hardware environment and software versions of the experiments are listed in Table 3.

Table 3. Experimental environment configuration.

Hardware and Software	Configuration Parameter
Computer	Operating System: Windows10
	CPU: Intel(R) Core (TM) i9-9900K CPU@3.60GHz
	GPU: NVIDIA GeForce RTX 3090
	RAM: 16 GB
	Video memory: 24 GB
Software version	Python3.9.12 + PyTorch1.9.1 + CUDA11.7 + cuDNN8.2.1 + Opencv4.5.5 +Visual Studio Code2022 (1.69.1)

In this study, the SGD (stochastic gradient descent) method was used to optimize the learning rate, and the epochs were determined by comparing the loss functions of the training set and validation set. The parameters of the training network are listed in Table 4.

Table 4. Training network parameters.

Parameter	Value
Batch size	64
Learning rate	0.01
Warm-up epochs	3
Number of iterations	120
Momentum parameter	0.937
Image size	640 × 640
Optimizer	SGD

2.4. Loss Function and Model Evaluation Metrics

The loss function of YOLOv5 consists of three components, which are confidence loss, bounding box regression loss, and classification loss. The expression of the YOLOv5 loss function is shown below:

$$Loss_{total} = \lambda_1 L_{obj} + \lambda_2 L_{box} + \lambda_3 L_{cls} \tag{1}$$

L_{obj}, L_{box}, and L_{cls} represent confidence loss, bounding box regression loss, and classification loss, respectively. λ_1, λ_2, and λ_3 are weight coefficients for the three losses, and changing these coefficients can adjust the emphasis on the three losses. In YOLOv5, L_{box} is calculated using L_{CIoU} [35], which can improve both the speed and accuracy of bounding box regression. The expression for L_{CIoU} is shown below:

$$\begin{cases} L_{CIoU} = 1 - IoU + \frac{\rho^2(b, b^{gt})}{c^2} + \alpha v \\ IoU = \frac{|b \cap b^{gt}|}{|b \cup b^{gt}|} \\ v = \frac{4}{\pi^2}\left(\arctan\frac{w^{gt}}{h^{gt}} - \arctan\frac{w}{h}\right)^2 \end{cases} \tag{2}$$

In the above expression, b and b^{gt} represent the predicted box and ground truth box, respectively; w^{gt}, h^{gt}, w, and h represent the width and height of the ground truth box and the predicted box, respectively; ρ represents the distance between the centers of the two boxes; c represents the maximum distance between the boundaries of the two boxes; and α is a weight coefficient. Both L_{obj} and L_{cls} use the BCEWithLogitsLoss, and their calculation formula is shown below:

$$Loss = -\frac{1}{n}\sum[y_n \ln x_n + (1 - y_n)\ln(1 - x_n)] \tag{3}$$

The BCEWithLogitsLoss function includes both the Sigmoid layer and the BCELoss layer and is suitable for multi-label classification tasks; y_n represents the ground truth label, and x_n represents the predicted label.

To verify the superior performance of the improved Yolov5 model, we measured the *mAP*, *FPS*, model volume, etc. Some commonly used metrics of precision (*P*), recall (*R*), average precision (*AP*), *F*1 Score (*F*1), and mean average precision (*mAP*) were selected to evaluate the model performance [36], and the metrics were defined as follows:

$$precision = \frac{TP}{TP + FP} \tag{4}$$

$$recall = \frac{TP}{TP + FN} \tag{5}$$

$$F1 = \frac{2 \cdot precision \cdot recall}{precision + recall} \tag{6}$$

$$AP = \int_0^1 p(r)dr \tag{7}$$

$$mAP = \frac{\sum_{i=1}^{N} AP_i}{N} \tag{8}$$

TP denotes the positive samples predicted to be correct, *FP* denotes the negative samples predicted to be incorrect, *FN* denotes the positive samples predicted to be incorrect, and *N* denotes the number of sample categories.

2.5. Improvement of YOLOv5 Network Architecture

2.5.1. YOLOv5 Network Architecture

YOLOv5 combines the characteristics of YOLOv1, YOLOv2, YOLOv3, and YOLOv4. YOLOv5 mainly contains four network models, namely, YOLOv5s, YOLOv5m, YOLOv5l, and YOLOv5x, and the model size and parameters increase sequentially in the four network structures. This study was based on the YOLOv5s network structure, as shown in Figure 3.

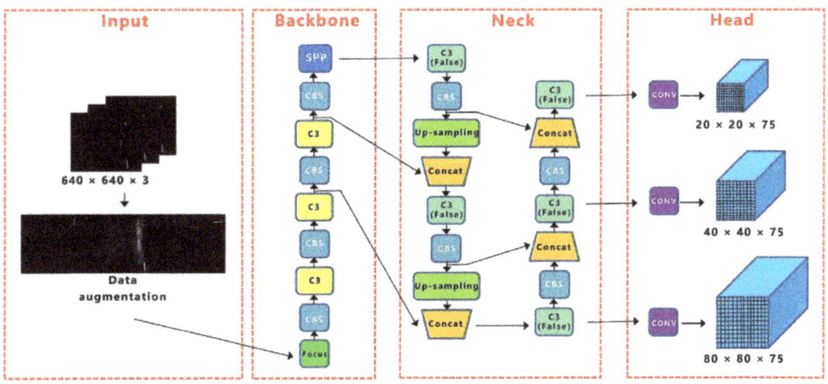

Figure 3. YOLOv5 network structure.

The YOLOv5 network structure consists of a backbone, neck, and head, and the image input first goes through the backbone for continuous feature extraction. The focus performs a slice operation on the input image; for example, if the input image size is 640 × 640 × 3, the slice operation will take a value for every other pixel on the image, and the result will be stacked on the channel to obtain a feature layer of 320 × 320 × 12. It is commonly understood to expand the image channel and compress the image height and width. The focus-module structure is shown in Figure 4.

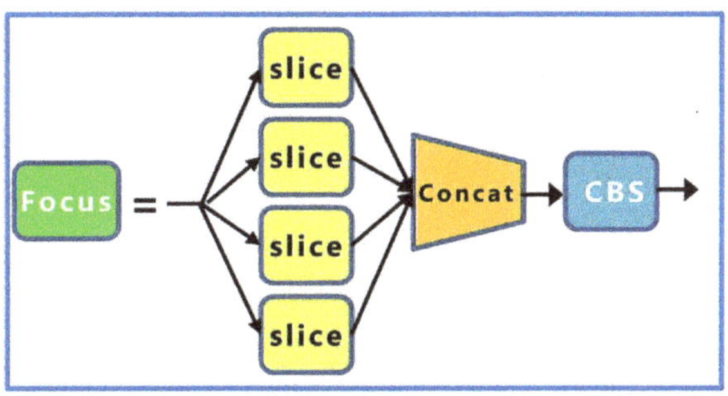

Figure 4. Focus network structure.

The second layer of the backbone is the CBS module with a convolution kernel size of 3 × 3, which performs convolution the calculation, batch standardization calculation, and SiLU activation function on the input data, adds nonlinearity to the network, and accelerates the convergence speed of the network. The third layer is the C3 module, which is mainly composed of n bottleneck modules, three CBS modules, and two convolution layers of size 1 × 1, it and is designed to better extract the deep features of the image. The structures of the bottleneck and C3 modules are shown in Figures 5 and 6, respectively.

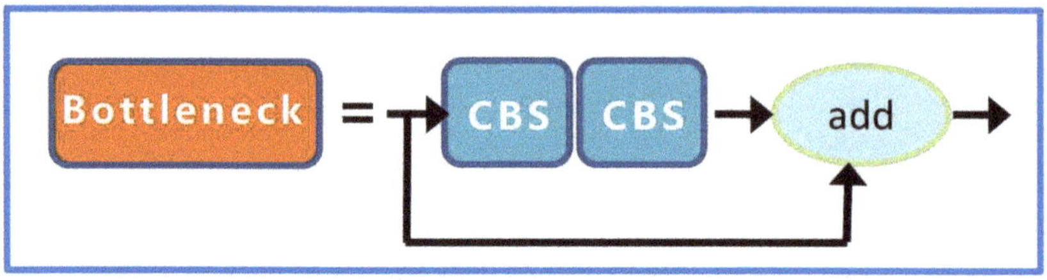

Figure 5. Bottleneck network structure.

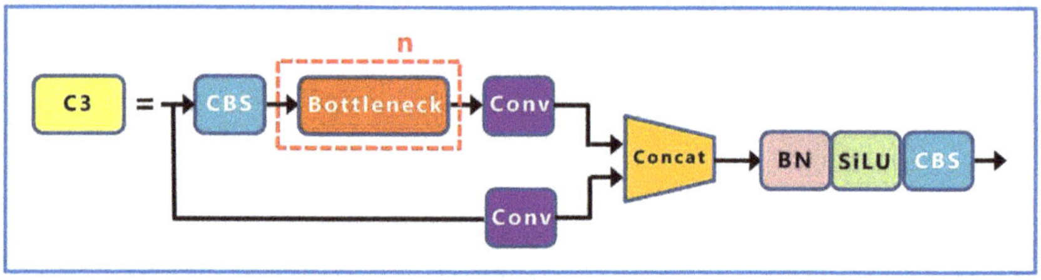

Figure 6. C3 network structure.

The last layer of the backbone is the SPP module. First, the number of channels of the input image is halved using the first CBS module, and then the feature map output from the first CBS module is passed through three maximum pool layers of different sizes (13 × 13, 9 × 9, and 5 × 5), and the residual edges constructed together with the output of the first CBS module are connected in parallel. Finally, the number of channels is halved by the second CBS module to ensure that the height and width of the feature map of different size inputs can be kept consistent after pooling; the structure of the SPP module is shown in Figure 7.

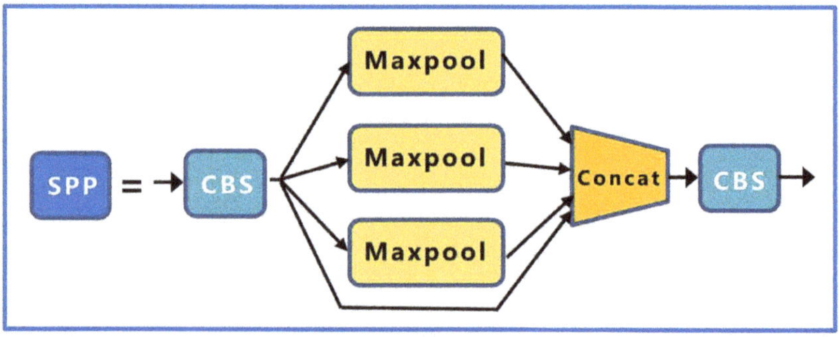

Figure 7. SPP network structure.

The neck network constructs feature pyramids for enhanced feature extraction to obtain more contextual information. Three feature maps are generated in the backbone network; the three feature layers are 80 × 80, 40 × 40, and 20 × 20 from shallow to deep. After the three effective feature layers are obtained, the FPN feature pyramid structure is constructed first, and the 20 × 20 feature layer is upsampled to obtain the 40 × 40 feature layer and then stacked with the corresponding 40 × 40 feature layers in the backbone network. A feature layer of 80 × 80 was obtained by upsampling twice in the FPN structure, and strong semantic features were transferred. Subsequently, the PAN structure was constructed to convey stronger localization features, and the 80 × 80 feature layer was downsampled to obtain a 40 × 40 feature layer, which was stacked with a 40 × 40 feature layer in the FPN network structure. The PAN network structure was downsampled twice, and the final outputs were 80 × 80, 40 × 40, and 20 × 20 enhanced effective feature layers, respectively. Finally, we used these three enhanced feature layers to input the Yolo Head to obtain the regression prediction and classification prediction results.

2.5.2. Proposed Approach: YOLOv5-CFD

This study made corresponding improvements to the backbone, neck, and head of YOLOv5. First, Shufflenetv2 was introduced as the backbone feature extraction network under the premise of ensuring detection accuracy. The weight parameter and volume of the network were reduced, and the lightweight design of the model was realized. Moreover, because the foreign fibers were mostly small-sized targets, the FPN + PAN structure was modified to obtain feature maps with more fine-grained information. Finally, the CA attention module was added to the front of the Yolo Head to improve the detection accuracy. The improved YOLOv5 (YOLOv5-CFD) network structure is illustrated in Figure 8.

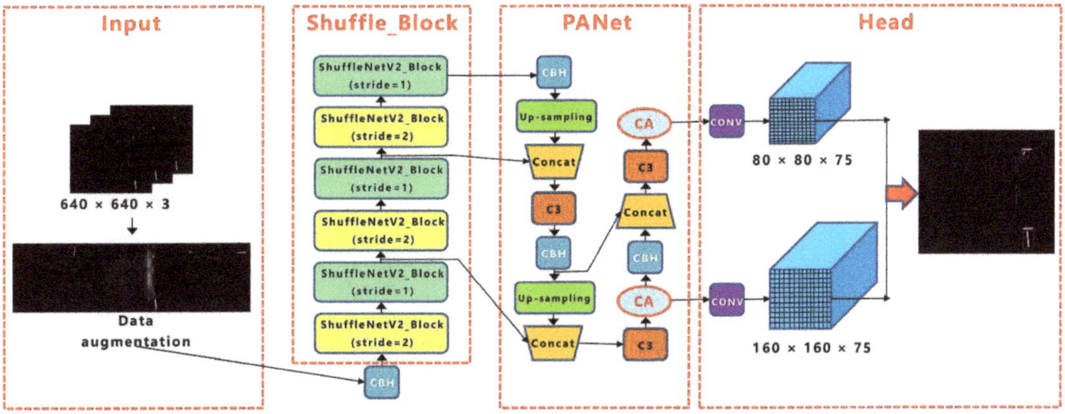

Figure 8. Network model of cotton foreign fiber detection.

Improvement of Backbone Network

ShufflenetV2 was proposed by Ma et al. [37] and was based on ShufflenetV1 and four efficient network design principles. The ShufflenetV2 model excels in speed and accuracy, making it an ideal lightweight network for deployment in mobile devices. First, ShufflenetV2 divides the input of the feature channel into two branches by the "Channel Split" operation. One branch has the same structure, and the other branch consists of three convolutions with the same input and output channels. The two branches are concatenated after convolution to keep the number of channels constant. Finally, the "Channel Shuffle" operation is used to ensure the information exchange between the two branches. ShufflenetV2 contains a basic unit and a unit for spatial downsampling (2×), as shown in Figure 9.

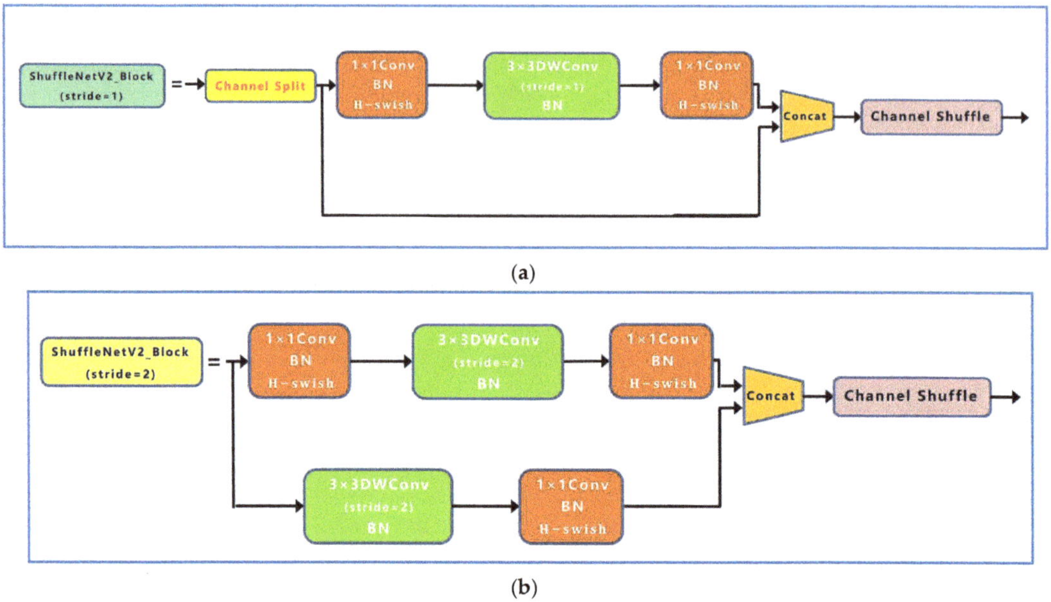

Figure 9. ShufflenetV2 network structure. (**a**) Basic unit; (**b**) unit for spatial downsampling (2×).

In this paper, ShufflenetV2 units with stride = 2 and stride = 1 were chosen to construct a new backbone network, and the output of each stage in the new backbone was connected to PANet. Moreover, we replaced the activation function in the ShufflenetV2 unit with the H-swish activation function, as shown in Equation (9):

$$H\text{-}swish(x) = \begin{cases} 0 & x \leq -3 \\ x & x \geq 3 \\ x \cdot (x+3)/6 & -3 < x < 3 \end{cases} \tag{9}$$

Improvement of PANet Network

Among the three effective features of the FPN + PAN structure output, the 20 × 20 and 40 × 40 feature maps were used to detect larger targets, whereas foreign fibers in cotton are mostly small-sized targets. Moreover, the image size of our input network was 2448 × 2048, and the grid pixels corresponding to the 20 × 20 and 40 × 40 feature maps were 128 × 108 and 64 × 54, respectively, when performing the bounding box regression. The k-means clustering statistics showed that nearly 75% of the foreign fiber target pixels were below 60, as shown in Figure 10, with 20 × 20 and 40 × 40 feature maps corresponding to anchors ([116, 90], [156, 198], [373, 326]) and ([30, 61], [62, 45], [59, 119]). The anchors were larger, and many operations were useless when performing the bounding box regression. The 20 × 20 and 40 × 40 feature maps and large target identification frames were discarded, making the bounding box regression more accurate and minimizing the model computational cost.

To solve the problem of an excessive number of small targets, the PANet network connection was improved to obtain a feature map with fine-grained information. A new 160 × 160 feature map was generated by upsampling the output of the backbone network twice and fusing it with the feature map of the corresponding size of the backbone. Because the improved backbone network generated three layers of feature mapping of 320 × 320, 160 × 160, and 80 × 80, the FPN did not require secondary upsampling; hence, the final YOLO detection heads were 160 × 160 and 80 × 80; Figure 11 shows the PANet network improvement schematic diagram of YOLOv5.

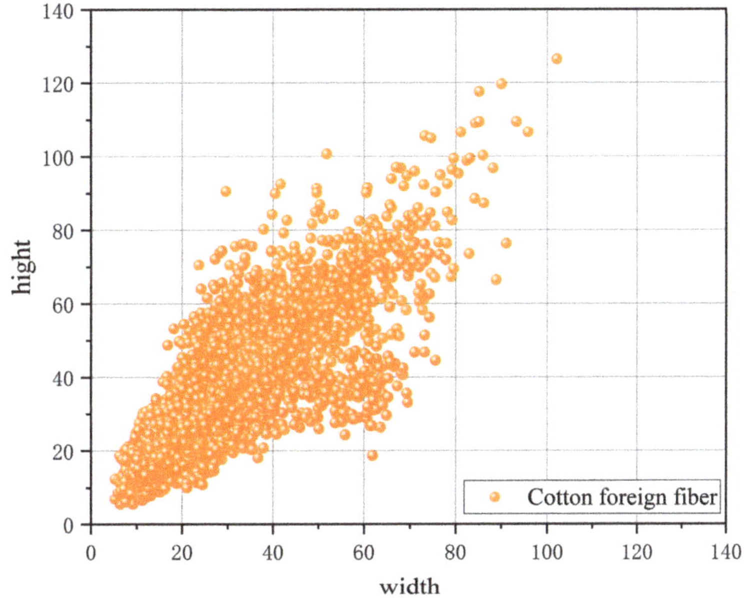

Figure 10. Cotton foreign fiber width and height distribution.

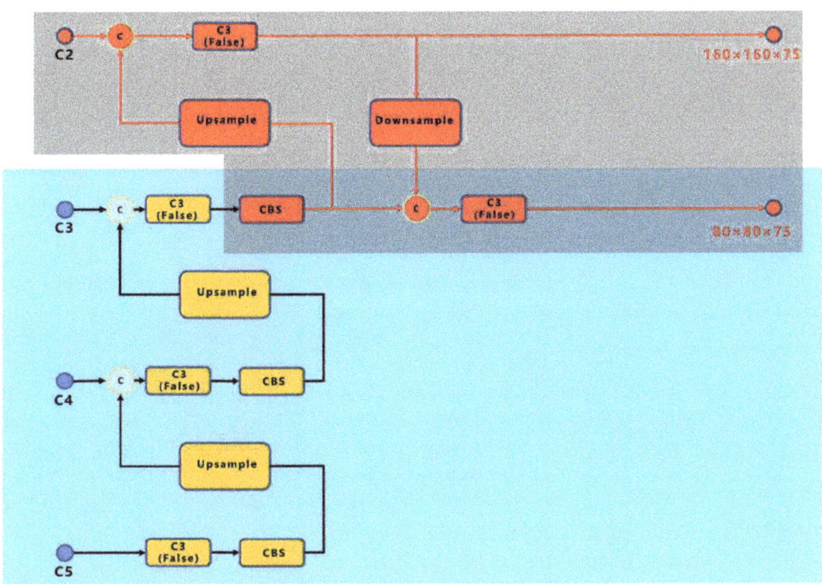

Figure 11. PANet network improvement schematic diagram of YOLOv5.

CA Module Design

Hou et al. [38] proposed a novel attention mechanism for mobile networks called "Coordinate Attention" by embedding location information into channel attention in 2021, as shown in Figure 12.

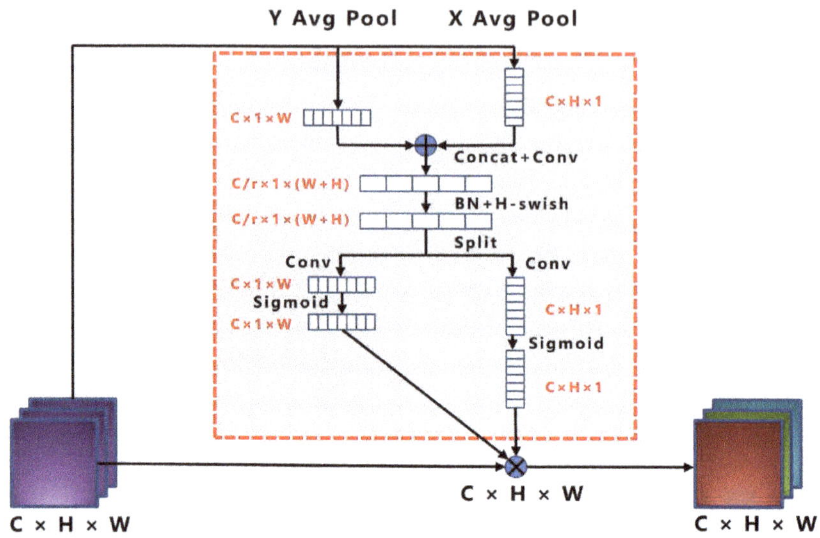

Figure 12. Coordinate Attention network structure.

Coordinate Attention focuses on the image width and height and encodes precise position information. First, the input feature map was divided into the width and height directions for global averaging pooling to obtain the feature maps in the width and height directions. The output of the c-th channel with the height and width is expressed as follows:

$$z_c^h(h) = \frac{1}{W} \sum_{0 \le i < W} x_c(h, i) \tag{10}$$

$$z_c^w(w) = \frac{1}{H} \sum_{0 \le j < H} x_c(j, w) \tag{11}$$

The above equation integrates the features from different directions and outputs a pair of feature maps with known directions. The module can capture long distance relationships in one direction while retaining spatial information in the other, helping the network locate targets more accurately.

Stitching together the feature maps in the width and height directions of the obtained global perceptual field, the channel is compressed to the original C/r using a 1×1 convolution. Subsequently, the BatchNorm and *H-swish* activation functions are used for encoding, followed by a 1×1 convolution to adjust the channels of the feature map to be equal to the number of channels of the input feature map. The attention weights g^h and g^w of the feature map on the height and width, respectively, are obtained after the sigmoid function, as shown below:

$$g^h = \sigma\left(F_h\left(f^h\right)\right) \tag{12}$$

$$g^w = \sigma(F_w(f^w)) \tag{13}$$

Finally, a weighted fusion is performed on the original feature map to obtain the final feature map with attention weights in the height and width directions, as shown in the following equation:

$$y_c(i,j) = x_c(i,j) \cdot g_c^h(i) \cdot g_c^w(j) \tag{14}$$

Based on the characteristics of multiple types and small targets with different fibers, this study added a CA module at the front end of each of the two detection heads of the

Yolo Head to improve the performance of the network at a low cost, thus improving the overall accuracy of target detection.

The flow chart of the foreign fiber detection method used in this study is shown in Figure 13.

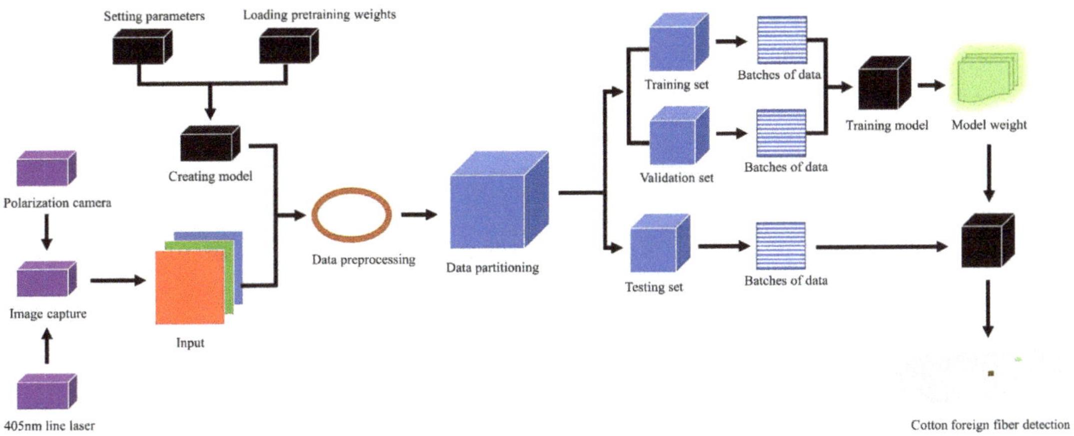

Figure 13. Flow chart of the proposed approach of foreign fiber detection.

3. Results and Discussion

Figure 14 shows the loss reduction curves of the YOLOv5-CFD model for the training and validation sets of foreign fiber images. As can be observed from the loss curve, the loss value dropped to a relatively small value when the number of training rounds was 20, and the network stabilized when the number of training rounds was 120.

The confusion matrix of the YOLOv5-CFD model is shown in Figure 15. It can be observed from the figure that most of the targets of different fiber types were correctly predicted with a low target miss rate, indicating that the model exhibited good performance.

Figure 16 shows the PR curve of YOLOv5-CFD test set and shows the change curve of the accuracy and recall of the detection results of twenty kinds of foreign fiber targets. According to statistics, the overall detection result mAP@0.5 was 96.9%.

3.1. Ablation Experiment

The effect of the improved method on the model performance was analyzed by ablation experiments. For comparison purposes, the experiment was divided into five groups. The first group was the original YOLOv5 network. In the second group, the ShufflenetV2 module was introduced into the backbone feature extraction network module of the YOLOv5. The third group modified the PANet network connection method using YOLOv5. In the fourth group, a CA module was added to the front of each of the two detection heads of YOLOv5. The last set of experiments was the result of the model used in this study. The experimental results are listed in Table 5.

As seen in Table 5, the use of the ShufflenetV2 module in the back-bone feature extraction network reduced mAP@0.5 and mAP@0.5:0.95 by 1.95% and 2.73%, respectively, but the model volume decreased by 5.89 MB, and the detection speed increased by 200 f/s. The introduction of the ShufflenetV2 module played an important role in reducing the model volume and improving the detection speed. The improvement of the PANet network reduced the model volume by 3.3 MB, increased the mAP@0.5 and mAP@0.5:0.95 by 0.27% and 4.63%, respectively, and increased the detection speed by 153 f/s. The introduction of the CA attention module improved the detection accuracy of the model and verified the effectiveness of the improved PANet and CA modules. In summary, the improved strategy based on YOLOv5 proposed in this study is important for facilitating the identification and detection of cotton foreign fibers in an actual production line.

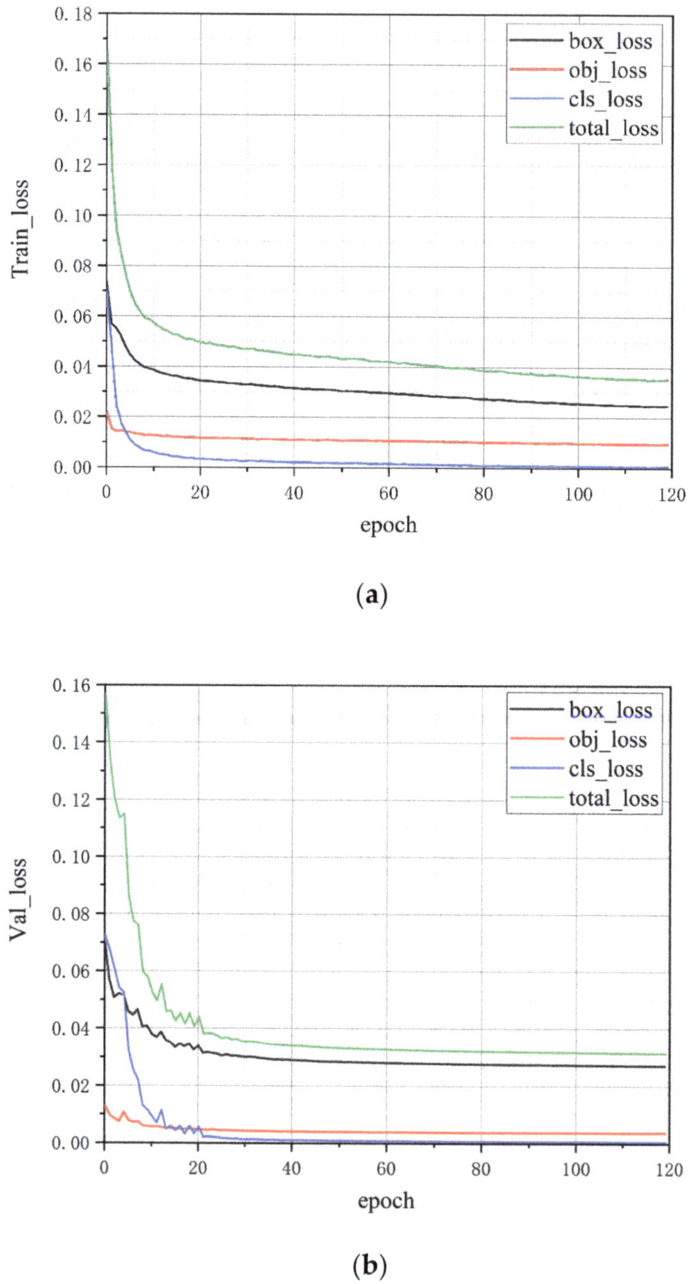

Figure 14. Four types of loss curves for each data set. (**a**) Training set; (**b**) Validation set.

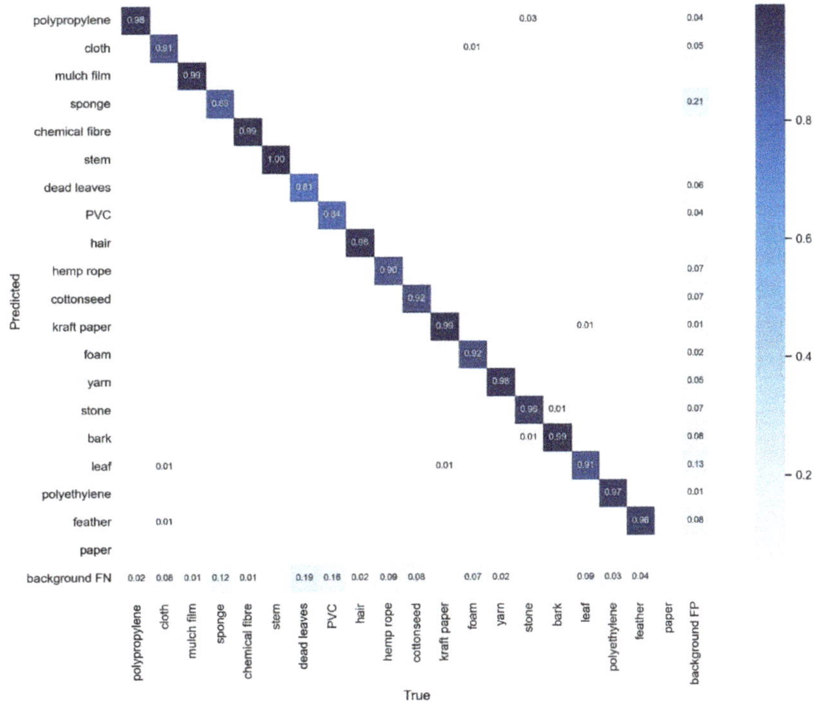

Figure 15. Confusion matrix of the YOLOv5-CFD model.

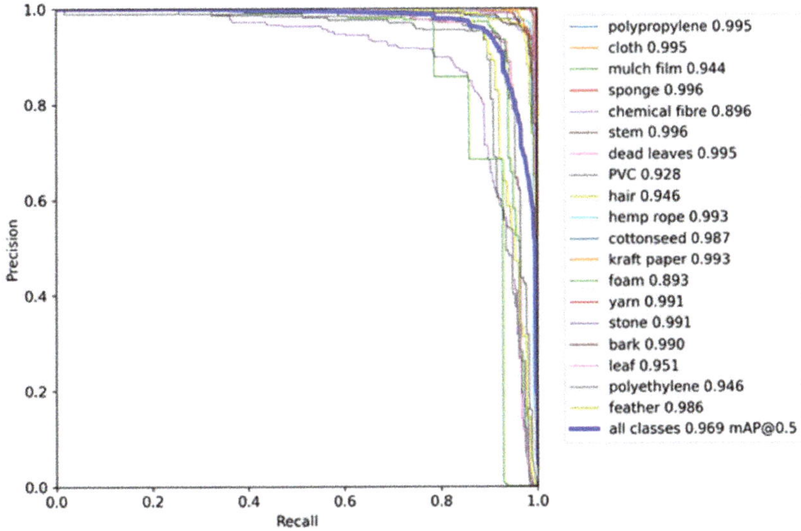

Figure 16. Test accuracy and recall curves of YOLOv5-CFD.

Table 5. Ablation experiments.

ShuffleNet V2	PANet	CA	H-Swish	Model Volume (MB)	mAP@0.5 (%)	mAP@0.5:0.95 (%)	FPS (f/s)
×	×	×	×	13.82	95.87	52.77	170
√	×	×	×	7.93	93.92	50.04	370
×	√	×	×	10.52	96.14	57.40	323
×	×	√	×	13.90	95.98	55.08	180
√	√	√	√	0.75	96.90	59.90	385

3.2. Comparison of Different Models

To verify the superiority of the YOLOv5-CFD model in cotton foreign fiber detection, we compared it with the most advanced foreign fiber detection models, YOLOv5, YOLOv4, SSD, and Faster-RCNN. The relevant parameters of the experiments were also strictly controlled using a uniform image size as the input and a uniform training and test set for experimental testing.

Comparing the overall test results of Faster-RCNN, SSD, YOLOv4, YOLOv5, and YOLOv5-CFD with mAP@0.5, as shown in Figure 17, it can be seen that YOLOv5-CFD model had better performance.

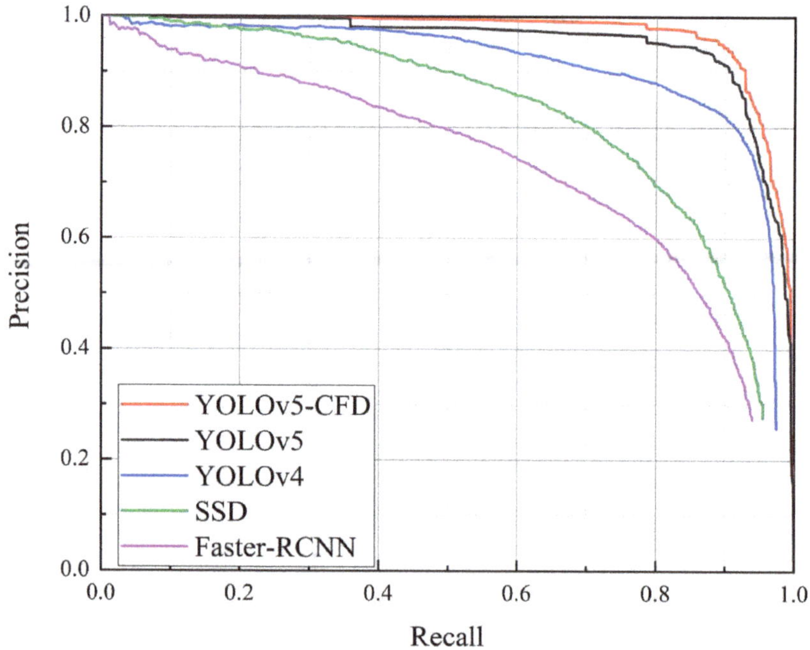

Figure 17. P–R curves of different detection models.

The pictures used in the comparative experiment in Figure 18 are from the test set of this paper [39]. Each experiment was conducted in the same environment. Figure 18 shows the detection effects of different models in different cases. The images contain complex light environments, small target foreign fibers, and multiple types of foreign fibers, so the problems of multiple types of small target foreign fibers in a complex light environment are fully considered, providing a convenient way to fully demonstrate the robustness and generalization ability of the model.

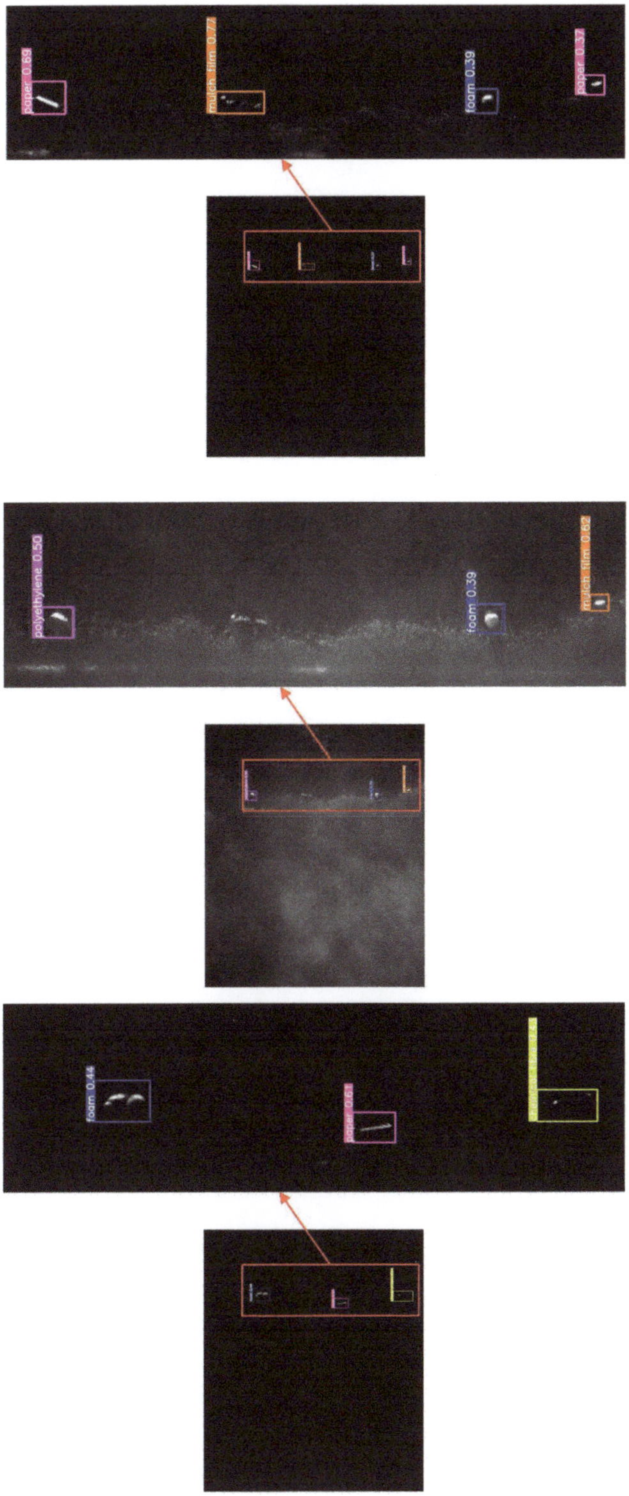

(a) YOLOv5-CFD

Figure 18. *Cont.*

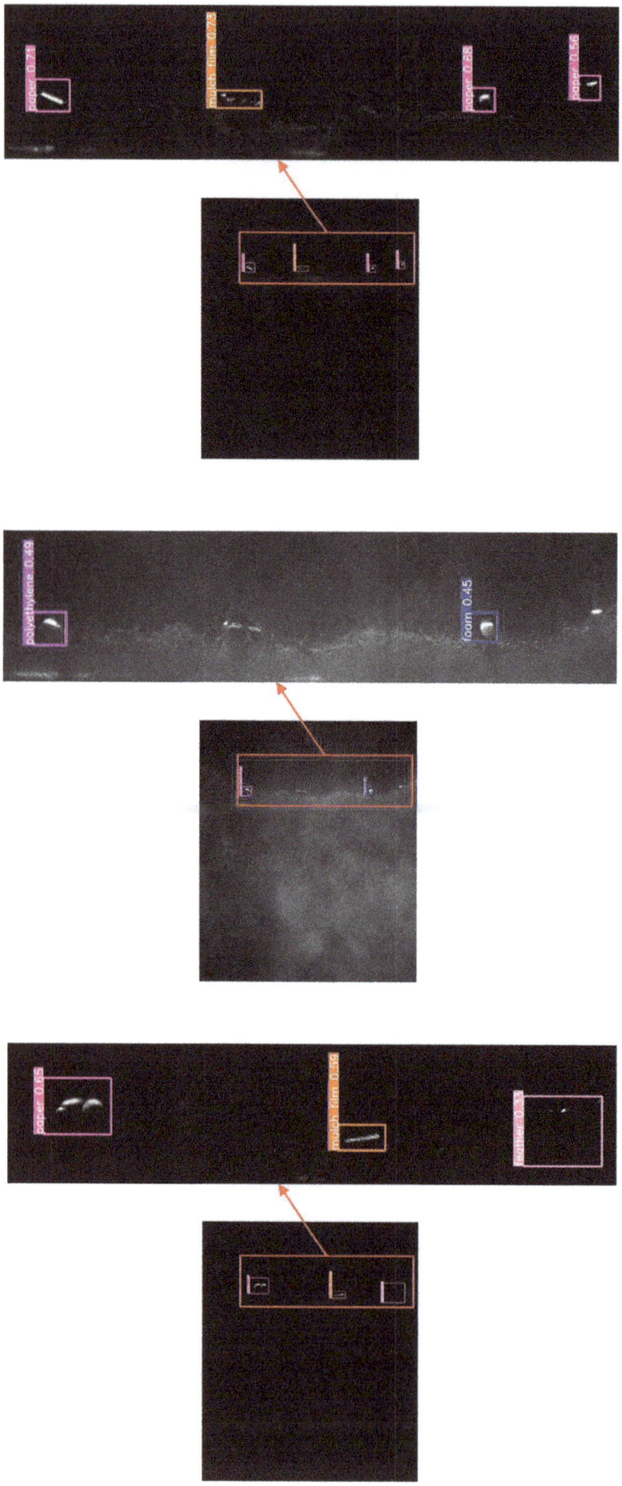

(**b**) YOLOv5

Figure 18. *Cont.*

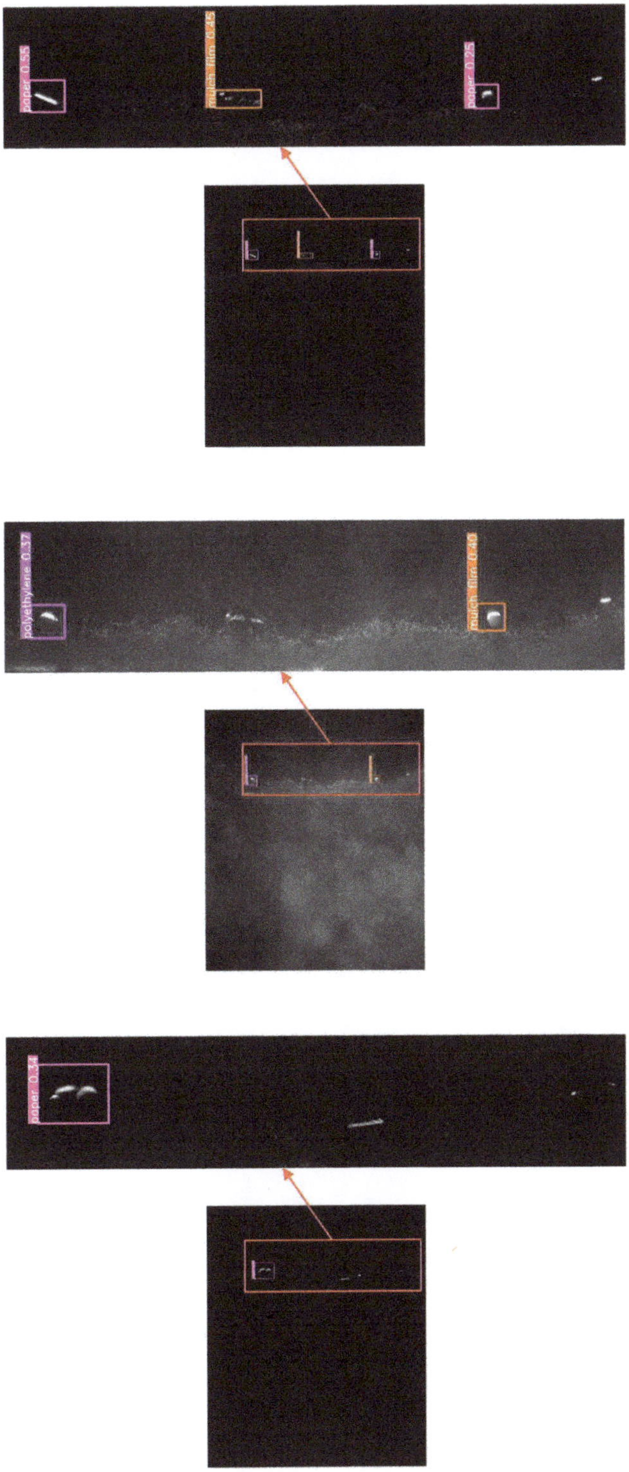

(c) YOLOv4

Figure 18. *Cont.*

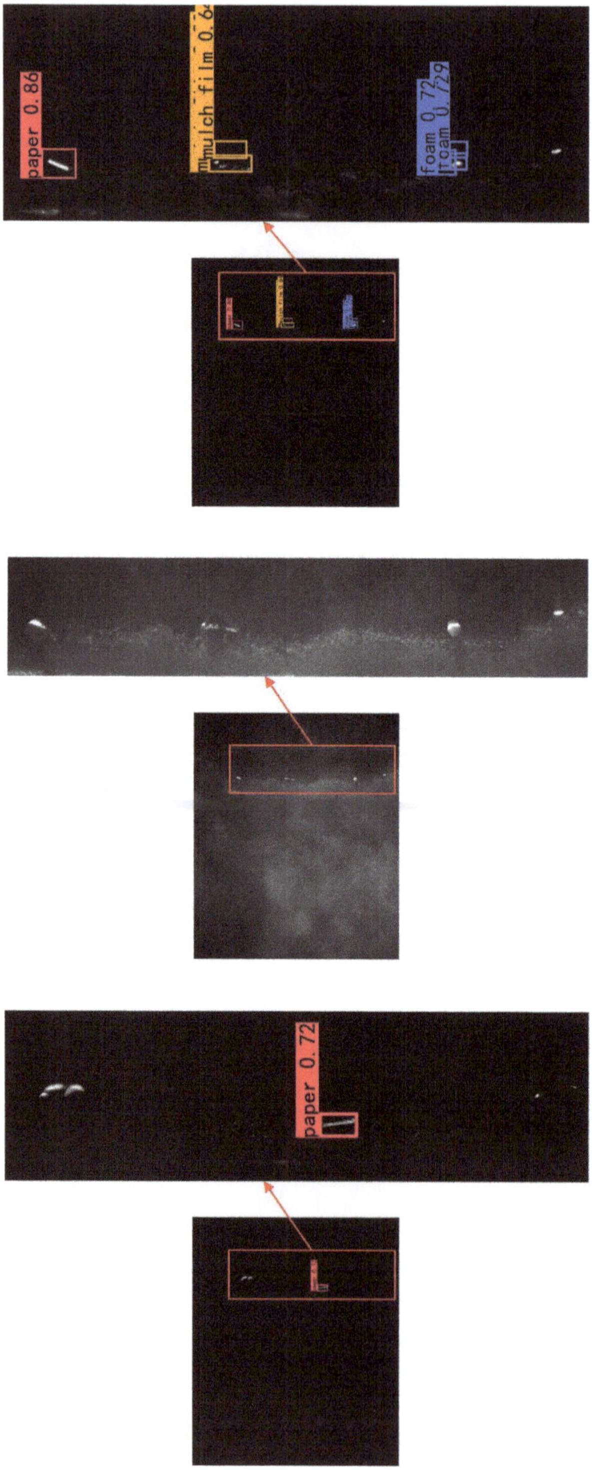

(d) SSD

Figure 18. *Cont.*

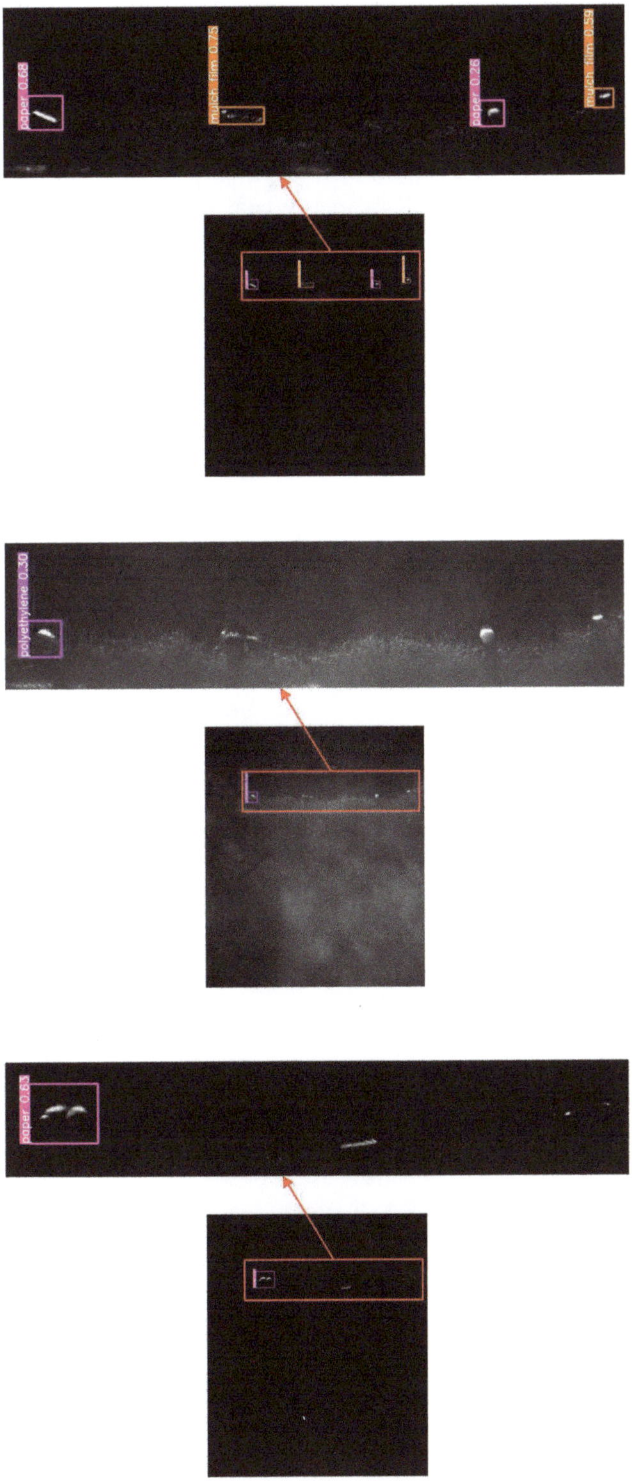

(e) Faster-RCNN

Figure 18. Visualization of different models in performance testing.

From the image detection results, it could be observed that for large foreign fibers, most of the five models were recognized, and YOLOv5-CFD had the highest correct classification rate. For small foreign fibers, YOLOv5-CFD had the highest recognition rate and correct classification rate. For the first image, YOLOv5-CFD was identified and classified correctly. In the second image, YOLOv5-CFD had the highest recognition rate with only one missed target, and YOLOv5 and Faster-RCNN had the highest correct classification rate. For the last image, YOLOv5-CFD, YOLOv5, and Faster-RCNN were all identified correctly, and only YOLOv5-CFD and SSD were classified correctly; however, the SSD model had multiple overlapping detection frames in the detection. In summary, the YOLOv5-CFD model outperformed the other four models in terms of the test results.

As shown in Table 6, the model volume, mAP@0.5, mAP@0.5:0.95, and FPS of the YOLOv5-CFD were up to 0.75 MB, 96.9%, 59.9%, and 385 f/s, respectively, which were better than the values of YOLOv5 (13.82 MB, 95.87%, 52.77%, and 170 f/s, respectively), followed by YOLOv4 (244.78 MB, 93.59%, 50.50%, and 88 f/s, respectively), and SSD (100.29 MB, 83.07%, 39.06%, and 128 f/s, respectively). Furthermore, the results of Faster-RCNN (108.91 MB, 75.68%, 33.60%, and 9 f/s, respectively) were worse. The results showed that the overall performance of the proposed YOLOv5-CFD was the best [40].

Table 6. Experimental results of different algorithms.

Model	Parameters	Model Volume (MB)	mAP@0.5 (%)	mAP@0.5:0.95 (%)	FPS (f/s)
YOLOv5-CFD	2.97×10^5	0.75	96.90	59.90	385
YOLOv5	7.28×10^6	13.82	95.87	52.77	170
YOLOv4	6.39×10^7	244.78	93.59	50.50	88
SSD	2.41×10^7	100.29	83.07	39.06	128
Faster-RCNN	2.84×10^7	108.91	75.68	33.60	9

The main improvement of the YOLOv5-CFD model is the volume size of the model and the detection speed; these enhancements meet the high requirements of the actual production line detection of cotton foreign fibers, and the detection accuracy of YOLOv5-CFD for small target foreign fibers is also the highest. Based on the above analysis, the YOLOv5-CFD object detection algorithm proposed in this study improves the detection speed and accuracy of foreign fiber targets and significantly reduces the model size.

3.3. YOLOv5-CFD Test Results

In order to test the robustness and anti-interference of the YOLOv5-CFD model, this paper repeatedly tested the miss-recognition rate, misjudgment rate, precision, recall, and F1 score of the model under different illumination, different incident angles, different cotton foreign fiber samples, different foreign fiber positions, different foreign fiber sizes, and different environments. Combined with the sampling frequency of the camera, the speed of the conveyor belt was set to 4 m/min. The misrecognition rate is the rate of failure to identify the presence of foreign fibers, and the misjudgment rate is the rate of judging the position where there is no foreign fiber as the presence rate. For each test condition, the precision and recall values for each category are first calculated, and then the averages of the precision and recall values for each category are taken. The test results of the YOLOv5-CFD model are shown in Table 7.

The experiments of foreign fibers (including mulch film, foam, feather, white paper, polyethylene, polypropylene, and chemical fiber) detection and classification were made. The results showed that the environmental light intensity changes had some influence on the foreign fiber classification, but little effect on the detection. The interference of strong light such as sunlight caused an increase in the misrecognition rate. The classification performance of the model was the best under dark conditions and the worst under sunlight conditions. Foreign fibers were difficult to identify with a small or large incidence angle such as 15° or 90°. When the incident angle was around 45°, the detection and classification of foreign fibers were optimal. For the different variety of samples, the YOLOv5-CFD model could generally detect foreign fibers well, and the average F1 score of the three numbered samples was about 0.69. Under the condition of different positions of foreign fibers, there were no omissions and misjudgments, and the classification results were the same. Under the condition of different sizes of foreign fibers, the minimum size of foreign fibers detected by the YOLOv5-CFD model was 0.5 mm^2. Smoke and dust almost had no interference of linear laser polarization imaging. In summary, the proposed method has good robustness and anti-interference, meets the basic detection of cotton foreign fibers on the actual production line, and has practical application value.

Table 7. YOLOv5-CFD model test results.

Different Conditions		Identification		Classification		
		Misrecognition Rate	**Misjudgment Rate**	**Precision**	**Recall**	**F1**
Illumination	dark	0	0	90.07%	97.03%	0.93
	lamplight	0	0	84.42%	92.35%	0.88
	sunlight	7.30%	0	72.20%	87.72%	0.79
Incidence angle	15°	100%	13.64%	0	0	0
	45°	0	0	90.07%	97.03%	0.93
	90°	3.93%	22.07%	67.85%	76.20%	0.72
Different varieties samples	115,549	0	5.54%	73.66%	66.23%	0.69
	114,835	0	7.92%	75.39%	70.03%	0.73
	114,712	12.87%	0	68.55%	60.34%	0.64
Different positions	upper edge	0	0	90.07%	97.03%	0.93
	middle	0	0	90.07%	97.03%	0.93
	lower edge	0	0	90.07%	97.03%	0.93
Foreign fiber size	1.5 mm^2	0	0	70.79%	77.05%	0.74
	1 mm^2	0	0	66.54%	78.21%	0.72
	0.5 mm^2	16.05%	0	60.10%	71.78%	0.65
	<0.5 mm^2	100%	0	0	0	0
Environment	smog	0	0	90.07%	97.03%	0.93
	dust	0	0	90.07%	97.03%	0.93

4. Conclusions

To address the problem of foreign fiber detection in cotton, a polarization imaging device of cotton foreign fiber was constructed using the difference in optical properties and polarization characteristics between cotton fibers and foreign fibers. Moreover, an object detection algorithm for cotton foreign fiber based on the improved YOLOv5 was proposed, which consisted of three key steps: The lightweight network Shufflenetv2 with the Hard-Swish activation function was used as the backbone feature extraction network, an improved PANet was added to YOLOV5, and a CA module was added before the Head of YOLOv5. The robustness and anti-interference of the improved YOLOv5 model under various conditions were also tested. Compared with the YOLOv5 foreign fiber detection model, the improved YOLOv5 foreign fiber detection model had better performance in mAP@0.5, mAP@0.5:0.95, and FPS, which increased by 1.03%, 7.13%, and 126.47%, respectively. The improved model is capable of performing online identification and classification of small foreign fiber targets of various types in cotton transportation.

Author Contributions: Conceptualization, R.W. and Z.-F.Z.; methodology, R.W.; software, R.W.; validation, R.W., Z.-F.Z. and B.Y.; formal analysis, R.W.; investigation, R.W.; resources, Z.-Y.C.; data curation, R.W.; writing—original draft preparation, R.W.; writing—review and editing, Z.-F.Z. and K.Y.; visualization, R.W. and H.-Q.X.; supervision, Y.-S.Z., R.-L.Z. and L.-J.G.; project administration, Z.-F.Z.; funding acquisition, Z.-F.Z., R.-L.Z. and L.-J.G. All authors have read and agreed to the published version of the manuscript.

Funding: This research was funded by the National Natural Science Foundation of China (No. 61905223, 11904327), and Henan Science & Technology Development Plan Project (No. 222102210085, 222102210319).

Institutional Review Board Statement: Not applicable.

Informed Consent Statement: Not applicable.

Data Availability Statement: Not applicable.

Acknowledgments: This research was funded by the National Natural Science Foundation of China (No. 61905223, 11904327), and Henan Science & Technology Development Plan Project (No. 222102210085, 222102210319).

Conflicts of Interest: The authors declare no conflict of interest.

References

1. Chen, B. Research on the testing technology of cotton foreign fiber content. *Mod. Bus. Trade Ind.* **2018**, *39*, 185–186.
2. Zhang, H.; Li, D. Applications of computer vision techniques to cotton foreign matter inspection: A review. *Comput. Electron. Agric.* **2014**, *109*, 59–70. [CrossRef]
3. Yang, W.; Li, D.; Zhu, L.; Kang, Y.; Li, F. A new approach for image processing in foreign fiber detection. *Comput. Electron. Agric.* **2009**, *68*, 68–77. [CrossRef]
4. Yue, X. Research on Multi-Parameter Optimization of Heterogeneous Fiber Sorting Machine Detection Rate Based on Neural Network. Master's Thesis, Tianjin Polytechnic University, Tianjin, China, 2020.
5. Shi, G. Remark upon foreign matter online detecting and clearing. *Shanghai Text. Sci. Technol.* **2007**, *35*, 24–25.
6. Chang, L. The Detecting System of Foreign Fibers in Cotton Based on DSP. Master's Thesis, Hefei University of Technology, Hefei, China, 2006.
7. Guo, S.; Kan, Z.; Zhang, R.; Guo, W.; Cong, T. Separation test of electrostatic separating device for machine-harvested seed cotton and plastic film residue. *Trans. Chin. Soc. Agric. Eng.* **2011**, *27*, 6–10.
8. Chen, Y.J.; Wu, T.R.; Shi, S.W.; Zhao, B.; Yang, S.H. Review of Cotton Foreign Fiber Detection Method Using Optical Imaging. *Laser Optoelectron. Prog.* **2021**, *58*, 1600007.
9. Pai, A.; Sari-Sarraf, H.; Hequet, E.F. Recognition of cotton contaminants via X-ray microtomographic image analysis. *IEEE Trans. Ind. Appl.* **2004**, *40*, 77–85. [CrossRef]
10. Luo, D.P.; Zhu, B.T.; Li, X. Fluorescent Effect of Ultra Violet and It's Application in Detection of Foreign Fibers in Cotton. *J. Henan Univ. Sci. Technol. Nat. Sci.* **2007**, *2007*, 63–66+2.
11. Cai, X.X.; Wu, L.L.; Liang, H.F.; Chen, J. Cotton foreign fiber detection based on near-infrared imaging technology. *Cotton Text. Technol.* **2021**, *49*, 6–10.
12. Hua, C.J.; Su, Z.W.; Qiao, L.; Shi, J. White foreign fibers detection in cotton using line laser. *Trans. Chin. Soc. Agric. Mach.* **2012**, *43*, 181–185.
13. Mustafic, A.; Jiang, Y.; Li, C.Y. Cotton contamination detection and classification using hyperspectral fluorescence imaging. *Text. Res. J.* **2016**, *86*, 1574–1584. [CrossRef]
14. Zhang, C.; Sun, S.L.; Shi, W.X.; Zeng, L.; Deng, D.X. Design and test of foreign fiber removal machine based on embedded system. *Trans. Chin. Soc. Agric. Mach.* **2017**, *48*, 43–52.
15. Kennedy, J.; Eberhart, R. Particle swarm optimization. In Proceedings of the ICNN'95-International Conference on Neural Networks, Perth, WA, Australia, 27 November–1 December 1995; pp. 1942–1948.
16. Goodman, E.D. Introduction to genetic algorithms. In Proceedings of the Fourteenth International Conference on Genetic and Evolutionary Computation Conference Companion-GECCO Companion'12, Philadelphia, PA, USA, 7–11 July 2012; pp. 671–692.
17. Dorigo, M.; Maniezzo, V.; Colorni, A. Ant system:optimization by a colony of cooperating agents. *IEEE Trans. Syst. Man Cybern. Part B* **1996**, *26*, 29–41. [CrossRef]
18. Otsu, N. A threshold selection method from gray-level histograms. *IEEE Trans. SMC* **1979**, *9*, 62–66. [CrossRef]
19. Liu, S.X.; Wang, J.X.; Zhang, H.; Li, W.; Li, D.L.; Zhang, X.H.; Fan, G.Q. Research on the multi-channel wavelet segmentation method of faint cotton foreign fibers. *Chin. J. Sci. Instrum.* **2016**, *37*, 60–66.
20. He, Y.; Wang, J.F. Rapid nondestructive identification of wood lacquer using Raman spectroscopy based on characteristic-band-Fisher-K nearest neighbor. *Laser Optoelectron. Prog.* **2020**, *57*, 013001. [CrossRef]
21. Jain, A.K.; Dubes, R.C. Algorithms for Clustering Data. *Technometrics* **1988**, *32*, 227–229.

22. Vapnik, V. Statistical Learning Theory. *DBLP* **1998**. [CrossRef]
23. He, X.Y.; Su, Z.W.; Deng, B.Y.; Pan, Y.F.; Chi, Z.Q. An artificial intelligence method for detecting foreign fiber in seed cotton. *Cotton Text. Technol.* **2018**, *46*, 49–52.
24. He, X.Y.; Wei, P.; Zhang, L.; Deng, B.Y.; Pan, Y.F.; Su, Z.W. Detection method of foreign fibers in seed cotton based on deep-learning. *J. Text. Res.* **2018**, *39*, 131–135.
25. Zhang, D. Research on Seed Cotton Foreign Fiber Sorting Recognition Algorithm Based on Deep Learning. Master's Thesis, Nanjing Forestry University, Nanjing, China, 2019.
26. Du, Y.H.; Dong, C.Q.; Zhao, D.; Ren, W.J.; Cai, W.C. Application of improved Faster RCNN model for foreign fiber identification in cotton. *Laser Optoelectron. Prog.* **2020**, *57*, 121007.
27. Dong, C.Q. Research on foreign fiber classification method based on improved Faster R-CNN model. Master's Thesis, Tianjin Polytechnic University, Tianjin, China, 2020.
28. Wu, M.X.; Wu, J.; Zhang, C.; Zhu, L. Detection of foreign fiber in cotton based on improved YOLOv3. *Chin. J. Liq. Cryst. Disp.* **2020**, *35*, 1195–1203. [CrossRef]
29. Wei, W.; Deng, D.X.; Zeng, L. Classification of foreign fibers using deep learning and its implementation on embedded system. *Int. J. Adv. Robot. Syst.* **2019**, *7*, 1729881419867600. [CrossRef]
30. Wu, Z.W.; Shi, H.Y. Recognition of Foreign Fiber in Cotton Based on DSConv and CBAM. *Cotton Text. Technol.* **2022**, *50*, 19–23.
31. Zhang, Y.; Zhang, S.J.; Feng, Z.Q. Improved Faster RCNN Target Detection Method for Foreign Fiber in Cotton. *Cotton Text. Technol.* **2022**, *5*, 37–41.
32. Zhang, C.; Li, T.; Li, J. Detection of Impurity Rate of Machine-Picked Cotton Based on Improved Canny Operator. *Electronics* **2022**, *11*, 974. [CrossRef]
33. Abayomi-Alli, O.O.; Damaševičius, R.; Misra, S.; Maskeliūnas, R. Cassava disease recognition from low-quality images using enhanced data augmentation model and deep learning. *Expert Syst.* **2021**, *38*, e12746. [CrossRef]
34. Ye, Y.; Li, Y.; Ouyang, R.; Zhang, Z.; Tang, Y.; Bai, S. Improving machine learning based phase and hardness prediction of high-entropy alloys by using Gaussian noise augmented data. *Comput. Mater. Sci.* **2023**, *223*, 112140. [CrossRef]
35. Zheng, Z.H.; Wang, P.; Liu, W.; Li, J.Z.; Ye, R.G.; Ren, D.W. Distance-IoU loss: Faster and better learning for bounding box regression. In Proceedings of the AAAI Conference on Artificial Intelligence, New York, NY, USA, 7–12 February 2020; pp. 12993–13000.
36. Zhao, W.; Wu, D.; Zheng, X. Detection of Chrysanthemums Inflorescence Based on Improved CR-YOLOv5s Algorithm. *Sensors* **2023**, *23*, 4234. [CrossRef]
37. Ma, N.; Zhang, X.; Zheng, H.T.; Sun, J. Shufflenet v2: Practical Guidelines for Efficient CNN Architecture Design. In Proceedings of the European Conference on Computer Vision (ECCV), Munich, Germany, 8–14 September 2018; pp. 116–131.
38. Hou, Q.; Zhou, D.; Feng, J. Coordinate attention for efficient mobile network design. In Proceedings of the IEEE/CVF Conference on Computer Vision and Pattern Recognition, Nashville, TN, USA, 19–25 June 2021; pp. 13713–13722.
39. Yu, J.; Zhang, W. Face Mask Wearing Detection Algorithm Based on Improved YOLO-v4. *Sensors* **2021**, *21*, 3263. [CrossRef] [PubMed]
40. Qiao, Y.L.; Guo, Y.Y.; He, D.J. Cattle body detection based on YOLOv5-ASFF for precision livestock farming. *Comput. Electron. Agric.* **2023**, *204*, 107579. [CrossRef]

sensors

MDPI

Review

Recent Advances in Wayside Railway Wheel Flat Detection Techniques: A Review

Wenjie Fu [1], Qixin He [1,*], Qibo Feng [1,2], Jiakun Li [1], Fajia Zheng [1] and Bin Zhang [1]

[1] Key Lab of Luminescence and Optical Information, Ministry of Education, Beijing Jiaotong University, Beijing 100044, China
[2] Dongguan Nannar Technology Co., Ltd., Dongguan 523050, China
* Correspondence: heqixin@bjtu.edu.cn; Tel.: +86-010-5168-8618

Abstract: Wheel flats are amongst the most common local surface defect in railway wheels, which can result in repetitive high wheel–rail contact forces and thus lead to rapid deterioration and possible failure of wheels and rails if not detected at an early stage. The timely and accurate detection of wheel flats is of great significance to ensure the safety of train operation and reduce maintenance costs. In recent years, with the increase of train speed and load capacity, wheel flat detection is facing greater challenges. This paper focuses on the review of wheel flat detection techniques and flat signal processing methods based on wayside deployment in recent years. Commonly used wheel flat detection methods, including sound-based methods, image-based methods, and stress-based methods are introduced and summarized. The advantages and disadvantages of these methods are discussed and concluded. In addition, the flat signal processing methods corresponding to different wheel flat detection techniques are also summarized and discussed. According to the review, we believe that the development direction of the wheel flat detection system is gradually moving towards device simplification, multi-sensor fusion, high algorithm accuracy, and operational intelligence. With continuous development of machine learning algorithms and constant perfection of railway databases, wheel flat detection based on machine learning algorithms will be the development trend in the future.

Keywords: wheel flat detection; wayside signal acquisition method; signal processing method

Citation: Fu, W.; He, Q.; Feng, Q.; Li, J.; Zheng, F.; Zhang, B. Recent Advances in Wayside Railway Wheel Flat Detection Techniques: A Review. *Sensors* **2023**, *23*, 3916. https://doi.org/10.3390/s23083916

Academic Editor: Ikhlas Abdel-Qader

Received: 18 March 2023
Revised: 9 April 2023
Accepted: 10 April 2023
Published: 12 April 2023

1. Introduction

Railways are considered to be one of the most important means of transportation at present. In China, the proportion of railway travel in people's daily life has increased year by year. In 2021, China's railway passenger volume has reached ~2.5 billion with an increase of 36.6 million over the previous year and a year-on-year increase of 16.9%, as shown in Figure 1a. In the same year, the total turnover of railway freight transportation reached ~2995.0 billion tons, with an increase of 255.2 billion tons over the previous year and a year-on-year increase of 9.3%, as shown in Figure 1b [1].

The demand for railway transportation has increased year by year, and the safety of railway transportation has become more and more important. According to the statistics of the United States Federal Railway Administration, 1234 train accidents were reported from January 2019 to February 2022 [2]. Among them, train derailments are the most common problems, with a total of 809 cases, accounting for 66% of the total number of failures, as shown in Figure 2.

As the most important running part of the train, the condition of railway wheels is closely related to the safety of train operation. Between 2005 and 2010, derailment accidents caused by rolling stock failures accounted for the highest proportion of accidents (38%) in 14 European countries (Austria, France, Germany, the UK, Sweden, Switzerland, Belgium, Bulgaria, Czech Republic, Hungary, Italy, the Netherlands, Poland, and Slovenia) [3].

Wheel flats are the most common local surface defect of railway wheels and are one of the important causes of train derailment [4]. Huge impact forces will be generated by the contact between flat wheels and rails, which will cause further damage to vehicles and rail components (wheel sets, bearings, rail ties, and so on) [5–12]. The excessive wheel–rail contact forces will increase the risk of train accidents and maintenance costs.

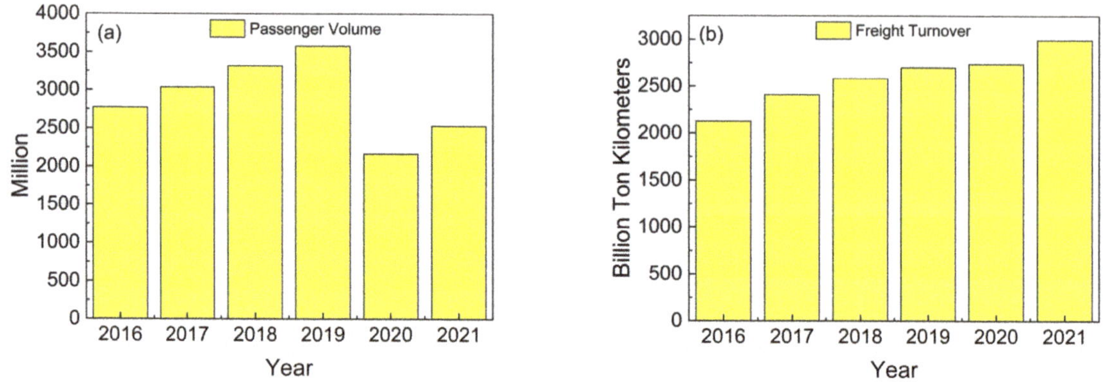

Figure 1. (**a**) Total railway passenger volume in China in 2021. (**b**) Total railway freight turnover in China in 2021.

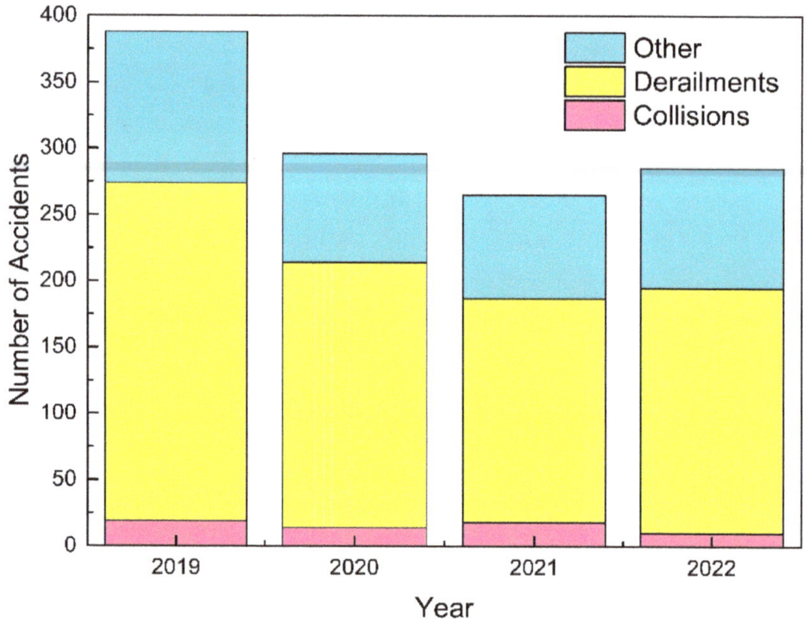

Figure 2. Main categories of railway failures in the United States from 2019 to 2022.

Wheel flats are caused by the loss of wheel tread material due to wheel slipping events [13]. Heat is generated as the wheels sliding on the rails, and the resulting temperature increase combined with the rapid cooling of the adjacent material can lead to the formation of brittle martensite on the wheel tread. Thermal impacts and phase transformations can generate large residual stresses, which will interact with the rolling contact stress to promote the formation and growth of cracks on the wheel tread [14]. There are many reasons for the wheels to slide on the rails, including (1) The sudden braking of the wheels when driving at high speed, thus causing the wheels to lock up and slide on the rail while the train is still moving [15]. (2) There is a local area where the wheel-rail friction becomes low. When there are foreign objects such as grease and leaves on the track, the wheel–rail adhesion is reduced, resulting in complete sliding between the wheel and rail [16–19]. In order to address the issue of wheel flats, the railway department has implemented various preventive measures, such as installing advanced anti-sliding systems on passenger trains. Despite these efforts, wheel flats cannot be completely eliminated. Therefore, wheel flat detection is considered to be an important measure of ensuring railway operation safety and reducing maintenance costs.

Human inspection has been the most common means of wheel flat detection in past decades. However, this method is time-consuming and prone to human errors. In-service testing methods can realize real-time detection of wheel flats without dismantling the wheelsets, and they have been studied by many researchers in recent years [20–23]. In-service testing methods can be divided into the on-board method and wayside method, according to different sensor installation positions. These two methods have their own advantages and disadvantages, and the selection of the method is based on a comprehensive consideration of the inspection duration, fault severity level, and so on [24–26]. In the on-board method, sensors are mounted on the axle box or wheels of a train, and different types of signals can be obtained by corresponding sensors such as accelerometers, microphones, etc., for analysis [27–35]. The flat signal obtained by the on-board method has strong periodicity and better robustness, which is conducive to signal processing and wheel flat detection. In the wayside method, sensors are usually installed on or nearby both sides of the track near train entrances or exit stations; the condition of all wheels can be evaluated as the train passes through the sensor system [36–38]. Unlike the periodic signals collected by the on-board method, each wheel can only be detected once by the wayside method, which leads to limited flat information and thus puts forward a high requirement for flat signal processing algorithms [39–42].

In 2017, Alemi and Corman reviewed the condition monitoring approaches for the detection of railway wheel defects [43]. In the recent few years, wheel flat detection has gained the attention of many researchers and flat detection techniques with new features were proposed, thus the summary and comparison of these techniques are of great significance. In this paper, the development of wheel flat detection techniques and corresponding signal processing methods with wayside deployment since 2016 are reviewed. The rest of this paper is organized as follows. The general structure of wayside in-service flat detection systems is demonstrated in Section 2. Wheel flat detection techniques proposed in the past five years is summarized in Sections 3 and 4. In Section 5, the merits and weaknesses of the mentioned techniques are discussed and concluded.

2. Structure of Wayside In-Service Flat Detection System

Wayside detection systems are commonly used in automated rolling stock inspection processes [43–46]. The general structure of wayside in-service flat detection systems for railway vehicles is shown in Figure 3. It can be divided into two main parts: the data acquisition module and the data analysis module [47]. The data acquisition module is usually installed in the work room beside the rail to acquire wheel flat information when the train passes. Cameras, microphones, pressure sensors and other kinds of sensors can be used in data acquisition the module to sense flat information.

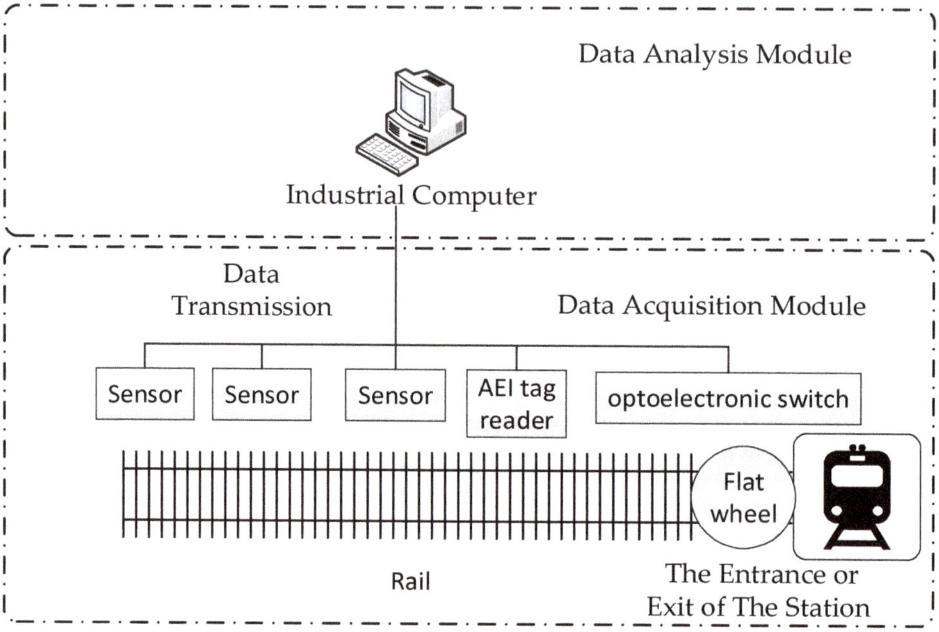

Figure 3. Wayside in-service flat detection system schematic.

When the train passes the wayside detection system, the sensor installed beside the rail, such as an optoelectronic switch, will generate a 'train coming signal' to start the system. Automatic equipment identification (AEI) tag readers are often mounted on rails along with sensors to detect the wagon IDs and identify each wheel. The signal processor and the control PC then record the data collected from each axle in a database, which are then transmitted to the railway control center or depot maintenance center for remote monitoring and diagnosis [48,49]. The data analysis module is usually installed in the monitoring room beside the track. The signals generated by the data acquisition module are sent to the data analysis module via wired or wireless data transmission [50]. The signals are pre-processed to reduce noise interference and processed by wheel flat algorithms to extract flat information.

3. Stress-Based Wheel Flat Signal Acquisition Method

The most commonly used wheel flat detection method is the stress-based method. In this method, the dynamic stress of the track when the train passes can be measured by different stress sensors such as strain gauges, accelerometers, and fiber Bragg gratings (FBG) [51–65].

In 2016, Matthias Asplund et al. checked the wheel profile parameters measured by the wayside wheel profile measurement system (WPMS) installed along the Swedish Iron Ore Line, and correlated them with the warning and alarm indications issued by the wheel defect detector (WDD) [66]. The WDD measures the force peak generated by static train load and dynamic load from the wheel defects [67]. WPMS can detect faults related to wheel profile, such as high flange, wide flange, thin flange, small flange angle, and abnormal wheel diameter. WDD can detect faults such as flats, large shelling, and severe wheel polygonization. Unfortunately, none of these systems can successfully capture surface cracks, spalling, small shelling, low levels of polygonization, or all subsurface defects, nor can they detect rolling contact fatigue (RCF) in wheels at an early stage [68,69].

In 2018, several vibration transducers were mounted on tracks by Tomasz Nowakowski et al., to detect the wheel flat as shown in Figure 4 [70]. A vibration signal processing method based on time-domain and frequency-domain was proposed. By analyzing the time signal envelopes of flat-wheel and ordinary-wheel trams, obvious peaks were found in the time signal envelopes of flat-wheel trams. They measured 15 tram passes with flawless wheels and 17 tram passes with flat wheels and found the time signal envelope of the flat wheel tram is characterized by an obvious peak with an amplitude higher than 35 m/s^2, while the time signal envelope of the flawless tram does not have this feature. Experimental results show that this method has high efficiency in wheel flat detection, and can be applied to a larger speed range.

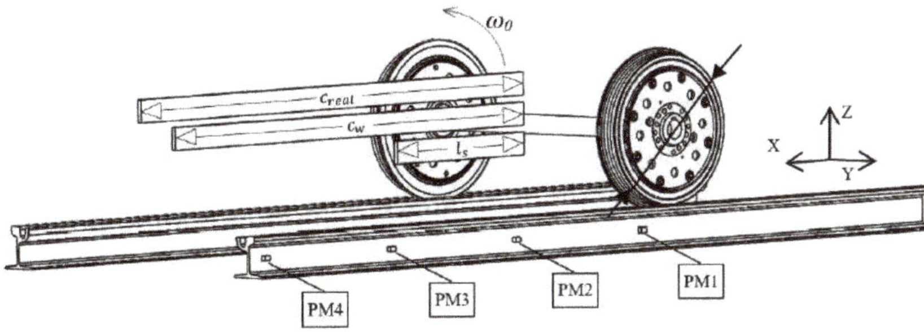

Figure 4. A view of the measurement points locations with a basic designation of dimensions and measurement realization. ω0: rotational velocity of the wheel, PM1~PM4: points of measurements [70].

Liu and Ni developed an FBG-based track-side wheel condition monitoring system for detecting wheel tread defects [71]. Two FBG strain gauge arrays mounted on the foot of the track were used to measure the dynamic strain of paired tracks excited by passing wheelsets. Each FBG array was approximately 3 m in length, slightly longer than the wheel circumference to ensure full coverage to detect any potential defects on the wheel tread. A defect detection algorithm was developed that utilizes online monitored rail responses to identify potential wheel tread defects. Data smoothing techniques were used to detrend and to pre-process the strain data and outlier analysis was used to perform diagnostics of responses to normalized data. Local defects can be identified by refined analysis of responses extracted in diagnostics. According to field tests, the proposed method can achieve satisfactory accuracy in wheel defect detection when the train running speed was higher than 30 kph, and some minor defects with a depth of 0.05 mm~0.06 mm were also successfully detected.

In the same year, Gabriel Krummenacher et al. proposed a wheel flat detection system based on the measurement of vertical force through wheel load checkpoints (WLC) installed on the rail [72]. Each WLC consists of four 1 m long measurement bars with four strain gauges per measurement bar. The strain gauges were installed perpendicular on the centerline of the rails. An automatic detection and classification method for railway wheel defects based on vertical force sensors was proposed. Wavelet features of time series data from sensors were designed and support vector machines were used as classifiers. Convolutional neural networks (CNN) for different wheel defect types were designed and trained through deep learning. A cyclic shift invariant artificial neural network was designed to detect flat and non-round wheels. The proposed method can be used to predict wheel defects without prior knowledge of how these defects will manifest in measurements.

In 2019, a railway wheel flat detection system based on a parallelogram mechanism was improved on the basis of the previous research of our group [73]. As the core component, the parallelogram mechanism is mainly composed of the measuring ruler, connecting rods, springs, hydraulic damper, limit block, and eddy current sensor, as shown in Figure 5. The wheel flat can be quantitatively detected by measuring the vertical displacement change of the measuring ruler. The depth of the wheel flat can be reflected by the amplitude of the sensor signal.

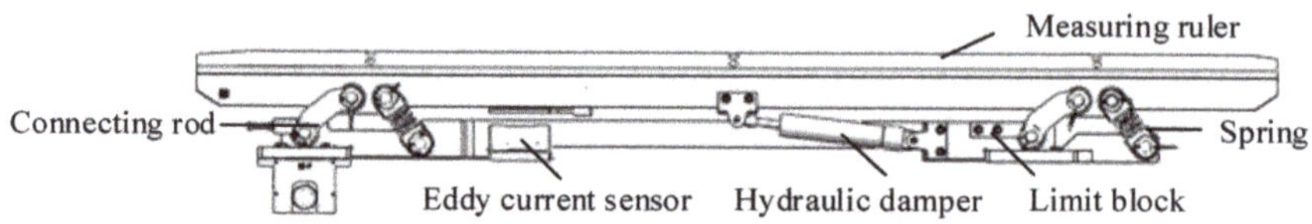

Figure 5. A schematic diagram of the parallelogram mechanism [73].

Wen-Jun Cao et al. used the dynamic signal of rail pad sensors (RPS) to identify the wheel flats [74]. A method to identify wheel flat dimensions using a dynamic measurement-based model update strategy was proposed. They used a model falsification approach to identify the size of the wheel flats, which can interpret high-dimensional time series in the context of inverse identification. This method has been successfully applied to process time series data and significantly reduces computation time. The system was field-tested on a test track at a train station in Singapore and the experimental results showed that the identified wheel flat size is within the real observation range.

In the same year, Alemi A. et al. proposed a fusion method for wheel defect recognition, which associates the collected samples with their positions on the wheel circumference coordinates [75]. As the magnitude of the contact force contains limited information about wheel defects, this study reconstructs defect signals from discrete samples collected by multiple sensors, such as WILDs. The obtained results show a considerable similarity between the contact force and the reconstructed defect signal, which can be used for further defect identification. The proposed method offers the potential to detect and identify defects at an early stage, including minor defects and long-wave defects. In addition to wheel defects, the reconstructed defect signal will also be influenced by other parameters, such as train velocity, axle load, number of sensors, and wheel diameter. In 2020, they carried out a parameter study to investigate the impact of these parameters [76]. The research shows that the fusion method can provide better performance when the signal-to-noise ratio (SNR) is high. Increasing the number of sensors can improve the results of the fusion process. Therefore, it is necessary to balance the cost of the interrogator supporting a large number of sensors with the accuracy and reliability of the fusion results.

In 2020, Chenyi Zhou et al. proposed a long-term monitoring method for wheel flats based on multi-sensor arrays, as shown in Figure 6 [77]. The dynamic strain response of the rail was captured efficiently by an array of sensors mounted on the rail web to ensure that all wheels were evaluated during the passage of the train. In order to realize accurate recognition and positioning of wheel flats, an algorithm based on multi-source data fusion was proposed. A vehicle–track system coupling dynamic model was established, and the sensitivity and reliability of different sensor layout schemes under different wheel flat conditions can be analyzed according to the model. By conjoint analysis of multi-sensor signals, the specific moment at which the wheel flat occurred can be precisely identified. By use of data fusion between multiple sensors, the specific location of wheel defects can be confirmed.

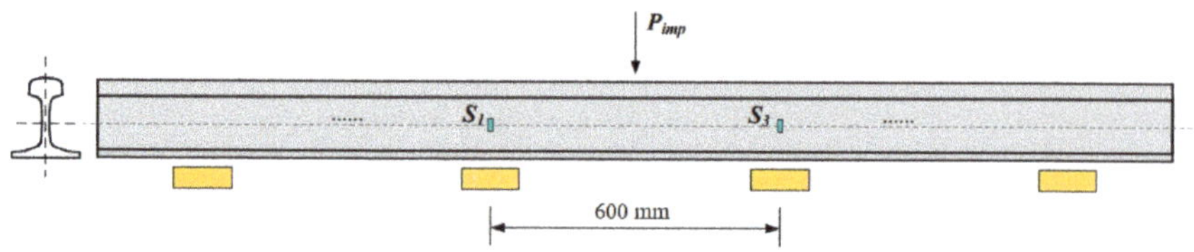

Figure 6. Plans for sensor arrangement [77].

Our group designed a new wheel flat detection system based on a self-developed reflective optical position sensor [78]. As shown in Figure 7, the sensor was composed of a tailed fiber laser, a four-quadrant detector, and a cube-corner prism. A in Figure 7a represents the light source and detection module composed of a tailed fiber laser and a four-quadrant detector. B in Figure 7a is a reflection module composed of cube-corner prism. The reflective optical position sensor is used to detect the vertical deformation of the rail under wheel–rail contact. The wheel–rail impact force of the entire circumference was measured by displacement detection of the collimated laser spot. The finite element method and multibody dynamics method were used to establish the vehicle–track coupled dynamic analysis model. A quantitative relationship between the sensor signal and the wheel flat length was established by the model. The system was assessed through simulation and laboratory investigation, and real field tests were conducted to certify its validity and correctness.

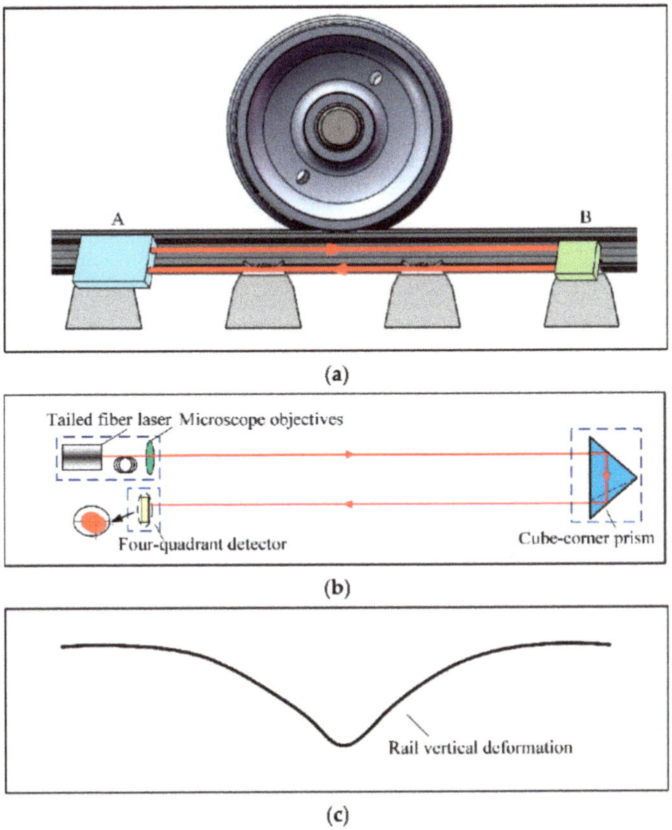

Figure 7. The schematic diagram of the sensor and the rail vertical deformation: (**a**) The installation location of the sensor; (**b**) The schematic diagram of the sensor; (**c**) The rail vertical deformation [78].

A vibration signal-based flat detection system was developed by Jyoti Barman and Durlav Hazarika [79]. In this system, the vibration signal was captured by an ADXL335 vibration sensor connected to the fish-plate of the track and linear time-frequency transform (wavelet transform) combined with a quadratic time-frequency transform (Wigner-Ville transform) were used to find the flat signal.

In 2021, Araliya Mosleh et al. proposed a wheel flat detection system based on strain gauges (SGS) mounted on the trackside which has been tested in several scenarios of varying complexity [80]. The layout scheme of the strain gauges is shown in Figure 8. The numbers in the figure represent the deployment position of the strain gauge on the rail. A method using envelope spectrum analysis to detect wheel flats was proposed. Through envelope spectrum analysis, wheel flats at different train speeds can be detected if a noticeable lag between the amplitudes of the envelope spectrum is observed. In 2021, they proposed a multi-sensory layout scheme to detect wheel flats on passenger and freight trains [81]. The effect of sensor type and installation location on the accuracy of the wheel flat detection system was analyzed and discussed. Experimental results show that using a layout scheme consisting of accelerometers is clearly more beneficial than using strain gauges to perform envelope spectrum analysis to detect defective wheels in situations where the signal is heavily contaminated by noise. Compared with the previous study, the number of sensors was reduced and the installation positions of sensors were optimized.

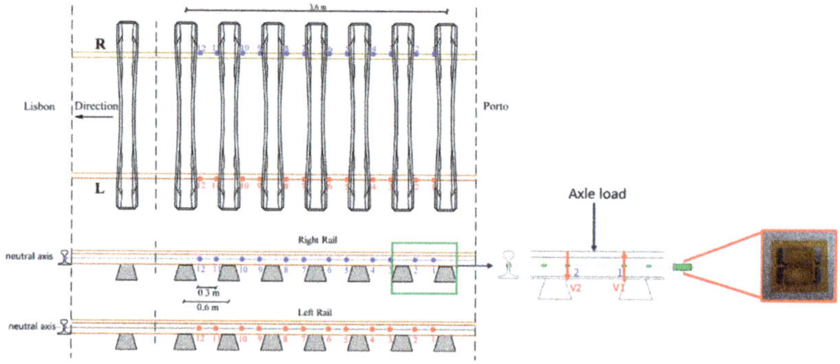

Figure 8. Strain gauges' positions [80].

Ni and Zhang established an FBG-based wayside monitoring system by deploying two arrays of FBG strain sensors beside the track [82]. Each sensor array includes 21 FBG gauges, and each gauge is evenly spaced at 0.15 m intervals on the rail foot of each single track. A Bayesian machine learning approach based on trackside strain-monitoring data was developed for online and quantitative assessment of railway wheel conditions. The cumulative density functions of the normalized Fourier amplitude spectra of the rail foot strain response under healthy wheel conditions were extracted as features. The probabilistic reference model was trained by Sparse Bayesian Learning (SBL). Due to the sparsity of SBL embedding, overfitting was avoided and the generalization ability was improved. As only a small number of basic functions were involved in the model, the computational efficiency of the model was competitive, and fast diagnosis can be realized in wheel condition assessment. The proposed method is verified by using the in-situ monitoring data collected by the wayside monitoring system during the train passing process. The proposed method is verified by comparing the diagnosis results obtained from the proposed online method and the offline wheel radius deviation measurement.

In 2022, Jian Mu et al. studied the dynamic behavior of the vehicle system and the contact force between the wheel and the rail [83]. The detection of wheel–rail vertical contact force was realized by the prototype through rail web strain gauges. Then a vehicle–rail coupling model considering the modal characteristics of the flexible wheelset and the track was established. The validity of the fitted curve to determine the flat length was checked by comparing the simulated plane length with the length calculated for the wheel–rail contact force.

Araliya Mosleh et al. proposed a data-driven machine learning method based on unsupervised learning, which can automatically differentiate between defective and healthy wheels of a train [84]. This method combines sets of acceleration and shear force records evaluated on the rail to enhance its sensitivity. By taking one or more train passing signals with different operating speeds, loading schemes, and track irregularities profiles as input, an artificial intelligence method was used to identify wheel flats. A continuous wavelet transform model was employed to extract features from multiple sensors, transforming time series measurements into alternative data. Then, the extracted features were normalized using Principal Component Analysis techniques, suppressing environmental and operational variations. Finally, data fusion was employed to merge the features from each sensor and enhance the sensitivity to detect wheel defects.

4. Sound- or Image-Based Wheel Flat Signal Acquisition Methods

In addition to stress-based methods, many sound- and image-based methods have also been used to detect wheel flats [85–88].

4.1. Sound-Based Method

When the wheel flat appears on the wheel tread, the wheel–rail contact force will become uneven and impact rolling noise will be generated. In the sound-based method, acoustic sensors such as microphones and acoustic emission (AE) sensors are installed on the side of the track to acquire flat signals. In 2017, Pawel Komorski et al. realized the identification of wheel flats by detecting collision noise generated by wheel–rail contact [89]. The layout scheme of the measurement system is shown in Figure 9. Three microphones were installed on one side of the track and spread along the track with a distance of 2.04 m, which is the length of the circumference of each tram wheel. A transmitter-receiver type photocell was placed between the tracks to measure the cross-section. The joint time-frequency analysis method (JTFA) was used to process acoustic signals and detect the flat wheels of trams, which consists of short-time Fourier transform (STFT) analysis and wavelet transforms. Subsequent studies have shown that the total cost can be reduced by reducing measurement points [90]. The improved system layout scheme is shown in Figure 10. The acoustic signal was analyzed according to the Fourier transform and the Hilbert transform. The novelty of this method is the use of acoustic signals instead of vibration signals to estimate the diagnostic parameters of the second and third rotational harmonic frequencies of the wheel. The highest sensitivity was obtained at the level of 13–15 dB.

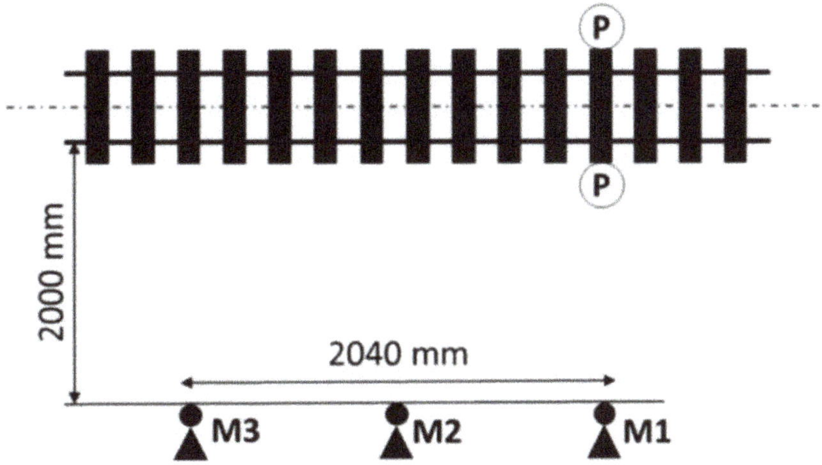

Figure 9. The scheme of measuring positions in the pass-by test; M—Microphones, P—photocells [89].

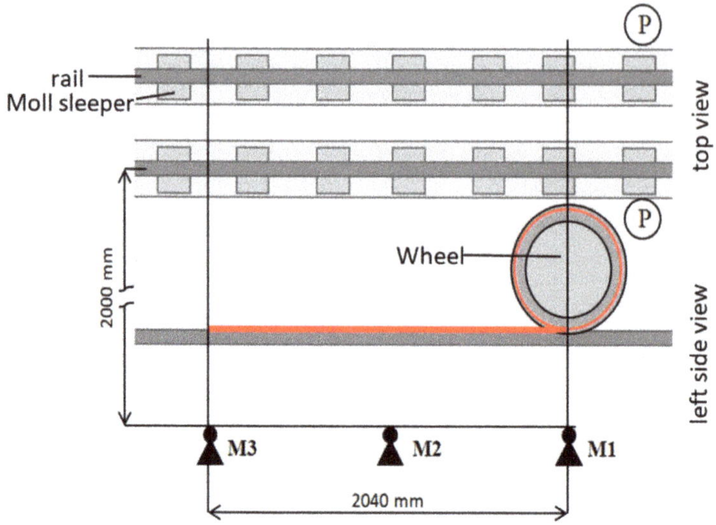

Figure 10. The scheme of measuring positions during acoustic pass-by tests; M1–M3—Microphones, P—Photocells [90].

In 2018, Metin Aktas et al. detected wheel flats based on acoustic emission (AE) technology and the parametric constraint optimization principle [91]. In the system, a single AE sensor was used and deployed on rails equipped with magnetic holders. As the train moves on the rail, the produced acoustic vibration can be measured continuously. A wheel flat detection algorithm based on parametric constraint optimization principles was proposed. This method compares the measured defect score curve with the predefined threshold curve to determine the wheel tread condition. Field tests were conducted and the results show that the system can effectively detect wheel flats at different train speeds with an accuracy rate of up to 90%. The challenge of this method is that the measured acoustic signal may contain the surrounding noise, which limits the method's accuracy.

4.2. Image-Based Method

With the development of computer vision technology, many image-based wheel flat detection systems have been designed [92,93]. In these systems, high-speed cameras were used to acquire a photo of the wheel tread when the train passes by. Then the wheel tread defects can be identified and localized by corresponding image processing algorithms.

In 2016, Hanieh Deilamsalehy et al. developed a computer-vision-based system for automatically detecting the sliding wheels and hot bearings from images taken by wayside thermal cameras [94]. From the acquired thermal images and sliding wheels, it can be seen that the sliding wheels possess a distinctive heat pattern at the wheel–track contact point. A method based on histograms of oriented gradients for identifying wheel flats and bearing parts was proposed. The feature descriptors were used by support vector machines to build fast classifiers with good detection rates. Simulated images of sliding wheels were used to train the algorithm. The monitoring method was tested with simulated images and a set of real thermal images taken on several trains of the Union Pacific Railroad (UPPR). 98% of the total number of defective wheels were detected without any false alarm. The model was improved by the author in the follow-up work, and the accuracy reached 100% [95].

In 2017, a system of collecting wheel tread defect images for online running trains was designed by Guangyu Guo et al. [96]. 16 high speed CCD cameras were used in the system to acquire the images of the entire wheel tread. By Support Vector Machine (SVM) classifier and Gaussian kernel, the wheel tread defect areas can be identified and located accurately. A wheel tread flat detection method was proposed to deal with wheel tread images captured by high-speed cameras. In this method, an SVM classifier was employed to recognize defective areas and a Gaussian kernel was used to locate defect areas. Three different types of descriptors were extracted to represent the defects, and experimental results show that compared with distribution vectors and area features, HOG features were considered to be better features for defect identification.

The image-based method uses vision cameras as flat sensors, which has the characteristic of non-contact measurement and fast response time. However, the accuracy of this method depends on image processing algorithms.

5. Summary

The advantages and disadvantages of the three signal acquisition methods are shown in Table 1. Currently, the sound-based method has been applied widely in non-destructive tests, but it is not a popular solution for wayside wheel flat detection [90]. The stress-based method is the most commonly used wheel flat detection method as its advantages of low cost, convenient installation, and high accuracy [89]. However, the traditional wheel flat detection system based on the stress method can only detect the wheel–rail contact area, which will cause more than 70% of the wheel tread area to be undetectable [96]. In addition, the size of the flat cannot be directly reflected by the acquired waveform, thus it is necessary to build models to acquire the relationship between flat size and the waveform. As models do not necessarily reflect the real field conditions, the quantitative measurement of wheel flats remains a challenge for the stress-based method. Generally, the image-based method is more expensive due to the use of lasers or cameras. In this method, the wheel flat size can be identified directly and quantitatively through image processing algorithms.

Table 1. Advantages and disadvantages of stress-based, sound-based, and image-based techniques.

Method	Advantages	Disadvantages
Stress-based method	Related technologies are more mature Simple installation and maintenance Low cost	It can only detect the condition of the wheel–rail contact area Quantitative measurement is difficult
Sound-based method	Acoustic emission method can realize repeated period measurement Low cost Easy to use	Quantitative measurement is difficult Relative technology application is less
Image-based method	Quantitatively measurable, Non-contact measurement, Long life	High cost

6. Conclusions

This paper discusses the research progress in wayside wheel flat detection since 2016 and compares the advantages and disadvantages of various wheel flat detection methods. During this period, the stress-based method is the most popular flat detection method. Strain gauges, accelerometers, and FBG are commonly used stress sensors to extract rail vibration signals. According to the Literature referenced in this paper, the application statistics of sensor types in stress-based wayside wheel flat detection systems in the past few years are shown in Figure 11. Different sensors have different characteristics, which can be flexibly selected according to the requirements. The strain gauge has the advantages of low price and high resolution, however it is non-linear and needs regular calibration. In addition, the measuring results are easily affected by external factors such as train speed, train weight, and electromagnetic interference. Therefore, more simulation studies are needed to evaluate the impact of train speed, train weight, and severity reflected in wayside measurement data. The FBG sensor has high precision and is immune to EMI induced noise. With wavelength multiplexing capability, multiple FBG sensors can be integrated, which is convenient for remote monitoring, deployment, and maintenance. The main challenges faced by FBG sensor applied in wheel flat detection include: ① Certain cross-sensitivity in strain and temperature response exists in FBG sensors; ② Currently, most of the systems based on FBG sensing have a measurement speed below 1 kHz. However, for high-speed measurement scenes such as wheel flat detection, the signal demodulation speed of FBG sensing-based schemes needs to be further improved. In addition, the study of new structure and packaging of FBG sensors has important significance. To address the issue of cross-sensitivity, researchers have put forth several temperature compensation models and algorithmic solutions [97]. In addition, it is also of great significance to study new structural designs and packaging options of FBG sensors to solve this problem. At present, the main factors restricting the detection speed of FBG signals are the modulation rate of the light source and modulator, and the response frequency of the detector. By improving the speed level of these devices, high-speed detection ability will be further improved. For FBG signal demodulation methods, spatial dispersion spectroscopy and time dispersion spectroscopy with higher sampling rates will be good choices. In addition, adopting the approach of collecting data first and processing them later can reduce the requirements for the signal analysis and processing capabilities. The research of effective installation methods and cost-effective high-performance demodulation systems for FBG sensors promote the application of FBG in the detection of wayside wheel flats. In addition to the above usual stress sensors, laser measurement technology has been introduced as a stress sensor which has brought many new ideas for wheel flat detection.

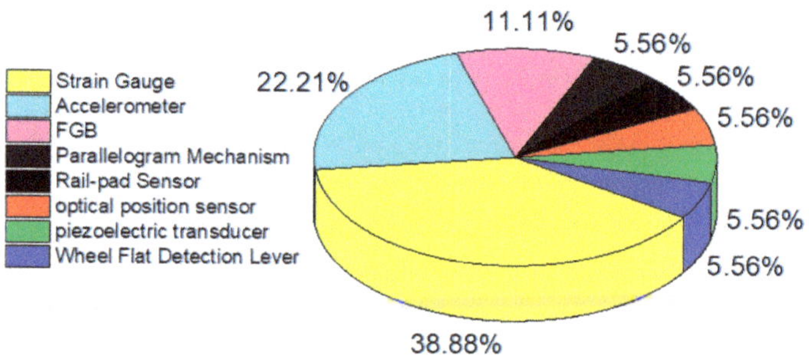

Figure 11. Usage statistics for different sensors.

Machine learning has been attracting more and more attention in the field of wheel flat detection. Machine learning algorithms, such as deep learning, can learn patterns from wheel flat signals, enabling them to realize self-feature extraction which can better capture flat signals. In addition, they are able to adapt to changing fault conditions, allowing them to perform more robustly when faced with new fault situations. Compared with the traditional threshold method, machine learning algorithms have higher accuracy but slightly lower precision [72]. This means that the threshold method has a low false alarm rate but is prone to miss detections, resulting in partially flat wheels being ignored after passing through the detection system. Compared with the traditional threshold method, machine learning algorithms have a high recall rate, indicating that they are less likely to miss detections of wheel flats. However, the application of machine learning also faces challenges. The performance of machine learning algorithms is mainly influenced by the model structure, the distribution of training data, and hyper parameters. A complex model structure and a large amount of training data greatly increase the cost of model training. In the real world, the amount of data for faulty wheels is severely imbalanced compared to that of normal wheels, with labeled data shortages decreasing the model's performance and robustness. Few-shot learning and unsupervised learning models have great potential in addressing these issues. Additionally, in order to ensure the generalization ability of machine learning algorithms and to make them applicable to in-service wheel flat detection on different railways, it is necessary to construct larger and more reasonable data sets. Appropriate model structures and well-distributed data sets can not only guarantee detection accuracy, but also ensure detection robustness by reducing overfitting. This may promote machine learning algorithms as the main development direction for wayside wheel flat detection.

In addition, by combining stress-based sensors, which are sensitive but have limited measurement range, and image-based sensors, which have a large measurement range but are easily affected by attached objects such as leaves and soil, multi-sensor fusion measurement can be achieved, which can further improve the sensitivity and accuracy of the measurement. Multi-sensor fusion can locate the specific position of wheel defects, but its results are affected by parameters such as the number of sensors and the length of the effective area [75–77]. By analyzing these parameters, multi-sensor fusion may have greater development potential in the field of wheel flat detection. The development direction of the wheel flat detection system may gradually tend to device simplification, multi-sensor fusion, algorithm accuracy and operation intelligence. In addition to single fault detection, the detection system is also gradually adopting multi-fault detection to be more efficiently. Apart from the wheel flat detection system, there are also wheel tread geometric parameter detection systems, wheel diameter detection systems, and so on [98,99]. Multi-system combined wheel status detection will become the future development trend.

Author Contributions: Writing—original draft preparation, W.F. and Q.H.; writing—review and editing, Q.H., J.L., F.Z., B.Z. and Q.F.; supervision, Q.F.; project administration, Q.F.; funding acquisition, Q.F. All authors have read and agreed to the published version of the manuscript.

Funding: This research was funded by the National Natural Science Foundation of China (no. 51935002).

Institutional Review Board Statement: Not applicable.

Informed Consent Statement: Not applicable.

Data Availability Statement: No new data were created or analyzed in this study. Data sharing is not applicable to this article.

Conflicts of Interest: The authors declare no conflict of interest.

References

1. Cilt China. Available online: http://www.ciltchina.cn/index.php?case=archive&act=show&aid=304 (accessed on 13 January 2023).
2. FRA Office of Safety Analysis. Available online: https://safetydata.fra.dot.gov/OfficeofSafety/publicsite/summary.aspx (accessed on 13 January 2023).
3. Ye, Y.G.; Shi, D.C.; Poveda-Reyes, S.; Hecht, M. Quantification of the influence of rolling stock failures on track deterioration. *J. Zhejiang Univ. Sci. A* **2020**, *21*, 938. [CrossRef]
4. Chong, S.Y.; Lee, J.R.; Shin, H.J. A review of health and operation monitoring technologies for trains. *Smart Struct. Syst.* **2010**, *6*, 1079–1105. [CrossRef]
5. Alexandrou, G.; Kouroussis, G.; Verlinden, O. A comprehensive prediction model for vehicle/track/soil dynamic response due to wheel flats. *Proc. Inst. Mech. Eng. F J. Rail Rapid Transit* **2016**, *230*, 1088–1104. [CrossRef]
6. Huang, L.W.; Li, Z.M.; Li, L.X.; An, Q. Methods to calculate accurate wheel/rail contact positions and static contact stress levels. *Proc. Inst. Mech. Eng. F J. Rail Rapid Transit* **2016**, *230*, 138–150. [CrossRef]
7. Kolonits, F. Analysis of the temperature of the rail/wheel contact surface using a half-space model and a moving heat source. *Proc. Inst. Mech. Eng. F J. Rail Rapid Transit* **2016**, *230*, 502–509. [CrossRef]
8. Manka, A.; Sitarz, M. Effects of a thermal load on the wheel/brake-block subsystem: The thermal conicity of railway wheels. *Proc. Inst. Mech. Eng. F J. Rail Rapid Transit* **2016**, *230*, 193–205. [CrossRef]
9. Jing, L.; Han, L.L. Further study on the wheel–rail impact response induced by a single wheel flat: The coupling effect of strain rate and thermal stress. *Veh. Syst. Dyn.* **2017**, *55*, 1946–1972. [CrossRef]
10. Ye, Y.G.; Shi, D.C.; Krause, P.; Tian, Q.Y.; Hecht, M. Wheel flat can cause or exacerbate wheel polygonization. *Veh. Syst. Dyn.* **2019**, *58*, 1575–1604. [CrossRef]
11. Xu, J.M.; Ma, Q.T.; Zhao, S.Q.; Chen, J.Y.; Qian, Y.; Chen, R.; Wang, P. Effect of wheel flat on dynamic wheel-rail impact in railway turnouts. *Veh. Syst. Dyn.* **2020**, *60*, 1829–1848. [CrossRef]
12. Wu, X.W.; Rakheja, S.; Ahmed, A.K.W.; Chi, M.R. Influence of a flexible wheelset on the dynamic responses of a high-speed railway car due to a wheel flat. *Proc. Inst. Mech. Eng. Part F J. Rail Rapid Transit* **2018**, *232*, 1033–1048. [CrossRef]
13. Han, L.L.; Jing, L.; Zhao, L.M. Finite element analysis of the wheel–rail impact behavior induced by a wheel flat for high-speed trains: The influence of strain rate. *Proc. Inst. Mech. Eng. Part F J. Rail Rapid Transit* **2018**, *232*, 990–1004. [CrossRef]

14. Bernal, E.; Spiryagin, M.; Cole, C. Wheel flat detectability for Y25 railway freight wagon using vehicle component acceleration signals. *Veh. Syst. Dyn.* **2019**, *58*, 1893–1913. [CrossRef]
15. Jergeus, J. Martensite formation and residual stresses around railway wheel flats. *Proc. Inst. Mech. Eng. C J. Mech. Eng. Sci.* **1998**, *212*, 69–79. [CrossRef]
16. Jergeus, J.; Odenmarck, C.; Lunden, R.; Sotkovszki, P.; Karlsson, B.; Gullers, P. Full-scale railway wheel flat experiments. *Proc. Inst. Mech. Eng. Part F J. Rail Rapid Transit* **1999**, *213*, 1–13. [CrossRef]
17. Wu, B.; An, B.Y.; Wen, Z.F.; Wang, W.J.; Wu, T. Wheel–rail low adhesion issues and its effect on wheel–rail material damage at high speed under different interfacial contaminations. *Proc. Inst. Mech. Eng. Part C J. Mech. Eng. Sci* **2019**, *233*, 5477–5490. [CrossRef]
18. Ling, L.; Cao, Y.B.; Xiao, X.B.; WEN, Z.F.; JIN, X.S. Effect of wheel flats on the high-speed wheel-rail contact behavior. *China Railw. Soc.* **2015**, *37*, 32–39.
19. Kumar, V.; Rastogi, V.; Pathak, P.M. Dynamic analysis of vehicle-track interaction due to wheel flat using bond graph. *Proc. Inst. Mech. Eng. Part K J. Multi-Body Dyn.* **2018**, *232*, 398–412. [CrossRef]
20. Brizuela, J.; Fritsch, C.; Ibanez, A. Railway wheel-flat detection and measurement by ultrasound. *Transp. Res. Part C Emerg. Technol.* **2011**, *19*, 975–984. [CrossRef]
21. Matsumoto, A.; Sato, Y.; Ohno, H.; Tomeoka, M.; Matsumoto, K.; Kurihara, J.; Ogino, T.; Tanimoto, M.; Kishimoto, Y.; Sato, Y.; et al. A new measuring method of wheel–rail contact forces and related considerations. In Proceedings of the 7th International Conference on Contact Mechanics and Wear of Rail/Wheel Systems, Brisbane, Australia, 30 September 2006; pp. 1518–1525. [CrossRef]
22. Matsumoto, A.; Sato, Y.; Ohno, H.; Shimizu, M.; Kurihara, J.; Saitou, T.; Michitsuji, Y.; Matsui, R.; Tanimoto, M.; Mizuno, M. Actual states of wheel/rail contact forces and friction on sharp curves—Continuous monitoring from in-service trains and numerical simulations. In Proceedings of the 9th Conference on Contact Mechanics and Wear of Rail/Wheel Systems, Chengdu, China, 15 August 2012; pp. 189–197. [CrossRef]
23. Lee, M.L.; Chiu, W.K. Determination of railway vertical wheel impact magnitudes: Field trials. *Struct. Health Monit.* **2007**, *6*, 49–65. [CrossRef]
24. Meixedo, A.; Goncalves, A.; Calcada, R.; Gabriel, J.; Fonscca, H.; Martins, R. Weighing in motion and wheel defect detection of rolling stock. In Proceedings of the 2015 3rd Experiment International Conference (exp.at'15), Ponta Delgada, Portugal, 2–4 June 2015; pp. 86–90. [CrossRef]
25. Kanehara, H.; Fujioka, T. Measuring rail/wheel contact points of running railway vehicles. In Proceedings of the 5th International Conference on Contact Mechanics and Wear of Rail/Wheel Systems, Tokyo, Japan, 25–28 July 2000; pp. 275–283. [CrossRef]
26. Mosleh, A.; Costa, P.; Calçada, R. Development of a Low-Cost Trackside System for Weighing in Motion and Wheel Defects Detection. *Int. J. Railw. Res.* **2020**, *7*, 1–9.
27. Shaikh, M.Z.; Ahmed, Z.; Chowdhry, B.S.; Baro, E.N.; Hussain, T.; Uqaili, M.A.; Mehran, S.; Kumar, D.; Shah, A.A. State-of-the-art wayside condition monitoring systems for railway wheels: A comprehensive review. *IEEE Access* **2023**, *11*, 13257–13279. [CrossRef]
28. Sanchis, R.; Cardona, S.; Martinez, J. Determination of the vertical vibration of a ballasted railway track to be used in the experimental detection of wheel flats in metropolitan railways. *J. Vib. Acoust.* **2019**, *141*, 021015. [CrossRef]
29. Dwyer-Joyce, R.S.; Yao, C.; Lewis, R.; Brunskill, H. An ultrasonic sensor for monitoring wheel flange/rail gauge corner contact. *Proc. Inst. Mech. Eng. Part F J. Rail Rapid Transit* **2013**, *227*, 188–195. [CrossRef]
30. Frankenstein, B.; Hentschel, D.; Pridoehl, E.; Schubert, F.; Meyendorf, N.; Baaklini, G.Y.; Michel, B. Hollow shaft integrated health monitoring system for railroad wheels. In Proceedings of the Conference on Advanced Sensor Technologies for Nondestructive Evaluation and Structural Health Monitoring, San Diego, CA, USA, 8–10 March 2005; pp. 46–55. [CrossRef]
31. Liang, B.; Iwnicki, S.D.; Zhao, Y.; Crosbee, D. Railway wheel-flat and rail surface defect modelling and analysis by time–frequency techniques. *Veh. Syst. Dyn.* **2013**, *51*, 1403–1421. [CrossRef]
32. Liang, B.; Iwnicki, S.; Ball, A.; Young, A.E. Adaptive noise cancelling and time–frequency techniques for rail surface defect detection. *Mech. Syst. Signal Process* **2015**, *54–55*, 41–51. [CrossRef]
33. Chen, Y.X.; Zhao, Z.X.; Kim, E.; Liu, H.Y.; Xu, J.; Min, H.; Cui, Y. Wheel fault diagnosis model based on multichannel attention and supervised contrastive learning. *Adv. Mech. Eng.* **2021**, *13*, 16878140211067024. [CrossRef]
34. Ye, Y.G.; Huang, C.H.; Zeng, J.; Zhou, Y.C.; Li, F.S. Shock detection of rotating machinery based on activated time-domain images and deep learning: An application to railway wheel flat detection. *Mech. Syst. Signal Process.* **2023**, *186*, 109856. [CrossRef]
35. Bernal, E.; Spiryagin, M.; Cole, C. Wheel flat analogue fault detector verification study under dynamic testing conditions using a scaled bogie test rig. *Int. J. Rail Transp.* **2022**, *10*, 177–194. [CrossRef]
36. Amini, A.; Entezami, M.; Huang, Z.; Rowshandel, H.; Papaelias, M. Wayside detection of faults in railway axle bearings using time spectral kurtosis analysis on high-frequency acoustic emission signals. *Adv. Mech. Eng.* **2016**, *8*, 1–9. [CrossRef]
37. Partington, W. Wheel impact load monitoring. *Proc. Inst. Civ. Eng. Transp.* **1993**, *100*, 243–245. [CrossRef]
38. Stratman, B.; Liu, Y.; Mahadevan, S. Structural health monitoring of railroad wheels using wheel impact load detectors. *J. Fail. Anal. Prev.* **2007**, *7*, 218–225. [CrossRef]
39. Palo, M.; Schunnesson, H.; Kumar, U.; Larsson-Kråik, P.-O.; Galar, D. Rolling stock condition monitoring using wheel/rail forces. *Insight Non-Destr. Test. Cond. Monit.* **2012**, *54*, 451–455. [CrossRef]

40. Palo, M.; Galar, D.; Nordmark, T.; Asplund, M.; Larsson, D. Condition monitoring at the wheel/rail interface for decision-making support. *Proc. Inst. Mech. Eng. Part F J. Rail Rapid Transit* **2014**, *228*, 705–715. [CrossRef]
41. Lee, K.Y.; Lee, K.K.; Ho, S.L. Exploration of using FBG sensor for derailment detector. *WSEAS Trans. Top. Syst.* **2004**, 3, 2433–2439.
42. Wei, C.L.; Lai, C.C.; Liu, S.Y.; Chung, W.H.; Ho, T.K.; Tam, H.Y.; Ho, S.L.; McCusker, A.; Kam, J.; Lee, K.Y. A Fiber Bragg Grating sensor system for train axle counting. *IEEE Sens. J.* **2010**, *10*, 1905–1912. [CrossRef]
43. Alemi, A.; Corman, F.; Lodewijks, G. Condition monitoring approaches for the detection of railway wheel defects. *Proc. Inst. Mech. Eng. F J. Rail Rapid Transit* **2017**, *231*, 961–981. [CrossRef]
44. Papaelias, M.; Amini, A.; Huang, Z.; Vallely, P.; Dias, D.C.; Kerkyras, S. Online condition monitoring of rolling stock wheels and axle bearings. *Proc. Inst. Mech. Eng. Part F J. Rail Rapid Transit* **2014**, *230*, 709–723. [CrossRef]
45. Wei, C.L.; Xin, Q.; Chung, W.H.; Liu, S.Y.; Tam, H.Y.; Ho, S.L. Real-time train wheel condition monitoring by Fiber Bragg Grating sensors. *Int. J. Distrib. Sens. Netw.* **2012**, *22*, 409048. [CrossRef]
46. Filograno, M.L.; Guillén, P.C.; Rodriguez-Barrios, A.; Martín-López, S.; Rodríguez-Plaza, M.; Andrés-Alguacil, Á.; González-Herráez, M. Real-time monitoring of railway traffic using Fiber Bragg Grating sensors. *IEEE Sens. J.* **2012**, *12*, 85–92. [CrossRef]
47. Li, C.S.; Luo, S.H.; Cole, C.; Spiryagin, M. An overview: Modern techniques for railway vehicle on-board health monitoring systems. *Veh. Syst. Dyn.* **2017**, *55*, 1045–1070. [CrossRef]
48. Gallikova, J.; Poprocky, P.; Volna, P. Implementation of FMEA method in maintenance of semi-trailer combination. *Diagnostvka* **2016**, *17*, 85–92.
49. Kasiar, L.; Zvolensky, P.; Barta, D.; Bavlna, L.; Mikolajčík, M.; Droździel, P. Diagnostics of electric motor of locomotive series 757. *Diagnostwka* **2016**, *17*, 95–101.
50. McGonigal, S. Defect Detectors Looking for Trouble in All the Right Places. Available online: https://trn.trains.com/railroads/abcs-of-railroading/2006/05/defect-detectors (accessed on 13 January 2023).
51. Liu, X.Z.; Xu, C.; Ni, Y.Q. Wayside detection of wheel minor defects in high-speed trains by a Bayesian blind source separation method. *Sensors* **2019**, *19*, 3981. [CrossRef]
52. Cui, K.; Qin, X.T. Numerical computation of wheel-rail impact noises with considering wheel flats based on the boundary element method. *J. Vibroeng.* **2016**, *18*, 3930–3940. [CrossRef]
53. Mishra, S.; Sharan, P.; Saara, K. Real time implementation of fiber Bragg grating sensor in monitoring flat wheel detection for railways. *Eng. Fail. Anal.* **2022**, *138*, 106376. [CrossRef]
54. Shi, D.C.; Ye, Y.G.; Gillwald, M.; Hecht, M. Designing a lightweight 1D convolutional neural network with Bayesian optimization for wheel flat detection using carbody accelerations. *Int. J. Rail Transp.* **2021**, *9*, 311–341. [CrossRef]
55. Goncalves, V.; Mosleh, A.; Vale, C.; Montenegro, P.A. Wheel Out-of-Roundness Detection Using an Envelope Spectrum Analysis. *Sensors* **2023**, *23*, 2138. [CrossRef]
56. Peng, X.Y.; Zeng, J.; Wang, J.B.; Wang, Q.S.; Li, D.D.; Liang, S.K. Wayside wheel-rail vertical contact force continuous detecting method and its application. *Measurement* **2022**, *193*, 110975. [CrossRef]
57. Erdozain, I.; Alonso, A.; Blanco, B. Wheel-track interaction in the presence of flats: Dynamic modelling and experimental correlation. In Proceedings of the International Conference on Noise and Vibration Engineering (ISMA)/International Conference on Uncertainty in Structural Dynamics (USD), Leuven, Belgium, 7–9 September 2020; pp. 2575–2584.
58. Mohammadi, M.; Mosleh, A.; Vale, C.; Ribeiro, D.; Montenegro, P.; Meixedo, A. An Unsupervised Learning Approach for Wayside Train Wheel Flat Detection. *Sensors* **2023**, *23*, 1910. [CrossRef]
59. Yang, K.; Peng, J.P.; Gao, X.R.; Peng, C.Y.; Zhang, Y.; Wang, Z.Y.; Zhao, Q.K.; Dai, L.X. Research on the detecting method of CRH wheel flat at low speed. In Proceedings of the 2016 18th International Wheelset Congress (IWC), Chengdu, China, 7–10 November 2016; pp. 126–129. [CrossRef]
60. Salehi, M.; Bagherzadeh, S.A.; Fakhari, M. Experimental detection of train wheel defects using wayside vibration signal processing. *Struct. Health Monit.* **2023**, 14759217221149614. [CrossRef]
61. Guedes, A.; Silva, R.; Ribeiro, D.; Vale, C.; Mosleh, A.; Montenegro, P.; Meixedo, A. Detection of wheel polygonization based on wayside monitoring and artificial intelligence. *Sensors* **2023**, *23*, 2188. [CrossRef]
62. Bureika, G.; Levinzon, M.; Dailydka, S.; Steisunas, S.; Zygiene, R. Evaluation criteria of wheel/rail interaction measurement results by trackside control equipment. *Int. J. Heavy Veh. Syst.* **2019**, *26*, 747–764. [CrossRef]
63. Kilinc, O.; Vágner, J. Vibration based diagnosis of wheel defects of metro train sets using one period analysis on the wayside. *Vibroeng. Procedia* **2017**, *11*, 13–18. [CrossRef]
64. Rivero, A.; Vanheeghe, P.; Grabe, H.; Duflos, E. Rolling stock monitoring with wayside train-monitoring systems using fiber optic cords. *J. Infrastruct. Syst.* **2021**, *27*, 04021040. [CrossRef]
65. Rombach, K.; Michau, G.; Ratnasabapathy, K.; Ancu, L.; Burzle, W.; Koller, S.; Fink, O. Contrastive feature learning for fault detection and diagnostics in railway applications. *arXiv* **2022**, arXiv:2208.13288. [CrossRef]
66. Asplund, M.; Palo, M.; Famurewa, S.; Rantatalo, M. A study of railway wheel profile parameters used as indicators of an increased risk of wheel defects. *Proc. Inst. Mech. Eng. Part F J. Rail Rapid Transit* **2016**, *230*, 323–334. [CrossRef]
67. Baeza, L.; Roda, A.; Carballeira, J.; Giner, E. Railway train-track dynamics for wheel flats with improved contact models. *Nonlinear Dyn.* **2006**, *45*, 385–397. [CrossRef]
68. Nielsen, J.C.O.; Johansson, A. Out-of-round railway wheels-a literature survey. *Proc. Inst. Mech. Eng. Part F J. Rail Rapid Transit* **2000**, *214*, 79–91. [CrossRef]

69. International Union of Railways. *Atlas of Wheel and Rail Defects*; International Union of Railways: Paris, France, 2004; pp. 36–37.
70. Nowakowski, T.; Komorski, P.; Szymanski, G.M.; Tomaszewski, F. Wheel-flat detection on trams using envelope analysis with Hilbert transform. *Lat. Am. J. Solids Struct.* **2019**, *16*, e148. [CrossRef]
71. Liu, X.Z.; Ni, Y.Q. Wheel tread defect detection for high-speed trains using FBG-based online monitoring techniques. *Smart Struct. Syst.* **2018**, *21*, 687–694. [CrossRef]
72. Krummenacher, G.; Ong, C.S.; Koller, S.; Kobayashi, S.; Buhmann, J.M. Wheel defect detection with machine learning. *IEEE Trans. Intell. Transp.* **2018**, *19*, 1176–1187. [CrossRef]
73. Gao, R.; He, Q.X.; Feng, Q.B. Railway wheel flat detection system based on a parallelogram mechanism. *Sensors* **2019**, *19*, 3614. [CrossRef] [PubMed]
74. Cao, W.J.; Zhang, S.L.; Bertola, N.J.; Smith, I.F.C.; Koh, C.G. Time series data interpretation for 'wheel-flat' identification including uncertainties. *Struct. Health Monit.* **2019**, *22*, 3–18. [CrossRef]
75. Alemi, A.; Corman, F.; Pang, Y.S.; Lodewijks, G. Reconstruction of an informative railway wheel defect signal from wheel–rail contact signals measured by multiple wayside sensors. *Proc. Inst. Mech. Eng. Part F J. Rail Rapid Transit* **2019**, *233*, 49–62. [CrossRef]
76. Alemi, A.; Corman, F.; Pang, Y.S.; Lodewijks, G. Evaluation of the influential parameters contributing to the reconstruction of railway wheel defect signals. *Proc. Inst. Mech. Eng. Part F J. Rail Rapid Transit* **2020**, *234*, 1005–1016. [CrossRef]
77. Zhou, C.Y.; Gao, L.; Xiao, H.; Hou, B.W. Railway wheel flat recognition and precise positioning method based on multisensor arrays. *Appl. Sci.* **2020**, *10*, 1297. [CrossRef]
78. Gao, R.; He, Q.X.; Feng, Q.B.; Cui, J.Y. In-service detection and quantification of railway wheel flat by the reflective optical position sensor. *Sensors* **2020**, *20*, 4969. [CrossRef] [PubMed]
79. Barman, J.; Hazarika, D. Linear and quadratic time-frequency analysis of vibration for fault detection and identification of NFR trains. *IEEE Trans. Instrum. Meas.* **2020**, *69*, 8902–8909. [CrossRef]
80. Mosleh, A.; Montenegro, P.; Costa, P.A.; Calcada, R. An approach for wheel flat detection of railway train wheels using envelope spectrum analysis. *Struct. Infrastruct. Eng.* **2021**, *17*, 1710–1729. [CrossRef]
81. Mosleh, A.; Montenegro, P.A.; Costa, P.A.; Calcada, R. Railway vehicle wheel flat detection with multiple records using spectral kurtosis analysis. *Appl. Sci.* **2021**, *11*, 4002. [CrossRef]
82. Ni, Y.Q.; Zhang, Q.H. A Bayesian machine learning approach for online detection of railway wheel defects using track-side monitoring. *Struct. Health. Monit.* **2021**, *20*, 1536–1550. [CrossRef]
83. Mu, J.; Zeng, J.; Wang, Q.S.; Sang, H.T. Determination of mapping relation between wheel flat and wheel/rail contact force for railway freight wagon using dynamic simulation. *Proc. Inst. Mech. Eng. Part F J. Rail Rapid Transit* **2022**, *236*, 545–556. [CrossRef]
84. Mosleh, A.; Meixedo, A.; Ribeiro, D.; Montenegro, P.; Calcada, R. Early wheel flat detection: An automatic data-driven wavelet-based approach for railways. *Veh. Syst. Dyn.* **2022**, 1–30. [CrossRef]
85. Sun, Z.L.; Lu, J.G. An ultrasonic signal denoising method for EMU wheel trackside fault diagnosis system based on improved threshold function. *IEEE Access* **2021**, *9*, 96244–96256. [CrossRef]
86. Kukenas, V.; Jasevicius, R.; Petrenko, V.; Kilikevicius, A.; Vainorius, D. Analysis of Sound Power Level Changes Caused by Wheel with Flat Spot Using Numerical and Physical Experiment. *Appl. Sci.* **2021**, *11*, 2141. [CrossRef]
87. Lv, S.; Zhou, F.Q.; Wei, Z.Z. Train wheel tread defects detection based on image registration. In Proceedings of the 2017 IEEE International Conference on Imaging Systems and Techniques (IST), Beijing, China, 18–20 October 2017. [CrossRef]
88. Onur, K.; Jakub, V. Condition monitoring of metro wheelsets using wayside vibration and acoustic sensors. *SSRN* **2022**, *1556–5068*, 20220296971.
89. Komorski, P.; Nowakowski, T.; Szymanski, G.M.; Tomaszewski, F. Application of Time-Frequency Analysis of Acoustic Signal to Detecting Flat Places on the Rolling Surface of a Tram Wheel. In Proceedings of the 14th International Conference on Dynamical Systems—Theory and Applications (DSTA)—Recent Developments in Mathematical Modelling of Dynamical Systems, Lodz, Poland, 11–14 December 2017; pp. 205–215. [CrossRef]
90. Komorski, P.; Szymanski, G.M.; Nowakowski, T.; Orczyk, M. Advanced acoustic signal analysis used for wheel-flat detection. *Lat. Am. J. Solids Struct.* **2021**, *18*, e338. [CrossRef]
91. Aktas, M.; Gunel, E.H.; Yilmazer, P.; Akgun, T. Detection of wheel flatten defect on the moving train with acoustic emission sensor. In Proceedings of the 29th IEEE Annual International Symposium on Personal, Indoor and Mobile Radio Communications (PIMRC), Bologna, Italy, 9–12 September 2018; pp. 386–390.
92. Liang, R.Y.; Ding, Y.Q.; Zhang, X.W.; Chen, J.S.; Guo, M.Z.; Zhao, L.; Wang, L.P. Copper strip surface defects inspection based on SVM-RBF. In Proceedings of the 4th International Conference on Natural Computation (ICNC 2008), Jinan, China, 18–20 October 2008; pp. 41–45. [CrossRef]
93. Zhang, X.G.; Xu, J.J.; Ge, G.Y. Defects recognition on X-ray images for weld inspection using SVM. In Proceedings of the International Conference on Machine Learning and Cybernetics, Shanghai, China, 26–29 August 2004; pp. 3721–3725.
94. Deilamsalehy, H.; Havens, T.C.; Lautala, P. Detection of sliding wheels and hot bearings using wayside thermal cameras. In Proceedings of the ASME 2016 Joint Rail Conference, Columbia, SC, USA, 12–15 April 2016; pp. 1–7.
95. Deilamsalehy, H.; Havens, T.C.; Lautala, P.; Medici, E.; Davis, J. An automatic method for detecting sliding railway wheels and hot bearings using thermal imagery. *Proc. Inst. Mech. Eng. Part F J. Rail Rapid Transit* **2017**, *231*, 690–700. [CrossRef]

96. Guo, G.Y.; Peng, J.P.; Yang, K.; Xie, L.M.; Song, W.W. Wheel tread defects inspection based on SVM. In Proceedings of the 2017 Far East NDT New Technology & Application Forum (FENDT), Xi'an, China, 22–24 June 2017; pp. 251–253.
97. Li, T.L.; Guo, J.X.; Tan, Y.G.; Zhou, Z.D. Recent Advances and Tendency in Fiber Bragg Grating-based Vibration Sensor: A Review. *IEEE Sens. J.* **2020**, *20*, 12074–12087. [CrossRef]
98. Ran, Y.F.; He, Q.X.; Feng, Q.B.; Cui, J.Y. High-Accuracy On-Site Measurement of Wheel Tread Geometric Parameters by Line-Structured Light Vision Sensor. *IEEE Access* **2021**, *9*, 52590–52600. [CrossRef]
99. Zheng, F.J.; Zhang, B.; Gao, R.; Feng, Q.B. A high-precision method for dynamically measuring train wheel diameter using three laser displacement transducers. *Sensors* **2019**, *19*, 4148. [CrossRef] [PubMed]

sensors

MDPI

Review

Advances of Research on Dual-Frequency Solid-State Lasers for Synthetic-Wave Absolute-Distance Interferometry

Mingxing Jiao *, Fei Jiang, Junhong Xing, Yun Liu, Tianhong Lian, Jianning Liu and Guangtao Li

School of Mechanical and Precision Instrumental Engineering, Xi'an University of Technology, Xi'an 710048, China
* Correspondence: jiaomx@xaut.edu.cn; Tel.: +13991960662

Abstract: Frequency-difference-stabilized dual-frequency solid-state lasers with tunable and large frequency difference have become an ideal light source for the high-accuracy absolute-distance interferometric system due to their stable multistage synthetic wavelengths. In this work, the advances in research on oscillation principles and key technologies of the different kinds of dual-frequency solid-state lasers are reviewed, including birefringent dual-frequency solid-state lasers, biaxial and two-cavity dual-frequency solid-state lasers. The system composition, operating principle, and some main experimental results are briefly introduced. Several typical frequency-difference stabilizing systems for dual-frequency solid-state lasers are introduced and analyzed. The main development trends of research on dual-frequency solid-state lasers are predicted.

Keywords: dual-frequency solid-state laser; frequency-difference tuning; frequency-difference stabilization; quadrature-demodulated Pound–Drever–Hall method; synthetic-wave absolute-distance interferometry

Citation: Jiao, M.; Jiang, F.; Xing, J.; Liu, Y.; Lian, T.; Liu, J.; Li, G. Advances of Research on Dual-Frequency Solid-State Lasers for Synthetic-Wave Absolute-Distance Interferometry. *Sensors* **2023**, *23*, 3206. https://doi.org/10.3390/s23063206

Academic Editors: Qibo Feng, Jiakun Li and Qixin He

Received: 9 February 2023
Revised: 10 March 2023
Accepted: 13 March 2023
Published: 17 March 2023

1. Introduction

The synthetic-wave absolute-distance interferometric measurement is a kind of high-accuracy no-guideway ranging technology based on the fraction-coincident method of the interference fringes, which provides an effective way to solve the technical problems of precision measurement and ultra-precision measurement of large-dimension workpieces. At present, the dual-frequency lasers [1,2] and optical-frequency combs [3–6] are normally used as the light sources for synthetic-wave absolute-distance interferometric systems, in which dual-frequency lasers have attracted great attention and strong research interests of scholars worldwide due to the advantages of simple structure and low cost. As the typical representatives of dual-frequency lasers, the Zeeman dual-frequency He-Ne laser at 632.8 nm [7], the two-longitudinal-mode He-Ne laser at 632.8 nm [8], and the birefringent dual-frequency He-Ne laser at 632.8 nm [9] have been successfully applied to the synthetic-wave absolute-distance interferometric system. However, the frequency differences of these dual-frequency lasers are generally <1 GHz due to the fact that the fluorescence linewidth of Ne atoms is relatively narrow (about 1500 MHz), and the corresponding synthetic wavelength is >300 mm; accordingly, it is difficult to improve the accuracy of the absolute-distance measurement. It is found that the fluorescence linewidth of solid-state crystals is much wider than that of the He-Ne gas medium, and a dual-frequency solid-state laser with a frequency difference larger than several tens of GHz or up to several THz can form much smaller synthetic wavelengths, so that the accuracy of the absolute-distance measurement can be significantly improved. In addition, the integer order of the synthetic-wave-interference fringes can be uniquely determined, provided that the preliminary measurement error of the measured distance is less than one fourth of the synthetic wavelength; additionally, when the frequency difference of the dual-frequency laser decreases, the corresponding synthetic wavelength becomes longer, and, accordingly, the preliminary measurement of the measured distance can be performed easily. Evidently different multistage synthetic wavelengths can be generated using a dual-frequency solid-state laser with tunable and large frequency difference as a light source. A step-by-step refined measurement method can be used to realize a high-precision absolute-distance measurement, and the final accuracy of the absolute-distance measurement mainly depends on the magnitude of the minimum synthetic wavelength and its stability, that is, it is determined by the maximum frequency difference of the dual-frequency solid-state laser source and its stability. In order to achieve the absolute-distance measurement accuracy to an order greater than 10^{-6}, the frequency-difference stability of the dual-frequency solid-state laser must be an order of 10^{-7} or better. Therefore, the frequency-difference-stabilized dual-frequency solid-state laser with tunable and large frequency difference can be used as an ideal light source for the synthetic-wave absolute-distance interferometric system [10–13].

Many experts and scholars worldwide have investigated the oscillating principles and key technologies of dual-frequency solid-state lasers with tunable and large frequency difference since 2000 [14–19], and they have successfully developed a series of new dual-frequency solid-state lasers, such as the birefringent dual-frequency solid-state laser [20–30], the biaxial dual-frequency solid-state laser [31,32], the two-cavity dual-frequency Nd:YAG laser (TCDFL) [33–40], and so on. Especially in the recent years, our research group has made important progresses in the research and development of the TCDFL and its quadrature-demodulated Pound–Drever–Hall (QD-PDH) frequency-stabilizing technology [41,42], and the QD-PDH frequency-stabilizing method has been applied to the frequency-difference stabilization of the TCDFL with a frequency difference of 24 GHz at 1064 nm [43,44]. This frequency-difference-stabilized TCDFL at 1064 nm has been successfully used as the light source for a synthetic-wave absolute-distance interferometric system [45].

2. Simultaneous Oscillation and Frequency-Difference Tuning of Orthogonally and Linearly Polarized Dual-Frequency Solid-State Laser

It is known that a dual-frequency laser output can be obtained by inserting a longitudinal-mode-splitting element into the laser cavity for the single longitudinal mode to be split into two orthogonally polarized components, provided that the laser meets the requirement of single-longitudinal-mode operation. At present, the commonly used methods of the laser single-longitudinal-mode selection for solid-state lasers mainly include the birefringent filter method, Fabry–Perot (F-P) etalon method, short-cavity method, twisted-mode-cavity method, ring traveling-wave-cavity method, etc. Additionally, the laser longitudinal-mode-splitting methods for solid-state lasers mainly include the birefringence splitting method and polarization splitting method. Due to the significant advantages of dual-frequency solid-state lasers, several different schemes of dual-frequency solid-state lasers have been designed and experimentally investigated by experts worldwide, which are based on the principles of single-longitudinal-mode selection by birefringent filter or intracavity F-P etalon, and the orthogonally and linearly polarized dual-frequency laser with tunable frequency difference has been simultaneously oscillated and output.

2.1. *Birefringent Dual-Frequency Solid-State Lasers*

2.1.1. Natural Birefringent Dual-Frequency Solid-State Lasers

Dual-frequency solid-state microchip lasers have the advantages of large frequency difference, high pumping efficiency, high beam quality, and narrow laser linewidth. To obtain a dual-frequency laser output with a tunable and large frequency difference, in 2009, the research group of Profs. A. Mckay and J. M. Dawes of Macquarie University reported a diode-pumped dual-frequency microchip Nd:YAG ceramic laser at 1064 nm [20], as shown in Figure 1. This laser consisted of a 0.25 mm-long highly doped (4%) ceramic Nd:YAG plate and two zeroth-order quarter-wave plates, which were optically bonded to BK7 glass substrates with planar 1064 nm resonator mirror coatings on each waveplate. The ceramic plate was glued to the input-waveplate mirror, and the output-waveplate mirror was positioned close to the ceramic. The overall cavity length was estimated to be 300 μm, so the laser at 1064 nm was forced to oscillate in single-longitudinal mode. Due to the natural birefringence effect, the single longitudinal mode was split and the orthogonally and linearly polarized dual-frequency laser at 1064 nm was oscillated and output. Figure 2 shows the experimentally observed typical spectrums of the two simultaneously lasing narrow-linewidth-longitudinal modes with approximately equal optical power, in which Figure 2a–c are the oscillating modes of the dual-frequency laser with different frequency difference, respectively. At a frequency difference > 120 GHz [see Figure 2c], the optical power in both polarizations decreased due to the limited gain near the edges of the spectral-gain linewidth, which ultimately limited the tuning range of the frequency difference. By tuning the relative angles between the principal axes of the two quarter-wave plates, the frequency difference was tuned linearly from a few gigahertz to over 150 GHz, as shown in Figure 3.

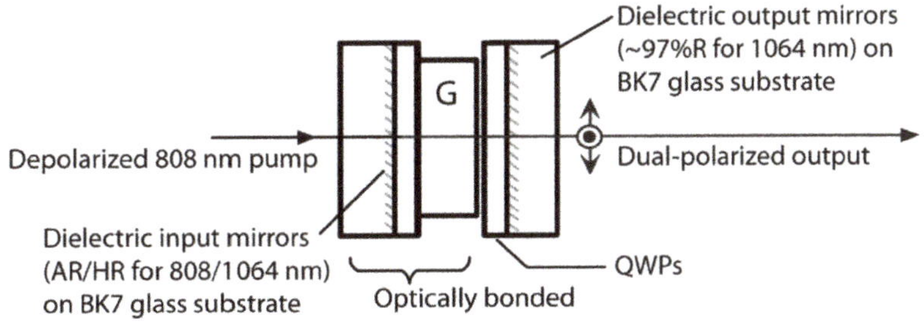

Figure 1. Schematic diagram of dual-frequency ceramic microchip Nd:YAG laser [20]. QWP: quarter-wave plate; G: glass.

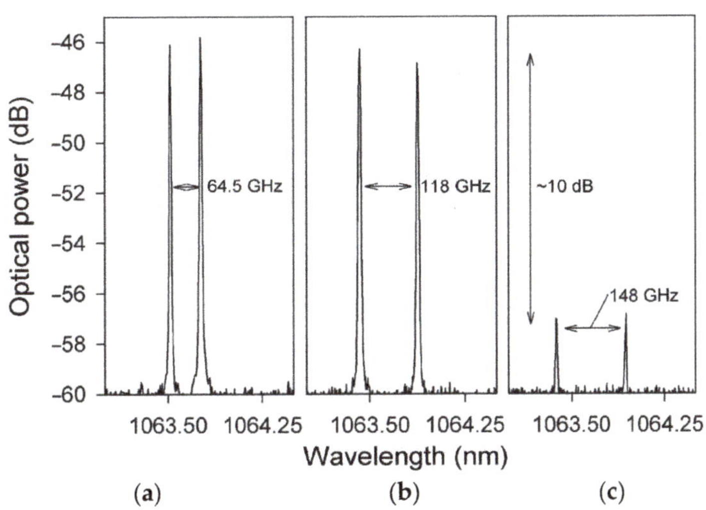

Figure 2. Oscillating spectrums of dual-frequency laser with difference frequency differences of (**a**) 64.5 GHz, (**b**) 118 GHz, and (**c**) 148 GHz [20].

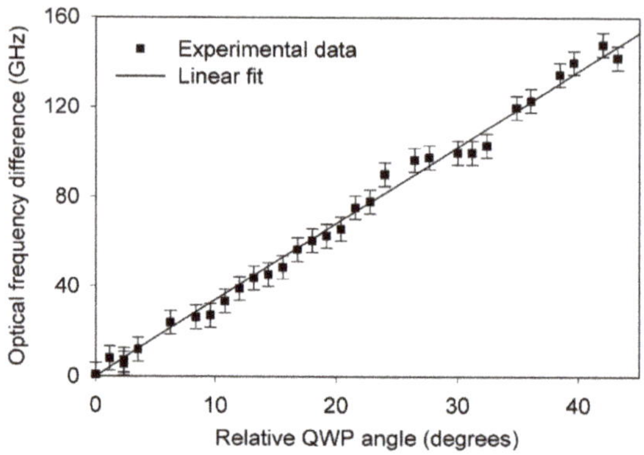

Figure 3. Tuning of the optical frequency separation between orthogonally polarized modes as a function of relative quarter-wave-plate angles [20].

In 2007, the research group of Prof. Zhao Changming of the Beijing Institute of Technology reported a diode-pumped tunable dual-frequency Nd:YAG laser at 1064 nm [21], as shown in Figure 4, in which an intracavity-fused quartz-made F-P etalon with a thickness of 0.5 mm was used as the laser longitudinal-mode selector, and two zeroth-order quarter-wave plates were vertically inserted into the laser cavity. As a result, an orthogonally polarized dual-frequency laser at 1064 nm was obtained, and the frequency difference of the dual-frequency laser was tuned by adjusting the relative angles between the principal axes of the two quarter-wave plates. The experimentally observed oscillating mode spectrum of the dual-frequency laser at 1064 nm was obtained by the use of a confocal scanning F-P interferometer, as shown in Figure 5, in which it can be seen that both longitudinal modes oscillated simultaneously in a single-longitudinal-mode, and the frequency difference of the dual-frequency laser at 1064 nm was tuned in a range from 50 MHz to 1.3 GHz. In 2010, the group also reported a diode-pumped coupled-cavity dual-frequency Nd:YAG laser at 1064 nm with tunable frequency difference [22], as shown in Figure 6. The single-longitudinal-mode oscillation was realized from a coupled-cavity setup consisting of the input-cavity mirror and a fused quartz-fabricated F-P etalon with a thickness of 1 mm and an effective reflectivity of 4%. The two quarter-wave plates were inserted into a cavity so as to split the laser-longitudinal mode, and the frequency difference was tuned in a range from 0 to 1.1 GHz by changing the angle between the principal axes of the two quarter-wave plates.

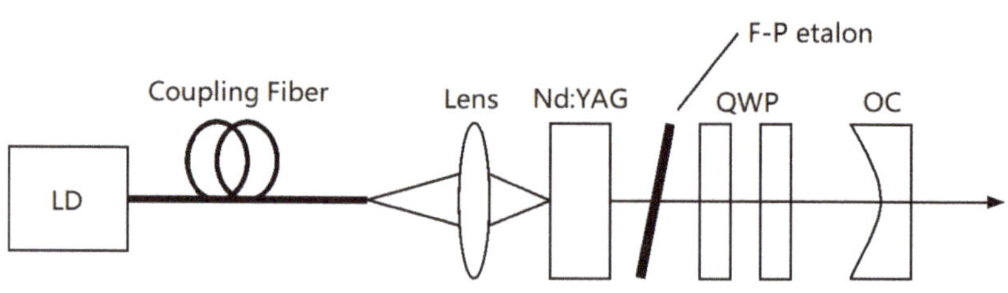

Figure 4. Schematic diagram of dual-frequency Nd:YAG laser [21]. LD: laser diode; F-P etalon: Fabry–Perot etalon; QWP: quarter-wave plate; OC: output coupler.

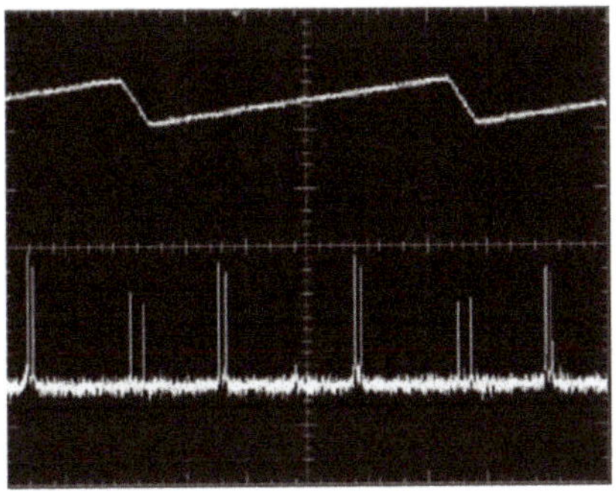

Figure 5. Oscillating spectrum of dual-frequency laser [21].

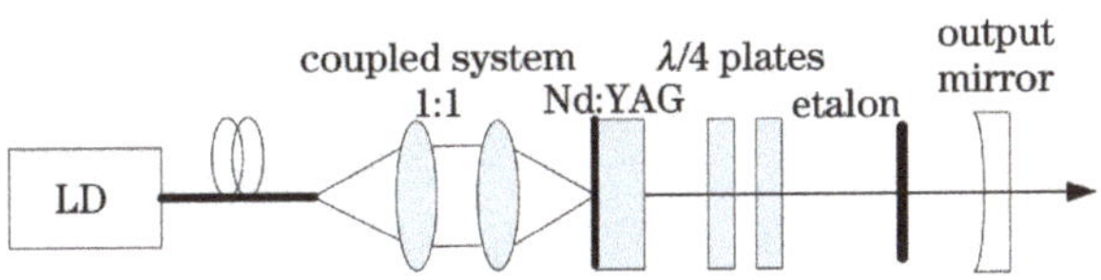

Figure 6. Diagram of the tunable two-frequency solid-state laser with coupled cavities [22]. LD: laser diode; $\lambda/4$ plate: quarter-wave plate.

In 2001, the research group of Prof. Zhang Shulian of the Tsinghua University reported a diode-pumped birefringent dual-frequency Nd:YAG laser at 1064 nm [23], as shown in Figure 7, in which the resonant cavity of the laser contained a piece of crystal-quartz-made birefringent F-P etalon (BFPE), serving as both a selector and a splitter of laser longitudinal modes. Because of the intracavity birefringent effect, each longitudinal mode was split into two linearly and orthogonally polarized components, i.e., ordinary mode (o-mode) and extraordinary mode (e-mode), and the unique transmission maximum of the etalon within the overall lasing bandwidth of the Nd:YAG laser was also split into two transmission maxima, namely ordinary peak (o-peak) and extraordinary peak (e-peak). The simultaneous operation of the two longitudinal modes could be obtained by making an o-mode and an e-mode coincide with the central positions of the o-peak and e-peak, respectively. A piece of BFPE with a geometrical thickness of 645 mm and a cut-angle of 10° was designed and fabricated, which was placed in a 40 mm-long cavity of the diode-pumped Nd:YAG laser. As a result, the orthogonally and linearly polarized dual-frequency laser at 1064 nm was output by slightly adjusting the tilt angle of the BFPE element, and a frequency difference of approximately 2 GHz was observed experimentally by a confocal F-P-scanning interferometer with a free spectral range of 4 GHz.

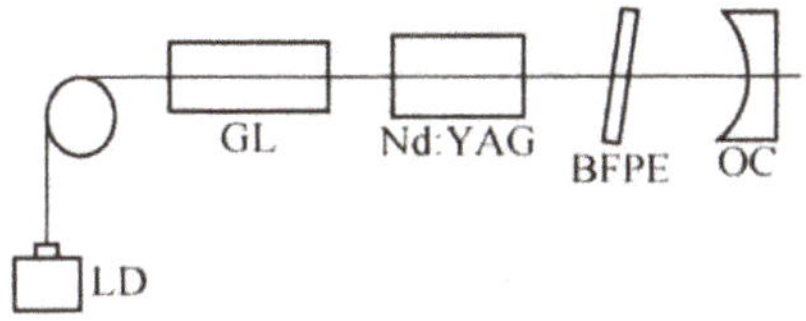

Figure 7. Experimental setup of diode-pumped birefringent dual-frequency Nd:YAG laser [23]. LD: laser diode; GL: gradient-index lens; BFFP: birefringent Fabry–Perot etalon; OC: output coupler.

In 2018, the research group of Prof. Jiao Mingxing of the Xi'an University of Technology designed and reported a diode-pumped dual-frequency Nd:YAG laser with a detuning twisted-mode cavity at 1064 nm [24], as shown in Figure 8, in which the polarizing beam splitter (PBS) was used as the polarizer and two quarter-wave plates were placed on each end of the Nd:YAG crystal. The laser oscillated in linearly polarized single-longitudinal-mode when the principal axes of the two quarter-wave plates were perpendicular to each other, and the single-longitudinal-mode was split into two orthogonally and linearly polarized modes when the twisted-mode cavity was detuned. As a result, an orthogonally polarized dual-frequency laser at 1064 nm was obtained, and the frequency difference was continuously tuned over the whole cavity free spectral range by rotating one of the quarter-wave plates in the plane perpendicular to the cavity axis. The oscillating-mode spectrum of the dual-frequency laser was observed experimentally by the use of a confocal scanning F-P interferometer, as shown in Figure 9. The dependence of the mode-splitting magnitude on the rotation angle of the quarter-wave plate was obtained experimentally, as indicated in Figure 10, and the largest frequency difference was about 3 GHz, determined by the laser cavity length.

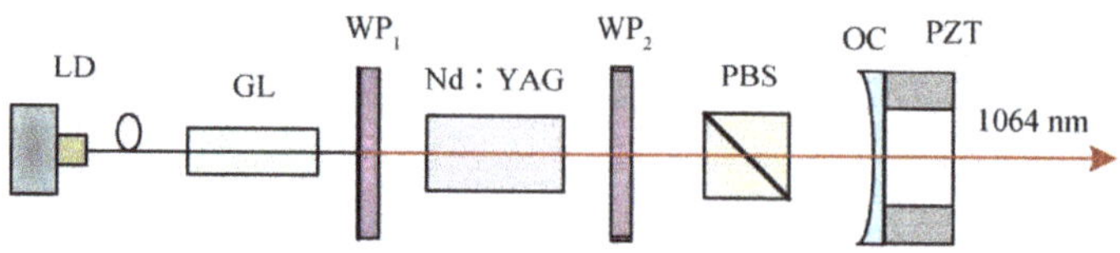

Figure 8. Schematic diagram of dual-frequency Nd:YAG laser with a detuning twisted-mode cavity [24]. LD: laser diode; GL: gradient-index lens; WP: wave plate; PBS: polarizing beam splitter; OC: output coupler; PZT: piezoelectric tube.

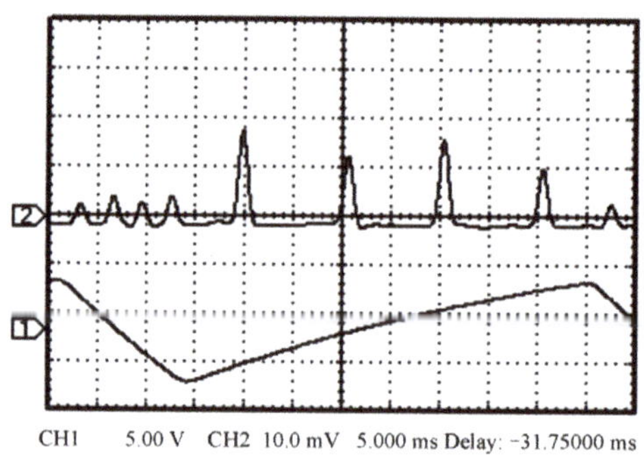

Figure 9. Oscillating spectrum of dual-frequency laser [24].

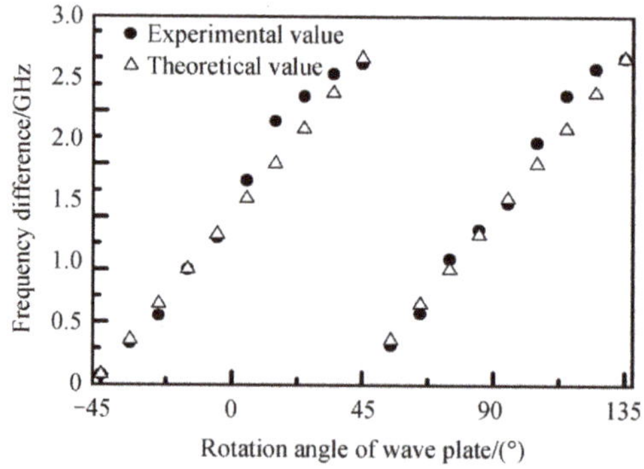

Figure 10. Dependence of mode-splitting magnitude on the rotation angle of quarter-wave plate [24].

2.1.2. Electro-Optical and Thermo-Optical Birefringent Dual-Frequency Solid-State Lasers

Compared to the crystal-quartz-made BFPE described above, an electro-optical birefringent F-P etalon (EO-BFPE) can also be used as not only the single-longitudinal-mode selector but also as the splitter of the laser-longitudinal mode. On this foundation, in 2007, the research group of Julien Le Gouet of Thales Research & Technology, France, reported an electro-optical birefringent dual-frequency Nd:YAG laser [25], as shown in Figure 11, in which a 400 μm-thick EO-BFPE made from lead zirconate tantalate ceramic was included in the laser cavity with an optical length of 14 mm. As a result, an orthogonally and linearly polarized dual-frequency laser at 1064 nm was obtained, and the frequency difference was discontinuously tuned in a range from 10 to 127 GHz when the direct voltage applied to the EO-BFPE was changed from 0 to 499 V. The typical oscillating spectrum of the dual-frequency laser with different frequency difference was observed experimentally, as shown in Figure 12.

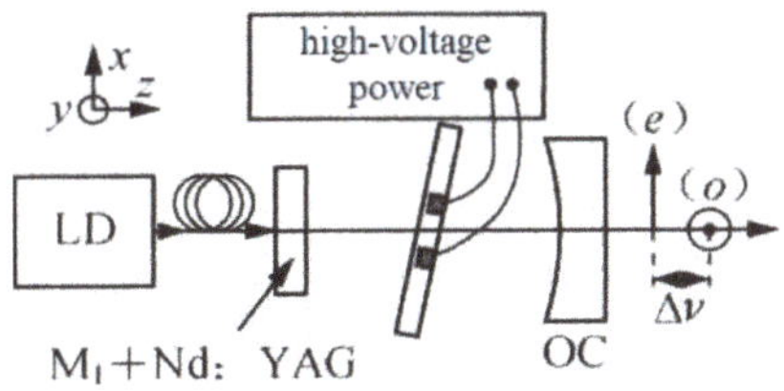

Figure 11. Schematic diagram of electro-optical birefringent dual-frequency Nd:YAG laser [25]. LD: laser diode; M: mirror; OC: output coupler.

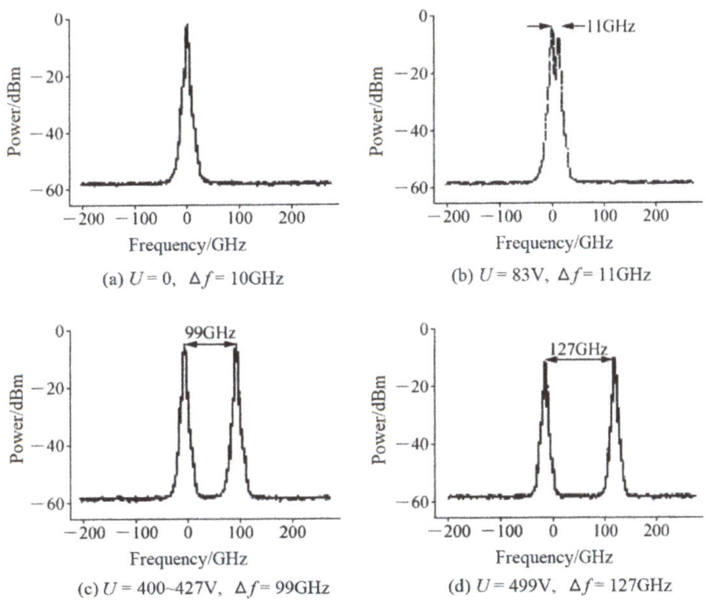

Figure 12. Oscillating spectrum of dual-frequency laser with different frequency difference [25].

The research group of Prof. M. Brunel of the Université de Rennes 1 has been devoted to dual-frequency solid-state laser technologies for a long time (since 1997). Using the thermo-optical birefringent effect of the intracavity $LiTaO_3$ crystal, in 2005, the group reported a microchip dual-frequency laser containing an erbium–ytterbium glass medium with a thickness of 190 μm and a $LiTaO_3$ crystal with a thickness of 130 μm [26], as shown in Figure 13. The cavity length was less than 0.5 mm; thus, the orthogonally and linearly polarized dual-frequency laser at 1530 nm was simultaneously oscillated and output, and the frequency difference of the dual-frequency laser was continuously tuned from a few

GHz to more than 50 GHz when the temperature of the $LiTaO_3$ crystal changed from nearly 45 to 10 °C, as shown in Figure 14.

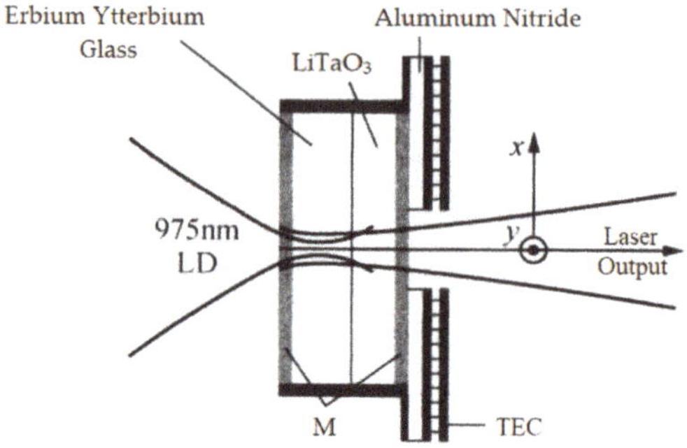

Figure 13. Schematic diagram of thermo-optical birefringent microchip dual-frequency laser [26]. LD: laser diode; M: mirror; TEC: thermo-electric cooler.

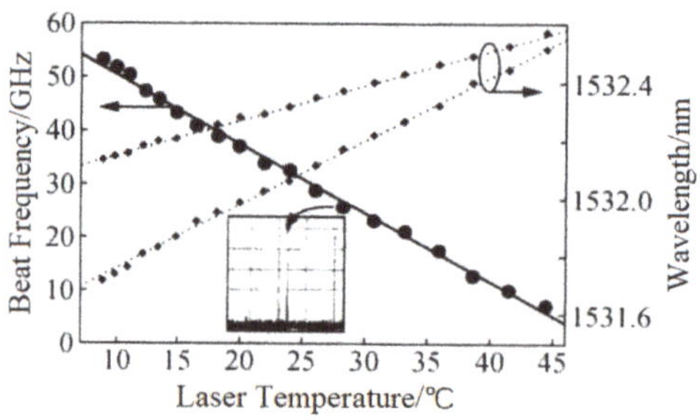

Figure 14. Dependence of oscillating wavelength and its beat frequency on laser temperature [26].

In 2008, the group also reported a diode-pumped dual-frequency solid-state laser [27], as shown in Figure 15, in which an erbium–ytterbium glass with a thickness of 0.75 mm was used as the gain medium. This laser also contained a 250 µm-thick F-P etalon acting as the single-longitudinal-mode selector and a 1 mm-thick $LiTaO_3$ crystal serving as the longitudinal-mode splitter. As a result, an orthogonally polarized dual-frequency laser at 1530 nm was obtained, and the frequency difference of the dual-frequency laser was tuned in a range from 100 MHz to 20 GHz when the temperature of the $LiTaO_3$ crystal varied from 22 to 42 °C, as shown in Figure 16.

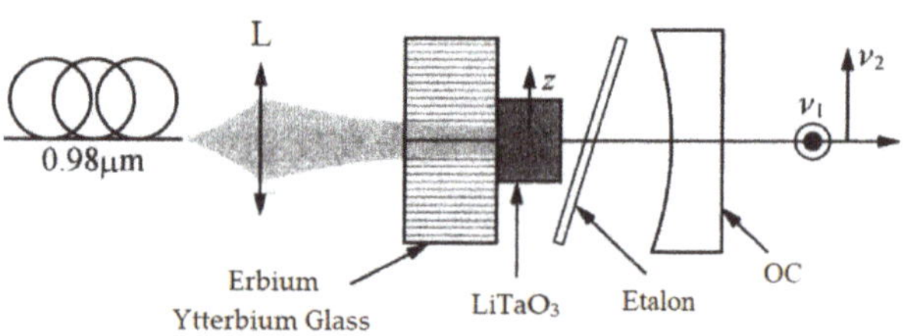

Figure 15. Schematic diagram of dual-frequency erbium–ytterbium glass laser [27]. L: lens; OC: output coupler.

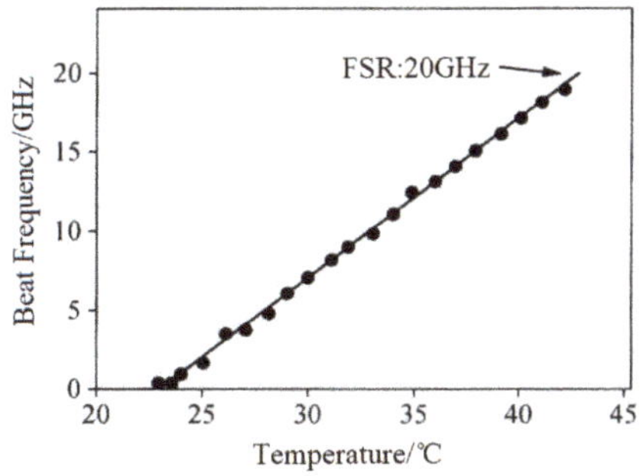

Figure 16. Dependence of frequency difference of dual-frequency laser on laser temperature [27].

2.1.3. Stress Birefringent Dual-Frequency Solid-State Lasers

In 2003, the research group of Prof. V. G. Gudelev of the National Academy of Sciences of Belarus reported a diode-pumped tunable dual-frequency Nd:YAG laser at 1064 nm with coupled resonators [28], as shown in Figure 17. The coupled resonators consisted of the two sides of the Nd:YAG crystal and the spherical output coupler of OC, and the fundamental cavity formed by the left side of the Nd:YAG crystal and the OC was forced to oscillate in either single-longitudinal mode or two longitudinal modes, depending on the geometrical cavity length of L. On the one hand, in the circumstance of single-longitudinal-mode oscillation, the laser mode was split due to the intracavity stress birefringence induced by the applied force of F to the OC element; therefore, an orthogonally and linearly polarized dual-frequency laser at 1064 nm was obtained, and the frequency difference was continuously tuned in a range from 50 MHz to 2.4 GHz when changing the applied force. The oscillating-mode spectrum of the dual-frequency laser with different frequency difference was observed experimentally, as shown in Figure 18. On the other hand, in the circumstance of two-longitudinal-mode oscillation of the fundamental cavity, each mode was split into two orthogonally and linearly polarized modes when the applied force existed, the mode-splitting magnitude increased with the applied force. When the mode splitting magnitude reached to a certain value, the system output a beam of orthogonally and linearly polarized dual-frequency laser at 1064 nm, the maximum frequency difference of which being 8.4 GHz, as shown in Figure 19.

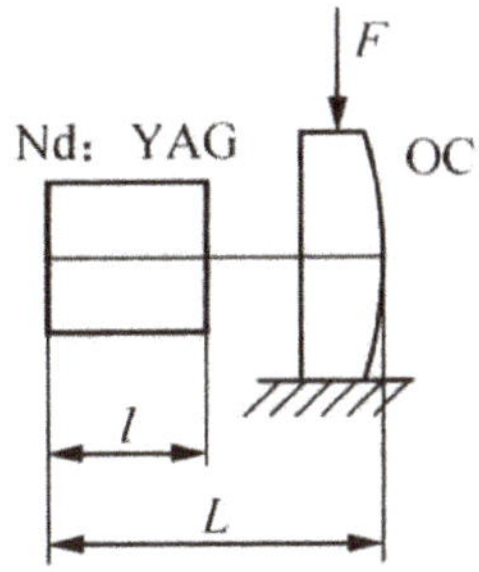

Figure 17. Schematic diagram of coupled-cavity dual-frequency Nd:YAG laser [28]. OC: output coupler.

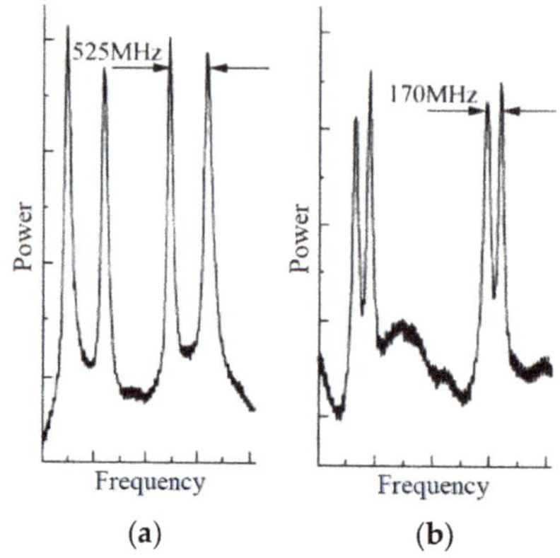

Figure 18. Oscillating spectrum of dual-frequency laser with different frequency difference of (**a**) 525 MHz and (**b**) 170 MHz [28].

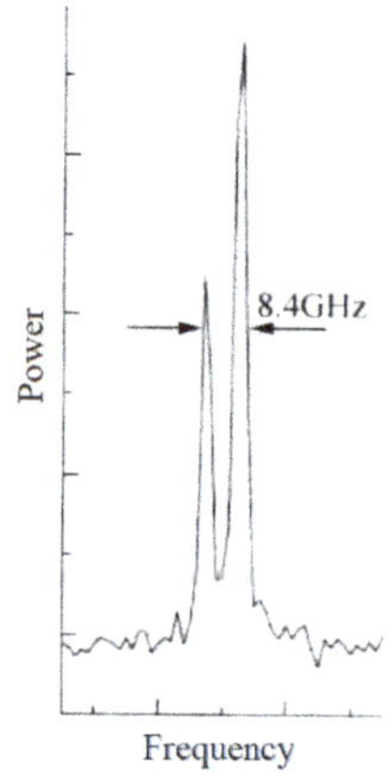

Figure 19. Oscillating spectrum of orthogonally polarized dual-frequency Nd:YAG laser [28].

In 2019, the research group of Prof. Zhang Shulian of the Tsinghua University reported a diode-pumped dual-frequency Nd:YAG microchip laser at 1064 nm [29], as shown in Figure 20, in which a <111>-cut Nd:YAG microchip with 1 $\pm$ 0.1% neodymium doping in a circular shape was included. The diameter of the crystal was 5 $\pm$ 0.05 mm and the thickness of the microchip was 1 $\pm$ 0.1 mm to reduce the multiple longitudinal modes inside the laser output. The two surfaces were parallel and coated with high-reflectivity dielectric films acting as the laser-cavity mirrors, so the laser at 1064 nm was forced to oscillate in single-longitudinal mode. Based on the effect of stress birefringence, the single-oscillated-laser mode was split into two orthogonally polarized components; thus, the orthogonally and linearly polarized dual-frequency laser output at 1064 nm was obtained, as shown in Figure 21. It can be seen that the frequency difference was in the scale of tens of MHz, determined by the internal stress inside the microchip laser. In 2021, the group also studied a microchip Nd:YAG dual-frequency laser with a frequency difference of 17.4 MHz [30], as shown in Figure 22, which was formed by the stress-induced birefringence in the microchip itself. The <111>-cut, 1%-doped and quasi-isotropic Nd:YAG crystal was processed into a plate with a diameter of 2.8 mm and a thickness of 1 mm. Both faces of the chip were dielectric-coated to form a monolithic-resonant cavity; only one longitudinal mode resonated in the cavity, which was split into two monofrequency components due to the stress-induced birefringence in the Nd:YAG crystal. Thus, the laser emitted an orthogonally polarized dual-frequency laser with an output power of 4.7 mW at 1064 nm. The polarization directions of these two components coincided with the principal stress directions, respectively, and the frequency difference between them was proportional to the difference between the magnitudes of two principal stresses.

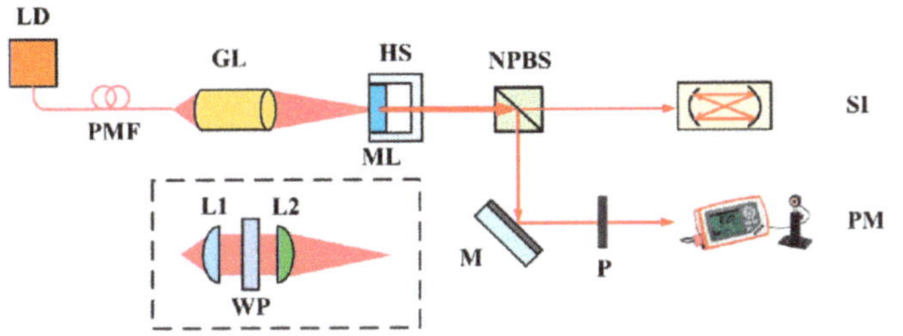

Figure 20. Schematic setup of the dual-frequency microchip laser [29]. LD: laser diode; PMF: polarization-maintaining fiber; GL: grin lens; L1, L2: lenses; WP: half waveplate; HS: heat sink; ML: Nd:YAG microchip laser; NPBS: nonpolarizing beam splitter; SI: scanning interferometer; PM: power meter; M: reflective mirror; P: polarizer.

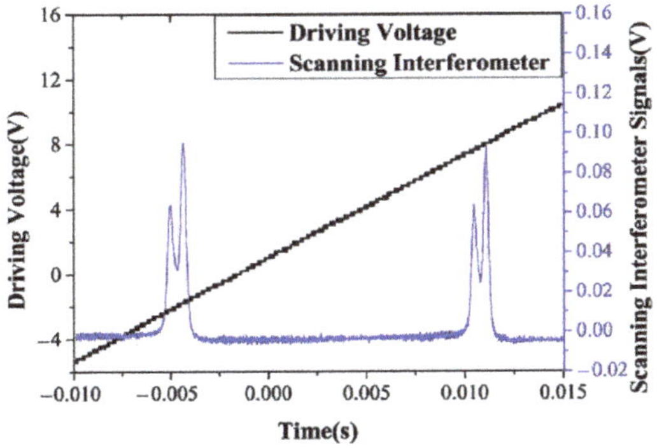

Figure 21. Oscillating spectrum of dual-frequency laser (blue trace) and triangular-wave voltage (black trace) [29].

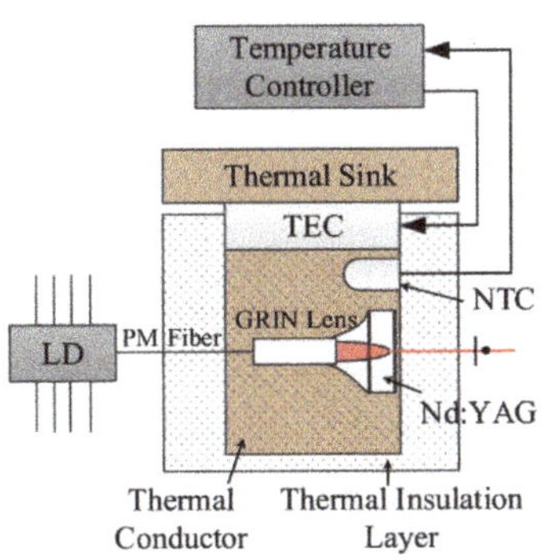

Figure 22. Schematic of microchip Nd:YAG dual-frequency laser [30]. LD: laser diode; TEC: thermo-electric cooler; NTC: negative-temperature-coefficient thermistor.

2.2. Biaxial Dual-Frequency Solid-State Lasers

In order to obtain a dual-frequency laser output with a tunable and large frequency difference, in 2004, the research group of R. Czarny of Thales Research & Technology,

France, reported a diode-pumped dual-frequency $KGd(WO_4)_2$ Yb^{3+}-doped laser for CW-THz generation [32], as shown in Figure 23, which included a 2 mm-long Yb:KGW active medium doped with 5% Yb. The 60 mm-long laser cavity was formed by two mirrors of M_1 and M_2. Spatial separation between the two eigenmodes inside a portion of the cavity was obtained using an AR-coated 10-mm long YVO_4 crystal cut at 45° of its optical axes. It led to two cross-polarized eigenmodes (o-mode and e-mode). The Yb:KGW crystal was oriented to maximize gain of eigenstate (e), and a quartz-made half waveplate $\lambda/2$ was inserted in the o-mode path. To independently tune the optical frequencies associated to each polarization eigenstate, two noncoated 150 μm-thick glass etalons of E_o and E_e were placed in the o-mode path and e-mode path, respectively. The single-longitudinal-mode operation of both eigenstates was obtained thanks to a quite short cavity (about 60 mm) and an additional 1 mm-thick glass etalon E_c inserted in the common path; thus, the orthogonally and linearly polarized dual-frequency laser at 1030 nm was obtained, as shown in Figure 24. The frequency difference between the two modes was step-tunable from DC to 3.1 THz. A maximum total optical output power of 120 mW was obtained with a beat-note linewidth narrower than 30 kHz.

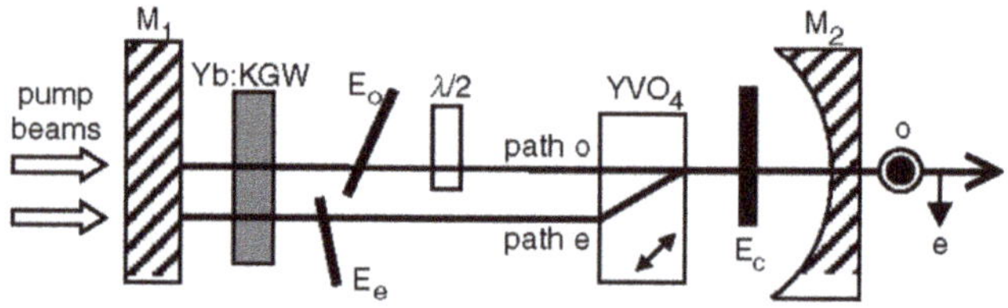

Figure 23. Schematic diagram of dual-frequency Yb:KGW laser [32]. E_o, E_e, E_c: etalon; M_1: plane-dichroic mirror; M_2: output coupler.

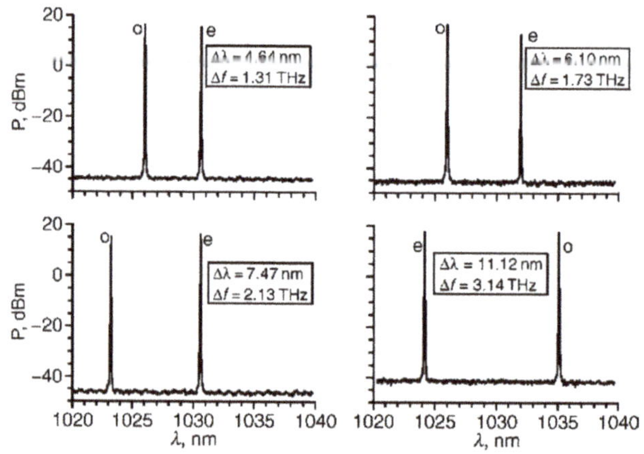

Figure 24. Measured optical spectra of dual-frequency laser when tuned for various frequency differences [32].

2.3. *Two-Cavity Dual-Frequency Solid-State Lasers*

According to the principle of single-longitudinal-mode selection by the use of a birefringent filter, the output of a high-power single-frequency laser is difficult to be obtained, due to the fact that the mode-selecting ability of the traditional birefringent filter is too low using a Brewster plate (BP) as a polarizer. Although the mode-selecting ability of the birefringent filter can be effectively enhanced by increasing the number of BP, some problems inevitably occur, such as complex laser structure, difficult adjustment of the laser cavity, etc. To overcome the shortcomings of the limited mode-selecting ability of the traditional birefringent filter, in 2006, Prof. Jiao Mingxing proposed a new birefringent filter in which a PBS was employed as the polarizer, and a TCDFL with a large frequency difference was invented and investigated [33,34]. Since then, our group have designed and researched several TCDFLs using different principles of single-longitudinal-mode selection, such as birefringent filter, F-P etalon, twisted-mode cavity, etc.

2.3.1. TCDFLs Using Birefringent Filter

In order to obtain a tunable dual-frequency laser with a large frequency difference, our group designed and investigated a diode-pumped dual-frequency Nd:YAG laser with a tunable frequency difference [35], as shown in Figure 25, using the light-splitting and polarizing functions of the PBS to form perpendicular linear and right-angle standing-wave cavities. Both cavities included a birefringent output coupler (BOC) fabricated from calcite crystal, and they employed a birefringent filter of PBS-BOC acting as laser-longitudinal-mode selectors. The p and s components of the 1064 nm laser light oscillating in single-longitudinal-mode were observed experimentally, as shown in Figure 26. The frequency difference of the dual-frequency laser was tuned by adjusting the tilt angles of the BOC, and the oscillating spectrums of the dual-frequency laser at 1064 nm were observed, as shown in Figure 27. It can be seen that the frequency difference was tunable in a range from 11.9 GHz to 148.4 GHz, and the maximum frequency difference reached the fluorescence line a width of Nd:YAG.

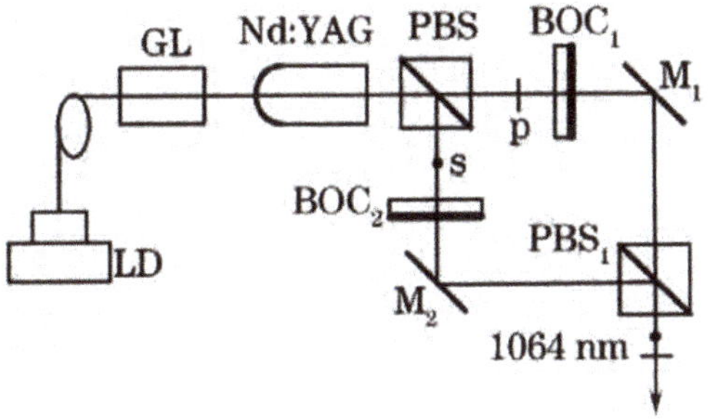

Figure 25. Schematic diagram of dual-frequency Nd:YAG laser system at 1064 nm [35]. LD: laser diode; GL: gradient-index lens; PBS: polarizing beam splitter; BOC: birefringent output coupler; M: mirror.

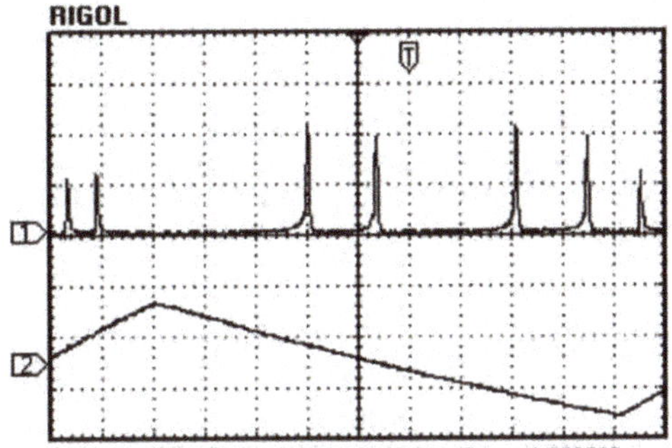

Figure 26. Mode patterns of simultaneous dual-frequency oscillation [35].

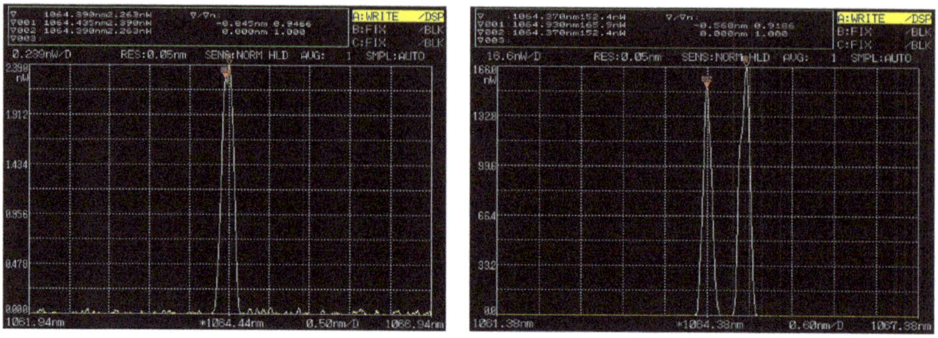

(a) (b)

Figure 27. Oscillating spectra of dual-frequency laser with different frequency differences of (**a**) 11.9 GHz and (**b**) 148.4 GHz [35].

Moreover, a diode-pumped TCDFL with a tunable and large frequency difference at 1064 nm was designed and investigated [36], as shown in Figure 28. Both cavities included a birefringent filter of PBS-BC, in which the birefringent crystal (BC) was a piece of a half-wave plate with a thickness of 0.7 mm. The p-polarized and s-polarized components of the laser at 1064 nm were forced to oscillate simultaneously in single-longitudinal mode in the linear and right-angle cavities, respectively. As a result, an orthogonally and linearly polarized dual-frequency laser at 1064 nm was obtained, and the oscillating-mode spectrum was observed experimentally by the use of a confocal scanning Fabry–Perot interferometer, as shown in Figure 29.

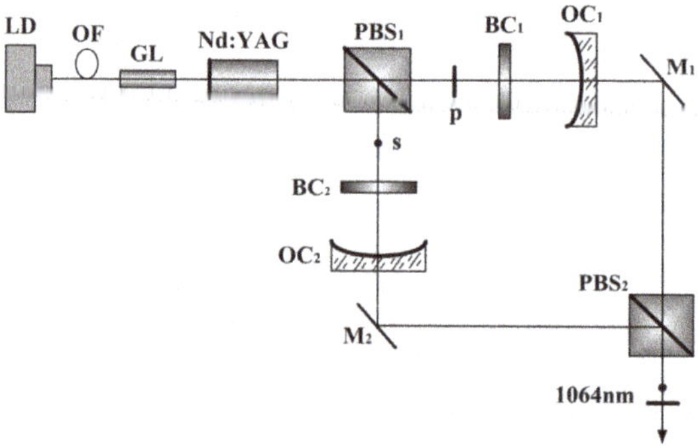

Figure 28. Schematic diagram of TCDFL based on the principle of longitudinal-mode selection by birefringent filter [36]. LD: laser diode; OF: optical fiber; GL: gradient-index lens; PBS: polarizing bean splitter; BC: birefringent crystal; OC: output coupler; M: mirror.

The tilt angle of the BC_2 was maintained at a constant 2.5°, and the frequency difference of the dual-frequency laser at 1064 nm was tuned by adjusting the tilt angles of the BC_1. The experimental relationship between the frequency difference and the tilt angles of the BC_1 was obtained, as shown in Figure 30. It can be seen that the frequency difference of the dual-frequency laser at 1064 nm was tuned in a range from 5.2 GHz to 147.3 GHz when the BC_1 was tilted from approximately 1.5° to 2.5°, and the maximum frequency difference reached nearly to the fluorescence linewidth of the Nd:YAG crystal.

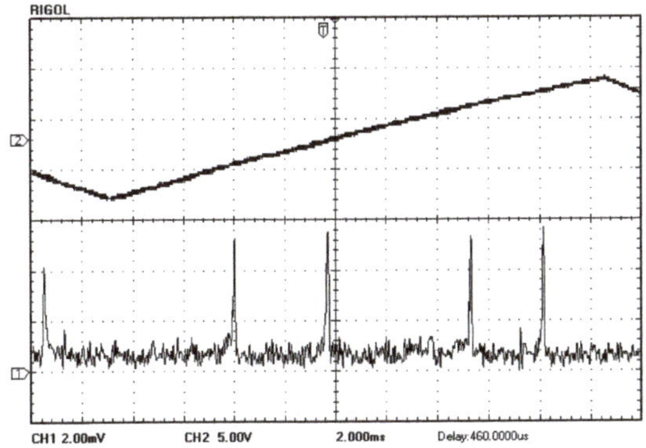

Figure 29. Oscillating mode spectrum of dual-frequency laser [36].

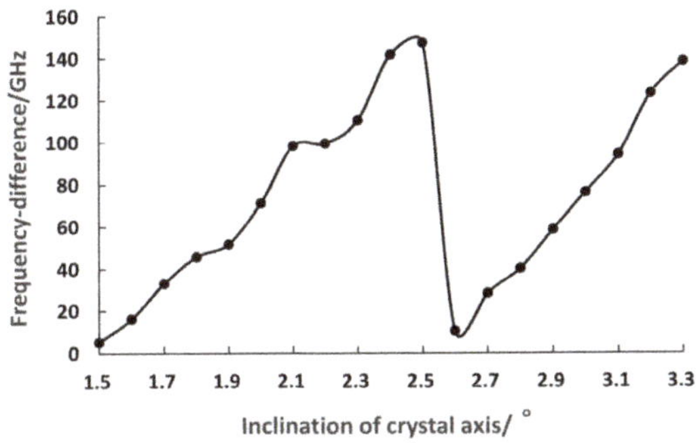

Figure 30. Relationship between frequency difference and tilt angle of the BC_1 [36].

In 2016, our group also reported a TCDFL with electro-optical modulators [37], as shown in Figure 31, which consisted of two standing-wave cavities that shared the same gain medium of Nd:YAG. An electro-optic birefringent filter consisting of PBS and lithium niobate (LN) was used not only to select longitudinal mode but also to tune frequency and frequency difference. As a result, the simultaneous operation of orthogonally and linearly polarized dual-frequency laser was obtained, as shown in Figure 32. The experimentally obtained results indicated that the frequency difference was tuned from 0 to 132 GHz by changing the DC voltages applied to LNs. When the pump power was 900 mW, the output powers from the linear and right-angle cavities were equal to 20 mW and 26 mW, respectively.

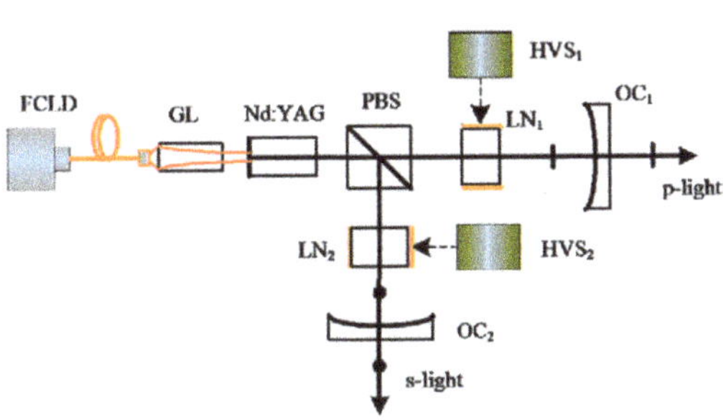

Figure 31. Schematic of two-cavity dual-frequency Nd:YAG laser [37]. FCLD: fiber-coupled laser diode; GL: gradient-index lens; PBS: polarizing bean splitter; LN: lithium niobate; HVS: high-voltage source; OC: output coupler.

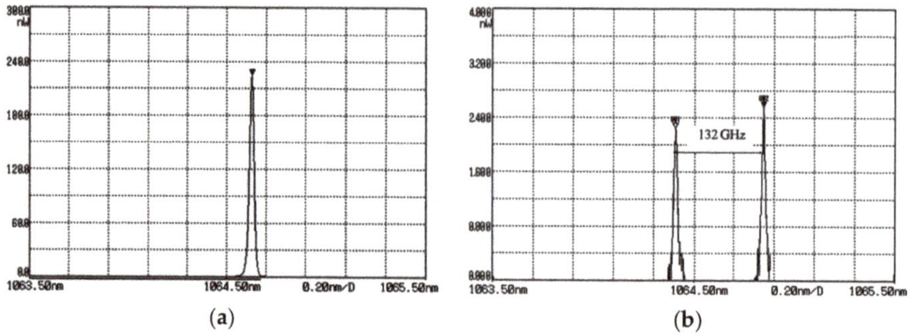

Figure 32. Oscillating spectrum of dual-frequency laser with different frequency differences of (**a**) 0 GHz and (**b**) 132 GHz [37].

2.3.2. TCDFL with a Twisted-Mode Configuration

In order to produce the dual-frequency laser with a tunable frequency difference at 1064 nm, in 2015, our group designed and investigated a diode-pumped TCDFL using a twisted-mode configuration [38], as shown in Figure 33, the two standing-wave cavities of which shared the same gain medium Nd:YAG, and the twisted-mode configuration reduced or even eliminated the spatial hole-burning effect of the gain so that the single-longitudinal-mode was oscillated in both standing-wave cavities. Thus, the orthogonally and linearly polarized dual-frequency laser at 1064 nm was obtained, and the oscillating mode spectrum was observed experimentally by the use of a confocal scanning F-P interferometer, as shown in Figure 34.

The experimental results showed that both cavities of the Nd:YAG laser oscillated steadily in single-longitudinal mode, and the frequency difference was tuned in a range from 0.3 GHz to 3 GHz by changing the cavity length. Theoretically, the maximum frequency difference was equal to one-half of the sum of both cavities' longitudinal-mode intervals.

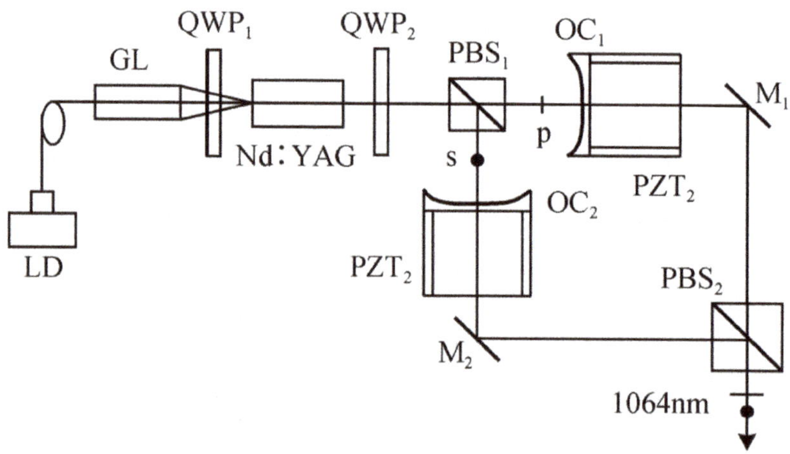

Figure 33. Schematic diagram of diode-pumped TCDFL using the twisted-mode configuration [38]. LD: laser diode; GL: gradient-index lens; QWP: quarter-wave plate; PBS: polarizing beam splitter; OC: output coupler; PZT: piezoelectric transducer; M: mirror.

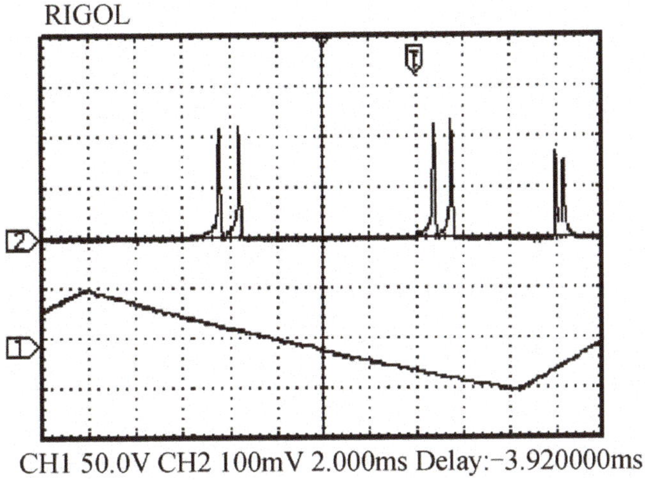

Figure 34. Oscillating-mode spectrum of dual-frequency laser [38].

2.3.3. TCDFL Using Intracavity F-P Etalon

The TCDFLs based on the principle of single-longitudinal-mode selection by the birefringent filter are not only simple but also have a low inserted cost. However, the spectral linewidths of the dual-frequency laser are not narrow enough due to the fact that the bandwidth of transmission peak of the birefringent filter is relatively wide, affecting the coherence lengths of the TCDFLs.

In 2022, our group reported a diode-pumped dual-frequency Nd:YAG laser with two standing-wave cavities sharing the common gain medium [40], which was based on the principle of single-longitudinal-mode selection by intracavity F-P etalon, as shown in Figure 35. With each of the cavities containing a piece of F-P etalon, the p-polarized and s-polarized components of the laser at 1064 nm were forced to oscillate simultaneously in single-longitudinal mode in both cavities, respectively. As a result, the orthogonally and linearly polarized dual-frequency laser at 1064 nm was output. Meanwhile, a coaxially propagating beam of the dual-frequency laser at 1064 nm escaped from the side of the intracavity PBS element because the PBS element had somewhat residual reflectance of the p-polarized beam and residual transmittance of the s-polarized beam.

The experimental results indicated that when both fused quartz F-P etalons with thicknesses of 0.5 mm and surface reflectivities of 90% were obliquely inserted into both cavities, respectively, the orthogonally and linearly polarized dual-frequency laser could be obtained by finely adjusting the tilt angles of the intracavity F-P etalons. The frequency difference tuning of the dual-frequency laser was realized by continually adjusting the tilt angles of both F-P etalons, and the typical oscillating spectrums of the dual-frequency laser were observed, as shown in Figure 36. It can be seen that the frequency difference was discontinuously tuned in a range from 16 to 76 GHz by finely adjusting the tilt angles of the intracavity F-P etalons, and theoretically, the maximum frequency difference was up to the oscillating bandwidth of the Nd:YAG laser. For the diode-pumped TCDFL with a frequency difference of 24 GHz, the threshold pump powers of the linear and right-angle cavities were equal to 1.9 W and 2 W, respectively, and the output powers of the linear and right-angle cavities were up to 229 and 190 mW, respectively.

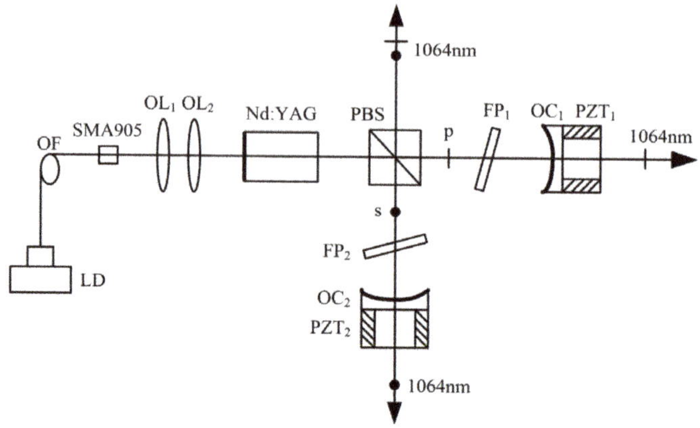

Figure 35. Schematic of TCDFL based on the principle of longitudinal-mode selection by single F-P etalon [40]. LD: laser diode; OF: optical fiber; SMA905: OF connector; OL: optical lens; PBS: polarizing beam splitter; FP: F-P etalon; OC: output coupler; PZT: piezoelectric transducer.

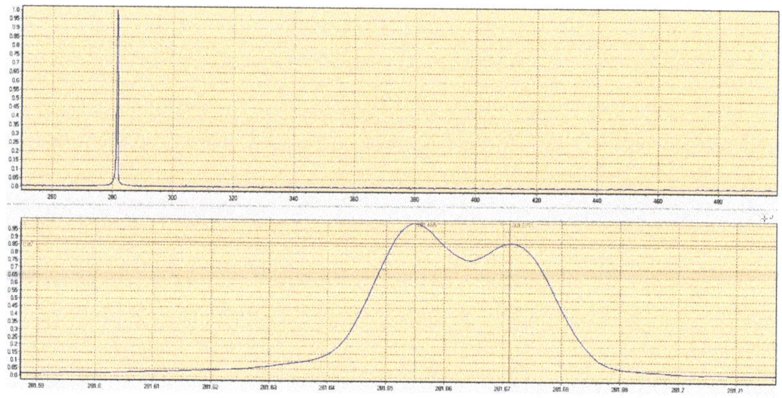

(a)

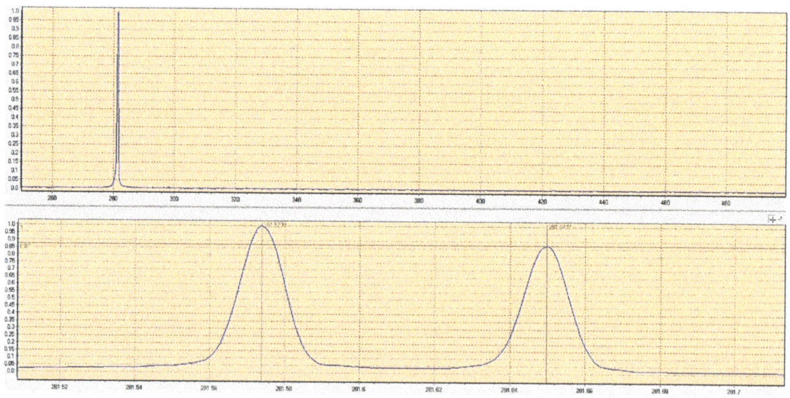

(b)

Figure 36. Oscillating spectrum of dual-frequency laser with different frequency differences of (a) 16 GHz and (b) 76 GHz [40].

3. Frequency-Difference-Stabilizing Systems for Dual-Frequency Solid-State Lasers

It is known that the accuracy of a synthetic-wave absolute-distance interferometric measurement ultimately depends on the maximum frequency difference of the dual-frequency laser and its stability. Therefore, the research on the frequency-difference-stabilizing technology of dual-frequency solid-state lasers is of great significance.

3.1. Frequency Difference Stabilization of Birefringent Dual-Frequency Solid-State Lasers

As described above, a kind of birefringent dual-frequency solid-state lasers has a single-axis configuration, and the frequency difference or beat-note stabilization has been investigated [46–50], the methods of which mainly include frequency-shifted optical feedback, saturable absorption, etc.

In 2007, the research group of L. Kervevan of France reported an original approach to stabilize the beat-note of a 1.53 μm dual-frequency Yb:Er glass laser via an optical self-injection process [51], as shown in Figure 37, in which the dual-frequency Yb:Er glass laser could output a dual-frequency laser at 1.53 μm with a beat-note of nearly 170 MHz. The optical self-injection process consisted of selecting one of the two linear modes as a master oscillator with a polarization filter, then frequency shifting the optical wave using an external acousto-optic modulator, and finally, using it to inject the other mode.

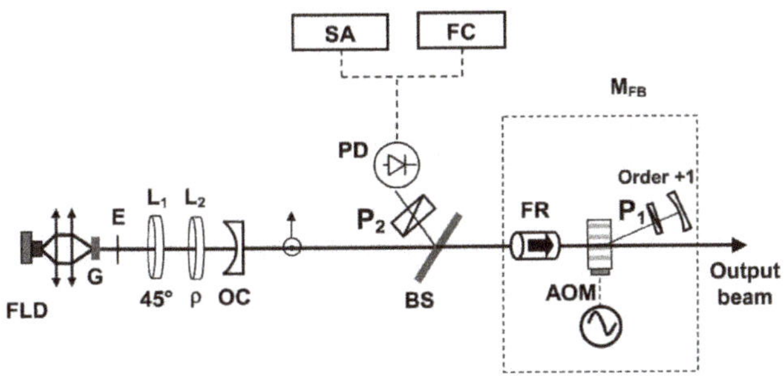

Figure 37. Schematic of self-injection stabilization process for 1.53 μm dual-frequency phosphate glass laser [51]. FLD: fiber-pigtailed laser diode; G: Yb:Er glass plate; E: intracavity etalon; L_1 and L_2: quarter-wave plates; P_1 and P_2: linear polarizers; OC: output coupler; BS: beam splitter; FR: Faraday rotator; AOM: acousto-optic modulator; PD: photodiode; SA: spectrum analyzer; FC: frequency counters; M_{FB}: optical feedback module; ρ: angular adjustment.

When the frequency-shifted optical beam was correctly reinjected into the oscillating mode of the laser cavity, the stability of the locking technique was tested by recording the beat note versus the synthesizer frequency, as shown in Figure 38a. It corresponded to a slow frequency deviation of about 0.27 mHz/s. A linear fit was applied to the measured synthesizer frequency [straight line in Figure 38a], and the difference between this reference line and the optical beat-note was observed in the histogram plotted curve in Figure 38b. It can be seen that a fitted Gaussian distribution curve allowed for estimating the stability of the frequency locking to be less than 0.25 Hz.

3.2. Frequency Difference Stabilization of TCDFL

Due to the fact that the TCDFL has both standing-wave cavities with two separate output couplers, it is easy to actively stabilize the resonant frequency of each cavity to a common frequency reference so that a high stability of the frequency difference can be obtained.

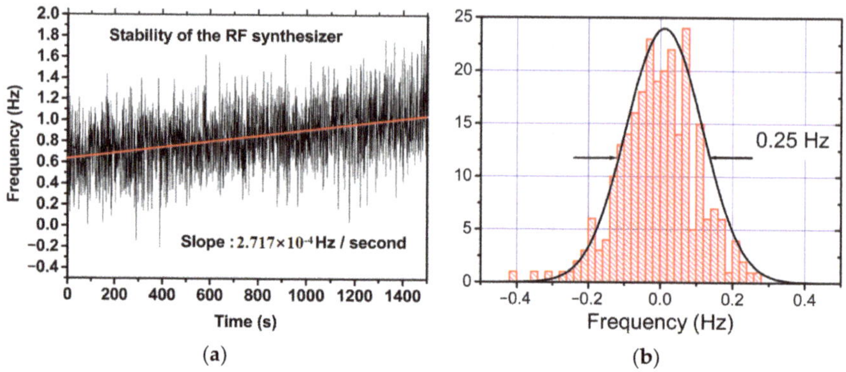

Figure 38. Schematic of self-injection stabilization process for 1.53 μm dual-frequency phosphate glass laser [51]. (**a**) Instantaneous fluctuations and long-term deviation of the synthesizer frequency versus time. (**b**) Histogram of the temporal stability of the beat note between the two orthogonal modes using the locking technique based on the frequency-shifted optical feedback loop.

As a commonly used method, the PDH frequency-stabilization method integrates with the technologies of both electro-optic phase modulation and optical heterodyne detection. In the past few decades, the laser-frequency-stabilizing technologies based on the PDH method have been widely investigated worldwide due to their advantages of fast servo response, low noise, and high-frequency stability [52–57].

3.2.1. Double-Modulator QD-PDH Frequency-Difference Stabilizing System for TCDFL

In 2022, our group reported a frequency-difference-stabilizing system for the diode-pumped TCDFL at 1064 nm using a double-modulator QD-PDH frequency-stabilizing method [43], as shown in Figure 39, which included two sets of QD-PDH frequency-stabilizing subsystems (see parts II and III) that shared the same F-P cavity as the frequency reference, and the magnitude of the frequency difference was required to be an integer number representing times of the free spectral range (FSR) of the referenced F-P cavity.

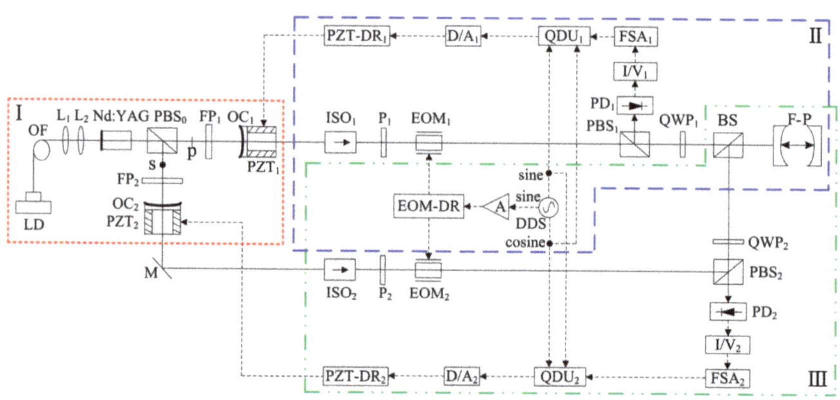

Figure 39. Schematic diagram of the frequency-difference-stabilizing system for the TCDFL using the double-modulator QD-PDH frequency-stabilizing method [43]. LD: laser diode; OF: optical fiber; L: lens; PBS: polarizing beam splitter; FP: F-P etalon; OC: output coupler; PZT: piezoelectric transducer; M: mirror; ISO: optical isolator; P: polarizer; EOM: electro-optic modulator; DDS: direct digital synthesizer; A: amplifier; EOM-DR: EOM driver; QWP: quarter-wave plate; BS: beam splitter; F-P: F-P reference cavity; PD: photodetector; I/V: current-to-voltage; FSA: frequency-selective amplifier; QDU: quadrature-demodulated unit; D/A: digital-to-analog; PZT-DR: PZT driver.

A QD-PDH method-based frequency-difference-stabilizing system for the diode-pumped TCDFL with a frequency difference of 24 GHz at 1064 nm was established and investigated, in which the free spectral range and the finesse of the referenced F-P cavity were equal to 375 MHz and 421, respectively. Both frequencies of the TCDFL were successfully frequency stabilized to the two different resonant frequencies of the F-P cavity during a period of about 1 h, the error signals of the two QD-PDH frequency-stabilizing subsystems were obtained experimentally, as shown in Figure 40, the maximum offset voltages of the QD-PDH error signals were equal to 77 mV and 74.2 mV, respectively, and correspondingly, the laser-frequency drifts of the linear and right-angle cavities were determined to be <0.35 MHz and 0.36 MHz, respectively. The frequency-difference fluctuations of the frequency-locked TCDFL are shown in Figure 41, and the maximum change in the frequency difference was <0.55 MHz. According to the Allan variance, the laser-frequency stabilities of the linear and right-angle cavities were better than 2.3×10^{-11} and 2.7×10^{-11}, respectively, corresponding to a frequency-difference stability better than 4.2×10^{-7}.

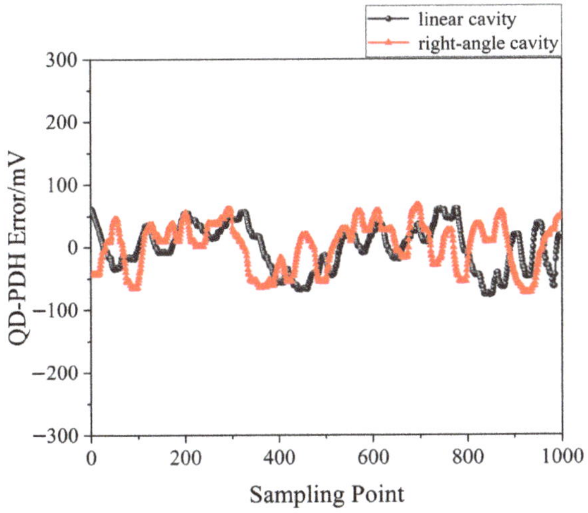

Figure 40. Error signals of the frequency-locked QD-PDH subsystems [43].

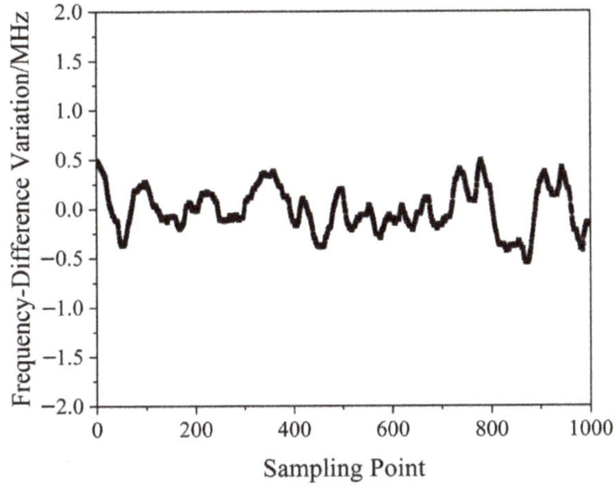

Figure 41. Frequency difference variation in the frequency-locked dual-frequency Nd:YAG laser [43].

3.2.2. Single-Modulator QD-PDH Frequency-Difference-Stabilizing System for TCDFL

In 2022, our group proposed a new scheme of the phase modulation of the orthogonally and linearly polarized dual-frequency laser using a single electro-optic modulator (EOM), and a simple frequency-difference-stabilizing system for the TCDFL using a single-modulator QD-PDH frequency-stabilizing method was designed [44], as shown in Figure 42, which included two sets of QD-PDH frequency-stabilizing subsystems (see parts II and III) that shared the same electro-optic phase modulation unit and the same frequency reference of the F-P cavity.

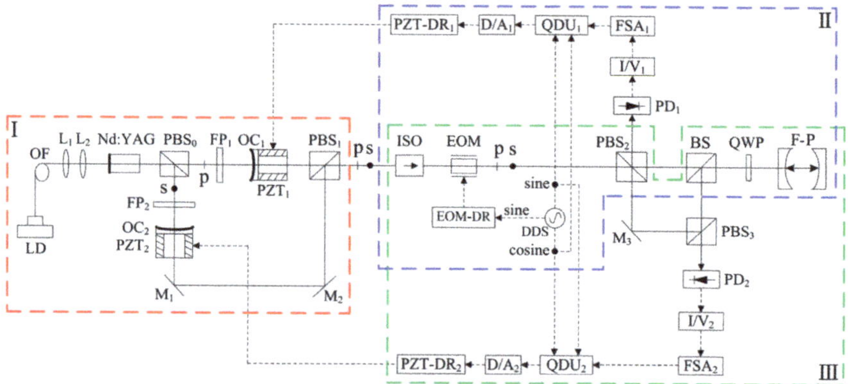

Figure 42. Schematic diagram of the frequency-difference-stabilizing system for the TCDFL using the single-modulator QD-PDH frequency-stabilizing method [44]. LD: laser diode; OF: optical fiber; L: lens; PBS: polarizing beam splitter; FP: F-P etalon; OC: output coupler; PZT: piezoelectric transducer; M: mirror; ISO: optical isolator; EOM: electro-optic modulator; DDS: direct digital synthesizer; EOM-DR: EOM driver; BS: beam splitter; QWP: quarter-wave plate; F-P: F-P reference cavity; PD: photodetector; I/V: current-to-voltage; FSA: frequency-selective amplifier; QDU: quadrature-demodulated unit; D/A: digital-to-analog; PZT-DR: PZT driver.

A QD-PDH frequency-difference-stabilizing system for the same diode-pumped TCDFL with a frequency difference of 24 GHz at 1064 nm was established and investigated. Both frequencies of the TCDFL at 1064 nm were successfully frequency-stabilized to the two different resonant frequencies of the referenced F-P cavity during a period of about 1 h, the error signals of the two QD-PDH frequency-stabilizing subsystems were obtained experimentally, as shown in Figure 43, the maximum offset voltages of the QD-PDH error signals were equal to 84.5 mV and 76.7 mV, respectively, and correspondingly, the laser-frequency drifts of the linear and right-angle cavities were determined to be <0.34 and 0.35 MHz, respectively. The frequency-difference fluctuations of the frequency-locked TCDFL are shown in Figure 44, and the maximum change in the frequency difference was <0.51 MHz. According to the Allan variance, the laser-frequency stabilities of the linear and right-angle cavities were better than 1.6×10^{-11} and 2.0×10^{-11}, respectively, corresponding to a frequency-difference stability better than 2.9×10^{-7}.

The experimental results obtained above indicate that compared with the double-modulator QD-PDH frequency-difference-stabilizing system shown in Figure 39, the single-modulator QD-PDH frequency-difference-stabilizing system shown in Figure 42 is not only simple, but also has better performances in the linear-dynamic range, frequency-discriminating sensitivity, frequency stabilization, and frequency-difference stabilization, as listed in Table 1.

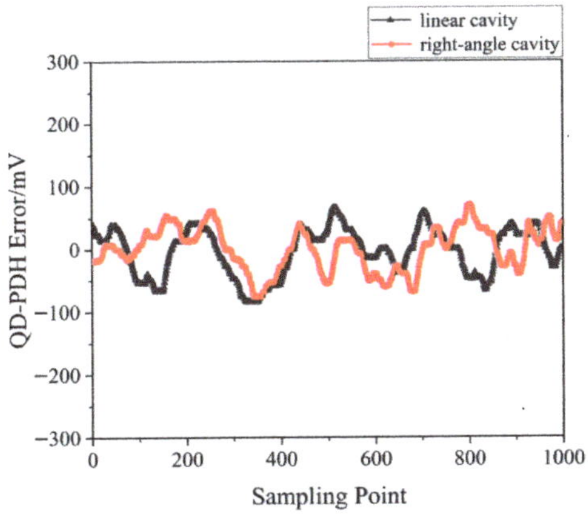

Figure 43. Error signals of the frequency-locked QD-PDH subsystems [44].

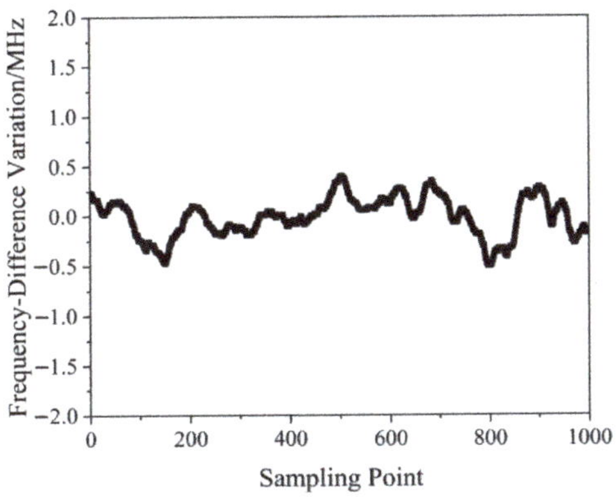

Figure 44. Frequency difference variation in the frequency-locked dual-frequency Nd:YAG laser [44].

Table 1. Comparison of experimental results of two QD-PDH frequency-difference-stabilizing systems [44].

Frequency-Stabilizing System	Resonant Cavity	Linear Dynamic Range/MHz	Frequency-Discriminating Sensitivity mV/MHz	Frequency Stability	Frequency-Difference Stability
Single-modulator QD-PDH	Linear cavity	5.36	251.71	1.6×10^{-11}	2.9×10^{-7}
	Right-angle cavity	5.25	222.34	2.0×10^{-11}	
Double-modulator QD-PDH	Linear cavity	5.08	222.72	2.3×10^{-11}	4.2×10^{-7}
	Right-angle cavity	4.87	208.98	2.7×10^{-11}	

4. Summary and Developing Trends

In order to achieve the high-accuracy synthetic-wave absolute-distance measurement, a frequency-difference-stabilized dual-frequency solid-state laser with a tunable and large frequency difference is needed to use as a light source to form a number of stable multistage synthetic wavelengths. We reviewed the advances in research on different kinds of dual-frequency solid-state lasers, including birefringent dual-frequency solid-state lasers, biaxial dual-frequency Yb:KGW lasers, and TCDFLs based on the different principles of single-longitudinal-mode selection. Additionally, several typical frequency-difference-stabilizing systems for dual-frequency solid-state lasers have been introduced and analyzed.

Taking into account of the requirements of synthetic-wave absolute-distance interferometry and the present research status of dual frequency, we may predict future development trends in dual-frequency solid-state lasers, mainly including the following aspects:

First of all, the new oscillating principles and methods of orthogonally and linearly polarized dual-frequency solid-state lasers will gain attraction for investigation. Especially, the different principles and methods of single-longitudinal-mode selection should be taken into account. In addition to the standing-wave TCDFLs described above, traveling-wave TCDFLs can also be designed and investigated.

Secondly, the new technologies for enlarging the frequency-difference-tuning range of dual-frequency solid-state lasers will be researched and developed. Intracavity second-harmonic generation (SHG) can be used to enlarge the frequency-difference-tuning range, due to the fact that the frequency difference of second-harmonic dual-frequency lasers is theoretically twice that of fundamental dual-frequency lasers.

Thirdly, the new technologies of the spectral-line narrowing of dual-frequency solid-state lasers will be researched and developed. The spectral linewidths of dual-frequency solid-state lasers determine the coherence lengths, which affect the measuring range of the synthetic-wave absolute-distance interferometry.

Finally, the new technologies of frequency difference stabilization of dual-frequency solid-state lasers will be researched and developed. The scheme of frequency difference stabilization depends on the dual-frequency solid-state laser configuration, and for TCDFL, the frequency difference stability of the dual-frequency laser will be further improved, provided that some special measurements be taken, such as using a highly stable F-P reference cavity with high finesse, optimizing the performances of the feedback control system, etc. It is worth mentioning that the femtosecond optical-frequency comb has a wide linewidth and a high frequency stability, which can be used as the frequency reference of the frequency-difference-stabilizing system.

Author Contributions: Conceptualization, M.J., J.X. and Y.L.; methodology, M.J.; software, M.J. and F.J.; validation, M.J., F.J., J.X., Y.L., T.L., J.L. and G.L.; formal analysis, M.J.; investigation, M.J., J.X., Y.L. and F.J.; resources, M.J.; data curation, M.J. and F.J.; writing—original draft preparation, M.J. and F.J.; writing—review and editing, M.J. and F.J.; visualization, M.J.; supervision, M.J.; project administration, M.J.; funding acquisition, M.J., J.X. and Y.L. All authors have read and agreed to the published version of the manuscript.

Funding: This research was funded by the National Natural Science Foundation of China, grant numbers 51875455 and 62075180, the Shaanxi Provincial Science and Technology Plan, grant number 2023-YBGY-400, and the Xi'an Science and Technology Plan, grant number 22GXFW0089.

Institutional Review Board Statement: Not applicable.

Informed Consent Statement: Not applicable.

Data Availability Statement: Data are available on request due to restrictions, e.g., privacy or ethical.

Conflicts of Interest: The authors declare no conflict of interest.

References

1. Deng, Y.; Chen, K.; Li, J.Y. He-Ne laser nanometer ruler system based on orthogonal polarization double longitudinal mode. *J. Appl. Opt.* **2017**, *38*, 316–320.
2. Liu, W.X.; Tang, N.; Ma, L.X.; Gao, K.F.; Sun, M.Z. Consistency of splitting frequency difference with longitudinal modes spacing variation in Zeeman dual-frequency laser. *Acta Phys. Sin.* **2021**, *70*, 146–152.
3. Rueda, A.; Sedlmeir, F.; Kumari, M.; Leuchs, G.; Harald, G. Resonant electro-optic frequency comb. *Nature* **2019**, *568*, 378–381. [CrossRef] [PubMed]
4. Karol, K.; Dorota, T.; Aleksandra, F.; Grzegorz, S. Fiber-based optical frequency comb at 3.3 μm for broadband spectroscopy of hydrocarbons. *Chin. Opt. Lett.* **2021**, *19*, 63–67.
5. Zhao, X.Y.; Qu, X.H.; Chen, J.W.; Zheng, J.H.; Wang, J.D.; Zhang, F.M. Method of measuring absolute distance based on spectral interferometry using an electro-optic comb. *Acta Phys. Sin.* **2020**, *69*, 96–108. [CrossRef]

6. Wang, G.C.; Li, X.H.; Yan, S.H.; Tan, L.L.; Guan, W.L. Real-time absolute-distance measurement by multi-wavelength interferometry synchronously multi-channel phase-locked to frequency comb and analysis for the potential non-ambiguity range. *Acta Phys. Sin.* **2021**, *70*, 138–149.
7. Tian, Z.G.; Zhang, L.; Zhang, S.L. Isocandela points frequency stabilization in He-Ne Zeeman-birefringence dual-frequency lasers. *Infrared Laser Eng.* **2016**, *45*, 53–58.
8. Zhou, Z.F.; Zhang, T.; Zhu, M.C.; Yin, B.B. Research on the laser beat-wave interferometry. *Chin. J. Lasers* **2005**, *32*, 101–104.
9. Xiao, Y.; Zhang, S.L.; Han, Y.M.; Li, Y. Full-inner-cavity birefringence dual-frequency He-Ne laser with fixed frequency-difference by angle-block and frequency-stabilization. *Chin. J. Lasers* **2001**, *28*, 509–512.
10. Willenberg, B.; Pupeikis, J.; Krüger, L.M.; Koch, F.; Keller, U. Femtosecond dual-comb Yb:CaF_2 laser from a single free-running polarization-multiplexed cavity for optical sampling applications. *Opt. Express* **2020**, *28*, 30275–30288. [CrossRef]
11. Cheng, R.; Yang, X.T.; Zhang, S.L. Cavity tuning characteristics of microchip Nd:YAG dual-frequency laser. *J. Appl. Opt.* **2012**, *33*, 1147–1151.
12. Kevin, A.; Abdelkrim, E.; Mehdi, A. Analytical modeling of dual-frequency solid-state lasers including a buffer reservoir for noise cancellation. *Opt. Express* **2018**, *26*, 8805–8820.
13. Loas, G.; Romanelli, M.; Mehdi, A. Dual-frequency 780-nm Ti:Sa laser for high spectral purity tunable cw THz generation. *IEEE Photonics Technol. Lett.* **2014**, *26*, 1518–1521. [CrossRef]
14. Villares, G.; Hugi, A.; Blaser, S.; Faist, J. Ultraviolet generation in a dual-periodic domain inverted structure in $LiTaO_3$crystal by frequency tripling a 1.064 μm laser. *Ferroelectrics* **2011**, *253*, 263–270.
15. Tao, L.P.; Li, L.; Yang, S.H.; Zhao, C.M. Laser diode pumped dual-frequency solid-state lasers. *Opt. Tech.* **2007**, *33*, 134–136.
16. Hu, M.; Huang, Q.F.; Zhang, H.; Xu, G.R.; Deng, J.; Liu, C.X.; Zhang, F. Spectral and frequency difference characteristics of the LD-pumped dual-frequency solid-state laser. *J. Optoelectron. Laser* **2014**, *25*, 472–477.
17. Gui, K.; Zhang, Z.L.; Xing, Y.X.; Zhang, H.Y. Frequency difference thermally and electrically tunable dual-frequency Nd:YAG/$LiTaO_3$ microchip laser. *Appl. Sci.* **2019**, *9*, 1969. [CrossRef]
18. Jin, T.; Hu, M.; Li, P.; Fan, H.D.; Han, N.; Feng, B.; Ou, J.; Zhou, X.F.; Yang, G.W.; Lu, Y.; et al. Experimental study of the dual-frequency laser based on the Nd:YVO_4/Nd:$GdVO_4$ combined crystal. *Laser Optoelectron. Prog.* **2018**, *55*, 283–287.
19. Danion, G.; Hamel, C.; Frein, L. Dual-frequency laser with two continuously and widely tunable frequencies for optical referencing of GHz to THz beat notes. *Opt. Express* **2014**, *22*, 17673. [CrossRef]
20. McKay, A.; Dekker, P.; Coutts, D.W.; Dawes, J.M. Enhanced self-heterodyne performance using a Nd-doped ceramic YAG laser. *Opt. Commun.* **2007**, *272*, 425–430. [CrossRef]
21. Li, L.; Zhao, C.M.; Zhang, P.; Yang, S.H. The study on diode-pumped two-frequency solid-state laser with tunable frequency difference. *Acta Phys. Sin.* **2007**, *56*, 2663–2669. [CrossRef]
22. Wu, X.; Yang, S.H.; Chen, Y.; Zhao, C.M.; Liu, Z.J. Tunable two-frequency solid-state laser with coupled-cavity configuration. *Acta Opt. Sin.* **2012**, *32*, 135–140.
23. Jiao, M.X.; Zhang, S.L.; Lian, J.W. Birefringent dual-frequency Nd:YAG laser with large frequency-difference. *Chin. J. Lasers.* **2001**, *28*, 100–102.
24. Xing, J.H.; Zhang, N.J.; Jiao, M.X.; Liu, Y. Dual-frequency Nd:YAG Laser with a detuning twisted-mode cavity. *Acta Photonica Sin.* **2018**, *47*, 7–13.
25. Julien, L.; Morvan, L.; Alouini, M.; Bourderionnet, J.; Dolfi, D.; Jean, P. Dual-frequency single-axis laser using a lead lanthanum zirconate tantalate (PLZT) birefringent etalon for millimeter wave generation: Beyond the standard limit of tunability. *Opt. Lett.* **2007**, *32*, 1090–1092.
26. Brunel, M.; Amon, A.; Vallet, M. Dual-polarization microchip laser at 1.53 μm. *Opt. Lett.* **2005**, *30*, 2418–2420. [CrossRef]
27. Pillet, G.; Morvan, L.; Brunel, M.; Bretenaker, F.; Dolfi, D.; Vallet, M.; Huignard, J.P.; Le, F.A. Dual-frequency laser at 1.5 μm for optical distribution and generation of high-purity microwave signals. *J. Lightw. Technol.* **2008**, *26*, 2764–2772. [CrossRef]
28. Gudelev, V.G.; Mashho, V.V.; Nikeenko, N.K.; Ryabtsev, G.I.; Stalmashonak, A.B.; Teplyashin, L.L. Diode-pumped CW tunable two-frequency YAG: Nd^{3+} laser with coupled resonators. *Appl. Phys. B* **2003**, *76*, 249–252. [CrossRef]
29. Li, J.Y.; Tan, Y.D.; Zhu, K.Y.; Lu, Y.Y.; Zhang, S.L. Dual-frequency solid-state microchip laser and its frequency difference control. *Opt. Eng.* **2019**, *58*, 116105. [CrossRef]
30. Chen, H.; Zhang, S.L. Microchip Nd:YAG dual-frequency laser interferometer for displacement measurement. *Opt. Express* **2021**, *29*, 6248–6256. [CrossRef]
31. Brunel, M.; Bretenaker, F.; Le, F.A. Tunable optical microwave source using spatially, resolved laser eigenstates. *Opt. Lett.* **1997**, *22*, 384–386. [CrossRef] [PubMed]
32. Czarny, R.; Alouini, M.; Larat, C.; Krakowski, M.; Dolfi, D. THz-dual-frequency Yb^{3+}:KGd $(WO_4)_2$ laser for continuous wave THz generation through photo mixing. *Electron. Lett.* **2004**, *40*, 20040597. [CrossRef]
33. Jiao, M.X. LD-pumped cogain two-cavity dual-frequency Nd:YAG laser with very large frequency-difference. *Laser Optoelectron. Prog.* **2008**, *60*, 0312025.
34. Jiao, M.X.; Xing, J.H.; Liu, Y.; Yang, Y.; Ma, S.H. Design and experimental study of two-cavity dual-frequency solid-state laser with large frequency difference. *Chin. J. Lasers* **2010**, *37*, 2784–2789. [CrossRef]
35. Xing, J.H.; Jiao, M.X. Design and experimental study of tunable dual-frequency Nd:YAG laser with large frequency difference. *Laser Optoelectron. Prog.* **2015**, *52*, 143–148.

36. Shan, D.J. Research on Frequency-Difference Tuning Techniques of Two-Cavity Dual-Frequency Nd: YAG Laser. Master's Thesis, Xi'an University of Technology, Xi'an, China, 2018.
37. Xing, J.H.; Jiao, M.X.; Liu, Y. T-shaped cavity dual-frequency Nd:YAG laser with electro-optical modulation. *Opt. Eng.* **2016**, *55*, 056115. [CrossRef]
38. Xing, J.H.; Jiao, M.X. Two-cavity dual-frequency Nd:YAG laser with a twisted-mode configuration. *Acta Photonica Sin.* **2015**, *44*, 142–147.
39. Zhou, Y.; Jiao, M.X.; Lian, T.H.; Xing, J.H.; Liu, Y.; Liu, J.N. Design and experimental investigation of passively Q-switched two-cavity dual-frequency Nd:YAG laser. *Chin. J. Lasers* **2018**, *45*, 54–61.
40. Jiang, F.; Jiao, M.X.; Wang, Y.D.; Xing, J.H.; Liu, Y. Two-cavity dual-frequency Nd:YAG laser based on the principle of longitudinal mode selection by Fabry-Perot etalon. *Opt. Eng.* **2022**, *61*, 046110. [CrossRef]
41. Su, J.; Jiao, M.X.; Ma, Y.Y.; Xing, J.H. Design of Pound-Drever-Hall laser frequency stabilization system using the quadrature demodulation. *Chin. J. Lasers* **2016**, *43*, 250–255.
42. Su, J.; Jiao, M.X.; Jiang, F. Pound-Drever-Hall laser frequency locking technique based on orthogonal demodulation. *Optik* **2018**, *168*, 348–354.
43. Jiang, F.; Jiao, M.X.; Su, J.; Xing, J.H.; Liu, Y.; Lian, T.H.; Liu, J.N. Design of frequency-difference stabilizing system for two-cavity dual-frequency Nd:YAG laser using quadrature-demodulated Pound-Drever-Hall method. *Opt. Eng.* **2022**, *61*, 046109.
44. Jiang, F.; Jiao, M.X.; Xing, J.H.; Liu, Y.; Lian, T.H.; Liu, J.N.; Li, H.Q. Design of frequency-difference stabilizing system for two-cavity dual-frequency Nd:YAG laser using single-modulator quadrature-demodulated Pound-Drever-Hall method. *Opt. Eng.* **2022**, *61*, 106106. [CrossRef]
45. Jiao, M.X.; Jiang, F.; Su, J.; Xing, J.H.; Liu, Y.; Lian, T.H.; Liu, J.N.; Wang, X.; Li, H.Q. Design of synthetic-wave absolute-distance interferometric system using two-cavity dual-frequency Nd:YAG laser with large frequency-difference. *Laser Optoelectron. Prog.* **2023**, *60*, L222685.
46. Brunel, M.; Vallet, M. Wavelength locking of CW and Q-switched Er3+ microchip lasers to acetylene absorption lines using pump-power modulation. *Opt. Express* **2007**, *15*, 1612–1620. [CrossRef]
47. Thévenin, J.; Vallet, M.; Brunel, M.; Gilles, H.; Girard, S. Beat-note locking in dual-polarization lasers submitted to frequency-shifted optical feedback. *J. Opt. Soc. Am. B* **2011**, *28*, 1104–1110. [CrossRef]
48. Rolland, A.; Brunel, M.; Loas, G.; Frein, L.; Vallet, M.; Alouini, M. Beat note stabilization of a 10-60 GHz dual-polarization microlaser through optical down conversion. *Opt. Express* **2011**, *19*, 4399–4404. [CrossRef]
49. Wang, X.H.; Xu, J.L.; Gao, S.F. Frequency stabilization of a dual frequency Yb^{3+}:$GdAl_3(BO_3)_4$ laser via nonlinear loss modulation in black phosphorus. *Laser Phys. Lett.* **2017**, *14*, 065802. [CrossRef]
50. Pique; Paul, J. Iodine-stabilized high-resolution dual-frequency Ti:sapphire laser. *Opt. Express* **2018**, *26*, 16402–16410. [CrossRef]
51. Kervevan, L.; Gilles, H.; Girard, S.; Laroche, L. Beat-note jitter suppression in a dual-frequency laser using optical feedback. *Opt. Lett.* **2007**, *32*, 1099–1101. [CrossRef]
52. Spencer, D.T.; Davenport, M.L.; Komljenovic, T. Stabilization of heterogeneous silicon lasers using Pound-Drever-Hall locking to Si_3N_4 ring resonators. *Opt. Express* **2016**, *24*, 13511–13517. [CrossRef] [PubMed]
53. Davila, R.J.; Baynes, F.N.; Ludlow, A.D.; Fortier, T.M.; Leopardi, H.; Diddams, S.A. Compact thermal-noise-limited reference cavity for ultra-low-noise microwave generation. *Opt. Lett.* **2017**, *42*, 1277–1280. [CrossRef] [PubMed]
54. Schmitz, J.; Meyer, H.M.; Köhl, M. Ultraviolet Fabry-Perot cavity with stable finesse under ultrahigh vacuum conditions. *Rev. Sci. Instrum.* **2019**, *90*, 063102. [CrossRef] [PubMed]
55. Yao, B.; Chen, Q.F.; Chen, Y.J.; Wu, B.; Mao, Q.H. 280 mHz linewidth DBR fiber laser based on PDH frequency stabilization with ultra-stable cavity. *Chin. J. Lasers* **2021**, *48*, 198–206.
56. Jang, Y.S.; Lim, J.K.; Wang, W.T.; Kim, S.W.; Savchenkov, A.; Andrey, B.M.; Wong, C.W. Measurement of sub-fm/Hz1/2 displacement spectral densities in ultrahigh-Q single-crystal microcavities with hertz-level lasers. *Photonics Res.* **2022**, *10*, 1202–1209. [CrossRef]
57. Yu, Y.; Zhang, H.; Wang, D.H.; Yuan, J.B.; Cao, J.; Huang, X.R. Review of precision measurements of the zero-thermal-expansion temperature of ultra-stable cavity. *Metrol. Meas. Technol.* **2022**, *42*, 30–37.

MDPI
St. Alban-Anlage 66
4052 Basel
Switzerland
www.mdpi.com

Sensors Editorial Office
E-mail: sensors@mdpi.com
www.mdpi.com/journal/sensors

www.ingramcontent.com/pod-product-compliance
Lightning Source LLC
Chambersburg PA
CBHW061621220926
51850CB00005B/566
9783036592527